Handbook of Plant Cell Biotechnology

Handbook of Plant Cell Biotechnology

Edited by Agatha Wilson

SYRAWOOD
PUBLISHING HOUSE
New York

Published by Syrawood Publishing House,
750 Third Avenue, 9th Floor,
New York, NY 10017, USA
www.syrawoodpublishinghouse.com

Handbook of Plant Cell Biotechnology
Edited by Agatha Wilson

International Standard Book Number: 978-1-68286-582-8 (Hardback)

Cataloging-in-Publication Data

Handbook of plant cell biotechnology / edited by Agatha Wilson.
p. cm.
Includes bibliographical references and index.
ISBN 978-1-68286-582-8
1. Plant cell biotechnology. 2. Plant biotechnology. I. Wilson, Agatha.
SB106.B56 H36 2018
630--dc23

Printed in China.

TABLE OF CONTENTS

PREFACE

Plant cells regulate all the vital processes occurring inside the plant. They cause cellular responses to external stimuli and determine the structure, growth and life cycle of the plant. Cells contain valuable evidence about the environment and the ecology that sustains the plant. Research in this field concerns with processes like phytoremediation, biofortification, etc. This book outlines the principles of plant cell biology. New technologies related to the field of plant breeding have been covered in it as well. It is an essential guide for both academicians and those who wish to pursue this discipline further.

The researches compiled throughout the book are authentic and of high quality, combining several disciplines and from very diverse regions from around the world. Drawing on the contributions of many researchers from diverse countries, the book's objective is to provide the readers with the latest achievements in the area of research. This book will surely be a source of knowledge to all interested and researching the field.

In the end, I would like to express my deep sense of gratitude to all the authors for meeting the set deadlines in completing and submitting their research chapters. I would also like to thank the publisher for the support offered to us throughout the course of the book. Finally, I extend my sincere thanks to my family for being a constant source of inspiration and encouragement.

Editor

1

Meta-Analysis of Gene Expression Signatures Reveals Hidden Links among Diverse Biological Processes in Arabidopsis

Liming Lai, Steven X. Ge*

Department of Mathematics and Statistics, South Dakota State University, Brookings, South Dakota, United States of America

Abstract

The model plant Arabidopsis has been well-studied using high-throughput genomics technologies, which usually generate lists of differentially expressed genes under various conditions. Our group recently collected 1065 gene lists from 397 gene expression studies as a knowledgebase for pathway analysis. Here we systematically analyzed these gene lists by computing overlaps in all-vs.-all comparisons. We identified 16,261 statistically significant overlaps, represented by an undirected network in which nodes correspond to gene lists and edges indicate significant overlaps. The network highlights the correlation across the gene expression signatures of the diverse biological processes. We also partitioned the main network into 20 sub-networks, representing groups of highly similar expression signatures. These are common sets of genes that were co-regulated under different treatments or conditions and are often related to specific biological themes. Overall, our result suggests that diverse gene expression signatures are highly interconnected in a modular fashion.

Editor: Francisco J. Esteban, University of Jaén, Spain

Funding: This work was supported by the National Institute of General Medical Studies at the National Institutes of Health (GM083226 to SXG, in part) and this material is based upon work supported partially by the National Science Foundation/EPSCoR Award No. IIA-1355423 and by the state of South Dakota. The funders had no role in study design, data collection and analysis, decision to publish, or preparation of the manuscript.

Competing Interests: The authors have declared that no competing interests exist.

* Email: xijin.ge@sdstate.edu

Introduction

Because of its small genome size, *Arabidopsis* thaliana has been a valuable model system for genetic mapping, sequencing and gene expression analysis [1]. Until March 2013, 1787 studies on gene expression of *Arabidopsis* were indexed in Gene Expression Omnibus (GEO) website in National Center for Biotechnology Information (NCBI) [2]. These studies investigated various biological processes by monitoring the gene expression level using the high-throughput genomics technologies such as DNA microarrays and RNA sequencing. The results were usually a set of genes associated with particular biological processes based on different experimental designs. Even though DNA microarrays suffer from noise and reproducibility issues [3], we believe that many of the noise could be filtered out by statistical analysis and that there are significant associations among these numerous results, or common modules in the transcriptional program.

Some studies have showed the relationships among gene lists in different species. Most researchers analyzed these gene lists using methodology of meta-analysis [4–7], which combines the results of studies that address a set of related research hypotheses, focusing on a special individual topic such as cancer or special treatment [8]. Several databases of gene lists have been created, such as L2L [9], LOLA [10], and MSigDB [11]. An network-based method was developed by Ge [12] to define associations among a large number of gene sets in human. Associations are defined as statistically significant overlaps between two gene lists. The method was applied successfully to a large number of human gene lists [12], and identified molecular links among diverse biological processes.

In this study, we used the methodology in [12] to analyze a set of *Arabidopsis* gene lists identified by genome wide expression studies. These lists were collected for AraPath [13], an *Arabidopsis* gene lists database we created recently. The objective was to systematically evaluate relationships among the gene lists and interpret the relationships. This process provides not only a new tool to uncover hidden links among vast amounts of gene lists, but a quantitative measure to describe the global gene expression of the *Arabidopsis* system under diverse conditions.

Materials and Methods

Data in this study was extracted from the AraPath [13], which is a gene lists database in *Arabidopsis* we created (Availability: http://bioinformatics.sdstate.edu/arapath/). As part of the database, the data contains a total of 1,065 co-expression gene lists, which were manually retrieved from published papers linked to GEO [2] before February, 2011.

Methodology of the analysis includes four steps. Step 1 is to evaluate overlapping genes among the 1,065 gene lists. A Perl programs was written to evaluate overlapping genes between all 566,580 pairs of lists. An overlap refers to a pair of gene lists, which has at least two common genes. And overlaps from the same paper were considered trivial and were removed. Because there are too much overlaps and microarray experiments tends to produce noisy data, we selected significant overlaps using stringent threshold. Step 2 computes p-values and q-values to identify significant overlaps. Based on the Hypergeometric distribution, we first calculate the likelihood (p-value) of observing the number of overlapping genes if these two gene lists are randomly drawn without replacement from a collection of 28,024 unique genes in terms of R program [14] we compiled. Then, p-values were translated into q-values based on the false discovery rate (FDR) [15] to correct that for multiple testing. Overlaps with very small q-value were significant overlaps. In this case, significant overlaps were identified with a q-value = 5.0E-9 as a cutoff. In step 3, network of significant overlaps was constructed based on outputs of the step 2 using Cytoscape[16]. Because this network includes too many nodes and edges, we need to further break the big clusters into smaller subclusters. In step 4, There are many algorithms that could decompose large networks into small, densely connected subnetworks such as those in [17,18]. We chose a simply algorithm that is available as a plug-in to Cytoscape. MCODE [19] is used to identify interconnected sub-networks and their clusters within the network of the step 3. To generally find locally dense regions (or clusters) of a graph is based on the clustering coefficient [19], C_i, which measures "clique" of the neighborhood of a vertex: $C_i = 2n/k_i\ (k_i - 1)$, where k_i is the vertex size of the neighborhood of vertex *i*, *n* is the number of edges in the neighborhood. According to the MCODE algorithm [19], however, clustering the main network into sub-networks is by means of vertex weighting, which is to weight all vertices based on their local network density using the highest *k*-core of the vertex neighborhood rather than the clustering coefficient C_i. A *k*-core is a graph of minimal degree *k*. The highest *k*-core of a graph is the central most densely connected sub-graph. Given a highly connected vertex, in a dense region of a graph, *v* may be connected to many vertices of degree one. These low degree vertices do not interconnect within the neighborhood of *v* and thus would reduce the clustering coefficient, but not the core-clustering coefficient (for detailed information about the MCODE algorithms, see the paper [19]). Here we created the sub-networks and found the modules and clusters using MCODE algorithms based on the following parameters: Node Score Cutoff = 0.15; k-core = 2; Degree Cutoff = 2; Max. Depth = 100. The DAVID web site [20,21] was applied to analyze the most significant functions of most frequently shared genes in each of sub-networks.

Results

A total of 1,065 gene lists were analyzed in this study. They include 277,349 gene entries corresponding to 28,024 unique genes. The average size of these gene lists is 87 genes, ranging from 1 to 2,952 genes. Its distribution is close to normality on a log10 scale (Figure 1). The results of analysis of the data are as follows.

Significant overlaps

By comparing all pairs of 1,065 gene lists using the Perl program, 16,261 significant overlaps were identified from a total of 192,642 overlaps. Based on the Hyper-geometric distribution, the probabilities (p-values) of observing the number of overlapping genes or more were first calculated if these two gene lists were randomly drawn without the replacement from a collection of 28,024 unique genes. The p-values were translated into q-values according to the false discovery rate (FDR) [15] to correct for multiple testing. Overlaps from the same paper were considered trivial and were removed. With a q-value $= 5 \times 10^{-}$ as a conservative cutoff, 16,261 significant overlaps were identified.

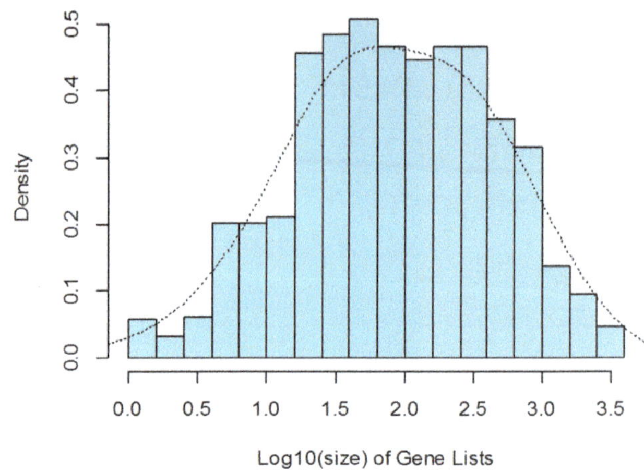

Figure 1. Histogram of log10 scale of size of gene lists.

Main network

The 16,261 significant overlaps are represented as an undirected network (Figure 2). In the network, nodes correspond to gene lists and edges indicate number of overlapping genes between two nodes within significant overlaps. This network highlights the correlation across gene expression signatures of diverse biological processes. It, thus, constitutes a "molecular signature map" in which the individual perturbations are placed in the context defined by others. This is a highly connected network with an average of 20.10 connections per gene list. The 809 nodes (75.96% of the 1,065 gene lists) and 16,261 edges are connected to a dominant main network. Most nodes are connected to a small number of gene lists. The network shows some different colors "cliques", which are some of the most connected graphs in terms of a vertex-weighting scheme based on the highest *k*-core of the vertex neighborhood. They are intuitively denser links within some neighborhoods.

Modules

To further explore these neighborhoods, we used the MCODE algorithm [19] to decompose the network into a total of 20 sub-networks. Of them, nine modules were further analyzed and their clustered information was shown in Table 1. The others were ignored because they are much lower score of density (less than 1.6) and have few nodes and edges (less than 6). The nine sub-networks are highly interconnected, suggesting that those genes are involved in common metabolic pathways or interact with each other under similar biological perturbations. The first three sub-networks are described in the following section. The remaining six sub-networks description and all the sub-networks figures and their composite outcomes tables are shown in Figures S1–S9 and Tables S1–S9 in File S1.

Sub-network 1

The sub-network 1 includes 46 nodes and 969 edges (Figure 3). The score of cluster density is 21.065, which is the highest score among nine sub-networks. This indicates it is the most densely connected. In Figure 3, the dark red nodes represent higher network density based on MCODE [19]. Dark green edges represent very small p-values. There are 31 nodes that represent up-regulated, eight down-regulated, and seven differently regulated. Most gene lists (67.39%) involving up-regulated nodes are related to seven biological themes and 25 treatments or conditions.

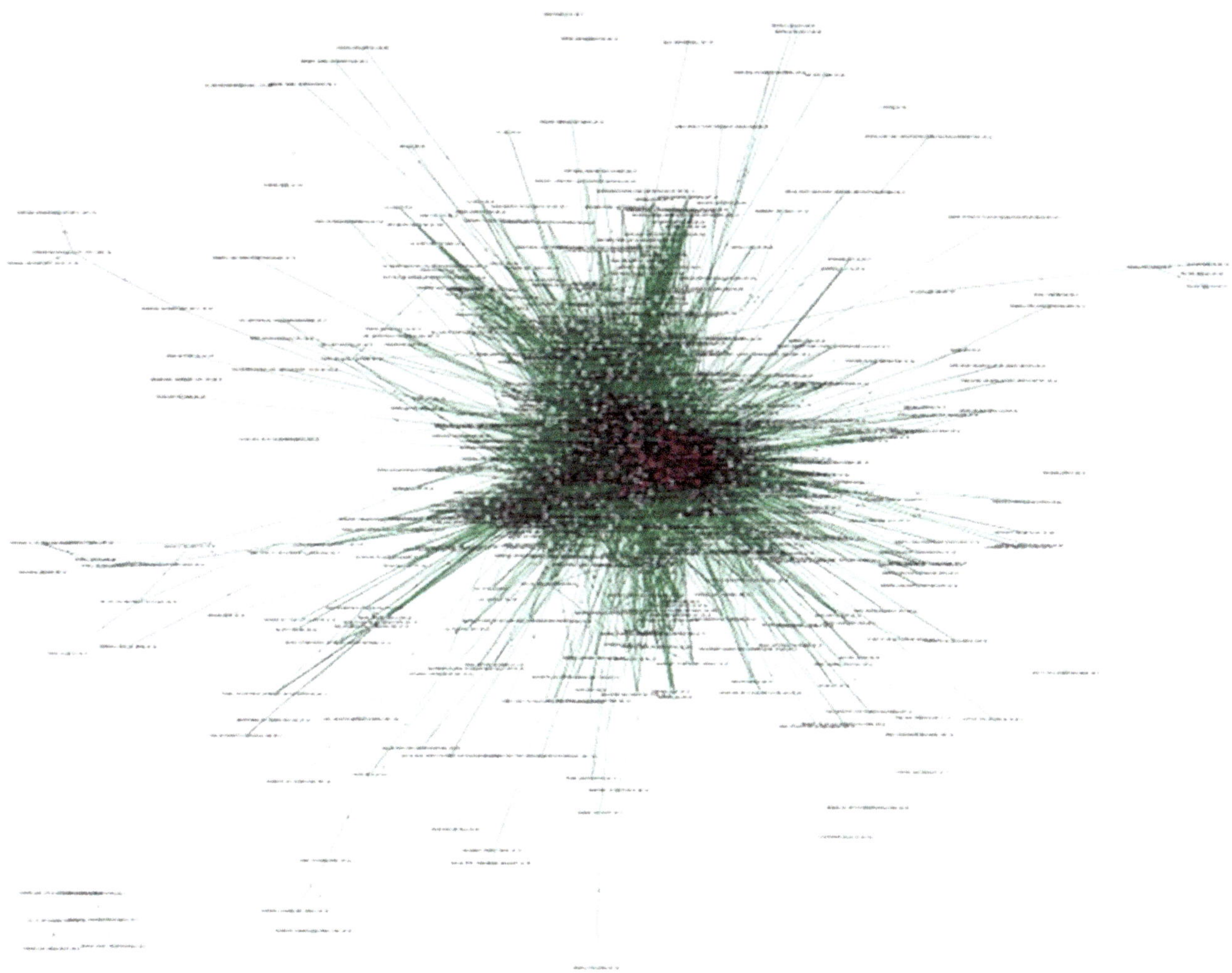

Figure 2. The main network created by Cytoscape. Node = name of gene list. Node Color = MCODE_Scores from small to large and corresponds to color from light green to dark red. Edge Color = p-values from large to small and corresponds to color from grey to dark green.

Table S1 in File S1 shows its composite outcomes. There are 46 gene lists and 256 frequently shared genes identified in this sub-network. They are regulated by 32 treatments or conditions from 32 publications related to 11 biological themes involving development, metabolism, disease, yield, function, genome analysis, immune, pathogen, mechanism, energy, virus, and photosynthesis in *Arabidopsis*. The top 10 most frequently shared genes with their gene descriptions were specifically listed corresponding to different gene lists, different biological themes, and treatments or conditions. For example, gene AT4G14365 ("putative E3 ubiquitin-protein ligase XBAT34") has the highest frequency of 35, which indicates it is the most active gene because it is regulated simultaneously under 35 gene lists in sub-network 1, namely, the gene connects directly 35 gene lists.

The most significant function of sub-network 1 is biological process in response to chitin (i.e. the most enriched term corresponding to the most frequently shared genes in sub-network 1 is "response to chitin") based on results of analysis of DAVID (Table 1). This indicates sub-network 1 is specifically associated with the chitin signaling pathway rather than by random chance. The other significant functions of sub-network 1 are responding to carbohydrate stimulus, organic substance, defense response, and bacterium based on the analysis of DAVID with a cutoff of 1.60×10^{-16} p-value. This suggests that sub-network 1 involves multiple signaling pathways.

Sub-network 2

Sub-network 2 is shown in Figure 4 and Table S2 in File S1. It includes 54 nodes (gene lists) and 168 frequently shared genes, which are regulated under 38 different treatments or conditions from 38 publications related to 10 biological themes. The score of cluster density is 9.907. There are 17 nodes to be up-regulated, 34 to be down-regulated, and three to be differently regulated. Most gene lists (62.96%) involving down-regulated nodes are related to nine biological themes and 23 treatments or conditions. Compared to sub-network 1, sub-network 2 has a lower cluster density score with even more treatments or conditions. Nine themes in sub-network 2 are common with sub-network 1: development, disease, function, genome analysis, mechanism, metabolism, photosynthesis, virus, and yield. This indicates the two sub-networks have relationships linked by same themes. No gene is common between the 256 frequently shared genes in sub-network 1 and the 168 frequently shared genes in sub-network 2. The two sub-networks have relatively independent functions. The most significant function of sub-network 2 is biological process of plastid thylakoid membrane based on results of DAVID (Table 1), suggesting that

Table 1. Summary of nine modules consisting of highly interconnected gene lists.

ID	Score Density†	#Nodes	#Edges	Unique genes	Shared genes*	Highest frequency	Most significantly enriched GO Term*	P-value*
1	21.065	46	969	10581	256	35	response to chitin	1.70E-38
2	9.907	54	535	12163	168	20	plastid thylakoid membrane	1.80E-59
3	9.545	33	315	9501	124	17	response to chitin	8.80E-13
4	9.062	48	435	13930	155	18	response to auxin stimulus	9.00E-10
5	2.559	34	87	9751	109	10	cell wall	5.30E-11
6	2.111	9	19	2565	34	6	External encapsulating structure organization	5.40E-07
7	2	15	30	6725	53	8	glycoside biosynthetic process	2.00E-05
8	2	6	12	5212	66	6	membrane-enclosed lumen	6.00E-14
9	1.846	13	24	3567	23	6	response to abiotic stimulus	6.80E-08
T	–	258	2426	–	988	–	–	–

Note: †Score Density = #Edges/#Nodes. *Most significantly enriched GO Term and p-value are the results of analysis of the DAVID in terms of the most frequently shared genes (i.e. Shared genes in Table 1) in each sub-network.

sub-network 2 is specifically associated with plastid thylakoid membrane, i.e. the lipid bilayer membrane of any thylakoid within a plastid. The other significant functions of sub-network 2 are response to chloroplast thylakoid membrane, thylakoid membrane, plastid thylakoid, and chloroplast thylakoid based on DAVID with cut-off p-value of 6.8×10^{-58}. The top 10 most frequently shared genes with their gene descriptions corresponding to gene lists, biological themes, and treatments or conditions were specifically listed in Table S2 in File S1. Gene AT4G27030 (fatty acid desaturase A), for example, has the highest frequency of 20, indicating it is the most active gene in sub-network 2.

Sub-network 3

The sub-network 3 includes 33 nodes (gene lists) and 124 most frequently shared genes (Figure 5 and Table S3 in File S1). There are 28 nodes to be up-regulated, four to be down-regulated, and one to be differently regulated. Most gene lists (84.85%) involving up-regulated nodes are related to nine biological themes and 20 treatments or conditions. By contrast to sub-networks 1 and 2, sub-network 3 is smaller in size, has a lower cluster density score, and less treatments or conditions for gene lists. Sub-network 3 are regulated by 25 treatments or conditions from 25 publications associated with 11 biological themes. Nine themes in sub-network 3 are common with sub-network 1: development, disease, energy, function, immune, mechanism, metabolism, photosynthesis, and virus. There are seven common themes (development, disease, function, mechanism, metabolism, photosynthesis, and virus) between sub-networks 3 and 2. These indicate sub-networks 3 and 1 or 2 have relationships linked by the same themes.

The top 10 most frequently shared genes with their gene descriptions corresponding to each gene list in sub-network 3 are specifically listed in Table S3 in File S1. Gene AT2G18690 has the highest frequency at 17, indicating it is the most active genes in sub-network 3. Other shared genes in sub-network 3 have lower frequency, which means they are less active than those in sub-networks 1 and 2. There is no common gene between the 168 most frequently shared genes in sub-network 2 and the 124 most frequently shared genes in sub-network 3. However, there are 72 genes of intersection between sub-networks 1 and 3. This indicates sub-networks 2 and 3 have relatively independent functions and sub-networks 1 and 3 have dependent relationship linked by the common shared genes. The most significant function of sub-network 3 is biological process in response to chitin based on results of analysis of DAVID (Table 1). It is specifically associated with the chitin signaling pathway, which is the same as sub-network 1 but different from sub-networks 2.

Discussion

The gene lists in our data are highly connected. Out of the 1,065 gene lists, 75.96% are connected in the main network in which many seemingly unrelated stimuli/perturbation may activate or deactivate the same molecular pathways. All the gene lists within each sub-network are highly connected by the most frequently shared genes. For example, in sub-network 8 (Figure S8 and Table S8 in File S1), AT1G56110 ("homolog of nucleolar protein NOP56"), AT3G05060 ("putative SAR DNA-binding protein"), and AT3G44750 (HDA3 histone deacetylase HDT1) are regulated by six different treatments or conditions from five publications corresponding to five special biological themes, which involves reproduction, photosynthesis, metabolism, development, and yield in *Arabidopsis* (Table S8 in File S1). And the most enriched term in sub-network 8 is membrane-enclosed lumen based on functional analysis of DAVID (Table 1). Therefore, the

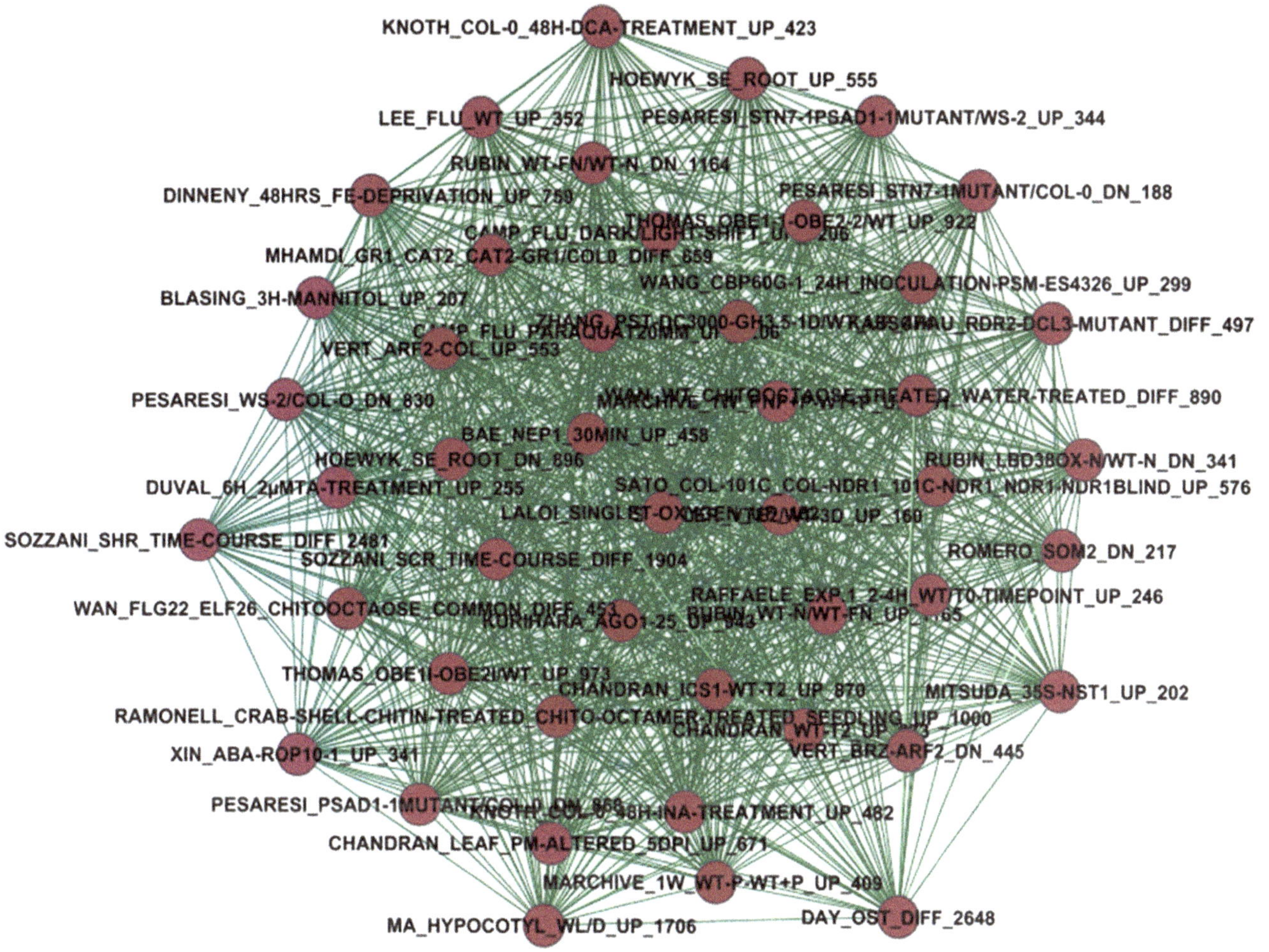

Figure 3. Sub-network 1 corresponding to the module 1/cluster 1 (see note in Figure 2 for meanings of node and edge and their color).

three genes not only connect the six gene lists, but have multiple functions by which we could propose a hypotheses that the three genes' interaction controls the reproduction, photosynthesis, metabolism etc. Furthermore, the three genes associated with the rapidly proliferating nature of the endosperm at 4 DA, with similar expression patterns to the early endosperm markers SUC5, PHE1, FWA, and FIS2 [22], were regulated by interploidy crosses, fis1X2x crosses at 5 DAP (two biological replicates of each), and unfertilized msi1 siliques at 7 DAF [23], kin10, starvation conditions, and sugar availability increase [24], sucrose [25], and 4h-carbon fixation [26]. These all became the organized links among the six gene lists because they associated with the three genes.

However,there are significant differences among the nine sub-networks. Based on Table 1, sub-networks 1–5 have more nodes than sub-networks 6–9. This does not mean the genes in sub-network 6–9 are less important than those in sub-networks 1–5. Generally, the most significantly enriched GO terms in sub-networks 1–9 are different except for sub-networks 1 and 3, which have the same theme "response to chitin". The second, third, and fourth most significantly enriched GO Terms in sub-network 1 are response to carbohydrate stimulus, organic substance, and defense response, respectively, which are different from those in sub-network 3. This indicates there are significantly different functions in sub-networks 1–9. Furthermore, "cliques" of all sub-networks are different in the whole network with 16,261 significant overlaps. These "cliques" are some of the most connected graphs in the NETWORK in terms of a vertex-weighting scheme based on the highest *k*-core of the vertex neighborhood. Therefore, they specify different meanings and information. Finally, all the most frequently shared genes of the sub-networks are different. For instance, sub-network 7 has the second smallest score of 2, with 15 nodes, and 30 edges, but possesses 53 most frequently shared genes, which are completely different from that of sub-network 1.

The most frequently shared genes are the strongest links within each of nine sub-networks and provide genomics research with important insights. They are the most important results we found in this study. For example, the gene AT5G39670 ("putative calcium-binding protein CML45", function as calcium ion binding) has the second highest frequency of 34 in the sub-network 1 (Table S1 in File S1). Its function as calcium ion binding indicates calcium-binding proteins participate in calcium cell signaling pathways by binding to Ca^{2+}. These proteins are expressed in many cell types during various growth stages in plants, and contribute to all aspects of the cell's functioning [24]. In the present study, we found that this gene responded to 24 treatments or conditions such as necrosis-ethylene, diurnal cycle, salicylic acid, iron deprivation, etc. according to various reports.

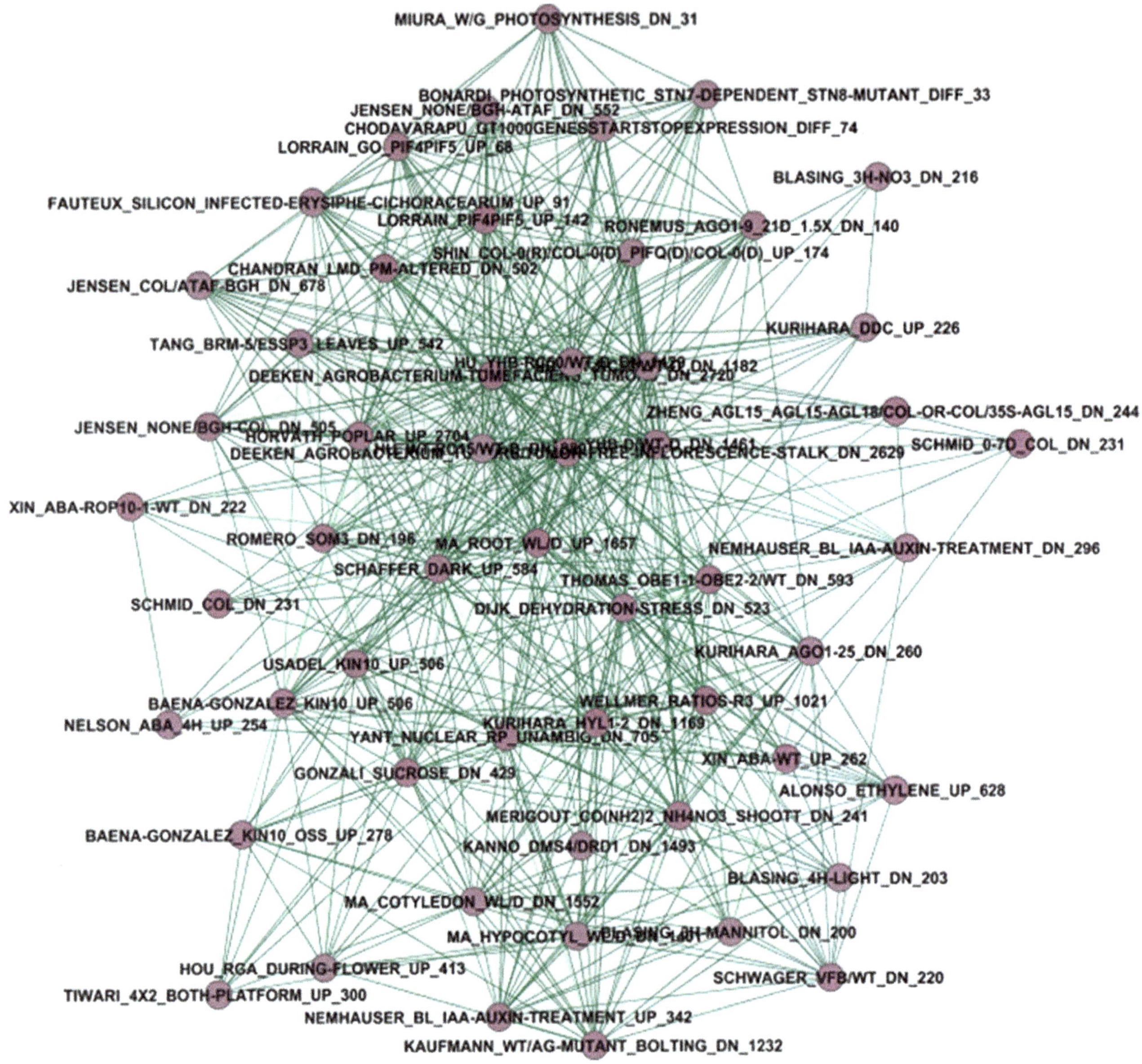

Figure 4. Sub-network 2 corresponding to the module 2/cluster 2 (see note in Figure 2 for meanings of node and edge and their color).

Furthermore, this gene mediates 10 crucial biological themes: immune, yield, disease, development, metabolism, function, photosynthesis, pathogen, energy, and virus. Also, the gene belongs to 24 up-regulated gene lists, seven down-regulated gene lists, and three different-regulated gene lists. In most cases, this gene is up-regulated, indicating that when the experimental treatments listed above were applied, there was an increased expression of the gene AT5G39670. However, some treatments or stimuli may cause a decreased expression of the above gene in order to protect its cells. Therefore, AT5G39670 is the second most active gene and second strongest link in sub-network 1. Similarly, gene AT4G14365 ("XBAT34", molecular functions: protein binding and zinc ion binding) is the strongest link as it is related to 35 gene lists in sub-network 1. Gene AT3G50930 ("BCS1", molecular functions: ATP binding and ATPase activity) is the third strongest link because it is associated with 33 gene lists in sub-network 1. Also, the most frequently shared genes are significantly different in these nine sub-networks. For instance, there are 124 most frequently shared genes in sub-network 3 and 155 most frequently shared genes in sub-network 4. However, there are only two common genes between these two sub-networks. These important results and their biological mechanisms need to be further addressed.

The results from this study, summarized as a molecular signature map, provide key insights into the underlying connections of diverse perturbations. More importantly, compared to previous reports that focused on specific themes, this study explored and established the hidden links among the gene lists on a global scale in *Arabidopsis*. These sub-networks will provide new putative gene targets for future research. For example, sub-network 4 shows that the top three genes AT1G74670, AT1G04240, and AT1G69530 are down-regulated by bioactive

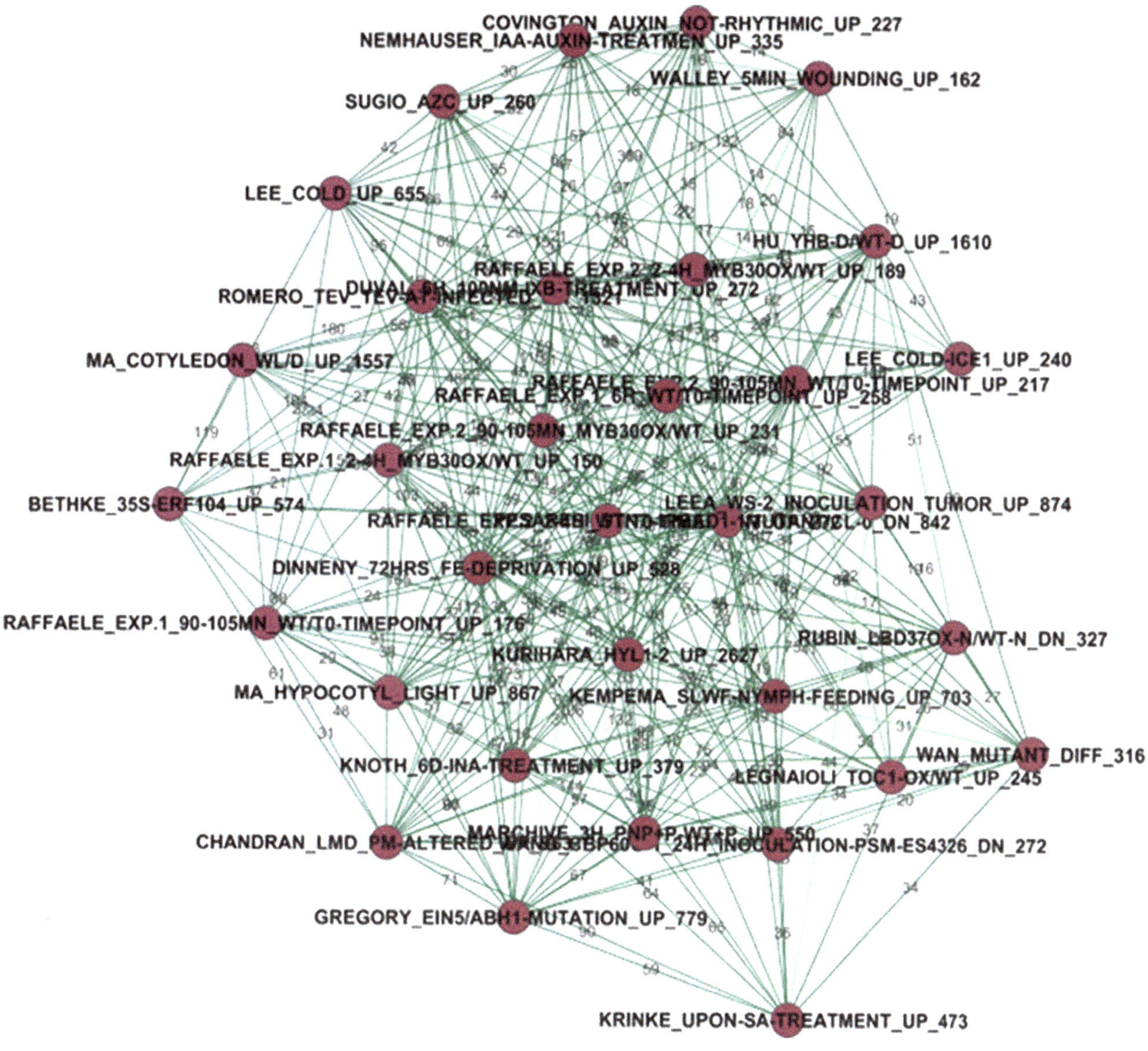

Figure 5. Sub-network 3 corresponding to the module 3/cluster 3 (see note in Figure 2 for meanings of node and edge and their color. edge label = number of overlapping genes between two nodes).

gibberellins corresponding to the gene list ZENTELLA_DEX_-VECTOR_DN_244 (Table S4 in File S1). This information provides some possible clues for future research regarding the mechanism of the regulation of plant growth by plant hormone gibberellins. Another example is gene AT4G27030, which has the highest frequency of 20 in sub-network 2. This gene is regulated by 14 treatments or conditions such as agrobacterium tumefaciens, kin10 and kin 11, dark, far-red light, etc. (Table S2 in File S1). This gene could be used to genetically modify crops for new and useful functions.

Conclusions

There are hidden links among the gene lists from the published papers concerning *Arabidopsis*. After performing systematic overlap analysis, we created 10 networks, including network and sub-networks 1–9 where there are a number of links among gene lists. Many seemingly unrelated stimuli/perturbation may activate or deactivate the same molecular pathways. These links are actually a set of overlapping genes. Of them, a total of 988 most frequently shared genes were identified from each sub-network. These genes are regulated by multiple treatments or conditions from different gene lists and related to different biological themes based on their sub-networks. They construct more active (stronger) links among the gene lists in our data.

Compared to previous reports focusing on specific themes, this study explored and established hidden links among the gene lists on a global scale in *Arabidopsis*. These results provide significant information about target genes or models for future research. In the future, it will be necessary for us to extend gene lists and develop more effective analysis methods to further explain the booming gene lists of microarray data.

Supporting Information Legends

File S1 Fig. S1. Sub-network 1 corresponding to cluster 1. Node = name of gene list. Node Color = MCODE_Scores from small to large and corresponds to color from light green to dark red. Edge Color = p-values from large to small and corresponds to color from grey to dark green. Fig. S2. Sub-network 2 corresponding to cluster 2. Fig. S3. Sub-network 3 corresponding to cluster 3. Fig. S4. The sub-network 4 correspoding to cluster 4. Fig. S5. The sub-network 5 corresponding to Cluster 5. Fig. S6. The sub-network 6 corresponding to Cluster 6. Fig. S7. The sub-network 7 corresponding to Cluster 7. Fig. S8. The sub-network 8 corresponding to Cluster 8. Fig. S9. The sub-network 9 corresponding to Cluster 9. Table S1. Results of sub-network 1 corresponding to cluster 1. Table S2. Results of sub-network 2 corresponding to cluster 2. Table S3. Results of sub-network 3 corresponding to cluster 3. Table S4. Results of sub-network 4 corresponding to cluster 4. Table S5. Results of sub-network 5 corresponding to cluster 5. Table S6. Results of sub-network 6 corresponding to cluster 6. Table S7. Results of sub-network 7 corresponding to cluster 7. Table S8. Results of sub-network 8 corresponding to cluster 8. Table S9. Results of sub-network 9 corresponding to cluster 9.

Author Contributions

Conceived and designed the experiments: SXG. Performed the experiments: LL. Analyzed the data: LL. Wrote the paper: LL SXG.

References

1. (2008) Arabidopsis thaliana. Wikipedia http://enwikipediaorg/wiki/Arabidopsis_thaliana: Wikipedia.
2. (2012). NCBI http://wwwncbinlmnihgov/geo/.
3. Ioannidis JP, Allison DB, Ball CA, Coulibaly I, Cui X, et al. (2009) Repeatability of published microarray gene expression analyses. Nat Genet 41: 149–155.
4. Miller BG, Stamatoyannopoulos JA (2010) Integrative meta-analysis of differential gene expression in acute myeloid leukemia. PLoS One 5: e9466.
5. Rogic S, Pavlidis P (2009) Meta-analysis of kindling-induced gene expression changes in the rat hippocampus. Front Neurosci 3: 53.
6. Yang XN, Sun X (2007) Meta-analysis of several gene lists for distinct types of cancer: A simple way to reveal common prognostic markers. BMC Bioinformatics 8.
7. Edwards YJK, Bryson K, Jones DT (2008) A Meta-Analysis of Microarray Gene Expression in Mouse Stem Cells: Redefining Stemness. PLoS One 3.
8. Glass GV Meta-analysis. Wikipedia http://enwikipediaorg/wiki/Meta-analysis. http://en.wikipedia.org/wiki/Meta-analysis: Wikipedia.
9. Newman JC, Weiner AM (2005) L2L: a simple tool for discovering the hidden significance in microarray expression data. Genome Biol 6: R81.
10. Cahan P, Ahmad AM, Burke H, Fu S, Lai YL, et al. (2005) List of lists-annotated (LOLA): A database for annotation and comparison of published microarray gene lists. Gene 360: 78–82.
11. Liberzon A, Subramanian A, Pinchback R, Thorvaldsdottir H, Tamayo P, et al. (2011) Molecular signatures database (MSigDB) 3.0. Bioinformatics 27: 1739–1740.
12. Ge SX (2011) Large-scale analysis of expression signatures reveals hidden links among diverse cellular processes. Bmc Systems Biology 5.
13. Lai LM, Liberzon A, Hennessey J, Jiang GX, Qi JL, et al. (2012) AraPath: a knowledgebase for pathway analysis in Arabidopsis. Bioinformatics 28: 2291–2292.
14. Team RC R: A language and environment for statistical computing. (2012) R Foundation for Statistical Computing, Vienna, Austria ISBN 3-900051-07-0, URL http://wwwR-projectorg/.
15. Benjamini Y, Hochberg Y (1995) Controlling the False Discovery Rate - a Practical and Powerful Approach to Multiple Testing. Journal of the Royal Statistical Society Series B-Methodological 57: 289–300.
16. Shannon P, Markiel A, Ozier O, Baliga NS, Wang JT, et al. (2003) Cytoscape: A software environment for integrated models of biomolecular interaction networks. Genome Research 13: 2498–2504.
17. Guimera R, Nunes Amaral LA (2005) Functional cartography of complex metabolic networks. Nature 433: 895–900.
18. Hsieh MH, Magee CL (2008) An algorithm and metric for network decomposition from similarity matrices: Application to positional analysis. Social Networks 30: 146–158.
19. Bader GD, Hogue CW (2003) An automated method for finding molecular complexes in large protein interaction networks. BMC Bioinformatics 4: 2.
20. Huang DW, Sherman BT, Lempicki RA (2009) Systematic and integrative analysis of large gene lists using DAVID bioinformatics resources. Nature Protocols 4: 44–57.
21. Huang DW, Sherman BT, Lempicki RA (2009) Bioinformatics enrichment tools: paths toward the comprehensive functional analysis of large gene lists. Nucleic Acids Research 37: 1–13.
22. Day RC, Herridge RP, Ambrose BA, Macknight RC (2008) Transcriptome analysis of proliferating Arabidopsis endosperm reveals biological implications for the control of syncytial division, cytokinin signaling, and gene expression regulation. Plant Physiology 148: 1964–1984.
23. Tiwari S, Spielman M, Schulz R, Oakey RJ, Kelsey G, et al. (2010) Transcriptional profiles underlying parent-of-origin effects in seeds of Arabidopsis thaliana. BMC Plant Biol 10: 72.
24. Baena-Gonzalez E, Rolland F, Thevelein JM, Sheen J (2007) A central integrator of transcription networks in plant stress and energy signalling. Nature 448: 938–942.
25. Gonzali S, Loreti E, Solfanelli C, Novi G, Alpi A, et al. (2006) Identification of sugar-modulated genes and evidence for in vivo sugar sensing in Arabidopsis. J Plant Res 119: 115–123.
26. Blasing OE, Gibon Y, Gunther M, Hohne M, Morcuende R, et al. (2005) Sugars and circadian regulation make major contributions to the global regulation of diurnal gene expression in Arabidopsis. Plant Cell 17: 3257–3281.

2

A Colloidal Singularity Reveals the Crucial Role of Colloidal Stability for Nanomaterials *In-Vitro* Toxicity Testing: nZVI-*Microalgae* Colloidal System as a Case Study

Soledad Gonzalo[1❡], Veronica Llaneza[2❡], Gerardo Pulido-Reyes[1,3❡], Francisca Fernández-Piñas[3], Jean Claude Bonzongo[2], Francisco Leganes[3], Roberto Rosal[1,4], Eloy García-Calvo[1,4], Ismael Rodea-Palomares[3*¤]

1 Departamento de Ingeniería Química, Universidad de Alcalá, Alcalá de Henares, Madrid, Spain, **2** Department of Environmental Engineering Sciences, University of Florida, Gainesville, Florida, United States of America, **3** Departamento de Biología, Facultad de Ciencias, Universidad Autónoma de Madrid, Madrid, Spain, **4** Instituto Madrileño de Estudios Avanzados (IMDEA) Agua, Alcalá de Henares, Madrid, Spain

Abstract

Aggregation raises attention in Nanotoxicology due to its methodological implications. Aggregation is a physical symptom of a more general physicochemical condition of colloidal particles, namely, colloidal stability. Colloidal stability is a global indicator of the tendency of a system to reduce its net surface energy, which may be achieved by homo-aggregation or hetero-aggregation, including location at bio-interfaces. However, the role of colloidal stability as a driver of ENM bioactivity has received little consideration thus far. In the present work, which focuses on the toxicity of nanoscaled Fe° nanoparticles (nZVI) towards a model microalga, we demonstrate that colloidal stability is a fundamental driver of ENM bioactivity, comprehensively accounting for otherwise inexplicable differential biological effects. The present work throws light on basic aspects of Nanotoxicology, and reveals a key factor which may reconcile contradictory results on the influence of aggregation in bioactivity of ENMs.

Editor: Vishal Shah, Dowling College, United States of America

Funding: This research was supported by the Comunidad de Madrid grants S-0505/AMB/0321 and S-2009/AMB/1511 (Microambiente-CM). The Spanish Ministry of Science and Innovation grants CGL2010-15675 sub-program BOS, and the CTM2008-04239/TECNO, and the Spanish Ministry of Economy and Competitiveness (MINECO) grants CTM2013-45775-C2-1-R, CTM2013-45775-C2-2-R and the US-National Science Foundation, grant number CBET-0853347.

Competing Interests: The authors have declared that no competing interests exist.

* Email: ismael.rodea@uam.es

❡ These authors contributed equally to this work.

¤ Current address: Departamento de Ingeniería Química, Universidad de Alcalá, Alcalá de Henares, Madrid, Spain

Introduction

Environmental health and safety (EHS) is one of the key challenges in the field of nanotechnology [1–3]. Despite research efforts, EHS development is hampered by a series of difficulties, from misconceptions in the field, to the lack of agreement on methodological aspects of EHS execution [1,4,5]. The latter is primarily due to the peculiar and complex intrinsic characteristics of nano-sized materials [1,3,4,6,7]. Overall, more than a dozen physicochemical properties could potentially contribute to hazardous interactions at the bio-nano interface [3,8,9]. However, it is hard to find comprehensive studies where key drivers of ENM bioactivity are clearly identified [1]. Currently, the influence of aggregation on ENM bioactivity is surrounded by controversy, mainly due to its broad methodological implications and the existence of contradictory results [1,4,10]. Despite the importance of the effects of aggregation in EHS [4,5], no international consensus exists on how to handle ENM aggregation [1,4,5]. Notably, Schorus & Lison [1] stated in their recent commentary "Focusing the research effort" that "the influence of SNPs (silica nanoparticles) aggregation in the biological response remains unclear and it remains impossible to state whether or not it is necessary to have a dispersion of SNPs before testing". Although they revised SNPs toxicity data, their statement may be generalized to ENMs. Aggregation is a symptom of a more general physicochemical condition of colloidal particles, *i.e.*, colloidal stability [11]. Colloidal stability has been considered a major issue in EHS assessment of ENMs to warrant exposure and dosimetry [2,4]. However, despite the growing body of evidence regarding biophysically mediated bioactivity of ENMs in a variety of model biological systems [8,12–14], the role of colloidal stability as its driver has been essentially overlooked. In the present work, the spontaneous occurrence of an nZVI speciation phenomenon [3,15], which we have termed *colloidal singularity*, showed that taking only colloidal stability into account allowed us to reconcile otherwise inexplicable biological effects.

Results

Physicochemical characterization of nZVI

The surface area and mean particle size of pristine nZVI powder was 43.9 m^2 g^{-1} and 20 nm, respectively, as calculated by the Nova 1200 BET (Brunauer–Emmett–Teller) method. Pristine nZVI powder showed a partial degree of surface oxidation as revealed by XRD analysis (**Table S1**). This is due to the high tendency of zero valent iron to oxidize in aerobic conditions [16,17]. nZVI suspensions ranging from 2.5 to 50 mgL^{-1} were prepared in Milli-Q water and OECD TG 201 algal growth medium [18]. Their main physicochemical properties (**Table 1**) were studied to select the best option for working as a stock suspension for subsequent EHS characterization. General trends were as follows: nZVI appeared aggregated with effective diameters of approximately 100 nm in Milli-Q water, and between 200–400 nm in OECD medium. The Polydispersity index was below 0.5 in Milli-Q water, and ranged from 0.5 to 0.8 in OECD TG 201 medium, indicating a high degree of polydispersity in particle size, especially in OECD TG 201 medium. nZVI appeared with a near neutral surface charge when dispersed in Milli-Q water, however it showed a negative surface charge in OECD TG 201 medium, ranging from −17.54 mV for 2.5 mgL^{-1} to −27.5 mV for 25 mgL^{-1}. The observed tendency of nZVI to acquire negative charge in saline medium has been associated with the high affinity of ions to be adsorbed onto the oxy-hydroxide surface shell of nZVI formed under oxidant conditions [17]. Primary particle size and morphology was investigated by Transmission electron microscopy (TEM) in nZVI suspensions prepared in Milli-Q water (**Figure 1**). 50 mgL^{-1} nZVI suspensions showed a high degree of agglomeration (**Figure 1.a**), however, the visualization of diluted suspensions (5 mgL^{-1}) allowed us to observe primary sized particles and small particle aggregates (**Figure 1.b, c**). The mean size, measured by TEM, of primary aggregates were 37.5±9.7 nm, and that of individual small and big nZVI primary particles was 4.0±1.4 and 12.7±1.7 nm, respectively (**Figure S1**). XEDS (X-Ray Energy Dispersive Spectroscopy) analysis confirmed Fe as the main constituent of the visualized particles (**Figure 1.d**). 25 mgL^{-1} of nZVI in Milli-Q water had the lowest polydispersity index (PDI) and effective diameter (**Table 1**). Therefore, it was selected as stock suspension.

nZVI working suspensions show a *colloidal singularity* in a narrow dose range

The stability of the nZVI working suspensions used in EHS was evaluated under shaking conditions by measuring the residual absorbance ($\lambda = 750$ nm) at relevant experimental lapse times [4,15] (0 h, 4 h, 24 h, 48 h and 72 h) (**Figure 1. e**). Extensive sedimentation of nZVI particles was observed within the first 4 h, in agreement with the high tendency of nZVI to homo-aggregate [19–21]. However, suspensions remained stable from 4 h to 72 h (**Figure 1.e**), confirming the relevant nZVI colloidal fractions in the exposure experiments [3,15]. The actual nZVI concentrations as stable colloidal fractions in equilibrated conditions were 20–40% of nominal concentrations. They maintained a linear correlation ($R^2 = 0.95$, $p<0.05$) with the initial dosage (**Figure 1. f**). Main physicochemical properties (e.g. size, surface charge and total surface area) of the stable colloidal nZVI fractions in the test system were monitored during 72 h (**Figure 2.a, b, c**). As a general trend, nZVI appeared in the form of aggregates of sizes around 150–500 nm, with ζ-potential around −15 mV and a total surface area of near 10^{-3} m^2 L^{-1}. However, nZVI presented a substantially different physicochemical state within a small dose range (0.1–0.5 mgL^{-1}). In this narrow interval nZVI appeared suspended in its primary size (4–12 nm) along the experimental lapse time (**Figure 2.a**), with a substantially more negative surface charge (ζ-potential = −28 mV) (**Figure 2.b**), and a nearly 10 times higher total surface area (**Figure 2.c**). To our knowledge, such an extreme asymmetry in behavior of colloidal suspensions without the interplay of any external factor has never been previously reported. For simplicity, we will herein refer to this phenomenon as *colloidal singularity*.

The *colloidal singularity* prevents nZVI toxicity

EHS assessment of nZVI suspensions was evaluated using an algal growth inhibition assay based on *Pseudokirchneriella subcapitata* (*P. subcapitata*) [18]. The assay consisted of monitoring the algal growth rates based on three different biomass end-points. Besides growth inhibition, a study of alterations in intracellular levels of reactive oxygen species (ROS) and cell cycle was also performed. All biomass related end-points tested correlated ($R>0.9$, $p<0.05$) (**Figure 3.b, c**) and showed non-linear dose-effect relationships (**Figure 3.a**) which mimicked the ζ-potential and particle size profiles showed in **Figure 2.a, b**. The biological response of the test model organism showed two statistically significant ($p>0.05$) regions of toxic response, bracketing a non-toxic region. The first toxic dose region occurred within nZVI concentrations of 0.05 to 0.075 mgL^{-1}, resulting in almost 40% maximum growth inhibition. The second window of concentrations resulting in adverse biological effects ranged from 1.0 to 2.5 mgL^{-1}, resulting in almost 60% maximum growth inhibition (**Figure 3.a**). The observed toxic response was consistently accompanied by a statistically significant ($p>0.05$) increase in intracellular ROS production, and cell cycle alterations (**Figure 4.a–e**). ROS production reached nearly 180% and 250% induction with respect to control levels for 0.075 mgL^{-1} and 2.5 mgL^{-1} nZVI, respectively (**Figure 4.a**). Cell cycle alterations consisted of two levels of abnormalities in cell cycle progression of cell populations. Firstly, nearly 50% of the cell population accumulated increasing intracellular DNA quantities. This means that cells accumulated in poorly defined S and G2/M phases (Black chart in **Figure 4.e**). Secondly, almost 30% of cell population showed DNA fragmentation (Blue spots in **Figure 4.d** and F region in **Figure 4.e**) which is usually linked to apoptosis [22,23]. ROS induction, cell cycle alterations and DNA fragmentation were clearly evident and statistically significant ($p<0.05$) from early stages of nZVI exposure (24 h) (**Figure 4.e, f**). Interestingly, between the two toxic end dose regions, there was a non-toxic dose region within a narrow range of nZVI concentrations (0.1 to 0.5 mgL^{-1}), coinciding with the *colloidal singularity* described earlier (**Figure 2**).

The nZVI-algal suspension can be regarded as a whole colloidal system and its stability properties can be studied as such. Destabilization of colloidal systems is typically accompanied by an increase in sedimentation rates, and a decrease (in absolute value) in ζ- potential which typically account for aggregation and sedimentation of colloidal particles [11,24,25]. **Figure 5.a** shows that the algae-nZVI systems exhibited clear signs of colloidal destabilization with the exception of the concentration range in which the colloidal singularity occurred. Furthermore, the decrease in algal growth rates clearly correlated with the sedimentation rates of the test systems ($R = -0.689$, $p<0.001$) (**Figure 5.b**). This suggests that the toxicity of nZVI is mediated by colloidal destabilization and hetero-aggregation between algal

Table 1. Main physicochemical parameters of nZVI suspensions prepared in Milli-Q water and OECD algal growth medium.

nZVI (mgL^{-1})	Dispersion Medium	Effective diameter (nm)[1]	Polydispersity Index (PDI)	ζ-potential (mV)	Total surface area ($m^2\ g^{-1}$)[3]	pH	Conductivity(μScm^{-2})
2.5	H_2O	124.5±21.0	0.471	−8.5±5.5	7.19	6	0.011
	OECD medium	284.2±47.0	0.609	−17.5±2.3	3.15	8.2	0.214
5	H_2O	82.1±13.5	0.407	−3.4±1.3	10.91	6	0.007
	OECD medium	183.4±27.4	0.816	−23.6±2.4	4.88	8.2	0.202
25	H_2O	108.9±16.4	0.406	−5.55±2.2	8.22	6	0.0032
	OECD medium	121.3±16.1	0.522	−27.50±2.7	7.38	8.2	0.207
50	H_2O	123.0±12.5	0.451	−7.0±3.3	7.28	6	0.012
	OECD medium	369.1±56.3	0.539	−22.2±2.5	2.43	8.2	0.206

Measurements were conducted at 25°C. Effective diameter and ζ-potential in milli-Q water were performed in the presence of 10 mM KCl. Data are expressed as mean ± standard deviation (SD) when applicable.
[1]: Effective diameter (hydrodynamic diameter measured by dynamic light scattering (DLS).
[3]: Total surface area (SSA) is calculated according to equation 1.

cells and nZVI. To test this hypothesis, events occurring at the bio-interface delimiting algal cells and the external microenvironment were studied by TEM and FTIR analysis (**Figure 6.**). As shown in **Figure 6.b, d**, the presence of electron-dense particles surrounding the external cell wall of the micro algae was identified at 0.075 mgL^{-1} and 2.5 mgL^{-1} of nZVI. However, they were absent in control cells and 0.5 mgL^{-1} nZVI treatment (**Figure 6.a**, **c**, respectively). XEDS analysis assigned Fe as the main constituent of those particles (**Figure 6.i, j**). Furthermore, FTIR analysis (**Figure 6.f and h**) revealed significant changes to the IR signals of algal cell where cell-nZVI attachment occurred (0.075 mgL^{-1} and 2.5 mgL^{-1} nZVI treatments). These changes consisted of a significant reduction of peak absorption at wavenumbers characteristic of asymmetric (2924 cm^{-1}) and symmetric (2853 cm^{-1}) sp^3 C-H stretching vibrations, O-H stretching (3222 cm^{-1}) and bending (1654 cm^{-1}) vibrations in hydroxyl functional groups and C=O stretching vibration (1743 cm^{-1}) of carbonyl functional groups [26]. Interestingly, IR signals remained almost unchanged in the 0.5 mgL^{-1} treatment (**Figure 6.g**) compared to control signal (**Figure 6.f**).

Colloidal destabilization reverts the colloidal singularity and induces nZVI toxicity

To further demonstrate the key role played by colloidal stability in the observed lack of toxicity in the colloidal singularity dose range, a destabilization experiment was performed. Destabilization of the *colloidal singularity* was induced by neutralizing the negative surface charge density of the 0.5 mgL^{-1} nZVI suspension by using a classical hydrolyzing metal coagulant [$Al_2(SO_4)_3$] [11,24]. Al^{3+}, like other polyvalent metal cations, is specifically adsorbed on negatively charge surfaces, thereby reducing the surface charge density [11]. Based on titration experiments with $Al_2(SO_4)_3$ (**Figure 7.a**), 0.01 μM $Al_2(SO_4)_3$ was chosen, as it proved to effectively reduce the ζ-potential of the 0.5 mgL^{-1} nZVI suspension from −28 mV to nearly −10 mV, resulting in partial colloidal destabilization. This $Al_2(SO_4)_3$ had no detrimental effect on the growth of *P. subcapitata* [27] (**Figure 7.b**), nor statistically significant effects on monitored parameters compared to control systems (**Figure 7.c,d**). The addition of $Al_2(SO_4)_3$ to the test system containing 0.5 mgL^{-1} nZVI caused 2-fold reductions in the growth rates of *P. subcapitata* ($p>0.05$) (**Figure 7.b**). This toxic response was accompanied by a statistically significant ($p>0.05$) increase in sedimentation rates (**Figure 7.c**), and a statistically significant ζ-potential reduction of the nZVI-algae colloidal system ($p>0.05$) (**Figure 7.d**).

Discussion

When suspensions of nZVI were prepared in OECD culture medium, we observed an ENM speciation phenomenon [3] as a function of dose. We denoted it as *colloidal singularity* since such a non-monotonic speciation phenomenon as a function of dose has never been previously reported. Interestingly, the ζ-Potential/particle size *vs* nZVI dosage diagrams resembled those of an inverted classical flocculation diagram [11,28]. In the classical flocculation diagram a fixed amount of suspended colloids is exposed to increasing concentrations of a hydrolyzing salt. The multivalent cations of the hydrolyzing salt (commonly Al^{3+} or Fe^{2+}) specifically adsorb on the negatively charged surface of the suspended colloids. This results in a characteristic non-linear behavior on the residual turbidity of the suspensions as a function of the resulting colloid/ions ratio. The classical flocculation diagram presents three differential stages: (1) in a first stage, the dose of coagulant is not enough to destabilize the colloids which

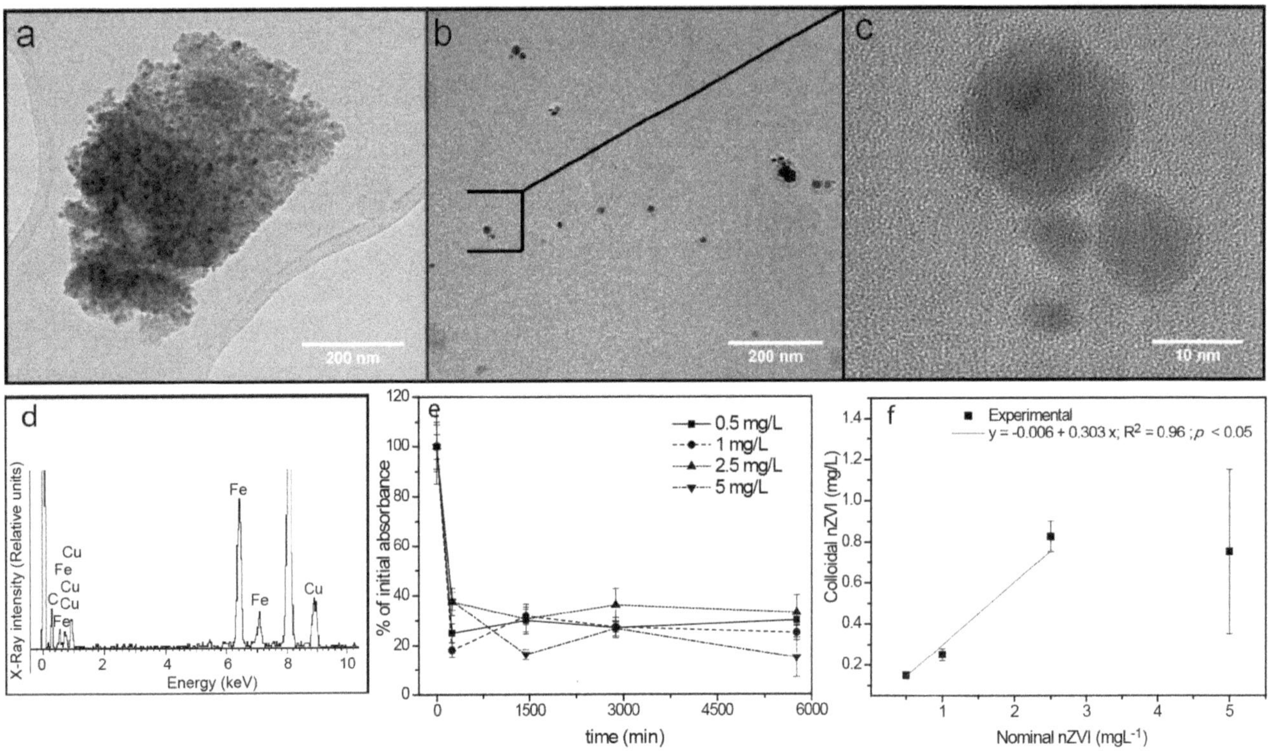

Figure 1. Characterization of nZVI working suspensions under experimental conditions. Representative TEM images of (a) nZVI aggregates in 50 mgL^{-1} nZVI stock suspensions, and (b) small nZVI aggregates and primary sized particles in diluted (5 mg/L) nZVI suspensions. (c) Detail of a small nZVI aggregate integrated by 4 agglomerated individual spherical particles. (d) XEDS spectra of (c), Fe is the main constituent. Ni and Cu intense signals in the XEDS spectra (when present), are artifacts due to the metallic grid supporting the samples, and of the sample holder of the TEM instrument, respectively. (e) Reduction of residual absorbance ($\lambda = 750$ nm) of nZVI suspensions as a result of aggregation and sedimentation at relevant experimental times (0 h, 4 h, 24 h, 48 h and 72 h). (f) Correlation between nominal nZVI concentration (mg/L) and nZVI concentration as relevant colloidal fraction (mg/L) in working suspensions at 72 h.

remained negatively charged and stable. (2) With increasing coagulant concentrations, the adsorbed cations effectively reduce the surface charge of suspended colloids, resulting in colloidal destabilization and removal of colloids. This typically happens in a very narrow coagulant dosage [11,24]. (3) A further increase in coagulant dose produces re-stabilization of colloidal particles by charge reversal resulting in a stable colloidal system. The success of the coagulation process depends entirely upon the adequate shaking [11,24]. In our case, two regions of colloidal destabilization bracketing a stable one appeared with increasing nZVI concentration. In contrast with the classical flocculation diagram, we have a fixed ion concentration (from the culture medium) and a variable number of colloidal particles. However, we also generated a variable colloid/ions ratio, which may result in a similar asymmetry in the colloidal stability of the suspensions. Interestingly, in the ζ-potential/particle size *vs* nZVI dose diagrams, we did not find charge reversal causing stabilization of nZVI in the colloidal singularity, but rather an increase in its electrophoretic mobility (ζ-potential). We hypothesized that the high affinity of anions to be adsorbed on nZVI surfaces [17] may promote an increased anion adsorption on the nZVI surface at the specific nZVI/ion ratios established in the *colloidal singularity* dose range This would result in a net increase in negative surface charge and may cause a reduction of its net surface energy, and of its tendency to interact with bio-interfaces.

The occurrence of the colloidal singularity is highly illustrative as a case study for EHS of ENMs. It clearly exemplifies the extreme relevance of ENMs speciation in their ability to interact with living organisms, resulting in differential biological effects [1,3]. When nZVI appeared stable in its primary size (in the *colloidal singularity*), no detrimental biological effects could be observed. This was true even when its total surface area was higher than that of any aggregated nZVI suspension. This is in apparent disagreement with previous reports [29,30] and the general consensus: a higher total surface implies a higher exposure, and hence bioactivity [2,4,5]. Here, the key point to reconcile the findings is to consider colloidal stability. The concomitant increase in the negative surface charge of nZVI suspensions implies stable colloidal systems. These systems do not cause biological effects due to their lack of tendency to interact with bio-interfaces [8,25,31]. This is true regardless of considerations on their surface area or their intrinsic bioactivity. It was further confirmed by the observed correlation between toxicity and hetero-aggregation tendency, and the destabilization experiment with $Al_2(SO_4)_3$.

Nanoparticle-induced oxidative stress caused by the formation of reactive oxygen species (ROS) has been established as one of the most common paradigms involved in the toxic responses induced by engineered nanoparticles in living organisms [7,8,32,33]. According to this paradigm, when the cellular antioxidant barriers are surpassed, ROS injury can ultimately induce cell death via alterations in cell cycle [33,34] and the activation of apoptotic signaling pathways [7,33]. Our results demonstrate that the toxicity of nZVI to *P.subcapitata* is linked to the production of ROS, cell cycle alterations and pre-apoptotic DNA fragmentation; and therefore can be assigned to the oxidative stress toxicity paradigm [5,7,33]. Furthermore, toxicity, ROS and cell cycle

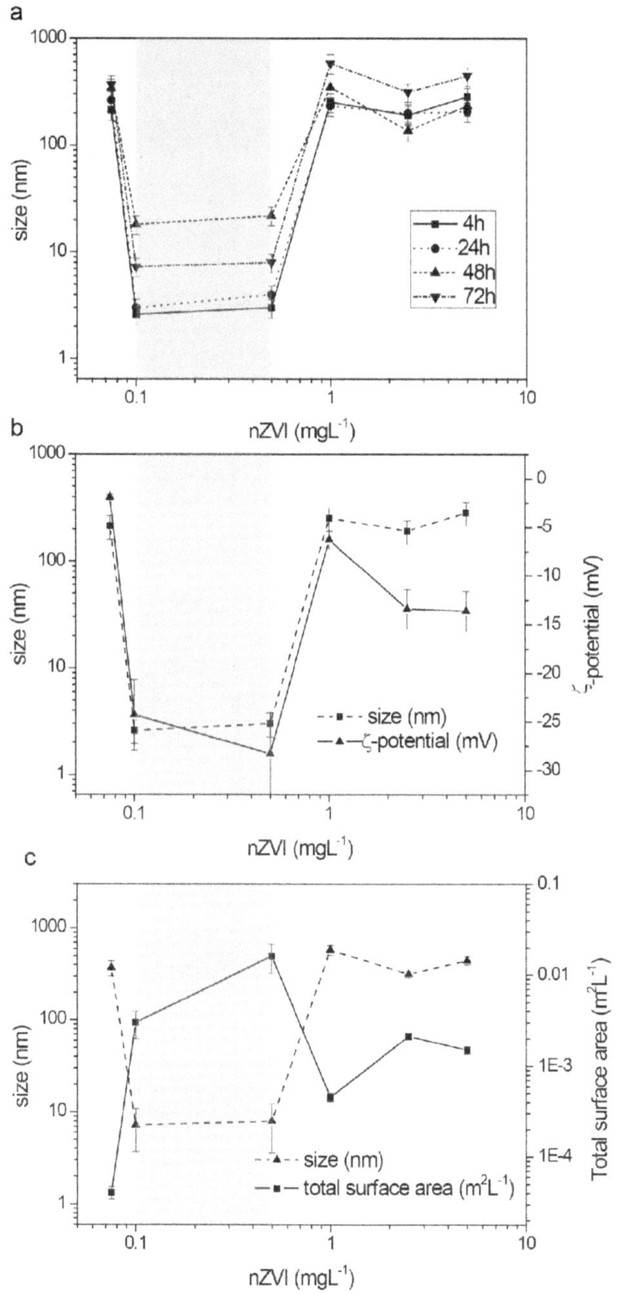

Figure 2. The occurrence of a *colloidal singularity*: Main physicochemical characteristics of colloidal nZVI fractions under experimental conditions. (a) Particle size of a range of nZVI concentrations at 4 h, 24 h, 48 h and 72 h. (b) ζ-potential and particle size for a representative exposure time (4 h). (c) Total surface area (m^2 L^{-1}) and particle size for a representative exposure time (4 h). The concentration range in which the *colloidal singularity* occurred is marked in blue.

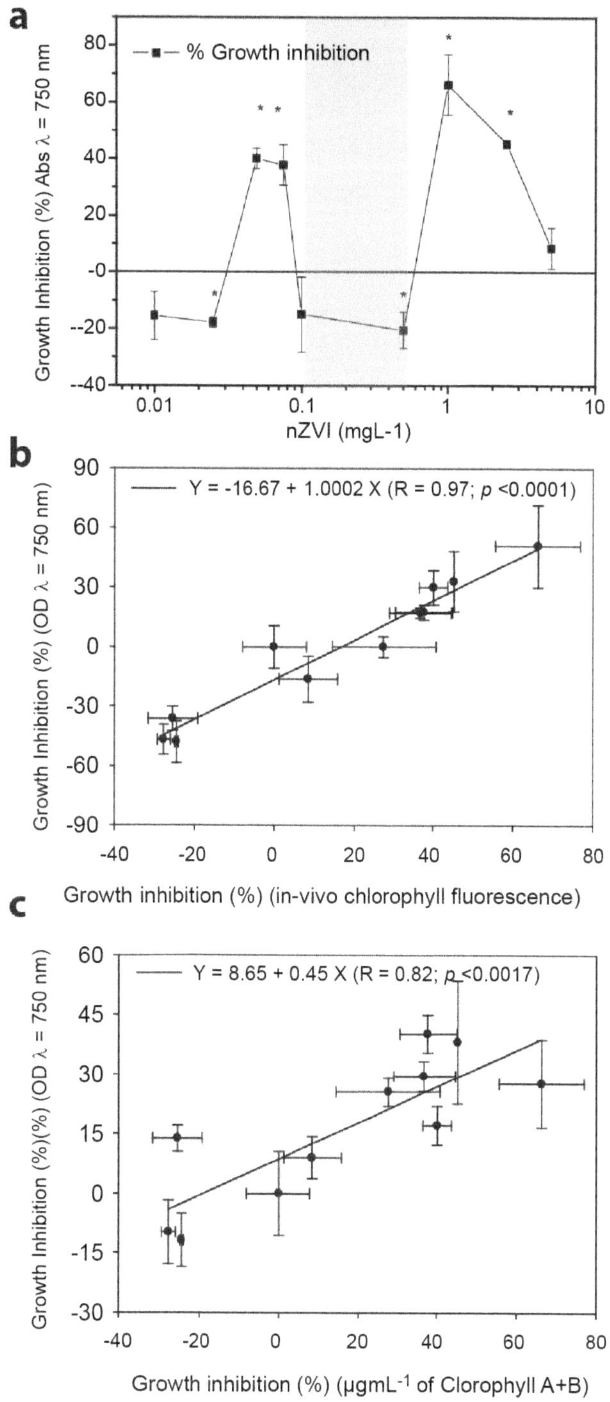

Figure 3. The *colloidal singularity* prevents nZVI toxicity. (a) Growth inhibition of *P. subcapitata* exposed to a linear nZVI dose gradient (0.025–5 mgL^{-1}) based on $OD_{\lambda=750\ nm}$ biomass surrogate. Typically, growth rate of control replicates was 1.4 d^{-1}. The coefficient of variance for 72 h control cultures was 9%. Statistically significant differences ($p<0.05$) are marked by an asterisk. (b,c) Correlations of biomass surrogates for growth inhibition of *P. subcapitata* exposed to an nZVI dose gradient (0.025–5 mgL^{-1}). The concentration range in which the *colloidal singularity* occurred is marked in blue.

alterations are mediated by direct attachment of nZVI particles to the outer cell envelopes of the microalgae, which was confirmed independently by TEM-XEDS and FTIR. To our knowledge, this is the first time that the mechanisms of toxicity of nZVI to a microalga have been elucidated. Keller et al. [35] described the toxicity of several commercial bare and polymer-coated nZVI to several marine and freshwater *microalgae*, including *P. subcapitata*. However, the investigation of the underlying toxic mechanism was beyond the scope of their study. ROS mediated toxicity of nZVI is consistent with previous studies with other biological

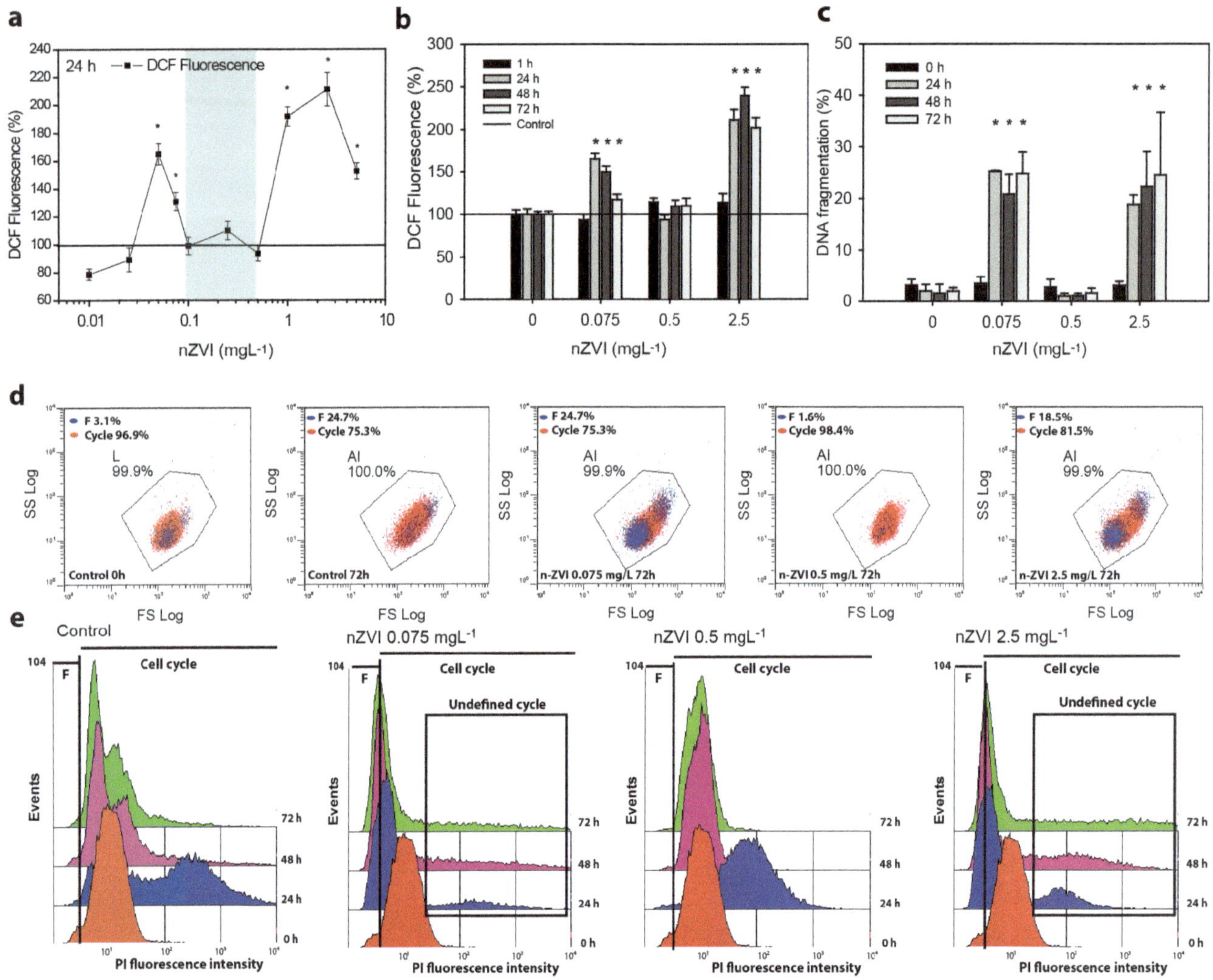

Figure 4. nZVI toxicity is consistently accompanied by ROS and cell cycle alterations. (a) Intracellular ROS formation as DCF fluorescence (% with respect to control levels) of *P. subcapitata* to an nZVI dose gradient (0.025–5 mgL^{-1}) at 24 h of exposure. The concentration range in which the *colloidal singularity* occurred is marked in blue. (b) Intracellular ROS formation as DCF fluorescence (% with respect to control levels) and (c) % of *P. subcapitata* cells showing DNA fragmentation, when exposed to representative nZVI concentrations (0, 0.075, 0.5, 2.5 mgL^{-1}) at relevant exposure times (0 h, 24 h, 48 h and 72 h). (d) Representative *FS-SS* distribution plots of *P. subcapitata* cells exposed to 0, 0.075, 0.5, 2.5 mgL^{-1} nZVI at relevant exposure times (0 h, 24 h, 48 h and 72 h). Red dots depict cells undergoing cell cycle progression, and blue dots depict cells showing DNA fragmentation (e) Representative cell cycle frequency histograms of *P. subcapitata* cells exposed to 0, 0.075, 0.5, 2.5 mgL^{-1} nZVI at relevant exposure times (0 h, 24 h, 48 h and 72 h). F: cell population containing less DNA than G0/G1 phase, which denoted a cell population showing DNA fragmentation. Black chart: cell population undergoing an abnormal cell cycle (poorly defined S and G2/M phases). Statistically significant differences ($p<0.05$) are marked by asterisks.

model systems, including bacteria, fish and mammals [8,16,36–40]. Similarly, toxicity mediated by direct attachment of nZVI to the cell surface is also well documented for bacteria [8,16,38,39,41]. FTIR spectra found in *P. subcapitata* control cells markedly resemble those of cyclic sugars [42–44] such as those found in algal cells walls, mainly cellulose, hemicellulose and pectins. The alterations in those peak vibrations when nZVI presented toxicity may confirm a direct alteration of the main chemical bonds of *P. subcapitata* cell wall. This suggests the presence of active binding sites of nZVI to *P. subcapitata* cell wall. The latter may imply that nZVI is able to induce intracellular specific toxic responses by damaging the structure and changing the physicochemical properties of surface biomolecules [45]. Jiang et al. [45] found that oxide nanoparticles (Al_2O_3, TiO_2 and ZnO) induced chemical and structural changes on bacterial cell envelopes, and that toxicity correlated with the severity of those changes. Similarly, other authors have found direct bond-interaction of oxide nanoparticles with algal and bacterial cell walls mediating toxicity and ROS [46–52]. The work by Gong et al. [46] on the toxicity of NiO nanoparticles to the microalgae *C. vulgaris* is especially interesting. They found that toxicity of NiO nanoparticles was mediated by hetero-aggregation. Furthermore, they found a decrease in the NiO signal and the appearance of a Ni° peak in the X-Ray diffraction spectra of NiO nanoparticles attached to the cell walls. This evidenced that a redox reaction was occurring between the NiO surface and the cell walls, resulting in the reduction of Ni (II) to Ni° [46]. The counterpart of the reduction is the oxidation of surface biomolecules of the organism resulting in chemical bonding. Our FTIR results and the fact that nZVI is covered by layers of iron oxy-hydroxides under oxidant

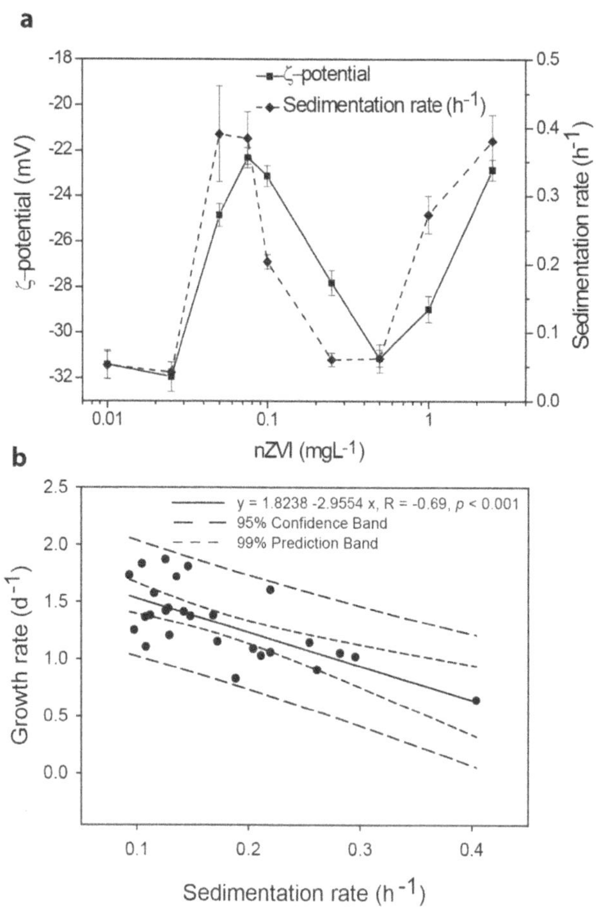

Figure 5. Toxicity of nZVI is mediated by colloidal destabilization and hetero-aggregation between algal cells and nZVI. (a) Sedimentation rates/ζ-potential of test systems exposed to increasing concentrations of nZVI. (b) Correlation of sedimentation rates *vs* algal growth rates.

conditions (Figure **S1**) [17], led us to think that the ROS production mechanism proposed for nZVI, *i.e.*, a Fenton/Haber–Weiss reactions [37–40], would not be the main mechanism of action responsible of ROS induction in *P. subcapitata*. We hypothesize that the spontaneous bio-reduction of some of the surface iron oxy-hydroxides by microalgal surface biomolecules, might be the origin of the attachment of nZVI to *P. subcapitata*. This may result in concomitant ROS induction and cell cycle alterations which is in agreement with the oxidative stress paradigm of redox cycling at the bio-interface [53]. However, this hypothesis requires further investigation.

From our results, it can be inferred that colloidal stability is a crucial regulator of the interaction of negatively charged ENMs with bio-interfaces. In this regard, there is a huge number of published research reporting toxicity of a variety of negatively charged metallic and carbonaceous ENMs on aquatic organisms [8,10,51,52]. A relevant part reported hetero-aggregation [8,10,15,51,52], despite the known negative surface charge of living cells [8,9]. This suggests a key role of negative-negative colloidal interactions in the fate and biological effects of ENMs in aquatic systems, which may be especially relevant for *microalgae* and other aquatic microorganisms in the "colloidal" size region. From our results, surface charge seems to play a major role in these negative-negative interactions because an increase (in absolute value) of nearly 10 mV (from −15 mV to −25 mV) seems to be enough to prevent hetero-aggregation. Similarly, a reduction from −25 mV to −14 mV was enough to induce it. However, would all ENMs and organisms have the same intrinsic tendency to generate hetero-aggregation? Logic seems to indicate that probably not. However, the rules governing the interactions between negatively charged colloids and organism interfaces remain unclear [8], and at present, no systematic studies concerning this issue have been reported.

From a basic methodological view-point, our findings have some broader implications for EHS evaluation of ENMs *in-vitro*. Firstly, when performing EHS screening [5], the need to achieve stable ENM suspensions, by stabilizing agents, polymers, etc., may yield false negative bioactivity results [5] if the ENM suspensions under evaluation are excessively stabilized. Secondly, overlooking colloidal stability may have severe implications in basic aspects of EHS characterization of ENMs. For example, characteristic-dependent ENM bioactivity, when comparing different sizes, shapes, aspect ratio, redox states, etc., can be masked due to a heterogeneous degree of colloidal destabilization of the working suspensions, resulting in misleading conclusions.

Conclusions

The influence of physicochemical state on the bioactivity and toxicity of ENMs is presently under discussion. In the present work, when a linear nZVI dose-range was prepared for EHS characterization, a spontaneous speciation phenomenon occurred which resulted in a qualitatively different physicochemical state of nZVI along the dose gradient. We called it *colloidal singularity*. Basically, nZVI generally appeared in the form of aggregates. However, there was a narrow dose-range (the *colloidal singularity*), between 0.1–0.5 mgL^{-1} nZVI where it appeared highly stable (low particle size and increased ς-potential). nZVI resulted in adverse biological effects when destabilized, but not when stable (the *colloidal singularity*). Furthermore, destabilization of those particular suspensions resulted in toxicity. Considering colloidal stability as a driver of bio-physical interactions allows us to explain our findings and previous apparently contradicting reports. In summary, our work demonstrates the role played by colloidal stability as a fundamental driver of ENM bioactivity, and opens up new perspectives on the relevant factors which may be considered in EHS and bioactivity testing of ENMs.

Experimental Section

nZVI and chemicals

All chemical reagents in the study were of reagent grade and used as purchased without further purification or pretreatment. Nano zero-valent iron (nZVI) powder was purchased from Quantum-Sphere (California, USA). The water used throughout the work was Milli-Q water. nZVI stock suspensions (25 mgL^{-1}) were prepared in Milli-Q water followed by a 10-minute sonication bath (Ultrason Selecta, Spain) to avoid particle agglomeration or aging. 25 mgL^{-1} of nZVI in Milli-Q water was selected as stock suspension since it had the lowest polydispersity index (PDI) and effective diameter (**Table 1**) and was sufficiently concentrated to be used as stock suspension for further EHS testing. Stock suspensions were always freshly prepared a few minutes prior to any physicochemical or biological experiment in order to ensure homogeneity of stock suspensions and prevent differences in aging and/or aggregation state throughout the different physicochemical and biological studies.

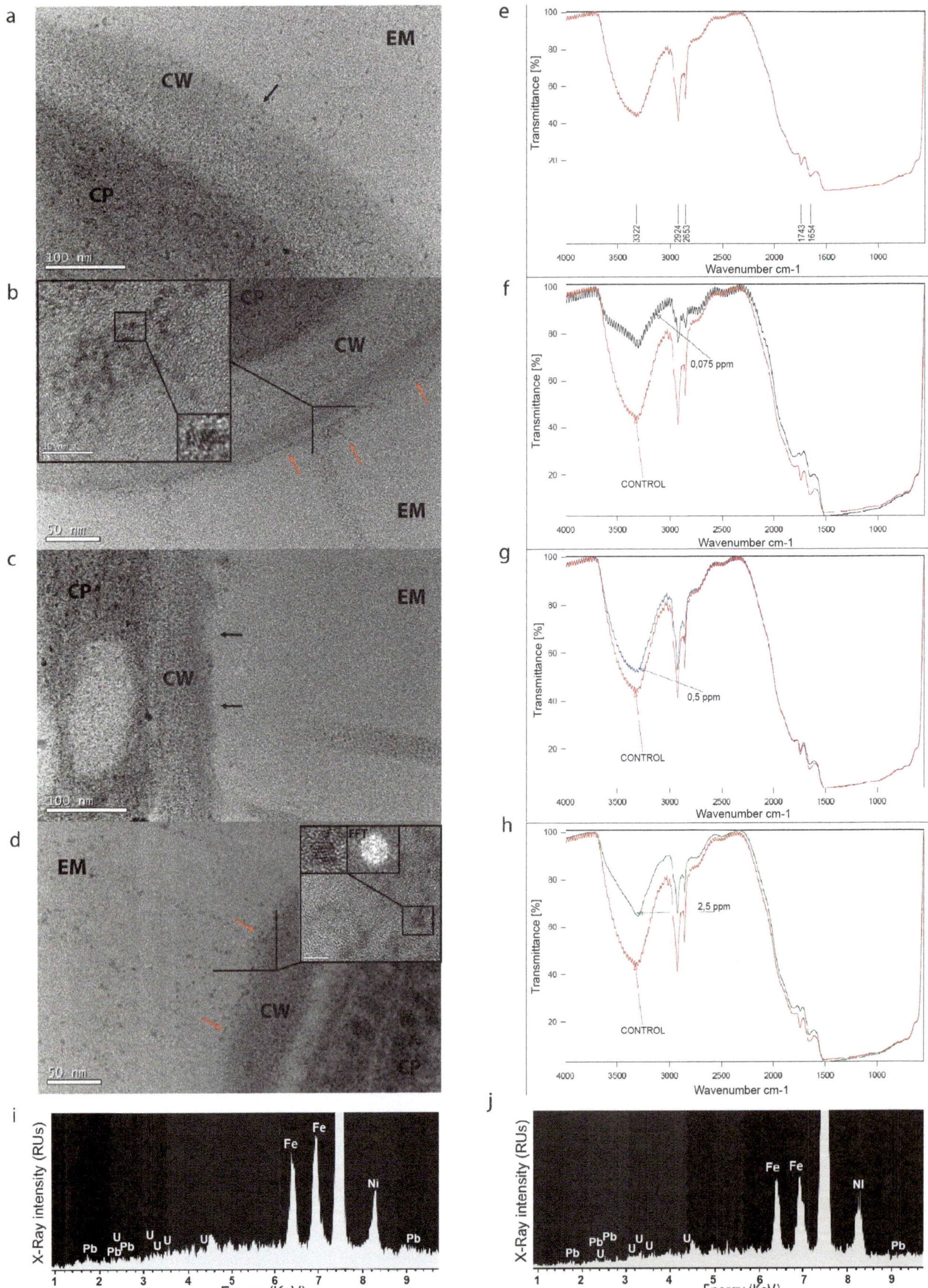
a
EM
CW
CP
100 nm
b
CP
CW
EM
50 nm
c
CP
EM
CW
100 nm
d
EM
FFT
CW
CP
50 nm
e
Transmittance [%]
3322
2924
2853
1743
1654
Wavenumber cm-1
f
0,075 ppm
CONTROL
Transmittance [%]
Wavenumber cm-1
g
0,5 ppm
CONTROL
Transmittance [%]
Wavenumber cm-1
h
2,5 ppm
CONTROL
Transmittance [%]
Wavenumber cm-1
i
X-Ray intensity (RUs)
Fe
Fe
Ni
Pb
U
Energy (KeV)
j
X-Ray intensity (RUs)
Fe
Fe
Ni
Pb
U
Energy (KeV)

Figure 6. Tracking nZVI physicochemical interactions at the bio-nano interface. Representative TEM images showing the interfacial space delimiting algal cells and the external microenvironment of untreated *P. subcapitata* control cells (a), and *P. subcapitata* cells exposed to 0.075 mgL^{-1} (b), 0.5 mgL^{-1} (c) and 2.5 mgL^{-1} (d) of nZVI during 72 h. EM: external microenvironment, CW: Cell wall, CP: Cytoplasm. Black arrows show the CW-EM interface and red arrows the nZVI attached to the outer region of the algal CW in (b) and (d). Insets in (b) and (d) show CW-EM interfacial regions where surface-bound nZVI accumulations were observed on the CW and where nZVI particles were identified. nZVI particles showed round shapes and near squared lattices. Interplanar spaces (2.8 A° and 3.2 A°) are consistent with those of hematite and iron (III) oxide-hydroxide. Fast Fourier Transform (FFT) pattern of the nanoparticle in inset of (d) revealed a truncated octahedral tridimensional structure [56]. Representative FTIR transmission profile of untreated *P. subcapitata* control cells (e) and *P. subcapitata* cells exposed to 0.075 mgL^{-1} (f), 0.5 mgL^{-1}, (g) and 2.5 mgL^{-1} (h) of nZVI during 72 h. Wavenumber (cm^{-1}) of absorption peaks vibrations analyzed in the study are marked in (e). (i, j) XEDS spectra of the insets in b and d, respectively. Fe is the main constituent; Ni and Cu intense signals in the XEDS spectra (when present), are artifacts due to the metallic grid supporting the samples, and of the sample holder of the TEM instrument, respectively.

The current use of nZVI particles in the remediation of water resources contaminated with recalcitrant organic pollutants is based on *in situ* injection of large quantities (up to 50 g nZVI/L) into the subsurface waters [40]. Following removal from solution via sedimentation and dispersal with water flow, a nZVI concentration gradient would develop downward from the point of injection and much lower concentrations as one moves away from the point of injection. While it is still unclear what the actual nZVI concentrations are along such a gradient, in this study, we used what could reasonably be concentrations in the low end of such a gradient.

Physicochemical characterization of nZVI stock suspensions

The Nova 1200 BET (Brunauer –Emmett –Teller) was used to determine the surface area of pristine nZVI. Morphology properties were studied by Transmission Electron Microscopy (TEM) at 200 KeV operating voltage in a JEOL JEM 2100 electron microscope with 0.25 nm point resolution coupled with XEDS (X-Ray Energy Dispersive Spectroscopy). ENMs size measurements, interplanar spacing of crystalline planes and Fast Fourier Transform (FFT) power spectrums were analyzed using Digital Micrograph software, 2.30.542.0 (Gatan Inc.). The size distribution of nZVI was obtained using dynamic light scattering (DLS, Malvern Zetasizer Nano ZS). Measurements were conducted performing 500 rounds measurements and in particle number mode. ς-potential was measured via electrophoretic light scattering combined with phase analysis light scattering in the same instrument. ζ-potential was derived from electrophoretic mobility by applying the Henry equation and the Smoluchowski approximation [11]. Measurements were conducted at ambient temperature (25°C) in Milli-Q water and/or OECD TG 201 assay media without any modification. DLS measurements were performed in particle number mode. The count rates obtained in DLS and ς-potential measurements were higher than 15 kilo counts per second (kcps) for all the measurements included in analysis. The quality of data was also evaluated by visual observation of autocorrelation curves.

Physicochemical characterization of stable nZVI fractions during the toxicity experiments

The suspension stability of nZVI particles was measured over time (0, 4, 24, 48 and 72 h) by spectrophotometry (Hitachi U-2000 spectrophotometer, Japan). 250 mL of the desired nZVI concentrations were prepared in OECD TG 201 algal growth medium in 500 mL Erlenmeyer flasks (chemical composition of OECD TG 201 algal growth medium can be found in **Table S2**.). Samples were maintained under identical experimental conditions as the bioassays. The temporal stability of nZVI particles was evaluated by measuring the residual absorbance at λ = 750 nm. 25 mL of samples were carefully taken from the supernatant of the test flasks to prevent catching sedimented non-colloidal nZVI fractions. Absorbance measurements were performed in 100 mm light path spectrophotometric glass cuvettes (100-OS, Hellma, Germany). Main physicochemical properties of nZVI colloidal fractions were evaluated by measuring ζ-potential and DLS of the supernatant of the test flasks. Total surface area (SSA) was calculated accordingly to the equation 1 proposed by Macé et al. [54] assuming spherical particles:

$$SSA(m^2g^{-1}) = \frac{\pi d^2}{\rho \frac{\pi}{6} d^3} \quad (1)$$

where ρ is the density of the solid particle (6700 Kg m^{-3} accordingly to producer) and *d* is the effective diameter of the particles.

Growth inhibition experiments using *P. Subcapitata*

The green alga *Pseudokirchneriella subcapitata* (*P. subcapitata*) was purchased from Microbiotests. Inc. (Denmark). *P. subcapitata* was routinely grown in 250 ml flasks at 28°C in light, ca. 65 µmol photons m^2 s^{-1} on a rotatory shaker at 135 rpm in OECD TG 201 standard algal culture medium [18] (**Table S2**.) (pH 8.2; Conductivity 0.214 µS/cm^2). Exposure experiments to nZVI suspensions were carried out in 12 mL of OECD TG 201 culture medium in 25 mL glass Erlenmeyer flasks. Before exposure to nZVI, cultures were washed once and re-suspended in fresh culture media to obtain a final optical density ($OD_{\lambda = 750\ nm}$) of 0.1. nZVI was added to achieve the desired concentrations, and cultures were exposed for up to 72 h in a rotary shaker at 135 rpm and 28°C under constant illumination. Growth inhibition experiments were performed in triplicate with serial dilutions essentially as described in the standard OECD TG 201 [18]. Three biomass surrogates were measured: $OD_{\lambda = 750\ nm}$, total chlorophyll content and *in-vivo* chlorophyll fluorescence. $OD_{\lambda = 750\ nm}$ was determined by spectrophotometry in a Hitachi U-2000 spectrophotometer (Japan) by measuring the culture absorbance at λ = 750 nm in 10 mm light path transparent plastic cuvettes. For nZVI nominal concentrations lower than 5 mg/L no interference of nZVI was observed in the determination of $OD_{\lambda = 750\ nm}$ of *P. subcapitata* (data not shown). For chlorophyll content determinations, culture aliquots (250 µL) were extracted in methanol at 4°C for 24 h in darkness. The total chlorophyll content of the extract (chlorophyll a+b) was determined by spectrophotometry as described elsewhere [52]. In-vivo fluorescence of chlorophyll was measured daily by transferring 100 µl of quadruplicate samples of cultures to an opaque black 96 well microtitter plate and by measuring fluorescence (485 nm/645 nm excitation/emission) on a Synergy HT multimode microplate reader (BioTek, USA).

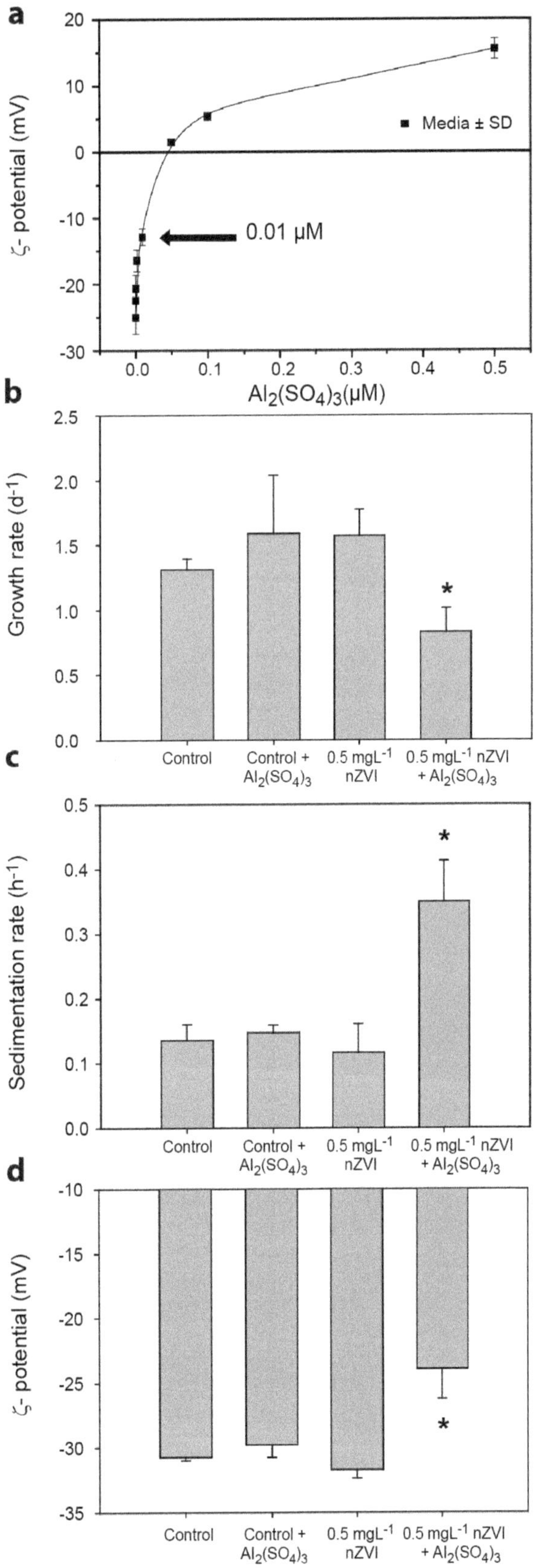

Figure 7. Colloidal destabilization reverts the *colloidal singularity* and induces nZVI toxicity. (a)Titration curve showing an exponential growth profile. Experimental ζ-potential of 0.5 mgL^{-1}+ nZVI $Al_2(SO_4)_3$ 0.01 μM is marked by an arrow. (b,c,d) Growth rates, sedimentation rates and ζ-potential (respectively) of colloidal test systems of *P. subcapitata* exposed to 0.5 mgL^{-1} of nZVI in the presence or absence of 0.01 μM $Al_2(SO_4)_3$. Statistically significant differences ($p<$ 0.05) are marked by an asterisk.

Assessment of ROS generation

Intracellular reactive oxygen species (ROS) produced by *P. subcapitata* was assessed by using the fluorescent probe 2′,7′-dichlorofluorescein diacetate (H_2DCFDA) (Invitrogen Molecular Probes; Eugene, OR, USA). The intracellular oxidation of H_2DCFDA generates 2,7-dichlorofluorescein (DCF), a fluorescent compound that serves as an indicator for hydrogen peroxide and other ROS, such as hydroxyl and peroxyl radicals. A 100 mM H_2DCFDA stock solution was freshly prepared in DMSO under dim light conditions to avoid degradation. Prior to analysis, the green alga was incubated for 30 min at room temperature (23°C) with a final concentration of 100 μM of H_2DCFDA. As a positive control for ROS formation, 3% H_2O_2 (v/v) was used. Fluorescence was monitored on a Synergy HT multimode microplate reader (BioTek,USA) with excitation and emission wavelengths of 488 and 530 nm, respectively. Results were normalized, for differences in cell numbers, by measuring chlorophyll content (as described previously) and expressed as arbitrary fluorescence units (AFU) per μg of chlorophyll a+b.

Cell cycle and cell population dynamics by flow cytometry

Alterations in cell cycle were assessed by propidium iodide (PI) staining and flow cytometry analysis using a Cytomix FL500 MPL flow cytometer (Beckman Coulter Inc., Fullerton, CA, USA). 1 mL of cell suspensions of the nZVI exposure experiments were collected in 1.5 mL eppendorf tubes and fixed by adding 3% paraformaldehyde (final concentration). Samples were kept in darkness at 4°C until further preparation. For PI staining, cells were gently washed with 1 mL PBS and were incubated for 3 h at 37°C in 1 mL of 0.05% Triton X-100, 0.24 mg/mL RNAase A (all from Sigma-Aldrich) in PBS buffer for membrane permeabilization. Cells were stained in 250 mL of staining buffer consisting of 0.05% Triton X-100, 0.24 mg/mL RNase A and 40 μg/mL PI and were incubated for 24 h in the dark at 4°C. Nuclear DNA content was then analysed using a Cytomics FC500 MPL flow cytometer with the MXP software (Beckman Coulter Inc., Fullerton, CA, USA) equipped with an argon-ion excitation wavelength (488 nm). The flow rate was set at 1 μL s^{-1} and at least 10,000 events (algal cells) were recorded. Light scattered by cells was collected at two angles: *FS* and *SS*, which were used in combination to distinguish between different cell populations. Non-algal particles were excluded from the analysis by setting an acquisition threshold value 1 for the FS parameter. Chlorophyll red autofluorescence was collected with a 675 nm long band pass filter (FL4), PI fluorescence was collected with a 575 nm long band pass filter (FL2), and the signal was adjusted to exclude the residual fluorescence signal coming from the chlorophyll red autofluorescence (FL4) (data not shown). Data acquisition was performed using MXP-2.2 software, and the analyses were performed using CXP-2.2 analysis software. Fluorescence was analysed in Log mode (chlorophyll fluorescence) and linear mode (PI fluorescence).

nZVI- algal interaction analyses by TEM microscopy and FTIR

For transmission electron microscopy (TEM) analysis, algal cell suspensions exposed to nZVI were collected by centrifugation using a swinging bucket rotor at low relative centrifugal forces

(RCFs = 1500 g) during 3 min in order to reduce the chance of artifacts [55], washed three times in phosphate buffer (0.1 M Na-phosphate, pH 7.2) to clean samples and remove unattached nZVI particles. After the final centrifugation step nearly 1 mm of cell pellet was formed. Supernatant was removed and 50 µl of 2% bacteriological ultrapure agar (CONDA, Spain) prepared in phosphate buffer was added to obtain 1 to 2 mm agar blocks. Cells were fixed in 3.1% freshly prepared glutaraldehyde in phosphate buffer (0.1 M Na-phosphate, pH 7.2) for 3 h at 4°C. Postfixation was in osmium tetroxide 1% prepared in phosphate buffer (0.1 M Na-phosphate, pH 7.2) for 2 h at 4°C. The samples were then gradually dehydrated in ethanol during 24 h and embedded in Durcupan resin (epoxy) (Fluka) sectioned (100 nm ultrathin films) in a Leica Reichert Ultracut S ultramicrotome and stained with uranyl acetate 2% in water. Samples were observed on Formvar/carbon Cu grids 300 mesh, at 200 KeV operating voltage in a JEOL JEM 2100 electron microscope with 0.25 nm point resolution coupled with XEDS (X-Ray Energy Dispersive Spectroscopy). All reagents used for TEM preparations were EM grade. Sample preparation for TEM microscopy involves centrifugation and washing of the cells, which may alter nZVI distribution on the algal cells [55]. Therefore, nZVI-cell interactions were further confirmed by IR spectroscopy. For Fourier Transformed Infrared (FTIR) analyses, 1 mL of exposed cells was centrifuged (5500 rpm, 5 min) and supernatant was removed. 5 µL of pelletized cells were transferred to 1 mm^2 glass cover-slips and were dried for 2 h at 25°C. Infrared spectra of the algal cell before and after nZVI contact experiments were obtained using a Bruker model IFs 66VFourier Transform Infrared (FTIR) spectrometer in transmission mode. This analysis allowed us to track changes in the chemical bonds and to elucidate the chemical groups involved in nZVI sorption processes.

Colloidal stability of nZVI-*algal* suspensions

Colloidal stability of nZVI-*algal* suspensions was evaluated by measuring ζ-potential, DLS and sedimentation rates of the colloidal system integrated by the algal cells and nZVI. For the determination of the sedimentation rates, a sedimentation system consisting of 15 mL falcon tubes was used. *Algae* exposed to the different nZVI concentrations were collected and $OD_{\lambda = 750\ nm}$ were determined by spectrophotometry (Hitachi U-2000 spectrophotometer, Japan). All colloidal systems were leveled to a homogeneous $OD_{\lambda = 750\ nm}$ of 0.350 with a final volume of 5 mL, therefore sedimentation rates were neither influenced by differences in initial optical densities nor test symmetry (mainly water column height). Residual turbidity ($OD_{\lambda = 750\ nm}$) of the test systems was measured spectrophotometrically as previously described by taking 250 µl of the near meniscus region of the test tubes at specific time intervals (0, 90, 180 min) and measuring the $OD_{\lambda = 750\ nm}$. Sedimentation rates were derived from the differences in residual turbidity of the test systems as a function of time according to equation 2.

$$\mu_{i-j}(h^{-1}) = \frac{\ln X_j - \ln X_i}{t_j - t_i} \qquad (2)$$

where μ_{i-j} is the average specific sedimentation rate (h^{-1}) from time i to j, X_j and X_i are the residual optical density at λ = 750 nm ($OD_{\lambda = 750\ nm}$) at times i to j; and t_j and t_i are time (h). ζ-potential of cells exposed to nZVI samples were measured at the end of the exposure experiments. For that, 2 mL of exposed cells were collected and transferred to 2.5 mL eppendorf tubes and ζ-potential was measured directly via electrophoretic light scattering combined with phase analysis light scattering in a Malvern Zetasizer Nano ZS as describe above.

Colloidal destabilization of nZVI suspensions with $Al_2(SO_4)_3$

nZVI Destabilization experiments were performed in 15 mL Erlenmeyer flask containing 12 mL OECD TG 201 culture medium. Algal cells were exposed to 0.5 mgL^{-1} of freshly prepared nZVI as described in the toxicity exposure experiments. 0.01 µM of $Al_2(SO_4)_3 \times 18\ H_2O$ (final concentration) was added to the test flask at the beginning of the experiments in order to induce colloidal destabilization and heteroaggregation of the nZVI-*algal* system. Growth inhibition, sedimentation rates and ζ-potential of the colloidal systems were measured as described above. The concentration of $Al_2(SO_4)_3$ used for the colloidal destabilization experiments was selected based on the results of a titration experiment of 0.5 mgL^{-1} of nZVI suspended in OECD TG 201 algal growth medium with increasing concentrations of $Al_2(SO_4)_3$ (**Figure 7.a**). 0.01 µM of $Al_2(SO_4)_3$ was selected due to its ability to partially neutralize the negative surface charges of nZVI (0.5 mgL^{-1}) without causing total neutralization or charge reversal.

Replication and Statistical Analysis

All the experimental procedures were performed at least in triplicate. The mean and standard deviations (SD) were calculated for each parameter. Statistical analyses were evaluated by R statistical program. A one-way ANOVA coupled with Student–Newman–Keuls *post-hoc* test was performed for comparing means of the toxicity effect and physicochemical characteristic among different nZVI concentrations. Statistically significant differences were considered to exist when $p<0.05$.

Supporting Information

Figure S1 X-Ray diffraction (XRD) of pristine nZVI powder. X-ray diffraction analysis of pristine nZVI powder showed peaks of Fe° (Fe) and iron oxide formation of magnetite/maghemite (Fe_3O_4/γ-FeOOH) (M) respectively.

Table S1 Size of nZVI primary particles and aggregates by TEM microscopy.

Table S2 Chemical composition of OECD TG 201 standard algal culture medium.

Acknowledgments

Authors want to acknowledge research support services (SIDI) of Universidad Autónoma de Madrid and the Centro Nacional de Microscopía Electrónica (ICTS) for their excellent technical support in TEM, flow cytometry and FTIR analyses. Thanks to Dr. Alberto Martín Molina and Dr. Arturo Moncho Jordá for their helpful discussion in the interpretation of the colloidal stability data.

Author Contributions

Conceived and designed the experiments: IRP SG VL GPR. Performed the experiments: SG VL GPR IRP. Analyzed the data: IRP SG GPR VL. Contributed reagents/materials/analysis tools: FFP FL RR JCB. Wrote the paper: IRP FFP FL RR JCB EGC.

References

1. Schrurs F, Lison D (2012) Focusing the research efforts. Nat Nanotechnol 7: 546–548.
2. Thomas CR, George S, Horst AM, Ji Z, Miller RJ, et al. (2011) Nanomaterials in the Environment: From Materials to High-Throughput Screening to Organisms. ACS Nano 5: 13–20.
3. Westerhoff P, Nowack B (2013) Searching for Global Descriptors of Engineered Nanomaterial Fate and Transport in the Environment. Acc Chem Res 46: 844–853.
4. Handy RD, Cornelis G, Fernandes T, Tsyusko O, Decho A, et al. (2012) Ecotoxicity test methods for engineered nanomaterials: Practical experiences and recommendations from the bench. Environ Toxicol Chem 31: 15–31.
5. Nel A, Zhao Y, Madler L (2013) Environmental Health and Safety Considerations for Nanotechnology. Acc Chem Res 46: 605–606.
6. (2012) Join the dialogue. Nat Nanotechnol 7: 545.
7. Nel A, Xia T, Madler L, Li N (2006) Toxic Potential of Materials at the Nanolevel. Science 311: 622–627.
8. Ma S, Lin D (2013) The biophysicochemical interactions at the interfaces between nanoparticles and aquatic organisms: adsorption and internalization. Environ Sci Process Impacts 15: 145–160.
9. Zhu M, Nie G, Meng H, Xia T, Nel A, et al. (2013) Physicochemical Properties Determine Nanomaterial Cellular Uptake, Transport, and Fate. Acc Chem Res 46: 622–631.
10. Kahru A, Dubourguier HC (2010) From ecotoxicology to nanoecotoxicology. Toxicology 269: 105–119.
11. Gregory J (2006) Particles in Water Properties and Processes. Boca Raton (Florida): Taylor & Francis. 180 p.
12. Nel AE, Madler L, Velegol D, Xia T, Hoek EM, et al. (2009) Understanding biophysicochemical interactions at the nano-bio interface. Nat Mater 8: 543–557.
13. Verma A, Stellacci F (2010) Effect of surface properties on nanoparticle-cell interactions. Small 6: 12–21.
14. Ivask A, ElBadawy A, Kaweeteerawat C, Boren D, Fischer H, et al. (2013) Toxicity Mechanisms in Escherichia coli Vary for Silver Nanoparticles and Differ from Ionic Silver. ACS Nano 8: 374–386.
15. Hartmann NB, Engelbrekt C, Zhang J, Ulstrup J, Kusk KO, et al. (2013) The challenges of testing metal and metal oxide nanoparticles in algal bioassays: titanium dioxide and gold nanoparticles as case studies. Nanotoxicology 7: 1082–1094.
16. Lee C, Kim JY, Lee WI, Nelson KL, Yoon J, et al. (2008) Bactericidal effect of zero-valent iron nanoparticles on Escherichia coli. Environ Sci Technol 42: 4927–4933.
17. Sun Y-P, Li X-Q, Cao J, Zhang W-X, Wang HP (2006) Characterization of zero-valent iron nanoparticles. Adv Colloid Interface Sci 120: 47–56.
18. OECD (2006) OECD TG 201. Guidelines for the Testing of Chemicals. Freshwater Algal and Cyanobacteria, Growth Inhibition Test.
19. Gilbert B, Ono RK, Ching KA, Kim CS (2009) The effects of nanoparticle aggregation processes on aggregate structure and metal uptake. J Colloid Interface Sci 339: 285–295.
20. Phenrat T, Saleh N, Sirk K, Tilton RD, Lowry GV (2006) Aggregation and Sedimentation of Aqueous Nanoscale Zerovalent Iron Dispersions. Environ Sci Technol 41: 284–290.
21. Rosicka D, Sembera J (2011) Assessment of Influence of Magnetic Forces on Aggregation of Zero-valent Iron Nanoparticles. Nanoscale Res Lett 6: 10.
22. Bortner CD, Oldenburg NBE, Cidlowski JA (1995) The role of DNA fragmentation in apoptosis. Trends Cell Biol 5: 21–26.
23. Zhang JH, Xu M (2000) DNA fragmentation in apoptosis. Cell Res 10: 205–211.
24. Henderson RK, Parsons SA, Jefferson B (2010) The impact of differing cell and algogenic organic matter (AOM) characteristics on the coagulation and flotation of algae. Water Res 44: 3617–3624.
25. Long Z, Ji J, Yang K, Lin D, Wu F (2012) Systematic and quantitative investigation of the mechanism of carbon nanotubes' toxicity toward algae. Environ Sci Technol 46: 8458–8466.
26. Crews P, Rodriguez J, Jaspars M (1998) Organic Structure Analysis. University of California, Santa Cruz.
27. Pitre D, Boullemant A, Fortin C (2014) Uptake and sorption of aluminium and fluoride by four green algal species. Chem Cent J 8: 8.
28. Duan J, Gregory J (2003) Coagulation by hydrolysing metal salts. Adv Colloid Interface Sci 100–102: 475–502.
29. Rabolli V, Thomassen LC, Uwambayinema F, Martens JA, Lison D (2011) The cytotoxic activity of amorphous silica nanoparticles is mainly influenced by surface area and not by aggregation. Toxicol Lett 206: 197–203.
30. Van Hoecke K, De Schamphelaere KA, Van der Meeren P, Lucas S, Janssen CR (2008) Ecotoxicity of silica nanoparticles to the green alga Pseudokirchneriella subcapitata: importance of surface area. Environ Toxicol Chem 27: 1948–1957.
31. Lin D, Ji J, Long Z, Yang K, Wu F (2012) The influence of dissolved and surface-bound humic acid on the toxicity of TiO2 nanoparticles to Chlorella sp. Water Res 46: 4477–4487.
32. Stone V, Donaldson K (2006) Nanotoxicology: signs of stress. Nat Nanotechnol 1: 23–24.
33. Xia T, Kovochich M, Liong M, Madler L, Gilbert B, et al. (2008) Comparison of the mechanism of toxicity of zinc oxide and cerium oxide nanoparticles based on dissolution and oxidative stress properties. ACS Nano 2: 2121–2134.
34. Burhans WC, Heintz NH (2009) The cell cycle is a redox cycle: Linking phase-specific targets to cell fate. Free Radic Biol Med 47: 1282–1293.
35. Keller AA, Garner K, Miller RJ, Lenihan HS (2012) Toxicity of Nano-Zero Valent Iron to Freshwater and Marine Organisms. PloS one 7: e43983.
36. Li H, Zhou Q, Wu Y, Fu J, Wang T, et al. (2009) Effects of waterborne nano-iron on medaka (Oryzias latipes): antioxidant enzymatic activity, lipid peroxidation and histopathology. Ecotoxicol Environ Saf 72: 684–692.
37. Phenrat T, Long TC, Lowry GV, Veronesi B (2009) Partial oxidation ("aging") and surface modification decrease the toxicity of nanosized zerovalent iron. Environ Sci Technol 43: 195–200.
38. Sevcu A, El-Temsah YS, Joner EJ, Cernik M (2011) Oxidative stress induced in microorganisms by zero-valent iron nanoparticles. Microbes Environ 26: 271–281.
39. Auffan Ml, Achouak W, Rose JR, Roncato M-A, Chanéac C, et al. (2008) Relation between the Redox State of Iron-Based Nanoparticles and Their Cytotoxicity toward Escherichia coli. Environ Sci Technol 42: 6730–6735.
40. Grieger KD, Fjordoge A, Hartmann NB, Eriksson E, Bjerg PL, et al. (2010) Environmental benefits and risks of zero-valent iron nanoparticles (nZVI) for in situ remediation: Risk mitigation or trade-off? J Contam Hydrol 118: 165–183.
41. Li Z, Greden K, Alvarez PJ, Gregory KB, Lowry GV (2010) Adsorbed polymer and NOM limits adhesion and toxicity of nano scale zerovalent iron to E. coli. Environ Sci Technol 44: 3462–3467.
42. Garside P, Wyeth P (2003) Identification of Cellulosic Fibres by FTIR Spectroscopy: Thread and Single Fibre Analysis by Attenuated Total Reflectance. Stud Conserv 48: 269–275.
43. Hadjoudja S, Deluchat V, Baudu M (2010) Cell surface characterisation of Microcystis aeruginosa and Chlorella vulgaris. J Colloid Interface Sci 342: 293–299.
44. Lugo-Lugo V, Barrera-Diaz C, Urena-Nunez F, Bilyeu B, Linares-Hernandez I (2012) Biosorption of Cr(III) and Fe(III) in single and binary systems onto pretreated orange peel. J Environ Manage 112: 120–127.
45. Jiang W, Yang K, Vachet RW, Xing B (2010) Interaction between Oxide Nanoparticles and Biomolecules of the Bacterial Cell Envelope As Examined by Infrared Spectroscopy. Langmuir 26: 18071–18077.
46. Gong N, Shao K, Feng W, Lin Z, Liang C, et al. (2011) Biotoxicity of nickel oxide nanoparticles and bio-remediation by microalgae Chlorella vulgaris. Chemosphere 83: 510–516.
47. Khan SS, Mukherjee A, Chandrasekaran N (2011) Studies on interaction of colloidal silver nanoparticles (SNPs) with five different bacterial species. Colloids Surf B Biointerfaces 87: 129–138.
48. Khan SS, Srivatsan P, Vaishnavi N, Mukherjee A, Chandrasekaran N (2011) Interaction of silver nanoparticles (SNPs) with bacterial extracellular proteins (ECPs) and its adsorption isotherms and kinetics. J Hazard Mater 192: 299–306.
49. Pakrashi S, Dalai S, T CP, Trivedi S, Myneni R, et al. (2013) Cytotoxicity of aluminium oxide nanoparticles towards fresh water algal isolate at low exposure concentrations. Aquat Toxicol 132–133: 34–45.
50. Sadiq IM, Dalai S, Chandrasekaran N, Mukherjee A (2011) Ecotoxicity study of titania (TiO(2)) NPs on two microalgae species: Scenedesmus sp. and Chlorella sp. Ecotoxicol Environ Saf 74: 1180–1187.
51. Rodea-Palomares I, Boltes K, Fernandez-Pinas F, Leganes F, Garcia-Calvo E, et al. (2011) Physicochemical characterization and ecotoxicological assessment of CeO2 nanoparticles using two aquatic microorganisms. Toxicol Sci 119: 135–145.
52. Rodea-Palomares I, Gonzalo S, Santiago-Morales J, Leganes F, Garcia-Calvo E, et al. (2012) An insight into the mechanisms of nanoceria toxicity in aquatic photosynthetic organisms. Aquat Toxicol 122–123: 133–143.
53. von Moos N, Slaveykova VI (2014) Oxidative stress induced by inorganic nanoparticles in bacteria and aquatic microalgae – state of the art and knowledge gaps. Nanotoxicology 8: 605–630.
54. Macé C, Desrocher S, Gheorghiu F, Kane A, Pupeza M, et al. (2006) Nanotechnology and groundwater remediation: A step forward in technology understanding. Remediation Journal 16: 23–33.
55. Schrand AM, Schlager JJ, Dai L, Hussain SM (2010) Preparation of cells for assessing ultrastructural localization of nanoparticles with transmission electron microscopy. Nat Protoc 5: 744–757.
56. Zheng RK, Gu H, Xu B, Fung KK, Zhang XX, et al. (2006) Self-Assembly and Self-Orientation of Truncated Octahedral Magnetite Nanocrystals. Adv Mater 18: 2418–2421.

3

Interacting Microbe and Litter Quality Controls on Litter Decomposition: A Modeling Analysis

Daryl Moorhead[1,2,3*¶], **Gwenaëlle Lashermes**[2,3¶], **Sylvie Recous**[2,3], **Isabelle Bertrand**[2,3]

1 Department of Environmental Sciences, University of Toledo, Toledo, Ohio, United States of America, **2** INRA, UMR614 Fractionnement des AgroRessources et Environnement, Reims, France, **3** Université Reims-Champagne Ardenne, UMR614 Fractionnement des AgroRessources et Environnement, Reims, France

Abstract

The decomposition of plant litter in soil is a dynamic process during which substrate chemistry and microbial controls interact. We more clearly quantify these controls with a revised version of the Guild-based Decomposition Model (GDM) in which we used a reverse Michaelis-Menten approach to simulate short-term (112 days) decomposition of roots from four genotypes of *Zea mays* that differed primarily in lignin chemistry. A co-metabolic relationship between the degradation of lignin and holocellulose (cellulose+hemicellulose) fractions of litter showed that the reduction in decay rate with increasing lignin concentration (LCI) was related to the level of arabinan substitutions in arabinoxylan chains (i.e., arabinan to xylan or A:X ratio) and the extent to which hemicellulose chains are cross-linked with lignin in plant cell walls. This pattern was consistent between genotypes and during progressive decomposition within each genotype. Moreover, decay rates were controlled by these cross-linkages from the start of decomposition. We also discovered it necessary to divide the Van Soest soluble (labile) fraction of litter C into two pools: one that rapidly decomposed and a second that was more persistent. Simulated microbial production was consistent with recent studies suggesting that more rapidly decomposing materials can generate greater amounts of potentially recalcitrant microbial products despite the rapid loss of litter mass. Sensitivity analyses failed to identify any model parameter that consistently explained a large proportion of model variation, suggesting that feedback controls between litter quality and microbial activity in the reverse Michaelis-Menten approach resulted in stable model behavior. Model extrapolations to an independent set of data, derived from the decomposition of 12 different genotypes of maize roots, averaged within <3% of observed respiration rates and total CO_2 efflux over 112 days.

Editor: Andrew C. Singer, NERC Centre for Ecology & Hydrology, United Kingdom

Funding: This study was supported by United States National Science Foundation grants DEB-0946257 and DEB-0918718, and Institut National de la Recherche Agronomique. The funders had no role in study design, data collection and analysis, decision to publish, or preparation of the manuscript.

Competing Interests: The authors have declared that no competing interests exist.

* Email: daryl.moorhead@utoledo.edu

¶ These authors are co-first authors on this work.

Introduction

Recent studies are challenging the ways that we have traditionally perceived and modeled decomposition. Decomposers were once thought to rapidly degrade labile fractions of plant litter, such as carbohydrates and proteins, leaving more recalcitrant compounds, like lignin, to become the foundation of stabilized soil organic matter [1,2,3]. However, microbial products rather than plant lignin actually may represent the larger fraction of SOM [4,5], and lignin *per se* may not persist throughout decay [6]. Unfortunately, many mathematical models are based on the traditional view of decomposition. In part this is because the chemical composition of litter has often been evaluated by proximate carbon analysis, typically yielding three, qualitatively different pools of compounds: polar and nonpolar extractives, acid hydrolysable materials, and acid non-hydrolysable materials. These pools provide a convenient structural framework for modeling, but lack the resolution to address finer scale biochemical transformations revealed by more contemporary studies [4,5,6]. Mathematical models must change to reflect these observations.

One of the changes needed in traditional decomposition models is to explicitly simulate the activities of microorganisms. If microbial contributions to SOM are more substantial than lignin, then the litter compounds fueling microbial activity may be more important to C stabilization than the plant lignin pool. This point was suggested by Smith et al. [7] who noted that the proportion of initial litter C remaining over time (125 days) was directly related to the initial rate of decomposition prior to lignin decay. In other words, the incorporation of C into microbial biomass increased its overall persistence despite an initially higher fractional loss through respiration [2,8]. A simple explanation is that the higher carbon use efficiency (CUE) of readily decayed materials, like sugars and proteins, transfers a larger fraction of substrate C to microbial biomass than lignin, and that these microbial products are more persistent [4,5]. These results suggest that the early stage of litter decay when microbial activities are greatest should be an integral component of mathematical models.

A second change needed in decomposition models is how they respond to differences and interactions between the chemical constituents of decaying litter. Despite the questionable role of plant lignin in decomposition, both the initial lignin content and the lignocellulose index (LCI) of litter are often the best predictors of decay rate [1,3]. The lignocellulose index is normally calculated as the ratio of the acid non-hydrolysable/(non-hydrolysable+ hydrolysable) products of proximate C analysis (see above), and assumed to represent the lignin (non-hydrolysable) and holocellulose (hydrolysable) fractions of litter. Although this assumption may be nearly true for fresh litter, both microbial and lignin degradation products increasingly contribute to proximate C fractions during decay [5,6]. Thus LCI and proximate C fractions are ambiguous metrics of litter chemistry and models using them to regulate decomposition conflate litter quality controls with the decomposition process. In reality, plant cell walls are usually the largest component of plant litter and have a specific biophysical structure of interconnected saccharide and phenolic polymers [9]. This arrangement partly explains the relatively consistent relationship between LCI and decay, but raises questions about the precise relationships between lignin and other litter constituents [10,11,12] needed to more accurately model substrate dynamics during decomposition.

Few empirical studies have examined litter chemistry during decay with sufficient resolution to improve models beyond the lignin or LCI controls normally calculated. However, Machinet et al. [11,13,14] examined the detailed litter chemistry of maize (*Zea mays* L.) roots for naturally occurring genotypes decomposing in laboratory incubations. These genotypes varied primarily in their lignin content. Long-term C losses (>200 days) were most often correlated with compounds associated with lignin and cross-linked between lignin and polysaccharides, whereas short-term (< 200 days) losses were more closely related to soluble compounds and cell wall sugars [11]. Different loss rates of different sugars (e.g., glucans, xylans and arabinans) suggested that the relationship between lignin and decomposition was partly mediated by the specific composition of the polysaccharide fraction of the litter, probably because arabinan substitution in xylan chains interferes with the degradation of hemicellulose [9]. In other words, the sugar composition of cell walls appears to define the transition from short- to long-term patterns of decay, providing a mechanistic explanations for patterns in microbial activity, as well as the empirical relationships between LCI and decay most often used in decomposition models [15].

The overall goal of this study was to simulate the relationships between litter decay, microbial production and litter chemistry at the early stage of decomposition, using data collected by Machinet et al. [11,13] to test and refine the Guild Decomposition Model (GDM) developed by Moorhead and Sinsabaugh [16]. Our two specific objectives were to simulate patterns of microbial productivity during decomposition as litter chemistry changed, and to simulate the relationship between measured changes in cell wall chemistry (sugars and lignin) and decay rate. We chose GDM because it calculates the decay of specific substrate pools as a combined function of both microbial activity and substrate characteristics, using the Michaelis-Menten equation of substrate saturation. The maximum velocity of decomposition (V_{max}) is thus a function of microbial biomass [16], so that estimated decay rates reflect both the changing chemical composition of the decomposing litter and microbial productivity.

Methods

Our general approach was to revise GDM to use the empirical data collected by Machinet et al. [11,13,14] to derive parameters needed to simulate decomposition. Then, we analyzed the data from a 112-day laboratory study of four maize genotypes that differed from one another in initial litter chemistry [13,14], to evaluate interactions between litter qualities likely controlling the decomposition process, and to obtain estimates for parameter values used to describe these processes. Next, we selected two key parameters for GDM, which preliminary analyses indicated varied with initial litter chemistry and had large effects on model behavior (below). We then optimized these parameters to produce the best possible fit between simulated and observed patterns of decomposition, with respect to CO_2 efflux and chemical transformations in decaying litter. Best-fit parameter values were in turn compared to initial litter chemical characteristics to determine possible relationships. These relationships were then used to derive parameter values needed to simulate CO_2 efflux during decomposition of maize roots for 12 additional genotypes [11].

Experimental data

The data used to drive the revised GDM model were obtained from Machinet et al. [13]. Briefly fine roots (diameter 2–3 mm) of four natural genotypes of maize (F2, F2bm1, F292 and F292bm3), which differed in their chemical composition (Table 1), were cut into 5 mm lengths, added to soil in laboratory microcosms and incubated at 15°C. Potassium-nitrate fertilizer was added to microcosms to insure no nitrogen limitation [17]. They monitored CO_2-C efflux on days 3, 7, 10, 14, 21, 29, 36, 42, 51, 57, 70, 80, 87, 95 and 112. Chemical characteristics of litter were determined on days 0 (initial), 14, 36, 57 and 112, on roots manually removed from soils. Machinet et al. [11] used the same experimental methods for 12 additional maize genotypes, but examined only the initial litter chemistry and measured CO_2 efflux during decomposition.

The suite of litter chemical characteristics reported by Machinet et al. [11,13] in root residues included C and N content, Van Soest soluble C (C-SOL) and N (N-SOL) contents, and cell wall polysaccharides: glucan (Glu), arabinan (Ara) and xylan (Xyl), which are the major polymer carbohydrates in graminaea, as well as galactan (Gal) and uronic acids contents (galacturonic (Ac Gal) and glucuronic (Ac Glu)). They also determined Klason lignin (KL) and the lignin monomers, guaiacyl (G) and syringyl (S), as well as ester-linked *p*-coumaric (pCA) and ferulic acids (FA_{ester}), and ether-linked ferulic acids (FA_{ether}). From these data we calculated the sum of cell wall sugars ($\sum$Sug), arabinan:xylan ratio (A:X), carbon:nitrogen ratio (C:N), lignocellulose index ($LCI = C_3/[C_2+C_3]$)) with C_1 (= C-SOL), C_2 (= $\sum$Sug) and C_3 (= KL), and non-detergent fiber (NDF) content.

Machinet et al. [13] estimated mass loss of litter based on cumulative CO_2 carbon efflux. Concentrations of litter chemical fractions at each date were multiplied by the estimated mass of remaining litter to estimate pool sizes during decomposition. Decay rate coefficients (k_i) were calculated for chemical constituents C_1, C_2 and C_3 for all 4 litter types over all 4 periods of observation (days 0–14, 14–36, 36–57 and 57–112), as the difference in the natural log of pool size between observation dates, divided by the time period.

Model revisions

This model was programmed in MATLAB (The MathWorks Inc., Natick, USA). We revised GDM to use Reverse Michaelis-Menten (RMM) functions to calculate decay rates as functions of

Table 1. Model parameters and state variables.

Factor	Description	Values[a] (genotype)	Unit
k_{1max}	Decay rate coefficient of substrate C_1	0.1	day^{-1}
k_{2max}	Maximum decay rate coefficient of substrate C_2	0.047[b]	day^{-1}
k_{3max}	Maximum decay rate coefficient of substrate C_3	0.001	day^{-1}
e_1	Substrate C_1 use efficiency	0.4	unitless
e_2	Substrate C_2 use efficiency	0.3	unitless
g	Coefficient for basal respiration	0.001	day^{-1}
BC_{max}	Maximum Biomass: total C ratio	0.05	unitless
LCI_T	LCI value at which C_3 decay starts	0.42 (F2)	unitless
		0.36 (F2bm1)	
		0.46 (F292)	
		0.45 (F292bm3)	
LCI_{max}	LCI value at which decay stops	0.7	unitless
K_{Bj1}	Half-saturation coefficient of all Guilds for substrate C_1	10	$mg\ C{\cdot}kg^{-1}$ soil
K_{B22}	Half-saturation coefficient of Guild 2 for	29 (F2)	$mg\ C{\cdot}kg^{-1}$
	substrate C_2	33 (F2bm1)	soil
		23 (F292)	
		19 (F292bm3)	
K_{B32}	Half-saturation coefficient of Guild 3 for	300	$mg\ C{\cdot}kg^{-1}$
	substrate C_2		soil
K_{B33}	Half-saturation coefficient of Guild 3 for substrate C_3	500	$mg\ C{\cdot}kg^{-1}$ soil
C_{1T}	Non-decomposable fraction of C_1	0.61[b] (F2)	unitless
		0.60 (F2bm1)	
		0.71 (F292)	
		0.90 (F292bm3)	
C_1	Litter C in C_1 pool (Van Soest soluble)	364 (F2)	$mg\ C{\cdot}kg^{-1}$
		419 (F2bm1)	soil
		361 (F292)	
		288 (F292bm3)	
C_2	Litter C in C_2 pool (acid hydrolysable fraction)	1238 (F2)	$mg\ C\ kg^{-1}$
		1166 (F2bm1)	soil
		1295 (F292)	
		1387 (F292bm3)	
C_3	Litter C in C_3 pool (acid non-hydrolysable	397 (F2)	$mg\ C{\cdot}kg^{-1}$
	fraction)	415 (F2bm1)	soil
		344 (F292)	
		325 (F292bm3)	
B_1	C in Guild 1	10	$mg\ C{\cdot}kg^{-1}$
			soil
B_2	C in Guild 2	10	$mg\ C{\cdot}kg^{-1}$
			soil
B_3	C in Guild 3	5	$mg\ C{\cdot}kg^{-1}$
			soil

[a]Initial values for variables.
[b]Empirically determined from experimental data.

microbial activity [18]: $dC_i/dt = k_i{\cdot}C_i{\cdot}B_j/(K_{Bji}+B_j)$, where C_i is the amount of substrate i, B_j is the amount of microbial biomass in guild j, and K_{Bji} is the half-saturation coefficient of guild j for substrate i. Note that we altered the more common RMM approach by using biomass (B) in place of an enzyme (E) pool, which assumes a constant ratio of E:B resulting from constitutive enzyme production [16,18]. We limited access to substrate pools by guilds: guild 1 was assumed to access only soluble resources (C_1); guild 2 was assumed to access both C_1 and C_2; guild 3 could access C_1, C_2 and C_3 (Figure 1).

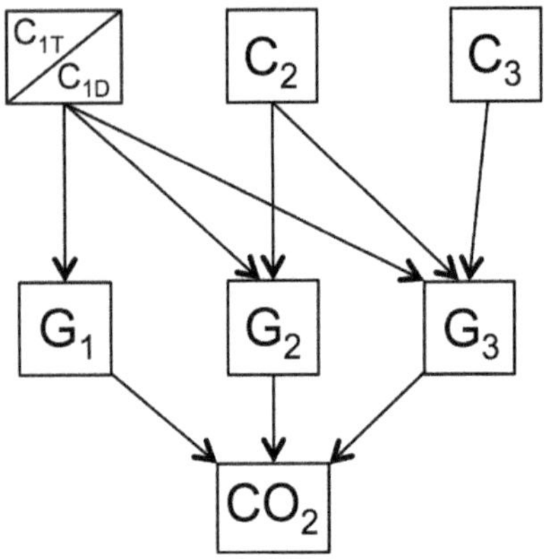

Figure 1. Carbon flow diagram for revised Guild Decomposition Model (GDM) simulating *Zea mays* root decomposition. Litter pools are two Van Soest soluble fractions, a decomposable fraction (C_{1D}) and a resistant fraction (C_{1T}), acid hydrolysable (C_2) and acid non-hydrolysable (C_3) fractions, and microbial guilds of opportunists (G_1), cellulolytic decomposers (G_2) and lignolytic decomposers (G_3).

A final revision to GDM was necessary to capture the dynamics of the soluble pool (C_1), a fraction of which persisted throughout the study by Machinet et al. [13]. This persistent fraction (C_{1T}) varied between genotypes and was considered to be non-decomposable during the time frame of the study (112 days).

Parameters and state variables

All parameters and state variables are reported in Table 1. Values of C_{1T}, the persistent fraction of the initial C_1 pool, were estimated from the average values of C_1 over time for each genotype (Table 1). We selected a decay rate coefficient (k_1) of 0.1 for the decomposable fraction of the C_1 pool (C_{1D}), due to its rapid loss.

The RMM approach made it possible to simplify the balance of C_2 and C_3 decay as functions of lignocellulose index (LCI) according to Moorhead et al. [19] (Text S1). In brief, the decay rate coefficients (k_2 and k_3) were described as linear functions of LCI: $k_i = m_i \cdot LCI + k_{imax}$, given empirically observed maximum values (k_{imax}) and slopes (m_i). This approach assumes that $k_3 = 0$ at a threshold level of $LCI = LCI_T$, wherein $LCI_T = 0.4$ [20], which defines the point at which LCI changes from being solely determined by C_2 decay ($LCI \leq LCI_T$) to also being determined by C_3 decay ($LCI > LCI_T$). The value of k_{2max} (maximum coefficient of C_2 decay) was estimated as the intercept of the linear regression of the observed decay rate coefficients for C_2 (k_2) against litter LCI for the four maize genotypes (Figure 2a; Table 1) over days 14–112, excluding values estimated over days 0–14 when we assumed that microorganisms had not begun to fully utilize pool C_2. Values of k_2 used during simulations were then estimated according to LCI of remaining litter. This revision to GDM provided a closer fit to observed patterns of C_2 decay (Figure 2a; $N = 12$, $R^2 = 0.97$, $P < 0.01$) than Moorhead and Sinsabaugh [16]. The value of k_{3max} was set to 0.001 because there was little evidence of C_3 decomposition for any litter type during incubations [13].

We selected identical values of K_{B11}, K_{B12} and K_{B13}, the half-saturation coefficients for the utilization of pool C_1 by all three guilds, assuming all microorganisms have similar affinities for soluble substrates (Table 1). We initially set K_{B22} and K_{B32} at the same values, again assuming that organisms capable of using the resource would have similar half-saturation coefficients, but higher than those for C_1, assuming that C_2 was generally less decomposable than C_1.

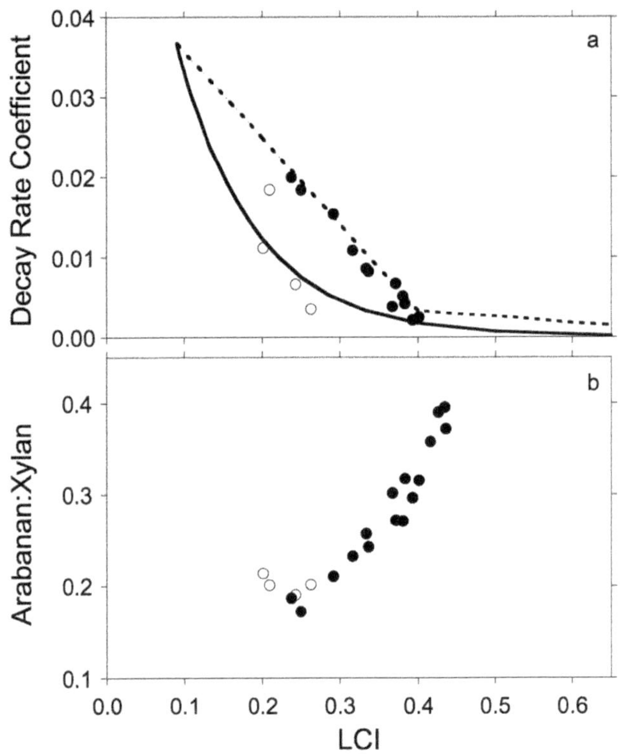

Figure 2. Observed and simulated decay rate coefficients for holocellulose and araban:xylan composition of remaining litter at residual values of lignocellulose index (LCI). Relationships between observed and simulated: a. decay rate coefficients (k_2) for litter pool C_2 and lignocellulose index (LCI) of litter (solid line is simulated according to Moorhead and Sinsabaugh [16], dashed line is simulated according to Moorhead et al. [19]), b. relationship between arabinan:-xylan (A:X) and LCI contents of decaying litter. Open circles are observations based on initial litter chemistry (day 0) and solid circles are from observations over time.

We then selected an even higher value for K_{B33} assuming that C_3 was even less decomposable than C_2. We are unaware of any published values for these parameters, and so followed the rationale of Moorhead and Sinsabaugh [16] in choosing values reflecting relative access to substrates of different qualities.

GDM requires an estimate of initial decomposer biomass, which it divides among three distinct microbial pools: early opportunists (guild 1), subsequent decomposition specialists (guild 2) and a final group of lignin degraders (guild 3) (Figure 1). Machinet et al. [13] estimated that microorganisms colonizing maize roots contributed 0–11% of the initial litter carbon content, and other studies suggest that microbial biomass rarely exceeds 2–3% of soil organic matter [21,22]. We set the initial microbial biomass pool at 1.25% (25 $mgC \cdot kg^{-1}$ soil) of total litter mass, assuming 10 $mgC \cdot kg^{-1}$ soil for guilds 1 and 2, and 5 $mgC\ kg^{-1}$ soil for guild 3 [16].

We estimated microbial production as the difference between the quantity of carbon released from decaying substrates and the amount mineralized through both growth- and maintenance-associated respiration. We assumed that this difference was immobilized in microbial biomass. GDM also calculates microbial turnover necessary to keep total microbial C less than 5% of the total system's organic C (microorganisms+substrates; Table 1). We estimated carbon use efficiency (CUE) as the difference between the amounts of C released from decaying substrates and mineralized through respiration, divided by the amount of C released by decomposition [23].

Optimizing parameter values

The values of two key model parameters that showed correlations with initial chemistry, LCI_T, and K_{B22}, were optimized for the four maize genotypes [13] using the *fmincon* function of the MATLAB Optimization Toolbox (The MathWorks, Inc., Natick, USA), which applies a sequential quadratic programming algorithm. The objective function was the sum of the root mean square errors calculated between the experimental and simulated CO_2 efflux rates, cumulative amount of carbon mineralized over 112-days and chemical evolution in the C_1 and C_2 pools during decomposition, normalized by the means of the observations.

We determined the best-fit estimates of parameters K_{B22} and LCI_T that most closely matched simulations to observed patterns of CO_2 efflux and both C_1 and C_2 pools over time (Table 1). We selected these parameters because the dynamics of pool C_2 were most closely related to cumulative CO_2 efflux and thus, estimated mass loss.

Sensitivity analysis

To evaluate the sensitivity of the model to parameter estimates (Table 1), we randomly varied model parameters e_1, e_2, k_{1max}, BC_{max} and K_{B11} within ±10% of their initial values (Table 1), 100 times for each litter type. We then calculated the differences between simulated and observed values of C_1, C_2, cumulative CO_2, and respiration rates on each date of observation [13] as a relative value = (observation-simulation)/observation. We summed these relative differences for each type of observation over all days of observation (i.e., through day 112), which produced a composite measure of the relative differences between observations and model output for each maize genotype. ANCOVA evaluated the contributions of variations in parameter values to variations in the relative differences between model output and observations by litter type. The type II sums of squares from ANCOVA were interpreted to represent the relative contribution of each parameter to model behavior.

Model extrapolation

The last set of simulations estimated decomposition for the 12 additional maize genotypes examined by Machinet et al. [13]. Relationships between values of C_{1T} and best-fit values of LCI_T and K_{B22} (Table 1) and the initial characteristics of litter chemistry were estimated for the four litter types described by Machinet et al. [13]: $LCI_T = -0.36 \cdot KL/AX + 0.70$ ($N = 4$; $R^2 = 0.923$; $P \leq 0.05$); $C_{1T} = 0.02 \cdot \sum Sug - 0.64$ ($N = 4$; $R^2 = 0.998$; $P \leq 0.05$); and $K_{B22} = 53.22 \cdot KL/AX - 13.65$ ($N = 4$; $R^2 = 0.975$; $P \leq 0.05$). These relationships were then used to estimate parameter values of C_{1T}, LCI_T and K_{B22} for simulations with each of the 12 additional genotypes reported by Machinet et al. [11], based on their initial chemistry. Principal components analyses evaluated relationships between the relative differences in observed and simulated values of respiration rates and cumulative CO_2 efflux, and initial litter chemistry characteristics for all 12 litters to determine if more detailed litter chemical characteristics than currently used in the model could provide additional insights to litter quality controls on decomposition.

Results

Decomposition dynamics

We discovered a close correspondence between the arabinan:-xylan ratio (A:X) and lignocellulose index (LCI) of the residue (Figure 2b) in the empirical data [11]. Between days 14 and 112, a linear regression of AX over LCI yielded an $R^2 = 0.91$ ($N = 12$, $P \leq 0.01$); we omitted days 0–14 because we expected decomposition to be limited by microbial activity rather than substrate [16,18]. Perhaps this relationship also explains why both AX and KL were related to best-fit model parameters LCI_T and K_{B22}.

Rates of CO_2 efflux for all four litter types rapidly increased to peak values within 14–21 days followed by gradual declines (Figure 3); cumulative CO_2 efflux rose rapidly during the first 36 days and then more slowly until day 112. For 3 of the 4 genotypes, the C_1 pool declined rapidly from day 1 to day 14, and then remained relatively constant during the rest of the incubation (Figure 4). For genotype F292bm3, the pool of C_1 remained essentially unchanged throughout the study. The C_2 pools of all four litters declined throughout the incubations, accounting for most of the litter mass loss over time (Figure 4). The C_3 fraction of the remaining litter showed a slight increase over the first 14 days of incubation (ca. 5–10%) for both F2 genotypes (not shown), but remained roughly constant for the two F292 genotypes [13].

Microbial production

Microbial biomass rapidly increased to peak values within 20–40 days, varying among genotypes (Figure 5a), declining most rapidly for those genotypes supporting the most rapid initial growth with the highest decay rates (F292 and F292bm3). Microbial turnover rates (Figure 5b), which should correlate with the generation of microbial products such as cell walls, were similar in shape but had lower peak values and lagged behind the patterns of rising and falling respiration rates (Figure 3). Genotypes with higher turnover rates also showed the greatest declines in biomass by day 112. Litter LCI increased over time in all litter types, coincident with declining biomass and turnover rates (Figure 5c), increasing most rapidly for litters decaying most rapidly (Figure 3). CUE declined over time for all litter types (Figure 5d), coincident with increasing LCI (Figure 5c), but started at higher values in litter types with larger pools of labile C_1 (i.e., C_{1D}), which had a higher C-assimilation efficiency (e_i) than C_2 (Table 1).

The total microbial production (C-immobilized into biomass including microbial turnover) by day 112 varied between litter types: 248, 195, 295 and 295 mgC for genotypes F2, F2bm1, F292 and F292bm3, respectively. These values were negatively related to initial litter KL and KL/AX contents, as well as final biomass and best-fit values of K_{B22}. They were positively related to total cumulative CO_2 efflux by day 112 and best-fit values of LCI_T (all $N = 4$, $P \leq 0.05$).

Model test

Our initial model parameter set, including best-fit estimates of K_{B22} and LCI_T, simulated rates of CO_2 efflux closely matching observations (all $N = 15$; omitting day 0), with R^2 values ranging from 0.87 for genotype F2bm3 to 0.98 for genotype F2bm1 (Fig. 3). Simulated rates peaked before observations for all litter types and at slightly lower values. However, simulated peak rates were within 10% of observed peaks for both F2 genotypes and 20% for both F292 genotypes. Simulated values of cumulative CO_2 efflux (to day 112) were even closer to observations, with R^2 values in excess of 0.99 for all genotypes (Figure 3). All simulations were within 5% of observed cumulative CO_2 efflux values. Simulated values of C_2 seldom differed by more than 5% of the observed values, with all $R^2 \geq 0.99$ (Figure 3). Finally, our model did not estimate any loss in pool C_3 during simulations and has no mechanism to generate an increase in this pool's size (Figure 1).

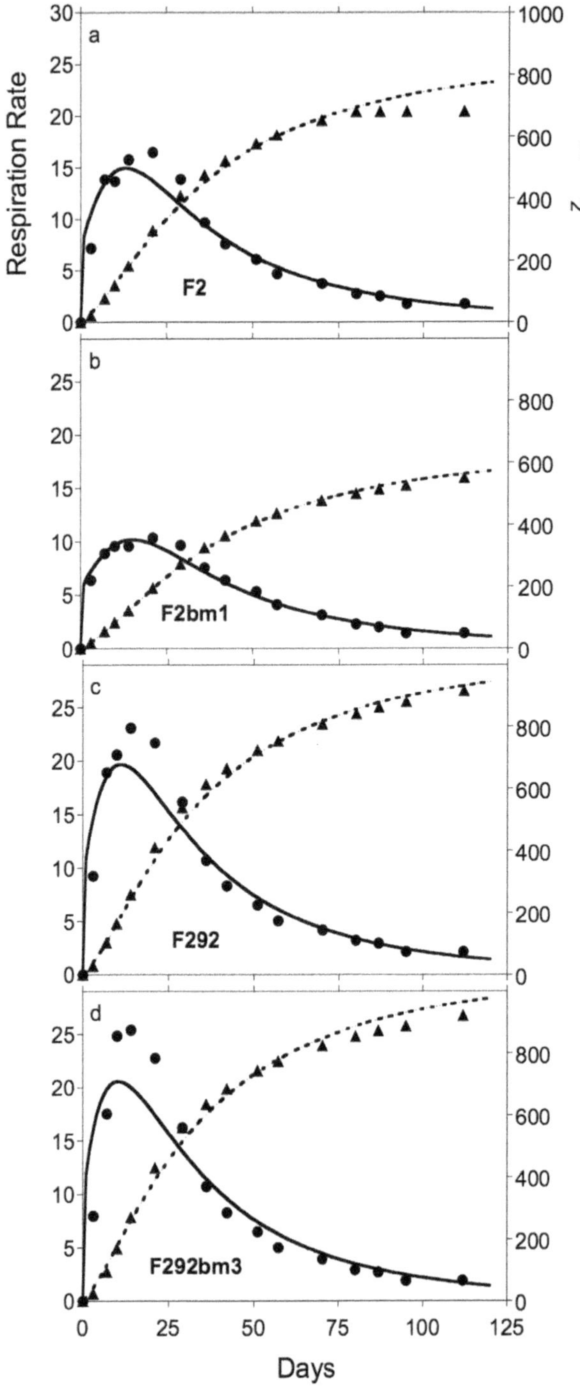

Figure 3. Observed and simulated respiration rates and cumulative CO_2 efflux during decomposition of *Zea mays* root litter. Observed (symbols) and simulated (lines) patterns of respiration rates (solid lines and circles, mgC·kg soil^{-1}·d^{-1}) and cumulative CO_2 efflux (dashed lines and triangles, mgC·kg soil^{-1}) for maize mutants, a. F2, b. F2bm1, c. F292 and d. F292bm3.

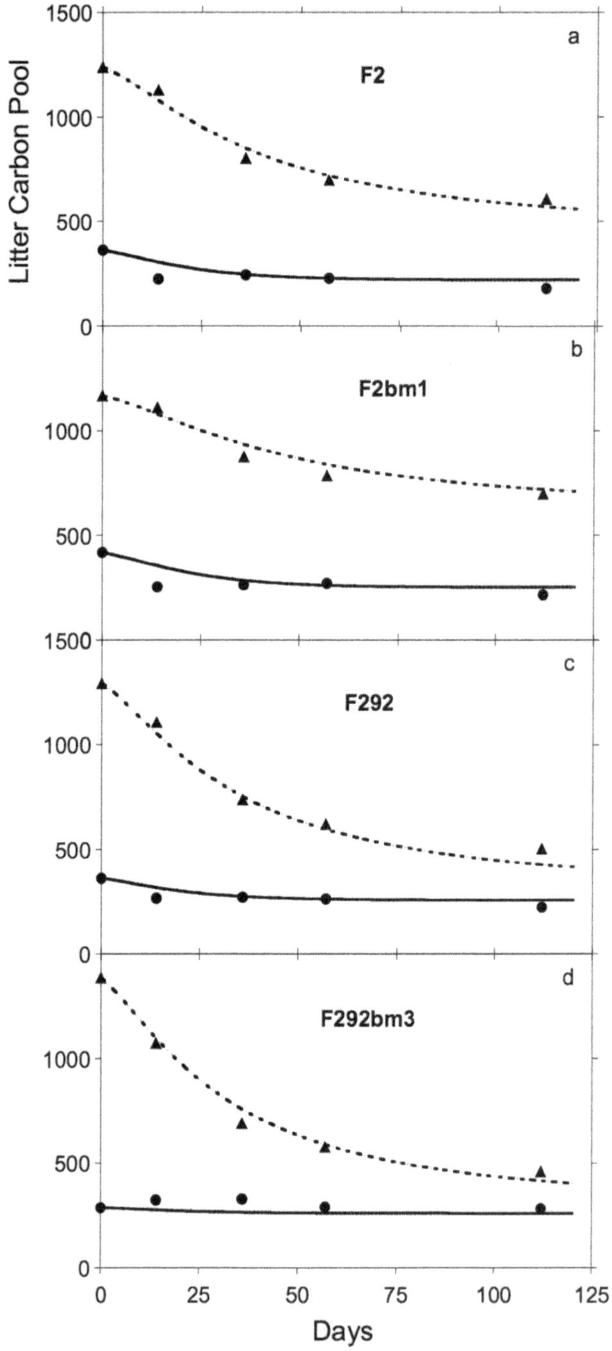

Figure 4. Observed and simulated mass remaining for soluble (C_1) and acid hydrolysable (C_2) chemical fractions of litter during decomposition of *Zea mays* roots. Observed (symbols) and simulated (lines) patterns of mass remaining for soluble C_1 pool (solid lines and circles, mgC·kg soil^{-1}) and acid hydrolysable C_2 pool (dashed lines and triangles, mgC·kg soil^{-1}) during decomposition of litter from maize mutants, a. F2, b. F2bm1, c. F292 and d. F292bm3.

Sensitivity analysis

The relative contributions of individual parameters to explaining variations in model behaviors differed by litter type and observation (Table 2). For example, the efficiency of C assimilation from substrate C_1 (e_1) rarely explained more than 1% of the variation in model behavior for any litter type. Moreover, no single parameter made a substantial contribution (≥10%) to explaining the variation in any model behavior when all four genotypes were pooled (not shown). When litter types were examined separately, variations in k_{1max} and K_{B11} often made their largest contributions to the same output variables for the

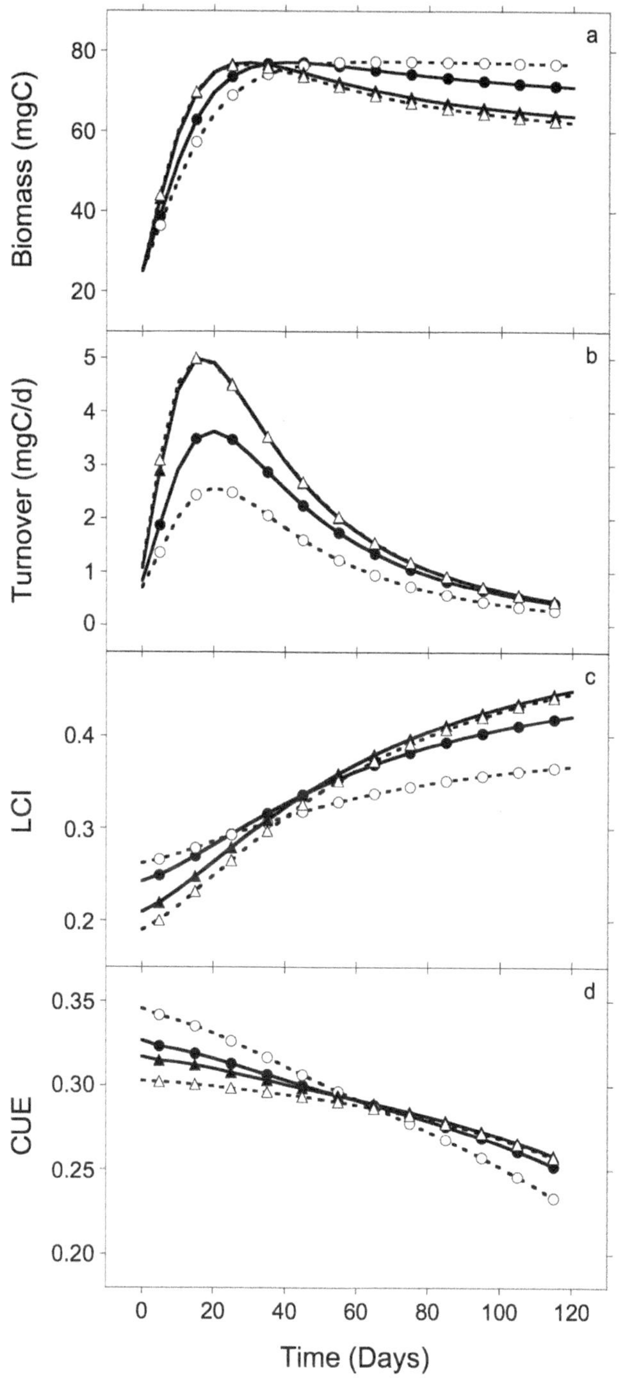

Figure 5. Simulated values of microbial biomass and turnover rate, remaining litter lignocellulose index (LCI), and carbon use efficiency (CUE), during decomposition of *Zea mays* root litter. Simulated values of a. total microbial biomass, b. daily microbial turnover rate, c. remaining litter lignocellulose index (LCI), and d. realized carbon use efficiency (CUE), over time, for litter genotypes F2 (filled circles), F2bm1 (open circles), F292 (filled triangles) and F292bm3 (open triangles).

same litter types. For example both parameters made substantial contributions (≥10%) to variations in C_1 for litter types F2, F2bm1 and F292bm3; cumulative CO_2 efflux for litter types F2, F2bm1 and F292; respiration rates for litter types F2bm1 and F292; and overall model fit for litter type F2bm1. Neither parameter made a substantial contribution to variation in C_2 for any litter type. In contrast, parameters e_2 and BC_{max} often made their largest contributions to model behaviors for litter types when k_{1max} and K_{B11} did not. For example, e_2 and BC_{max} made substantial contributions to variations in C_2 for litter types F2, F2bm1 and F292bm3; cumulative CO_2 efflux for litters F292 and F292bm1; respiration rates for F2, F292 and F292bm3; and overall model fit for F2.

Model extrapolation

When we simulated decomposition of the 12 different maize genotypes [11], we found that overall simulated rates of respiration were strongly related to observations ($N = 180$, $R^2 = 0.69$, $P \leq 0.01$), and that simulated rates of respiration averaged 1±17% lower than observed rates (data not shown). However, the relative differences ([observations-simulations]/observations) varied over time (Text S2). The first two axes of the principal components analysis explained 54% of the variation in the relative differences between observed and simulated respiration and litter chemical characteristics (Figure 6a). Differences in rates between days 10–21 were more strongly related to axis 2, along with litter LCI, KL/AX and $\sum$Sug. In contrast, differences in respiration for most days ≥50 were more strongly associated with axis 1, along with several chemical characteristics, the strongest being arabinans, AX, galactose, NDF and pCA (Figure 6a).

In contrast to the daily respiration rates, there was no significant relationship between simulated and observed peak respiration rates (means = 17.5±1.8 and 21.4±9.9 mgC·kg $soil^{-1} \cdot d^{-1}$, respectively; not shown). The PCA showed that relative differences between peak rates (PR) were more closely related to axis 2 and opposite those of rates between days 10–21 (Figure 6a), with significant positive correlations with KL and KL/AX.

Simulated values of cumulative CO_2 efflux were also strongly related to observations ($N = 180$, $R^2 = 0.923$, $P \leq 0.01$), and averaged only 3±18% greater than observations. The first two axes of the PCA explained 69% of the variation in the relative differences between observed and simulated cumulative CO_2 efflux and initial litter chemistry characteristics (Figure 6b). All values for CO_2 efflux were tightly clustered and closely associated with the first axis. Litter chemical characteristics, NDF, arabinans, galactose and AX, were also closely related to the first axis. The relative differences between observations and simulations at day 36 (d36) showed a significant, positive correlation with the initial soluble content of litter (C_1 = C-SOL) and negative correlation with galactan (GAL) content. Differences by day 59 were also related to C_1 and galactan. Although there were no significant differences between simulations and observations at day 112, these variations were significantly related to several aspects of initial litter chemistry, including C_1.

Estimates of microbial production by day 112 for these 12 genotypes ranged 203–303 mgC (mean = 270±25 mgC), and were negatively correlated with KL/AX and final microbial biomass, as well as initial LCI. Production was positively related to total cumulative CO_2 efflux by day 112 and best-fit values of LCI_T, as well as initial litter concentrations of Glu, $\sum$Sug, and C_2 (all $N = 12$, $P \leq 0.05$). A stepwise regression explained nearly all variation in microbial production as a function of total CO_2 efflux and initial concentrations of xylans (Xyl) and guaiacyl (G) in litter ($N = 12$, $R^2 = 0.998$, $P \leq 0.01$).

Table 2. Results of sensitivity analysis quantifying the relative contributions (%) of random variations in model parameters (column headings) to resulting variations in model behaviors (row labels), based on sums of squares from ANOVA relating model output to parameters.

Source	Genotype	e_1	e_2	k_{1max}	BC_{max}	K_{B11}
C_1-Carbon	F2	0.2*	1.3*	47.5**	3.9*	30.6**
C_1-Carbon	F2bm1	2.0*	0.5*	63.7**	0.7*	38.9**
C_1-Carbon	F292	0.1*	1.7	3.9*	42.5**	1.2
C_1-Carbon	F292bm3	0.0	2.3*	65.6**	7.0*	22.1**
C_1-Carbon	All Litters	0.2	0.0	2.1*	0.1	6.3*
C_2-Carbon	F2	1.1*	38.4**	0.2*	53.9**	0.1
C_2-Carbon	F2bm1	1.2	13.2**	0.1*	71.4**	0.0
C_2-Carbon	F292	0.0	3.5*	0.8*	73.0**	0.1
C_2-Carbon	F292bm3	0.0	40.8**	0.0	40.9**	0.0
C_2-Carbon	All Litters	1.0*	1.6*	0.2	3.3*	0.4
Cumulative CO_2	F2	0.2*	1.2*	48.1**	3.8*	30.1**
Cumulative CO_2	F2bm1	0.4*	0.1	64.4**	0.2*	40.3**
Cumulative CO_2	F292	0.7*	11.7**	35.7**	12.3**	38.1**
Cumulative CO_2	F292bm3	0.5*	48.8**	6.6*	21.2**	4.5*
Cumulative CO_2	All Litters	0.2	0.0	0.6	0.3	1.6*
Respiration Rate	F2	1.0	28.7**	9.6*	17.8**	7.6*
Respiration Rate	F2bm1	0.3	0.1	48.5**	0.3	30.2**
Respiration Rate	F292	2.1*	40.8**	20.6**	11.4**	14.0**
Respiration Rate	F292bm3	0.2*	45.3**	1.8*	34.7**	0.5*
Respiration Rate	All Litters	0.0	0.2	0.0	0.4	0.1
Overall Fit	F2	3.0*	34.8**	4.0*	31.9**	2.1*
Overall Fit	F2bm1	0.6*	0.2	57.9**	0.6*	36.3**
Overall Fit	F292	2.6*	46.5**	4.6*	16.8*	7.9*
Overall Fit	F292bm3	0.3*	49.2**	0.9*	31.3*	0.7*
Overall Fit	All Litters	0.0	0.2	0.6	0.4	0.6
Maximum		30.4	49.2	65.6	73.0	40.3
Count>10%		0	11	9	12	9

*significant parameter contributions (P≤0.05) to variation in model behavior.
**contributions of parameters representing at least 10% of the explained variation in model behavior (P≥0.05).

Discussion

Lignin-cellulose interaction

Simulations provided a close match to observed patterns of holocellulose (C_2) decay (Figure 4). This pool represented the largest fraction of litter (Table 1), also explaining why simulations closely fit patterns of CO_2 efflux (Figure 3). Long-term decomposition (months to years) has often been negatively correlated to initial lignin content of litter [3], partly because lignin decays slowly and partly because some products of decomposition increase the size of the non-hydrolysable pool often interpreted as lignin [6]. However, Machinet et al. [11] also demonstrated the effects of biochemical connections between cell wall polysaccharides and lignin on the pattern of cumulative CO_2 efflux from decomposing maize roots. The fit between k_2 and LCI (Figure 2a) reveals an interaction between C_2 and C_3 in early stages of litter decay [19], well before C_3 begins to decline [13,20]. In other words, there is unlikely to be qualitatively separate pools of lignin-shielded and unshielded holocellulose, as is sometimes implied [1,3]. Instead, biochemical linkages between cellulose, hemicellulose and lignin components of cell walls influence patterns of decomposition throughout the process.

A brief explanation of these relationships is that the structural composition of hemicellulose is a primary chain consisting mainly of xylans with branching arabinan side chains that interact with other cell wall polymers. The level of arabinoxylan substitution (represented by A:X) increases with the progressive enzymatic degradation of plant material, during both digestion in the rumen [9,24] and decomposition in soils [12,13], as the more exposed elements of xylan chains are more readily hydrolyzed than those near arabinan branches. Gunnarsson et al. [25] found that initial xylan and arabinan concentrations in litter were as important as the total amount of hemicellulose in describing C mineralization over the first 9 days of laboratory incubations, and that arabinan was the single most important factor. These results were consistent with those of Machinet et al. [11], who identified initial arabinan content as an early predictor of C mineralization (days 3–7), followed by AX (days 7–14). In addition, hydroxycinnamic acids (ferulic and *p*-coumaric acids) play a key role in cross-linking arabinoxylans with lignin and this cohesive network also hampers decomposition [10,11,12]. These cross-linkages explain the

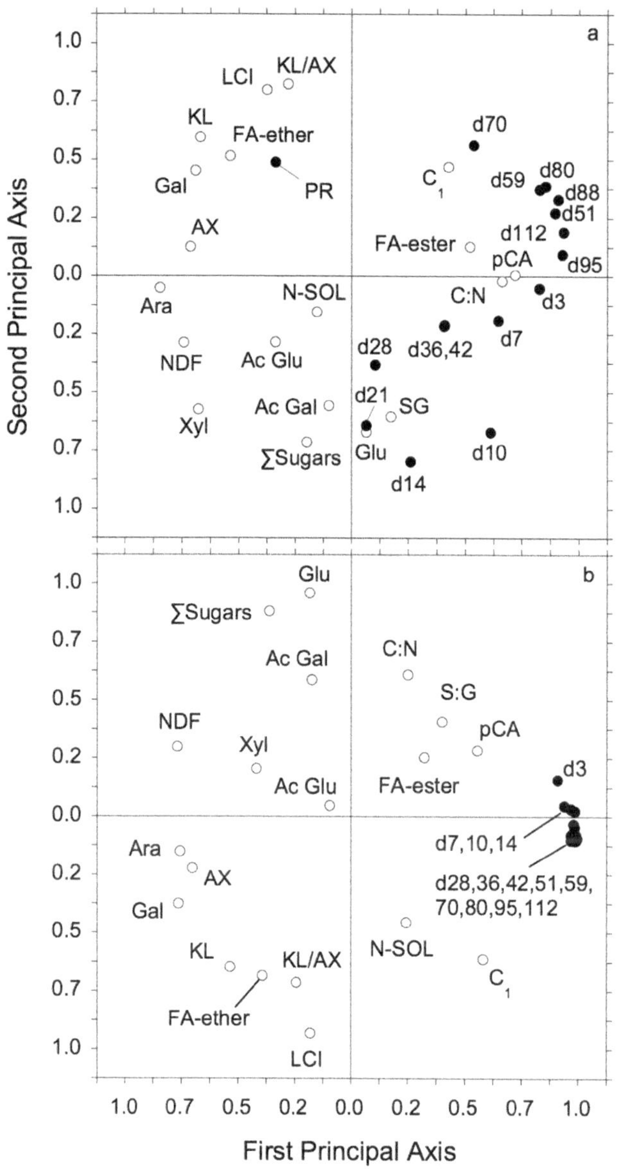

Figure 6. Principal components analysis of variations in initial litter chemistry characteristics and relative differences between observed and simulated rates of microbial respiration and cumulative CO_2 efflux during decomposition of *Zea mays* roots. Results of principal components analysis of variations in initial litter chemistry characteristics and relative differences = (observations − simulations)/observations from simulated decomposition of 12 novel maize mutants [11] for: a. rates of respiration on days 3–112 (e.g., d3 = rate on day 3) and peak respiration (PR), and b. cumulative CO_2 efflux by days 3–112 (parameter definitions given in text).

negative effect of lignin on holocellulose decay long reported in the literature [1,3] and also why k_2 decreased as AX increased early in decomposition (Figure 2).

Our model revision approximated this complex control with a straightforward relationship between LCI and k_2 (Figure 2) that included interactions between C_2 and C_3 (based on empirical sugar and lignin determinations). Moreover, key model parameters LCI_T and K_{B22} were most closely related to initial KL/AX ratios of litter, emphasizing the interaction between holocellulose and lignin throughout the decomposition process.

Soluble dynamics

Data from Machinet et al. [13] also revealed that the soluble C_1 pool included a fraction (C_{1T}) that persisted throughout the study (Figure 4). Although the soluble fraction of litter is usually considered to be highly decomposable, the persistence of a relatively large soluble pool is common in both soils and decaying litter [26,27]. The soluble pool usually contains a hydrophilic fraction including sugars and amino acids that are readily used by microorganisms [28,29], and a more hydrophobic portion, including soluble polyphenols (e.g., tannins) that are less rapidly utilized [26,30]. This differential use of compounds in the C_1 pool by decomposer microorganisms changes the pool's overall quality and decomposability with time [1]. Although the soluble pool is replenished with degradation products from non-soluble substrates [1,31], products of C_2 hydrolysis would enter the more labile fraction of the soluble pool, which cycles much more rapidly than the more persistent fraction [28,29]. In addition, Machinet et al. [13] found no decrease in the C_3 pool over time (not shown), so that it's degradation products could not have increased the more persistent fraction of the C_1 pool (C_{1T}). Simulating the persistence of a sizeable pool of C_1 with GDM required dividing the pool into labile and persistent fractions (Figure 1), with the labile fraction (C_{1D}) rapidly declining during decay and the persistent C_{1T} pool remaining intact (Figure 4). Moorhead and Sinsabaugh [16] found that simply routing the products of C_2 degradation through the whole C_1 pool GDM could not explain the size of this composite pool.

Respiration patterns

Simulations tended to overestimate rates of CO_2 efflux between days 0–14, and underestimate rates between days 14–36 (Figure 3). A positive relationship between the most labile components of litter and early respiration (hours to weeks) or decay rate is commonly reported [32,33], and most models, including GDM, assume a greater decay rate for soluble substrates [15,16]. In terms of substrate dynamics, GDM tended to underestimate early losses (day 14) of C_1 and overestimate C_2 losses (Figure 4), suggesting that values of k_{1max} should be slightly increased and K_{B1i} (i = 1,2,3) reduced (Table 1), to stimulate decay of C_{1D} and possibly initial CO_2 efflux. However, GDM underestimated respiration to a greater extent for litter types with lower amounts of C_{1D}, i.e., with higher C_{1T} (F292 and F292bm3). We found that C_{1T} was most closely related to the total sugar content of cell walls (∑Sug; N = 4; $R^2 = 0.998$; $P \leq 0.05$) and that xylan, arabinan and glucan concentrations were highly correlated with each other (not shown). Further resolution of the chemical composition of the soluble fraction (C_{1D} and C_{1T}) and its dynamics during decomposition are needed to improve the mathematical descriptions of these relationships.

Microbial production

Moorhead and Sinsabaugh [16] argued that litter decay is initially limited by microbial action, because there is a time lag in the colonization of fresh litter by decomposer microorganisms (Figure 2). Whether this lag was a numerical (biomass) or functional (physiological) response in the study by Machinet et al. [13] is unknown because they did not monitor microbial biomass. Therefore, the most speculative part of this study was our simulation of microbial dynamics. Nonetheless, the patterns and magnitudes of simulated microbial biomass and turnover rates were consistent with observed patterns of respiration (Figures 3,

5b), reported limits to microbial biomass concentration in soils [21,22], and changes in litter chemistry likely controlling microbial activities (Figures 2, 4 and 5c; [1]) as well as CUE (Figure 5d) [23]. Parameter estimates for C-assimilation efficiencies (e_i) for various substrate pools, coefficient of microbial basal respiration rate (g), and maximum biomass:total system C ratio (BC_{max}), most directly affected the relationships between substrate decomposition and biomass dynamics (Table 1). However, variations in any or all of these parameter values generate similar patterns with respect to different litter chemistry, although amplitudes and temporal regimes vary [16].

Our primary reason for simulating biomass dynamics was to determine if litter quality affected microbial production consistent with the idea that more decomposable chemical fractions of litter not only decay more rapidly but also generate more microbial products likely to enter stable soil organic matter pools [4,5]. In fact, simulated microbial production was significantly and negatively correlated to KL and KL/AX, which are inversely related to decomposability, but production showed no significant positive relationship to any simple measure of litter quality. However, it was positively related to the sum of the two most decomposable carbon pools, i.e., C_2+C_{1D}, (N = 4, rho = 0.982, P≤ 0.05). The contributions of the relatively larger pools of C_{1D} in litter types F2 and F2bm1 to simulated production were small compared to the larger pools of C_2 in the other litters, despite the lower C assimilation efficiency of C_2 versus C_1 (Table 1). Thus our results were consistent with the observations of Smith et al. [7] and others who found that higher initial rates of C incorporation into biomass coincided with higher losses of litter through respiration [2,8].

Sensitivity analysis

The most interesting result of our sensitivity analysis was the lack of any simple, overall interpretation. No single parameter consistently explained>10% of the observed variability in any model behavior (Table 2). Usually a parameter important to explaining one model behavior, such as the contributions of k_{1max} to simulated C_1 dynamics or cumulative CO_2 efflux (Table 2), made little contribution to other model outputs. The largest discrepancies between simulations and observations were in early respiration and C_1 loss (discussed above). Of the tested parameters, e_2, the C assimilation efficiency for C_2, appeared to be most important to respiration rates, which seems reasonable because C_2 was the largest substrate pool and provided most of the C respired (Table 1, Figure 4). As for the dynamics of the C_1 pool, parameter k_{1max} appeared to be most important, followed by K_{B11}; thus parameters controlling the degradation rate of this pool explained differences between simulations and observations (Figure 4). These results are consistent with our earlier conclusion that greater resolution of the C_1 pool composition and dynamics might provide a better understanding of decomposition.

Extrapolations

We simulated the decomposition of 12 additional maize genotypes in the second set of incubations conducted by Machinet et al. [11] based on the assumption that the relationships between key parameters, LCI_T, K_{B22} and C_{1T}, and litter chemical characteristics were consistent with those for the four genotypes examined by Machinet et al. [13]. The assumption seemed reasonable because all 16 genotypes were naturally occurring varieties of *Zea mays*, and likely to be more similar in chemistry and tissue architecture than unrelated species more commonly used in comparative decomposition experiments [6].

In general, simulations were within a few percent of observed values of respiration rates and cumulative CO_2 efflux over the entire period of incubation. Analyses of these differences between simulations and observations provided relatively little additional insight to patterns of cumulative CO_2 efflux (Figure 6b). The importance of the initial C_1 pool (as both C_{1T} and C_{1D} estimates) to cumulative CO_2 efflux through time again emphasized the importance of initial decay rate to longer-term patterns [7,8]. The relationships between CO_2 efflux on day 112 and initial arabinan and *p*-coumaric acid concentrations and AX suggest that the cross-linkages among hemicellulose and lignin became increasingly important with progressive decay. The tight cluster of cumulative CO_2 efflux along the first axis of our PCA also underscored the importance of cross-linkages among cell wall constituents (e.g., arabinan, AX, NDF and *p*-coumaric acid) to litter decay (Figure 6b), consistent with Machinet et al. [11].

The differences between simulated and observed respiration rates were more variable, suggesting temporally shifting controls on decomposition. Peak rates and those on days 14–28 were related to initial KL and KL/AX, as well as glucan and S:G (Figure 6a). In contrast, rates on days>42 were more closely associated with arabinan, AX, *p*-coumaric acid and ester-linked ferrulic acids, perhaps because polysaccharide-ester linked ferulic acids can form ether-links with lignin [34,35] and the syringyl units of lignin can be esterified by *p*-coumaric acids, which is typical of grass cell walls [11]. However, the biggest differences between simulated and observed rates were on days 36–42, which were most closely related to KL, ether-linked ferrulic acids and galactan (Figure 6b). The frequent importance of KL and AX (or their chemical constituents) to these patterns was surprising, because KL (in LCI) was used to estimate k_2, and KL/AX to estimate LCI_T and K_{B22} (previously discussed) used in simulations. Clearly, simple linear relationships were insufficient to capture the subtleties of these controls. The relationships between respiration and galactan and C:N ratio (Figure 6b) hint at a microbial control [14], in part because galactan is sometimes used as an index to microbial contributions [36], but it is also a hemicellulosic sugar, along with arabinan, rhamnan, and xylan [37].

Conclusions

We found that the level of arabinan substitution in xylan chains (AX) was an important control in early stages of decomposition, and was also linearly related to LCI calculated on the basis of fine scale cell wall chemistry. These relationships provide a plausible, mechanistic explanation for earlier, empirical descriptions of LCI effects on decomposition as a result of biochemical cross-linkages between polysaccharides and lignin. Thus lignin and LCI serve as convenient, negative proxies for the decomposability of litter even at the start of decay. However, additional research is needed to determine the chemical composition and dynamics of the non-hydrolysable product of proximate C analysis that is typically termed "lignin" if we are to discover the mechanistic relationships between LCI and latter stages of decomposition.

We also found that dividing the soluble pool of litter (C_1) into separate persistent (C_{1T}) and labile (C_{1D}) pools, was necessary to accurately simulate the dynamics of the composite soluble pool during early decomposition because not all soluble compounds are equally decomposable. The finer scale chemical composition and dynamics of the soluble component of litter is needed to determine the possible sources and fates of these compounds.

These relationships between simulated patterns of litter decay, litter chemistry (LCI, AX, C_{1T} and C_{1D}) and microbial productivity were consistent with the notion that the more rapid utilization of substrates with high carbon use efficiency generates greater amounts of microbial products that can contribute to stable soil organic matter pools than more persistent substrates with lower C-assimilation, like lignin. Our assumption that lignin decay provides little to no net C-acquisition by microorganisms is also consistent with recent observations that little lignin C enters stable SOC pools.

Finally, our extrapolations with litter types that differed in initial chemistry demonstrated that relationships we found between key model parameters, LCI_T, K_{B22} and C_{1T}, and decomposition were robust across a large range of maize litter types, highlighting the importance of access (K_{B22}) to the largest, rapidly decaying pool of substrate (C_2) by microorganisms able to use this resource (G_2), as well as negative controls imposed by the less-accessible substrate pools (KL, C_{1T}). Moreover, differences between simulations and observations indicated that temporal controls on decay rates shifted from relative substrate pool sizes (both accessible and persistent) dominating at the start to factors related to cross-linkages between structural polysaccharides and lignin with progressive decomposition.

In closing, our results suggest that interactions between decomposer microorganisms and litter quality characteristics at the earlier stages of decomposition may provide more insights to soil organic C stabilization than later stages dominated by the more persistent chemical characteristics of the cell wall, if indeed microbial products comprise a large fraction of stable soil C pools.

Supporting Information

Data S1 Data used to test and refine model. Observed respiration rates and chemical composition of residues of maize roots during decomposition [13].

Figure S1 Differences over time between observed and simulated respiration rates and cumulative CO_2 efflux from the decomposition of *Zea mays* roots. Relative differences between observations and simulations of litter decomposition from 12 novel maize genotypes [11] for: a. respiration rates (mgC·kg soil^{-1}·d^{-1}) over time (means±95% confidence intervals, all N = 12), b. cumulative CO_2 efflux (mgC·kg soil^{-1}) over time (means±95% confidence intervals, all N = 12).

Text S1 Lignocellulose controls. Relationships between decay rate coefficients for polysaccharides (C_2) and polyphenolics (C_3) as functions of lignocellulose index ($LCI = C_3/[C_2+C_3]$)).

Text S2 Sensitivity analysis. Patterns of differences between observed and modeled patterns of respiration associated with decomposition of decaying maize roots.

Author Contributions

Conceived and designed the experiments: DM GL SR IB. Performed the experiments: DM GL. Analyzed the data: DM GL SR IB. Contributed reagents/materials/analysis tools: DM GL SR IB. Wrote the paper: DM GL SR IB.

References

1. Berg B, McClaugherty C (2008) Plant litter: Decomposition, humus formation, carbon sequestration, Second edition. New York: Springer. 338 p.
2. McMahon SK, Williams MA, Bottomley PJ, Myrold DD (2005) Dynamics of microbial communities during decomposition of carbon-13 labeled ryegrass fractions in soil. Soil Sci Soc Am J 69: 1238–1247.
3. Melillo JM, Aber JD, Muratore JF (1982) Nitrogen and lignin control of hardwood leaf litter decomposition dynamics. Ecology 63: 621–626.
4. Wickings K, Grandy AS, Reed SC, Cleveland CC (2012) The origin of litter chemical complexity during decomposition. Ecol Lett 15: 1180–1188.
5. Cotrufo MF, Wallenstein MD, Boot CM, Denef K, Paul E (2013) The microbial efficiency-matrix stabilization (MEMS) framework integrates plant litter decomposition with soil organic matter stabilization: do labile plant inputs form stable soil organic matter? Glob Change Biol 19: 988–995.
6. Preston CM, Nault JR, Trofymow JA (2009) Chemical changes during 6 years of decomposition of 11 litters in some Canadian forest sites. Part 2. 13C abundance, solid-state 13C NMR spectroscopy and the meaning of "Lignin". Ecosystems 12: 1078–1102.
7. Smith JL, Bell JM, Bolton H Jr, Bailey VL (2007) The initial rate of C substrate utilization and longer-term soil C storage. Biol Fertil Soils 44: 315–320.
8. Blagodatskaya EV, Blagodatsky SA, Anderson T -H, Kuzyakov Y (2009) Contrasting effects of glucose, living roots and maize straw on microbial growth kinetics and substrate availability in soil. Eur J Soil Sci 60: 186–197.
9. Chesson A (1988) Ligninopolysaccharide complexes of the plant cell wall and their effect on microbial degradation in the rumen. Anim Feed Sci Tech 21: 219–228.
10. Grabber JH, Mertens DR, Kim H, Funk C, Lu FC, et al. (2008) Cell wall fermentation kinetics are impacted more by lignin content and ferulate cross-linking than by lignin composition. J Sci Food Agr 89: 122–129.
11. Machinet GE, Bertrand I, Barriere Y, B Chabbert B, Recous S (2011) Impact of plant cell wall network on biodegradation in soil: Role of lignin composition and phenolic acids in roots from 16 maize genotypes. Soil Biol Biochem 43: 1544–1552.
12. Amin BAZ, Chabbert B, Moorhead D, Bertrand I (2014) Impact of fine litter chemistry on lignocellulolytic enzyme efficiency during decomposition of maize leaf and root in soil. Biogeochemistry 117: 169–183.
13. Machinet GE, Bertrand I, Chabbert B, Recous S (2009) Decomposition in soil and chemical changes of maize roots with genetic variations affecting cell wall quality. Eur J Soil Sci 60: 176–185.
14. Machinet GE, Bertrand I, Chabbert B, Watteau F, Villemin G, et al. (2009) Soil biodegradation of maize root residues: Interaction between chemical characteristics and the presence of colonizing micro-organisms. Soil Biol Biochem 41: 1253–1261.
15. Manzoni S, Porporato A (2009) Soil carbon and nitrogen mineralization: theory and models across scales. Soil Biol Biochem 41: 1355–1379.
16. Moorhead DL, Sinsabaugh RL (2006) A theoretical model of litter decay and microbial interaction. Ecol Monogr 76: 151–174.
17. Recous S, Robin D, Darwis D, Mary B (1995) Soil inorganic N availability: Effect on maize residue decomposition. Soil Biol Biochem 27: 1529–1538.
18. Schimel JP, Weintraub MN (2003) The implications of exoenzyme activity on microbial carbon and nitrogen limitation in soil: A theoretical model. Soil Biol Biochem 35: 549–563.
19. Moorhead DL, Lashermes G, Sinsabaugh RL, Weintraub MN (2013) Calculating co-metabolic costs of lignin decay and their impacts on carbon use efficiency. Soil Biol Biochem 66: 17–19.
20. Herman J, Moorhead D, Berg B (2008) The relationship between rates of lignin and cellulose decay in aboveground forest litter. Soil Biol Biochem 40: 2620–2626.
21. Anderson T-H, Domsch KH (1989) Ratios of microbial biomass carbon to total organic carbon in arable soils. Soil Biol Biochem 21: 471–479.
22. Wardle DA (1998) Controls of temporal variability of the soil microbial biomass: A global-scale synthesis. Soil Biol Biochem 30: 1627–1637.
23. Sinsabaugh RL, Manzoni S, Moorhead DL, Richter A (2013) Carbon use efficiency of microbial communities: stoichiometry, methodology and modeling. Ecol Lett 16: 930–939.
24. Vailhé MAB, Provan GJ, Scobbie L, Chesson A, Maillot MP, et al. (2000) Effect of phenolic structures on the degradability of cell walls isolated from newly extended apical internode of tall fescue (Festuca arundinacea Schreb.). J Agr Food Chem 48: 618–623.
25. Gunnarsson S, Marstorp H, Dahlin AS, Witter E (2008) Influence of non-cellulose structural carbohydrate composition on plant material decomposition in soil. Biol Fertil Soils 45: 27–36.
26. Cleveland CC, Neff JC, Townsend AR, Hood E (2004) Composition, dynamics, and fate of leached dissolved organic matter in terrestrial ecosystems: Results from a decomposition experiment. Ecosystems 7: 275–285.
27. Bengtson P, Bengtsson G (2007) Rapid turnover of DOC in temperate forests accounts for increased CO2 production at elevated temperatures. Ecol Lett 10: 783–790.
28. Kalbitz K, Schmerwits J, Schwesig D, Matzner E (2003) Biodegradation of soil-derived dissolved organic matter as related to its properties. Geoderma 113: 273–291.

29. Bertrand I, Prevot M, Chabbert B (2009) Soil decomposition of wheat internodes of different maturity stages: Relative impact of the soluble and structural fractions. Bioresource Technol 100: 155–163.
30. Nishimura S, Maie N, Baba M, Sudo T, Sugiura T, et al. (2012) Changes in the quality of chromophoric dissolved organic matter leached from senescent leaf litter during the early decomposition. J Environ Qual 41: 823–833.
31. Sinsabaugh RL (1994) Enzymatic analysis of microbial pattern and process. Biol Fertil Soils 17: 69–74.
32. Lagomarsino A, Moscatelli MC, De Angelis P, Grego S (2006) Labile substrates quality as the main driving force of microbial mineralization activity in a poplar plantation soil under elevated CO2 and nitrogen fertilization. Sci Total Environ 372: 256–265.
33. Hernandez DL, Hobbie SE (2010) The effects of substrate composition, quantity, and diversity on microbial activity. Plant Soil 335: 397–411.
34. Jung HG, Deetz DA (1993) Cell wall lignification and degradability. In: Jung HG, Buxton DR, Hatfield RD, Ralph J, editors. Forage cell wall structure and digestibility. Madison: ASA-CSSA-SSSA. Pp. 315–346.
35. Jacquet G, Pollet B, Lapierre C (1995) New ether-linked ferulic acid-coniferyl alcohol dimers identified in grass straws. J Agr Food Chem 43: 2746–2751.
36. Oades JM (1984) Soil organic matter and structural stability: mechanisms and implications for management. Plant Soil 76: 319–337.
37. Merila P, Malmivaara-Lamsa M, Spetz P, Stark S, Vierikko K, et al. (2010) Soil organic matter quality as a link between microbial community structure and vegetation composition along a successional gradient in a boreal forest. Appl Soil Ecol 46: 259–267.

4

Multiple Complexes of Nitrogen Assimilatory Enzymes in Spinach Chloroplasts: Possible Mechanisms for the Regulation of Enzyme Function

Yoko Kimata-Ariga*, **Toshiharu Hase**

Institute for Protein Research, Osaka University, Suita, Osaka, Japan

Abstract

Assimilation of nitrogen is an essential biological process for plant growth and productivity. Here we show that three chloroplast enzymes involved in nitrogen assimilation, glutamate synthase (GOGAT), nitrite reductase (NiR) and glutamine synthetase (GS), separately assemble into distinct protein complexes in spinach chloroplasts, as analyzed by western blots under blue native electrophoresis (BN-PAGE). GOGAT and NiR were present not only as monomers, but also as novel complexes with a discrete size (730 kDa) and multiple sizes (>120 kDa), respectively, in the stromal fraction of chloroplasts. These complexes showed the same mobility as each monomer on two-dimensional (2D) SDS-PAGE after BN-PAGE. The 730 kDa complex containing GOGAT dissociated into monomers, and multiple complexes of NiR reversibly converted into monomers, in response to the changes in the pH of the stromal solvent. On the other hand, the bands detected by anti-GS antibody were present not only in stroma as a conventional decameric holoenzyme complex of 420 kDa, but also in thylakoids as a novel complex of 560 kDa. The polypeptide in the 560 kDa complex showed slower mobility than that of the 420 kDa complex on the 2D SDS-PAGE, implying the assembly of distinct GS isoforms or a post-translational modification of the same GS protein. The function of these multiple complexes was evaluated by in-gel GS activity under native conditions and by the binding ability of NiR and GOGAT with their physiological electron donor, ferredoxin. The results indicate that these multiplicities in size and localization of the three nitrogen assimilatory enzymes may be involved in the physiological regulation of their enzyme function, in a similar way as recently described cases of carbon assimilatory enzymes.

Editor: Douglas Andrew Campbell, Mount Allison University, Canada

Funding: This work was supported by grants-in-aid for Scientific Research on Priority Areas (23570165) from the Japan Society for the Promotion of Science (to YKA), www.jsps.go.jp. The funders had no role in study design, data collection and analysis, decision to publish, or preparation of the manuscript.

Competing Interests: The authors have declared that no competing interests exist.

* Email: a-yoko@protein.osaka-u.ac.jp

Introduction

Intracellular enzymes are pertinently distributed and/or co-localized with functionally related proteins rather than evenly dispersed within cells or organelles, and dynamically change their states in response to environmental changes, for the biological reactions to proceed efficiently in a highly controlled fashion [1,2]. Various chloroplast enzymes are subjected to light/dark modulation of their activity through redox modulation [3], whose rate is adjusted by other factors such as specific metabolites [4]. In some cases, the reversible changes in the redox and activation states are accompanied by oligomerization and re-dissociation of transient complexes. For the enzymes of photosynthetic carbon assimilation, several lines of evidence have suggested that enzymes of the Calvin cycle associate to form multiprotein complexes. A multiprotein complex including two major Calvin cycle enzymes and a small protein (CP12) has been identified [5], and its reversible dissociation allowing for rapid regulation of enzyme activity was shown to be mediated by thioredoxin in response to changes in light availability [6]. Another recent finding for the key photosynthetic enzyme, ferredoxin-NADP$^+$ oxidoreductase (FNR), is its reversible association with thylakoid binding proteins (Tic 62 and TROL) in response to light signal and stromal pH, which regulates the stability and dynamic light-dependent membrane tethering of FNR [7,8]. Assimilation of nitrate is another major biological process in photosynthetic organisms, and has a large effect on plant growth and development [9]. Nitrate transported into cells is reduced to nitrite by nitrate reductase present in the cytosol and is further reduced to ammonium by nitrite reductase (NiR) in the chloroplast. The resulting ammonium is fixed as the amine group of Gln by glutamine synthetase (GS), and then two molecules of Glu are synthesized from Gln and 2-oxoglutarate by glutamate synthase (also called as glutamine oxoglutarate aminotransferase; GOGAT). The holoenzyme of NiR contains a siroheme and a 4Fe-4S cluster, and the GOGAT holoenzyme contains an FMN and a 3Fe-4S cluster. NiR, GS and GOGAT are nuclear-encoded and known to be located in the stromal fraction of chloroplasts in higher plants. NiR and GOGAT require reducing powers for their reactions, and their physiological electron donor is ferredoxin (Fd) (or NADH in the case of GOGAT), which is reduced by light-dependent reactions of the photosynthetic electron-transfer chain. These enzymes are known to be highly regulated during development and by external conditions such as light availability and nitrogen availability at the level of protein expression (reviewed in [9,10]). Findings of the multiprotein complexes of the Calvin cycle enzymes and FNR suggest that similar post-translational mechanisms may also exist for the regulation of nitrogen assimilatory enzymes. A recent study of protein profiles of GS gene products in the legume *Medicago truncatula* showed that the GS polypeptides assembled into organ-specific protein complexes with different molecular mass, implying organ-specific post-translational modifications under defined physiological conditions [11]. In this study, the possibility for the formation of protein complexes involving three nitrogen assimilatory enzymes, NiR, GS and GOGAT, was investigated using spinach leaves, by the method of blue native PAGE (BN-PAGE). BN-PAGE is a powerful tool for separating protein complexes from biological membranes and the soluble fraction under native conditions [12], which was also used for the above analyses of the Calvin cycle enzymes [6] and FNR [7,8]. In spinach, NiR is encoded in a single gene per haploid genome, and expressed as a single polypeptide of 65 kDa protein, whose X-ray crystal structure was recently solved [13]. GOGAT in plant plastids is present as two distinct classes, with different physiological electron donors, Fd and NADH. Fd-dependent GOGAT (Fd-GOGAT) is encoded in a single gene in spinach, expressed as a large single polypeptide of 165 kDa, and accounts for more than 95% of the total GOGAT activity in photosynthetic plant tissue [14]. NADH-GOGAT in plants is present as a 240 kDa polypeptide containing an additional NADH-binding region at the C-terminus. GS in higher plants also exists as two distinct types, cytosolic GS (GS1) and plastidic GS (GS2), with different polypeptide sizes of 38–40 kDa and 42–45 kDa, respectively [15]. Plant GS1 and GS2 polypeptides assemble into decameric complexes to be active holoenzymes [16,17]. In this study, we found each of NiR, GS and GOGAT assembled into discrete protein complexes in the chloroplasts of spinach leaves, and the dissociation profile of these complexes varied in response to the changes in the stromal conditions. The function of these complexes was evaluated, thus suggesting the involvement of the multiplicities in size and localization of these nitrogen assimilatory enzymes in the physiological regulation of enzyme function, in a similar way as the cases of the carbon assimilatory enzymes.

Materials and Methods

Isolation of stroma and thylakoid proteins

Chloroplasts were prepared essentially as described by Mach [18] from mature leaves of spinach (*Spinacia oleracea*) purchased at a local market, and then ruptured in 20 mM Hepes-KOH, pH 7.5 by repeated freeze-thawing at −20°C, unless otherwise specified. Following centrifugation at 10,000×g for 5 min at 4°C, the supernatant was filtered through a 0.2-μm filter membrane, and reserved as a stromal fraction. The pellet was re-suspended in the same buffer and centrifuged as above; this washing step was repeated twice in order to remove residual stromal components. The resulting pellet of the membrane fraction was re-suspended for solubilization in a Native PAGE sample buffer (50 mM Bis-Tris-HCl, pH 7.2, 50 mM NaCl and 10% glycerol) (life technologies) containing 1% n-dodecyl-ß-D-maltoside (DDM), at a chlorophyll concentration of 0.5 mg/ml, and then incubated for 30 min at 4°C with constant rotation, followed by centrifugation at 20,000×g for 15 min at 4°C. The supernatant was reserved as a fraction of thylakoid proteins (possibly containing proteins from other plastid membranes such as envelop), adjusted to 0.25% Coomassie brilliant blue (CBB) G-250, and used for BN-PAGE analysis. For the preparation of proteins from whole chloroplasts, isolated chloroplasts were directly suspended in the Native PAGE sample buffer containing 1% DDM, and the solubilized proteins were recovered by the same procedure as the membrane fraction described above.

BN-PAGE, Native PAGE, two-dimensional SDS-PAGE and western blot analysis

BN-PAGE was performed at 4°C, by using a 4–16% linear polyacrylamide gradient of Native PAGE Novex Bis-Tris Gel system (life technologies) according to the manufacture's instruction. Native PAGE was performed according to the same protocol as BN-PAGE except that CBB was omitted from the running buffer and the sample loading buffer. Unless otherwise specified, samples loaded on the gel were derived from 10 μg on a chlorophyll basis of the chloroplasts whose proteins were extracted. For two-dimensional SDS-PAGE after BN-PAGE (2D BN/SDS-PAGE), the lanes were cut out after the run, and incubated in a denaturing solution composed of 1% SDS and 1% 2-mercaptoethanol for 15 min before applying on top of a 12% SDS-PAGE gel. For western blot analysis, proteins on the gels were transferred to an immobilon PVDF membrane (Millipore) and probed with polyclonal antibodies raised against maize recombinant proteins (Fd-GOGAT, GS1 and FNR) and spinach NiR; proteins were detected by using horseradish peroxidase conjugated to Protein A (life technologies) with enhanced chemiluminescence (Western ECL substrate, Bio-Rad).

pH treatment of chloroplasts or stromal fraction

For the preparation of stroma and thylakoid proteins under different pH conditions, chloroplasts were suspended in 20 mM potassium phosphate buffer at pH 6.0 or pH 8.0, followed by repeated freeze-thawing, and incubated on ice for 30 min before the isolation of the stroma and thylakoid proteins. For the pH shift assay of stromal proteins, the isolated stromal fraction prepared by using 20 mM Hepes-KOH, pH 7.5 buffer was diluted with three-times volume of the same buffer, or 20 mM potassium phosphate buffer at pH 6.0 or pH 8.0, and incubated for 30 min at room temperature before BN-PAGE analysis. For the pH shift assay of whole chloroplasts, chloroplasts disrupted in 20 mM Hepes-KOH, pH 7.5 by repeated freeze-thawing were diluted with three-times volume of the same buffer, or 20 mM potassium phosphate buffer at pH 6.0 or pH 8.0, and incubated for 30 min at room temperature before extraction of stromal proteins for the analysis.

DTT treatment of thylakoid membranes

Thylakoid membranes prepared by using 20 mM potassium phosphate buffer at pH 6 was solubilized in the Native PAGE sample buffer containing 1% DDM for 30 min, and then incubated for 15 min in the absence or presence of DTT at 2, 10 and 50 mM at 4°C with constant rotation. After further incubation for 15 min at room temperature, the supernatant was recovered by centrifugation at 20,000×g for 15 min at 4°C, and used for BN-PAGE and 2D BN/SDS-PAGE analyses.

GS activity assays in gel and in solution

GS enzyme activity assays in gel and in solution were performed essentially as described by Seabra et al. [11]. For an in-gel GS activity assay, Native PAGE was performed as described above in this section, and the gel was incubated in GS activity solution and then in a stop solution containing $FeCl_3$ for chelating the product [11], allowing the detection of reddish brown color corresponding to the activity. GS activity in solution was determined by quantification of γ-glutamyl hydroxamate produced by the transferase reaction of glutamine and hydroxylamine, per minute per mg of total protein [11].

Fd-binding assay

Fd-affinity resin was prepared by using recombinant Fd from *Leptolyngbya boryana* [19] and CNBr-activated Sepharose 4B (GE Healthcare), following the manufacture's directions. For a Fd-binding assay, stromal proteins prepared by using 20 mM potassium phosphate buffer at pH 6.0 or pH 8.0 were mixed with the Fd resin, and the unbound fraction was recovered by centrifugation. The resulting resin was washed with each potassium phosphate buffer (at pH 6.0 or pH 8.0), and the bound proteins were eluted with the same buffer containing 0.2 M NaCl and analyzed by western blots after BN-PAGE.

Results and Discussion

Multiple complexes of nitrogen assimilatory enzymes in spinach chloroplasts

Using stroma and thylakoid proteins isolated from the chloroplasts of mature spinach leaves, native protein profiles of GOGAT, NiR and GS were analyzed by western blots under BN-PAGE (Fig. 1B-D). Proteins from the two chloroplast fractions were well-separated as seen in their CBB-stained patterns (Fig. 1A). Multiple bands with various patterns among the three enzymes were detected as follows. In the analysis of GOGAT (Fig. 1B) using antibody which specifically recognizes Fd-GOGAT, two bands with different mobilities were detected exclusively in the stromal fraction (lane S). The faster-migrating band corresponds to the size of the holoform of Fd-GOGAT (approx. 190 kDa), and the size of the other band was estimated to be 730 kDa, which is about four times of the holoform. NiR is also present almost exclusively in stroma (Fig. 1C), and in addition to the major band at around 70 kDa of the holoenzyme, smear and faint bands at around 120, 170 and 340 kDa were observed. On the other hand, in GS (Fig. 1D), besides the band at 420 kDa of the decameric holoenzyme in stroma, a slower-migrating band at 560 kDa was detected in the thylakoid fraction. This 560 kDa protein was relatively resistant against the elution from thylakoid membranes under the condition of 0.1% DDM by which complexes of FNR with Tic62/TROL were mostly eluted (our preliminary experiment), indicating that its binding to thylakoid membranes is relatively tight under the current experimental conditions. In the analysis of FNR as a control (Fig. 1E), monomer size of the holoenzyme at 40 kDa was detected in both fractions, and additional bands around 400 kDa in thylakoids and 220 kDa in stroma are thought to be the complexes of FNR with Tic 62, Trol or some other proteins [7,8]. Next, the identities of the multiple bands observed for each enzyme were analyzed by western blots under 2D BN/SDS-PAGE (Fig. 1F–I) as follows, and their physiological significances were further analyzed.

Analysis of GOGAT complex

As shown in Figure 1F of GOGAT analysis, the two bands detected by BN-PAGE analysis exhibited the same mobility on the 2D SDS-PAGE, showing a polypeptide size of 170 kDa. Together with the previous reports that Fd-GOGAT is a single copy gene in spinach, and that NADH-GOGAT shows a very different molecular weight of 240 kDa, the 730 kDa protein was shown to be either a homomultimer (most likely a tetramer) of 190 kDa holoenzyme or its complex with other protein(s) in chloroplasts. In order to further investigate the identity and physiological significance of the 730 kDa complex, its response to the changes in the conditions of stromal solvent was analyzed (Fig. 2A). Stromal proteins were extracted from isolated chloroplasts using either a slightly acidified solvent (at pH 6) to mimic the nightly stromal environment or a slightly alkalized solvent (at pH 8) simulating the changes in the stromal pH following illumination. As shown in Figure 2A, distribution of the two bands is clearly different between the two solvent conditions (lanes 1 and 5); the 730 kDa is the major band at pH 6, on the other hand at pH 8, it mostly disappeared, and the intensity of the 190 kDa band was somewhat increased. No GOGAT signal was detected in the thylakoid fraction. This suggests that the 730 kDa complex is converted into the 190 kDa holoenzyme, and the latter form may be relatively unstable at pH 8 under the current experimental conditions. By addition of NaCl at 0.1 M (lane 2) or DTT at 10 mM (lane 3) into the stromal solvent at pH 6, the 730 kDa complex was mostly converted into the 190 kDa protein, suggesting the involvement of electrostatic interactions and disulfide bonds for the assembly of the complex. Addition of ascorbic acid at 10 mM (lane 4), which was shown to affect the redox regulation of FNR-Tic62 complex [20], had no significant effect under the current conditions. Next, whether the pH-dependent conversion between the two forms (730 kDa and 190 kDa) of GOGAT is reversible or not was investigated (Fig. 3). Stromal proteins extracted by using a standard extraction buffer at pH 7.5 (lane 1) were diluted by a buffer at pH 6 or 8 (lanes 3 and 4). Treatment with pH 8 buffer increased the 190 kDa monomer and decreased the 730 kDa complex (lane 4), while no significant change was seen by the treatment with pH 6 buffer (lane 3), in other words, re-association was not observed. Basically the same results were obtained by the same treatments of whole chloroplasts (lanes 5–7) instead of the stromal fraction, indicating that the effect of thylakoid components for the conversion is negligible. Thus, Fd-GOGAT forms a dissociable 730 kDa complex (probably a tetramer). In this regard, there is a report that bacterial NADPH-GOGAT, which consists of two subunits (α and β) and shares considerable homology with Fd-GOGAT throughout its sequence of α subunit, appears to be a $(\alpha\beta)_4$ tetramer in an analytical gel filtration experiment [21] and by X-ray small angle scattering measurements [22]. Although Fd-GOGAT is generally considered to be monomeric in solution, both NADPH-GOGAT α subunit and Fd-GOGAT form similar dimers in crystals [23,24]. Therefore, a tetramer of Fd-GOGAT may be present under physiological conditions. This 730 kDa complex dissociates into the 190 kDa monomer enzyme in response to DTT and alkalized pH (Figs. 2&3), suggesting the involvement in the redox regulation of the function of GOGAT, which will be addressed in the latter section.

Analysis of NiR complex

The situation of NiR turned out to be partly similar to that of GOGAT. In the western analysis of 2D BN/SDS-PAGE (Fig. 1G), the multiple bands observed by BN-PAGE analysis showed the same mobility as the monomer (65 kDa), indicating that they are derived from the same NiR gene product, known to be a single copy in spinach. Therefore, the multiple bands other than the monomer NiR detected on BN-PAGE analysis were either homo-multimeric forms of NiR enzyme or its complexes with other components in the chloroplasts. Responses to the different conditions in the stromal solvent such as salt and pH (Fig. 2B) have some common features with those of GOGAT, but there also are a few intriguing differences. As in the case of GOGAT in Figure 2A, protein distributions are largely different between the treatments of pH 6 and pH 8 (lanes 1 and 5). Intensive shifts of the bands toward higher molecular weight up to around 800 kDa were observed at pH 6 (lane 1); monomer band

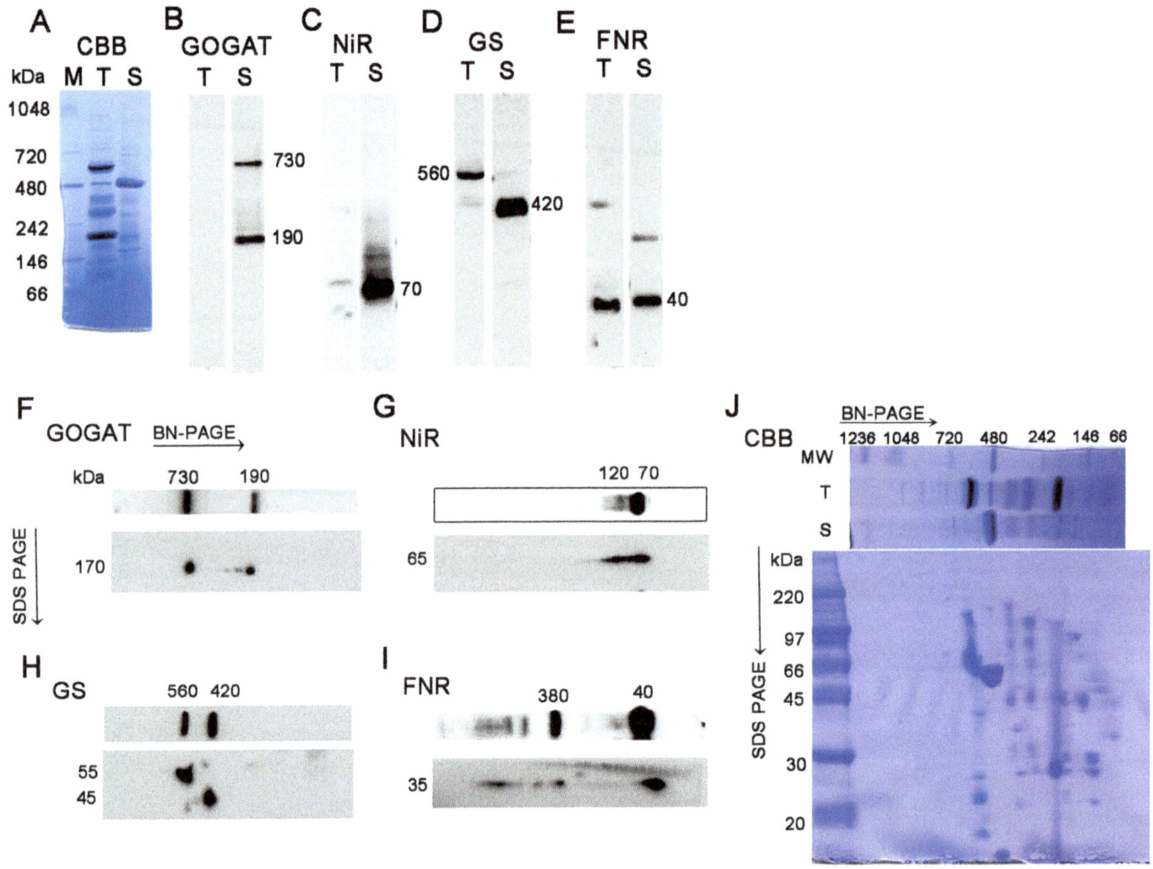

Figure 1. Protein complexes in spinach chloroplasts analyzed by blue-native PAGE (BN-PAGE) and SDS-PAGE. Upper panel; CBB-staining (A) and western blots (B–E) of BN-PAGE analysis of thylakoid (T) and stroma (S) proteins extracted from spinach chloroplasts. Lower panel; western blots (F–I) and CBB-staining (J) of 2D SDS-PAGE after separation by BN-PAGE of stromal proteins (F and G) and whole chloroplast proteins (H, I and 2D SDS-PAGE of J). Western blots were probed with polyclonal antibodies against each protein indicated. The numbers beside each band stand for the estimated molecular weights. All the samples loaded were derived from 10 µg on a chlorophyll basis of chloroplasts whose proteins were extracted.

at around 70 kDa became very faint, and instead, most abundant band of 170 kDa and less abundant, smear bands at around 340 kDa and 800 kDa are detected. The identity of the 50 kDa band is not clear but may be either a partial degradation product of NiR or a non-specifically detected band. Addition of DTT at 10 mM reduced the above shift and increased the 70 kDa monomer at both pH 6 and 8 (lanes 3 and 7), which is similar to the case of GOGAT. Unlike GOGAT, addition of NaCl at 0.1 M largely reduced the overall signals (lanes 2 and 6), and addition of ascorbic acid at 10 mM increased the monomer protein (lanes 4 and 8), implying that these agents affect the stability and/or dissociation of NiR. Another difference from GOGAT is the apparent reversible feature between NiR monomer and multiple complexes in response to the pH shifts (Fig. 3B); dilution of the stromal extract with pH 6 buffer increased the higher molecular-weight complexes, especially the 170 kDa band (lane 3 as compared to lane 1), while dilution with pH 8 buffer almost diminished the signals for those complexes (lane 4). Dilution of whole chloroplast suspension instead of the stromal extract conferred basically the same results (lanes 5–7) except that more intensive shifts of the bands were seen at lower pH, as observed in Figure 2B. Thus the conversion between monomer NiR and the higher molecular-weight complexes upon the pH shifts appears to be reversible, and their distribution varies in response to the relatively moderate changes in the solvent conditions, which may be related to the regulation of physiological function of NiR, addressed in the latter section.

Analysis of GS complex

Another intriguing feature implying a regulatory mechanism of enzyme function was found by the analysis of the GS complex. Unlike the cases of GOGAT and NiR, the 2D SDS-PAGE analysis of the 420 kDa and 560 kDa proteins (Fig. 1H) exhibited polypeptides of different molecular weights (around 45 and 55 kDa, respectively), suggesting either that they are derived from different GS isoproteins or that a post-translational modification of the GS polypeptide occurred. The 45 kDa polypeptide corresponds to the size of plastidic GS2 protein, but the 55 kDa is rather large for the conventional plant GS proteins. In this connection, Seabra et al. [11] reported that some of the organ-specific complexes of *M. truncatula* GS2 protein, detected by western blots after Native PAGE, appeared to contain a polypeptide of higher molecular mass (50 kDa) in addition to a 42 kDa conventional GS2 polypeptide as analyzed by western blots after SDS-PAGE. Because basically only one GS2 isoprotein was shown to be expressed in this plant except in seeds, this additional polypeptide (50 kDa) was indicated to represent a post-translational modification of 42 kDa *M. truncatula* GS2 protein.

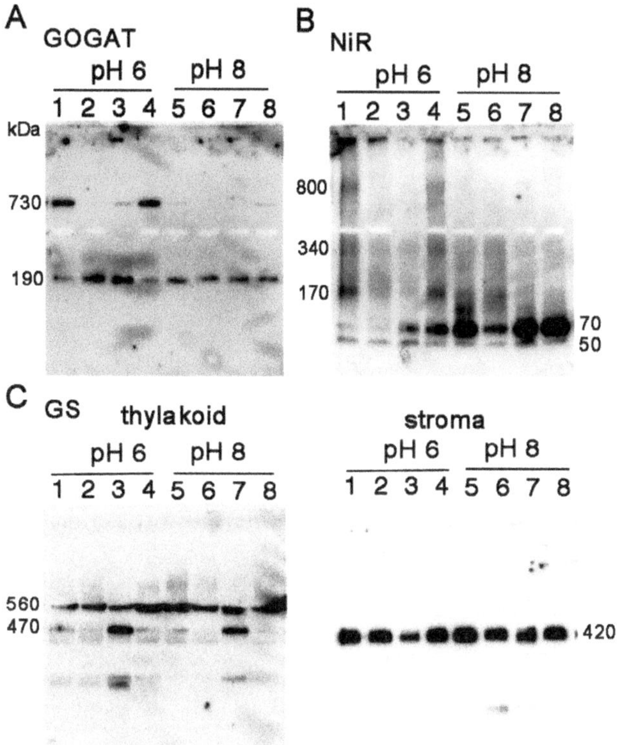

Figure 2. Protein distribution in response to the changes in solvent conditions of chloroplasts. Stroma and thylakoid proteins extracted from spinach chloroplasts under different solvent conditions of potassium phosphate buffer at pH 6 (lane 1) or pH 8 (lane 5) containing 0.1 M NaCl (lanes 2 and 6), 10 mM DTT (lanes 3 and 7) or 10 mM ascorbic acid (lanes 4 and 8) were analyzed by western blots after BN-PAGE. Results of stromal proteins for GOGAT (A) and NiR (B), and stroma and thylakoid proteins for GS (C) were presented. The numbers beside each band stand for the estimated molecular weights.

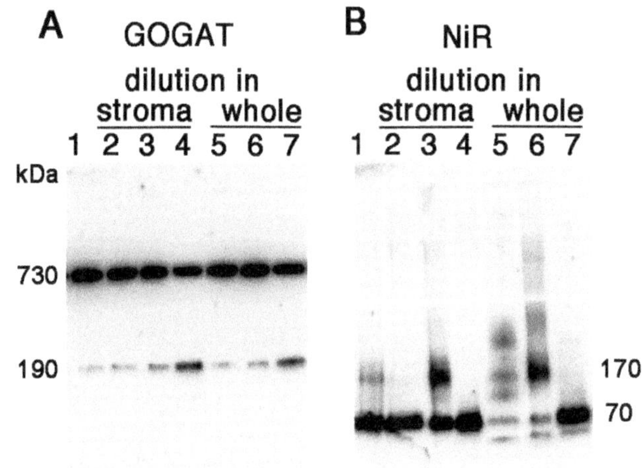

Figure 3. Protein distribution in response to the pH shift of stroma and whole chloroplasts. Stromal proteins extracted by using Hepes-KOH buffer at pH 7.5 (lane 1) were four-fold diluted with the same buffer (lane 2), potassium phosphate buffer at pH 6 (lane 3) or pH 8 (lane 4), incubated for 30 minutes at room temperature and analyzed by western blots for GOGAT (A) and NiR (B) after BN-PAGE. Stromal proteins extracted from whole chloroplasts treated with above three conditions were also analyzed (whole; lanes 5–7).

Since only one GS2 isoprotein has been reported in spinach (*S. oleracea*), it is not clear whether multiple GS2 isoproteins are expressed in spinach leaves or not. Concerning membrane-bound GS proteins, a slower migrating form of the GS2 protein was shown to accumulate in the fraction of plastid membranes in root nodules of *M. truncatula* [25], and the interaction of cytosolic GS1 protein with a symbiosome membrane protein (nodulin 26) in soybean root nodules was reported recently [26].

In contrast to the cases of GOGAT and NiR, no significant differences in the GS profile were observed between the two pH conditions in both the stroma and thylakoid fractions (Fig. 2C). Instead, addition of DTT produced an additional 470 kDa band below the 560 kDa band in thylakoids (lanes 3 and 7 in the left panel), together with the decrease in the amount of the stromal 420 kDa holoenzyme especially at pH 6 (lane 3 in the right panel). These results suggest either that the stromal decameric GS (420 kDa) may be recruited into thylakoid membranes under the conditions at 10 mM DTT via anchoring protein(s), or that the 470 kDa protein may be a cleaved or dissociated product of the 560 kDa protein complex. In order to seek out the identity of this 470 kDa band, 2D BN/SDS-PAGE analysis of thylakoid proteins treated with DTT was performed (left panel in Fig. 4). The 470 kDa protein on BN-PAGE exhibited the same mobility as that of the 560 kDa protein, suggesting that the 470 kDa band is probably derived from the 560 kDa complex although the subunit composition and integrity of the two complexes is not clear. Our preliminary experiment shows that the 470 kDa protein is more easily released from thylakoid membranes compared to the 560 kDa protein by the mild extraction with 0.1% DDM. This 470 kDa protein may be an intermediate complex to be released into stroma as a decameric holoenzyme by unknown mechanism. The identities of these higher molecular-weight forms are not clear, but in the root nodules of *M. truncatula*, the slower migrating form of GS2 protein as described above was shown to localize in the fraction of the plastid membrane as a catalytically inactive form [25]. Also, among the multiple GS complexes (420~620 kDa) observed in the leaves of the same plant [11], only the lowest 420 kDa band was detected by an in-gel GS activity assay, implying the involvement of this multiplicity in the regulatory mechanism for GS enzyme activity. Thus, the activity of the GS complex in spinach chloroplasts was addressed in the following section.

Native PAGE analysis and GS activity assays

Native PAGE without the CBB dye was performed for an in-gel GS activity assay because the dye interferes the activity staining in BN-PAGE. Native PAGE without detergents is limited to the separation of acidic proteins with a pI below the pH of the gel and is often characterized by protein aggregation and broadening of protein bands as compared with BN-PAGE. However, it is even milder than BN-PAGE and thought to offer advantages for retaining physiological protein assemblies and oligomerization. As shown in Figure 5A, the migration pattern of molecular weight markers (lane M) was very similar to that of BN-PAGE (lane M in Fig. 1A). Minor differences were seen in the pattern of stromal proteins with slower mobility of a major band (ribulose 1,5-bisphosphate carboxylase/oxygenase complex) compared to that in BN-PAGE (lane S in Figs. 1A&5A). In contrast, the pattern of thylakoid proteins was rather different between the two PAGEs especially for the slower mobility of chlorophyll binding protein complexes (around 200, 300 and 600 kDa in BN-PAGE; lane T in Figs. 1A&5A). For the patterns of the three nitrogen assimilatory enzymes, no large difference was seen except for the apparent lack

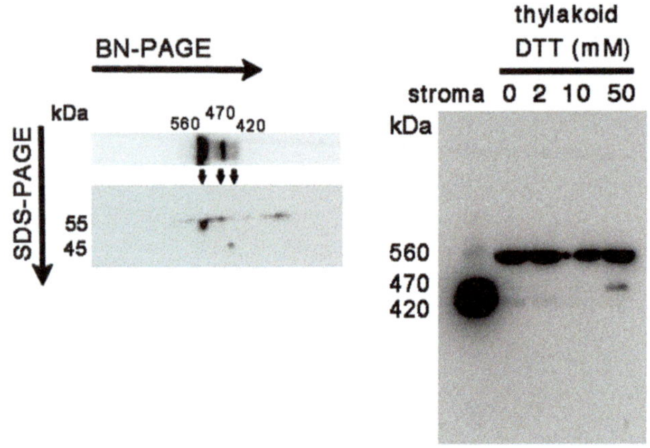

Figure 4. Western analysis of GS distribution in thylakoid proteins treated with DTT. Thylakoid proteins treated with DTT were analyzed by western blots of BN-PAGE (right) and 2D SDS-PAGE after BN-PAGE (left) using the GS polyclonal antibody. In the pattern of 2D SDS-PAGE (left), a faint spot at about 45 kDa is probably derived from the contamination of the stromal GS holoenzyme which is fairly abundant in the stromal fraction.

of the 190 kDa GOGAT monomer, and instead, smear bands migrating around the 480 kDa marker were observed in addition to the major band around 720 kDa marker on Native PAGE (Fig. 5B). Thus, the behavior of the monomer form of GOGAT appears to be different between the two native PAGEs. For NiR (Fig. 5C), more intensive shifts, as compared to that on BN-PAGE (Fig. 1C), were observed even at neutral pH. These results may represent more physiological situation than those in BN-PAGE. As for GS (Fig. 5D), the distribution pattern was basically the same as those on BN-PAGE (Fig. 1D) except that the mobility of the band in the thylakoid fraction was slightly slower. The identity of a very faint band observed above the major band in stroma is not clear at present. Thus, multiple protein complexes common to any of the three nitrogen assimilation enzymes were not observed also by this Native PAGE analysis. Since the profile of the bands detected by the anti-GS antibody remained basically the same between the analyses of BN-PAGE and Native PAGE, an in-gel GS activity assay using Native PAGE was performed in order to evaluate whether the molecular species in each fraction are enzymatically active or not (Fig. 5E). After the run of the Native PAGE, chlorophyll-binding proteins were detected as a green color in the lanes of thylakoids (T) and whole chloroplasts (W). Subsequent GS activity staining caused yellowish background of the whole gel and reddish staining of only the 420 kDa protein in the lanes of stroma (lane S) and whole chloroplasts (lane W), in agreement with the results of the leaf GS activity staining of *M. medicago* plants described in the previous section. The GS transferase activity in solution was also measured to be 2.1 µmol min^{-1} mg^{-1} total protein in the stromal extract. The activity in the thylakoid extract was about 0.07 µmol min^{-1} mg^{-1} total protein, which was barely above the background level. Taken together, the 420 kDa GS holoenzyme present in stroma was shown to be a catalytically active form while the slower migrating protein complex (560 kDa) detected by the anti-GS antibody in thylakoids appeared to have no significant GS activity. The 560 kDa protein in thylakoids may represent an inactive form caused by a post-translational modification as described in the previous section. In this regard, the higher molecular weight form could result from the covalent binding of ubiquitin or ubiquitin-like proteins of approx. 10 kDa; Arabidopsis GS2 has been identified as a potential substrate for SUMO (small ubiquitin-like modifier) by the yeast two-hybrid system [27] although SUMOylation has not been tested in the plants. Another possibility has been postulated for the slower migrating form of the root nodule GS2 protein to be a non-processed precursor form which contains the transit peptide [25]. Analytical studies such as a proteomic analysis of the isolated protein are required to verify these possibilities. Reactivity of the anti-GS antibody we used was examined by additional western analyses (Figure S1 and S2), but we can't totally deny the possibility that the higher molecular weight forms (the 560 kDa protein and the 55 kDa polypeptide) detected by the GS antibody are a non-GS protein that may contain some antigenic structures common to GS proteins.

Fd-binding ability of multiple forms of GOGAT and NiR proteins

GOGAT and NiR are known to interact with Fd as a physiological electron donor. In order to evaluate the function of the multiple forms of GOGAT and NiR, the ability of each form for binding with Fd was investigated using the stromal extracts prepared under different pH conditions (Fig. 6). FNR as another major Fd-dependent enzyme and GS as a non-Fd dependent enzyme were also analyzed as controls. We routinely use an affinity resin with recombinant Fd from the cyanobacteria *L. boryana* [19] for the purification of Fd-dependent enzymes because of its stability on an affinity resin and because *L. boryana* Fd functions as an electron donor for Fd-dependent enzymes in higher plants to an extent comparable to that achieved by plant Fds. At pH 8, where GOGAT, NiR and FNR are mostly present as monomers in the stromal extracts, profiles of the Fd-binding and elution were quite similar among these three enzymes (lanes 6–10 in panels A, B and D); part of the enzymes were bound and released with first one-column volume of the elution buffer containing 0.2 M NaCl. The relative intensity of the signals in the elution fraction (lane 9) was somewhat higher for NiR than for FNR and GOGAT. This is consistent with the binding strength of these enzymes in the leaf extracts of maize and potato on Fd-affinity chromatography; in the order of GOGAT<FNR<NiR (T. Hase and Y. Chikuma, personal communications). In contrast, no detectable band was seen in the elution fraction of GS (lane 9 in

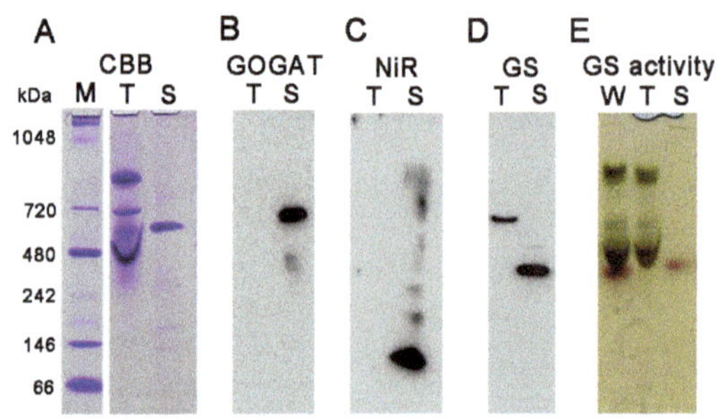

Figure 5. Protein complexes in spinach chloroplasts analyzed by using Native PAGE. CBB staining (A), western blots (B–D) and in-gel GS activity staining (E) after Native PAGE analysis of thylakoid (T) and stroma (S) proteins extracted from spinach chloroplasts. For in-gel GS activity assay, proteins extracted from whole chloroplasts (W) were also used. Lane M stands for the protein standard markers, but the migration distance does not necessarily correlate with the molecular mass in this Native PAGE. Samples loaded were derived from 10 µg (A–D) and 20 µg (E) on a chlorophyll basis of chloroplasts whose proteins were extracted.

panel C), indicating no significant binding. Thus, monomer forms of GOGAT and NiR, as well as FNR, at least partly bind with Fd in the stromal extracts under the conditions of pH 8. On the other hand, at pH 6, higher molecular weight forms of GOGAT (730 kDa) and NiR (170 kDa) in addition to the monomer forms were clearly observed in the extracts (lanes 1 in panels A and B). For NiR protein, the 170 kDa band together with the 70 kDa monomer band was greatly diminished in the unbound fraction (lane 2 in panel B), indicating their efficient binding with Fd. The reason for the detection of only the monomer band in the elution fraction (lanes 4,5 in panel B) is assumed to be due to the dissociation of the 170 kDa molecule under the conditions of 0.2 M NaCl (as implied from the results in lane 2 of Fig. 2B). FNR as a control also showed similar binding profile with NiR, although only the monomer was detected under this condition (lanes 1–5, panel D). On the other hand in GOGAT, significant reduction in the intensity of the bands was not observed in the unbound fraction (lane 2 in panel A), but partial binding was indicated from the band in the elution fraction (lane 5 in panel A). Because the 730 kDa molecule appears to dissociate into the 190 kDa monomer under 0.1 M NaCl condition (lane 2 in Fig. 2A), it is not clear which molecules, either or both of the 730 kDa and the 190 kDa proteins, actually bound with Fd. Under pH 6 conditions, GS protein also appears to partly bind with Fd resin (lane 5 in panel C); whether this is due to non-specific binding or transient complex formation of GS with other Fd-binding proteins is not clear at present.

Taken together, both the monomer (70 kDa) and the higher molecular weight complex (170 kDa) of NiR exhibited the binding ability with Fd; oligomerization of NiR may be related to its other function such as enzyme activity and protein stability. On the other hand in GOGAT, the monomer binds with Fd at least pH 8, and the 730 kDa complex appears to have either negligible or low Fd-binding ability under pH 6 conditions. As for Fd-GOGAT from *Synechococcus* sp. PCC 6301, there is a report that two protein bands appeared after non-denaturing PAGE of purified GOGAT proteins, both presenting GOGAT activity *in situ* using methyl viologen as a small artificial electron donor although the data were not shown [28]. Thus, the higher molecular weight complex of GOGAT in spinach chloroplasts may itself retain catalytic activity but may assume conformation which reduces the binding with Fd for the purpose to regulate the GOGAT reaction in the stromal fraction.

To sum up, individual complexes with larger molecular sizes of holoenzyme were detected for all the three nitrogen assimilatory enzymes, GOGAT, GS and NiR, studied in this article. Figure 7 shows schematic models for the pH-dependent conversion of monomers and oligomers (or multi-protein complexes) of GOGAT and NiR, and possible localization of GS inferred from this study, together with a previously presented model for FNR [7]. The pH-dependent conversion of monomers and multimeric complexes of GOGAT and NiR, which appears to be accompanied by differential binding to Fd and possibly by differential enzyme activity or protein stability, indicates the physiological significance of these complexes for the regulation of enzyme function in response to the conditions such as redox state in stroma and light availability. To our surprise, GS signal was observed as another inactive complex in thylakoid fraction (or some other plastid membranes) with a larger polypeptide, postulating a different regulatory mechanism from those of the above two enzymes, which involves different localization, similarly to the case of FNR, and possibly a novel post-translational modification.

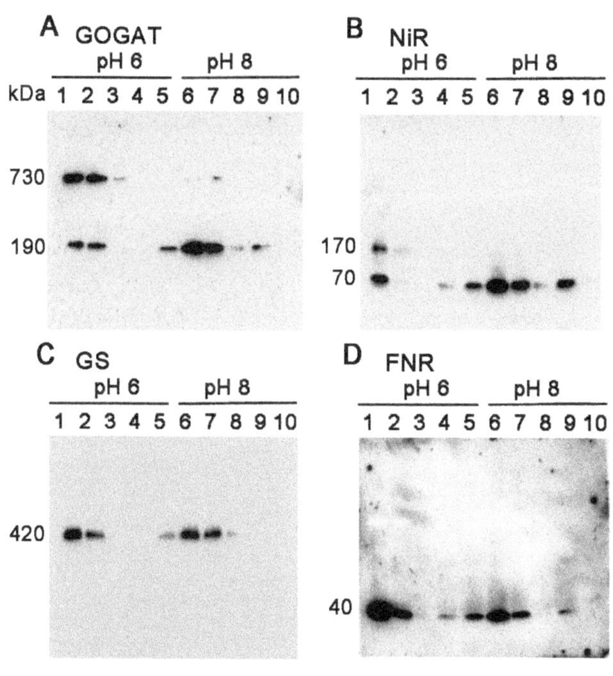

Figure 6. Fd-binding assay of protein complexes in the stromal fraction of spinach chloroplasts. Stromal proteins extracted from spinach chloroplasts by using potassium phosphate buffer at pH 6 (lane 1) or pH 8 (lane 6) were mixed with Fd resin, and the resulting unbound fractions (lanes 2&7), washed fraction (lanes 3&8) and fractions eluted with first (lanes 4&9) and second (lanes 5&10) one-column volume of the same buffer containing 0.2 M NaCl were collected and analyzed by western blots after BN-PAGE.

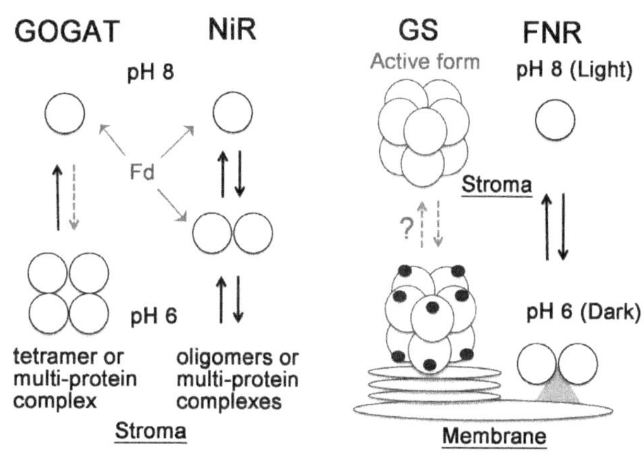

Figure 7. Models of pH-dependent complex formation and relocation of carbon- and nitrogen assimilatory enzymes in chloroplasts. Complex formation (possibly oligomerization) of GOGAT and NiR and localization of GS signals observed in this study are depicted together with a previously presented model for FNR [7]. GOGAT and NiR display pH-dependent conversion of monomers and oligomers (or protein complexes with unknown proteins) in the stromal fraction, which appears to be accompanied by differential binding to Fd (Fig. 6) and possibly by differential enzyme activity or protein stability. GS is postulated to exist both as an active enzyme of decamer in stroma and as a modified, probably inactive form (depicted with black dots) in the plastid membranes (such as thylakoids and envelop), which may interconvert by an unknown mechanism. FNR is known to relocate between stroma and thylakoids (possibly as a dimer for the stabilization of this enzyme) in the light, redox and pH-dependent manners [7,8].

Supporting Information

Figure S1 Western analysis of spinach proteins by SDS-PAGE (A) and BN-PAGE (B). A: Total proteins from spinach leaves (lane L), whole chloroplasts (W), stroma (S) and thylakoid fraction (T) were analyzed by western blot after 7.5% SDS-PAGE probed with anti-GS antibody (GS) and preimune serum (PI). In spinach leaves, only GS2 isoprotein is considered to be expressed [ref. S1 in Figure S1] and is thought to be detected by the polyclonal antibody against maize GS1a proteins [ref. S2 in Figure S1] that shares 75% identity with spinach GS2, through the consequence of the cross-reactivity of the antibody by recognizing the epitopes common to the two GS isozymes. As expected from the western analysis of the whole chloroplasts by 2D BN/SDS-PAGE (Figure 1H), stroma and thylakoid fractions contained the polypeptides with distinct sizes of approx. 41 kDa and 51 kDa, respectively. The deduced MWs are somewhat different from those in the 2D analysis (45 kDa and 55 kDa), but the size in the stromal fraction of the 1D SDS-PAGE (41 kDa) appears to be closer to the estimated size of spinach GS2 protein (372 residues) and would be more accurate than the 2D analysis. B: Proteins extracted from whole chloroplasts were analyzed by western blot after BN-PAGE, probed with anti-GS antibody (GS) and preimune serum (PI). The numbers beside each band stand for the estimated molecular weights. All the samples loaded were derived from 10 µg on a chlorophyll basis of chloroplasts whose proteins were extracted.

Figure S2 Western analysis of maize proteins by BN-PAGE (A) and SDS-PAGE (B). A: Thylakoid (T) and stroma (S) proteins were extracted from maize chloroplasts using the same protocol as described for spinach in the 'Materials and Methods' section, and were analyzed by western blot after BN-PAGE, probed with anti-GS antibody (GS). The pattern is quite similar to that of spinach leaves (Figure 1D), although the mobility of the major band detected in the lane of thylakoid proteins is slightly faster (approx. 540 kDa) than that of spinach (560 kDa). B: Total proteins from maize leaves (lane L), stroma (S) and thylakoid fraction (T) were analyzed by western blot after 7.5% SDS-PAGE, probed with anti-GS antibody (GS). The analysis showed the bands, 41 kDa for stroma and 49 kDa for thylakoids, which are similar to those of spinach (Figure S1A). The relative content of the maize GS2 protein in the stromal fraction appears to be lower compared to that of spinach. Additional lowest band detected in the lane of the whole leaves is thought to be a cytosolic GS1 protein (39 kDa), which is known to be absent in spinach leaves, leading to the confirmation that the anti-GS antibody reacts with both GS1 and GS2. Thus, the patterns of the GS signals detected in the stroma and thylakoid fractions of maize chloroplasts were basically the same as those of spinach on both BN-PAGE and SDS-PAGE analyses. The numbers beside each band stand for the estimated molecular weights.

Acknowledgments

The authors thank Dr. Shingo Kikuchi for helpful advices with experimental methods and techniques.

Author Contributions

Conceived and designed the experiments: YKA. Performed the experiments: YKA. Analyzed the data: YKA TH. Contributed reagents/materials/analysis tools: YKA. Wrote the paper: YKA TH.

References

1. Kuriyan J, Eisenberg D (2007) The origin of protein interactions and allostery in colocalization. Nature 450: 983–90.
2. Winkel BSJ (2004) Metabolic channeling in plants. Annu Rev Plant Biol 55: 85–107.
3. Buchanan BB (1984) The ferredoxin/thioredoxin system: a key element of the regulatory function of light in photosynthesis. Bioscience 34: 378–383.
4. Scheibe R (1991) Redox-modulation of chloroplast enzymes. A common principle for individual control. Plant Physiol 96: 1–3.
5. Wedel N, Soll J (1998) Evolutionary conserved light regulation of Calvin cycle activity by NADPH-mediated reversible phosphoribulokinase/CP12/glyceraldehyde-3-phosphate dehydrogenase complex dissociation. Proc Natl Acad Sci U S A 95(16): 9699–704.
6. Howard TP, Metodiev M, Lloyd JC, Raines CA (2008) Thioredoxin-mediated reversible dissociation of a stromal multiprotein complex in response to changes in light availability. Proc Natl Acad Sci U S A 105(10): 4056–61.
7. Benz JP, Lintala M, Soll J, Mulo P, Bolter B (2010) A new concept for ferredoxin–NADP(H) oxidoreductase binding to plant thylakoids. Trends Plant Sci 15: 608–613.
8. Juric S, Hazler-Pilepic K, Tomasic A, Lepedus H, Jelicic B, et al. (2009) Tethering of ferredoxin: NADP+ oxidoreductase to thylakoid membranes is mediated by novel chloroplast protein TROL. Plant J 60: 783–794.
9. Xu G, Fan X, Miller AJ (2012) Plant nitrogen assimilation and use efficiency. Annu Rev Plant Biol 63: 153–82.
10. Masclaux-Daubresse C, Daniel-Vedele F, Dechorgnat J, Chardon F, Gaufichon L, et al. (2010) Nitrogen uptake, assimilation and remobilization in plants: challenges for sustainable and productive agriculture. Ann Bot 105(7): 1141–57.
11. Seabra AR, Silva LS, Carvalho HG (2013) Novel aspects of glutamine synthetase (GS) regulation revealed by a detailed expression analysis of the entire GS gene family of *Medicago truncatula* under different physiological conditions. BMC Plant Biol 13: 137.
12. Schägger H, von Jagow G (1991) Blue native electrophoresis for isolation of membrane protein complexes in enzymatically active form. Anal Biochem 199(2): 223–31.
13. Swamy U, Wang M, Tripathy JN, Kim SK, Hirasawa M, et al. (2005) Structure of spinach nitrite reductase: implications for multi-electron reactions by the iron-sulfur: siroheme cofactor. Biochemistry 44(49): 16054–63.
14. Suzuki A, Knaff DB (2005) Glutamate synthase: structural, mechanistic and regulatory properties, and role in the amino acid metabolism. Photosynth Res 83(2): 191–217.
15. Forde BG, Cullimore JV (1989) The molecular biology of glutamine synthetase in higher plants. In: Miflin, BJ, editor.Oxford Surv Plant Mol Cell Biol Vol. 6.London: Oxford Univ Press. pp. 247–296.
16. Unno H, Uchida T, Sugawara H, Kurisu G, Sugiyama T, et al. (2006) Atomic structure of plant glutamine synthetase: a key enzyme for plant productivity. J Biol Chem 281(39): 29287–96.
17. Torreira E, Seabra AR, Marriott H, Zhou M, Llorca O, et al. (2014) The structures of cytosolic and plastid-located glutamine synthetases from *Medicago truncatula* reveal a common and dynamic architecture. Acta Crystallogr D Biol Crystallogr 70(Pt 4): 981–93.
18. Mach J (2002) Organelle preparations. In: Weigel D, Glazebrook J, editors.Arabidopsis a Laboratory Manual.New York: Cold Spring Harbour Laboratory Press. pp. 217–218.
19. Sakakibara Y, Kimura H, Iwamura A, Saitoh T, Ikegami T, et al. (2012) A new structural insight into differential interaction of cyanobacterial and plant ferredoxins with nitrite reductase as revealed by NMR and X-ray crystallographic studies. J Biochem 151(5): 483–92.
20. Stengel A, Benz P, Balsera M, Soll J, Bölter B (2008) TIC62 redox-regulated translocon composition and dynamics. J Biol Chem 283(11): 6656–67.
21. Stabile H, Curti B, Vanoni MA (2000) Functional properties of recombinant *Azospirillum brasilense* glutamate synthase, a complex iron-sulfur flavoprotein. Eur J Biochem 267(9): 2720–30.
22. Petoukhov MV, Svergun DI, Konarev PV, Ravasio S, van den Heuvel RH, et al. (2003) Quaternary structure of *Azospirillum brasilense* NADPH-dependent glutamate synthase in solution as revealed by synchrotron radiation x-ray scattering. J Biol Chem 278(32): 29933–9.
23. Binda C, Bossi RT, Wakatsuki S, Arzt S, Coda A, et al. (2000) Cross-talk and ammonia channeling between active centers in the unexpected domain arrangement of glutamate synthase. Structure 8(12): 1299–308.
24. van den Heuvel RH, Ferrari D, Bossi RT, Ravasio S, Curti B, et al. (2002) Structural studies on the synchronization of catalytic centers in glutamate synthase. J Biol Chem 277(27): 24579–83.
25. Melo PM, Lima LM, Santos IM, Carvalho HG, Cullimore JV (2003) Expression of the plastid-located glutamine synthetase of *Medicago truncatula*. Accumula-

tion of the precursor in root nodules reveals an in vivo control at the level of protein import into plastids. Plant Physiol 132(1): 390–9.
26. Masalkar P, Wallace IS, Hwang JH, Roberts DM (2010) Interaction of cytosolic glutamine synthetase of soybean root nodules with the C-terminal domain of the symbiosome membrane nodulin 26 aquaglyceroporin. J Biol Chem 285(31): 23880–8.
27. Elrouby N, Coupland G (2010) Proteome-wide screens for small ubiquitin-like modifier (SUMO) substrates identify Arabidopsis proteins implicated in diverse biological processes. Proc Natl Acad Sci U S A. 107(40): 17415–20.
28. Marqués S, Florencio FJ, Candau P (1992) Purification and characterization of the ferredoxin-glutamate synthase from the unicellular cyanobacterium *Synechococcus* sp. PCC 6301. Eur J Biochem. 206(1): 69–77.

5

The Shade Avoidance Syndrome in Arabidopsis: The Antagonistic Role of Phytochrome A and B Differentiates Vegetation Proximity and Canopy Shade

Jaime F. Martínez-García[1,2*], Marçal Gallemí[2], María José Molina-Contreras[2], Briardo Llorente[2], Maycon R. R. Bevilaqua[2,3¤], Peter H. Quail[4,5]

1 Institució Catalana de Recerca i Estudis Avançats, Barcelona, Spain, **2** Centre for Research in Agricultural Genomics (CRAG), Consortium CSIC-IRTA-UAB-UB, Barcelona, Spain, **3** CAPES foundation, Ministry of Education of Brazil, Brasilia - DF, Brazil, **4** Department of Plant and Microbial Biology, University of California, Berkeley, California, United States of America, **5** US Department of Agriculture/Agriculture Research Service, Plant Gene Expression Center, Albany, California, United States of America

Abstract

Light limitation caused by dense vegetation is one of the greatest threats to plant survival in natural environments. Plants detect such neighboring vegetation as a reduction in the red to far-red ratio (R:FR) of the incoming light. The low R:FR signal, perceived by phytochromes, initiates a set of responses collectively known as the shade avoidance syndrome, intended to reduce the degree of current or future shade from neighbors by overtopping such competitors or inducing flowering to ensure seed production. At the seedling stage these responses include increased hypocotyl elongation. We have systematically analyzed the Arabidopsis seedling response and the contribution of phyA and phyB to perception of decreased R:FR, at three different levels of photosynthetically active radiation. Our results show that the shade avoidance syndrome, induced by phyB deactivation, is gradually antagonized by phyA, operating through the so-called FR-High Irradiance Response, in response to high FR levels in a range that simulates plant canopy shade. The data indicate that the R:FR signal distinguishes between the presence of proximal, but non-shading, neighbors and direct foliar shade, via a intrafamily photosensory attenuation mechanism that acts to suppress excessive reversion toward skotomorphogenic development under prolonged direct vegetation shade.

Editor: Enamul Huq, University of Texas at Austin, United States of America

Funding: MG and MJMC received FPI fellowships from the Spanish Ministry of Economy and Competitivity (MINECO). BL received a Marie Curie IIF fellowship (CarotenActors) from the European Union FP7. JFMG received a 1-year fellowship from the Spanish Ministry of Education for a sabbatical stay in PHQ laboratory. The authors' research is supported by grants from the Generalitat de Catalunya and Spanish MINECO - Fondo Europeo de Desarrollo Regional (FEDER) to JFMG (2009-SGR697, Xarba and BIO2011-23489) and by National Institutes of Health (2R01 GM-047475), Department of Energy (DEFG03-87ER13742), and United States Department of Agriculture Agricultural Research Service Current Research Information System (5335-21000-032-00D) to PHQ. The funders had no role in study design, data collection and analysis, decision to publish, or preparation of the manuscript.

Competing Interests: The authors have declared that no competing interests exist.

* Email: jaume.martinez@cragenomica.es

¤ Current address: UEM Universidade Estadual de Maringá, Departamento de Biotecnologia, Biologia Celular e Genética, Maringá, Paraná, Brazil

Introduction

The shade avoidance syndrome (SAS) refers to a set of plant responses aimed at adapting plant growth and development to high plant density environments, like those found in forests, prairies or orchard communities. Two related but different situations can occur in these environments: plant proximity (without direct vegetative shading) and direct plant canopy shade [1–3]. Because vegetation preferentially reflects far-red (FR) light compared to other wavelengths, plant proximity generates a reduction in the red (R, about 600–700 nm) to far-red (FR, between 700–800 nm) ratio (R:FR) in the light impinging on neighbors. By contrast, under a plant canopy, light from the visible region (called photosynthetically active radiation or PAR, between 400–700 nm) is strongly absorbed by the chlorophyll and carotenoid photosynthetic pigments whereas FR, which is poorly absorbed by the leaves, is transmitted through (or reflected from) vegetation. As a consequence, under direct plant canopy shade both the amount of PAR (light quantity) and R:FR (light quality) are greatly reduced, in the latter case mostly by the selective depletion of R light caused by the filtering of sunlight through the leaves [1,2,4–6].

This low R:FR signal is perceived by the phytochrome (phy) photoreceptors. Phys detect the R and FR part of the spectrum and have a major role in controlling several adaptive responses such as seed germination, stem elongation, leaf expansion and flowering time. In Arabidopsis (*Arabidopsis thaliana*), a small gene family of five members encodes the phys (*PHYA-PHYE*) [7]. Although phyB is the major phy controlling the SAS, genetic and physiological analyses have shown that other phys act redundantly with phyB in the control of some aspects of SAS-driven development, such as flowering time (phyD, phyE), petiole elongation (phyD, phyE) and internode elongation between rosette leaves (phyE) [2,4]. The photolabile phyA has the unique capacity to function as a FR-light sensor through a mechanism termed the FR-High Irradiance Response (HIR) [2,8,9]. In contrast to the other phys, phyA has an antagonistic negative role in the SAS hypocotyl response, although varying degrees of regulation have been reported: *phyA* mutant seedlings growing under low R:FR light showed from moderately [10] to extremely long hypocotyls [11]. An antagonistic activity between phyA and phyB has also been shown in seedlings exposed to varying ratios of monochromatic R and FR [12,13]. However, it has been argued that the adaptive significance of this phyA antagonism is limited and may instead be an inevitable consequence of the intrinsic properties of phyA selected for their role in seedling deetiolation [2].

Phys exist in two photoconvertible forms, an inactive R-absorbing Pr form and an active FR-absorbing Pfr form. In light-grown plants, the steady-state ratio of Pr and Pfr conformers depends on the R:FR ratio. Under high R:FR the photoequilibrium is displaced towards the active Pfr form and the SAS is suppressed. Under low R:FR the photoequilibrium is displaced towards the inactive form and SAS is induced. This induction is regulated at least partly by the interaction of active phys with various PHYTOCHROME INTERACTING FACTORs (PIFs) [14–16], which results in rapid changes in the expression of dozens of *PHYTOCHROME RAPIDLY REGULATED* (*PAR*) genes, postulated to be instrumental in implementing the SAS responses [17–21]. Because many of these *PAR* genes encode transcriptional regulators, it is assumed that shade responses are a consequence of the phy regulation of a complex transcriptional network, as postulated for seedling de-etiolation [22,23], that seems to be organized in functional modules [24]. Genetic analyses demonstrated positive and negative roles in SAS regulation for several *PAR* genes encoding transcriptional regulators, including members of the homeodomain leucine zipper class II (*ATHB2*, *ATHB4*, *HAT1*, *HAT2* and *HAT3*), basic-helix-loop-helix (*BEE1*, *BEE2*, *BIM1*, *BIM2*, *HFR1*, *PAR1*, *PAR2* and *PIL1*) and B-BOX CONTAINING (BBX) families of proteins [17–19,24–30]. Most of these studies were done analyzing hypocotyl elongation. Therefore, low R:FR perception rapidly changes the balance of positive and negative factors, resulting in the appropriate SAS responses, i.e., eventually causing hypocotyls to elongate. Evidence for the involvement of several of these factors in controlling auxin levels and sensitivity in mediating this elongation response has been reported [16,31,32].

The light treatments used to induce the SAS vary among laboratories, resulting in differences in the extent of the responses (usually hypocotyl length) reported for the same genotype. For instance, a review of several papers in the field reported that Arabidopsis Col-0 hypocotyls elongate in response to low R:FR (under laboratory conditions usually provided by white light supplemented with FR light, W+FR) from a minimum of about 2.5 mm to a maximum of ca. 9 mm [15,20,26,28,33–35]. In addition to media composition, variations in the timing and nature of the W+FR treatment might also explain some of the observed differences. For instance, the reported effect of the negative SAS regulator HFR1 on the shade-induced hypocotyl elongation ranged from mild [25] to very strong [27]. We have noted previously that the very strong phenotype of *hfr1* mutant seedlings was observed under shade conditions that reduced both R:FR and PAR (400–700 nm), whereas the mild phenotype occurred under shade conditions where only the R:FR ratio was reduced without significantly affecting the PAR [25]. Indeed, although the SAS is generally considered to be mainly induced by light of reduced R:FR, other light parameters, such as low-intensity light of the whole PAR spectrum and low blue light (which is part of the PAR spectrum), are also known to contribute to these responses [6,36–38]. Together, these observations highlight the fact that different shade conditions (such as variable PAR and/or R:FR) employed by different labs might account for some of the observed variability in the SAS response.

In this paper we have investigated the effect of both the level of PAR and supplemental FR (which results in different R:FR ratios without altering PAR) in the incoming light on hypocotyl elongation. To address the contribution of the two major phys in this response, we have systematically analyzed wild-type, *phyA* and *phyB* mutant seedlings. We observe that, independently of the PAR level employed, the R:FR ratio strongly and differentially affects elongation of wild-type, *phyA* and *phyB* hypocotyls. Our results indicate that quantitative variation in the R:FR ratio provides a dual signal with a likely different meaning in nature: when the R:FR is moderately lowered, it mimics plant proximity without direct shading, whereas when it is very low, it mimics direct plant canopy shade. In addition, the effects of these two environmental conditions can be distinguished genetically, with phyA and phyB having different roles in transducing the signals, as shown previously for seedling de-etiolation.

Materials and Methods

Plant material and growth conditions

Arabidopsis (*Arabidopsis thaliana*) plants for seed production were grown in the greenhouse as described [39]. The *phyA-501* (SALK_014575) [40] and *phyB-9* [41] mutant lines are in Col-0 ecotype. The *phyA* and *phyB* mutant lines in *Ler* have been described previously [42]. Homozygous *phyA-501* plants were genotyped as indicated in Figure S1 by using specific oligos: MSO31 (5′–TAG-AGC-ACC-GCA-CAG-CTG-CC-3′), MSO32 (5′– GAA-GCT-ATC-TCC-TGC-AGG-TGG– 3′) and LBb1 (5′-GCG-TGG-ACC-GCT-TGC-TGC-AAC-T-3′).

All the experiments were performed with seeds surface-sterilized and sown on Petri dishes with solid growth medium without sucrose (GM–; 0.215% (w/v) MS salts plus vitamins, 0.025% (w/v) MES pH 5.8) [17]. After stratification (3–6 days), plates were incubated in growth chambers at 22°C under continuous W that was provided by 2–4 cool-white horizontal fluorescent tubes (Figures 1–3), unless otherwise stated. These tubes delivered different amounts of photosynthetically active radiation (PAR), and a R:FR of about 2.5. In Figures 4 and 5, plates were incubated in growth chambers at 22°C under continuous W provided by 4 cool-white vertical fluorescent tubes (PAR of 20–25 $\mu mol \cdot m^{-2} \cdot s^{-1}$, R:FR of about 2.5). Simulated shade (W+FR) was generated by enriching W with supplementary FR provided by LED lamps (www.quantumdev.com; or www.philips.com/horti). Unless otherwise stated, fluence rates and PAR were measured with a LI-1800 spectroradiometer (Li-Cor Inc., www.licor.com); to calculate the R:FR, windows of 30 nm around the R (640–670 nm) and FR (720–750 nm) peaks were employed. For Figure 4, 5, and S4, fluence rates were measured with a Spectrosense2 meter associated with a 4-channel sensor (Skye Instruments Ltd., www.skyeinstruments.com), which measures PAR (400–700 nm) and 10 nm windows in the R (664–674 nm) and FR (725–735 nm) regions.

Hypocotyl length measurements

The National Institutes of Health ImageJ software (Bethesda, MD, USA; http://rsb.info.nih.gov/) was used on digital images to measure the length of hypocotyl seedlings (after laying out seedlings flat on agar plates) as indicated elsewhere [26]. At least 25 seedlings were used for each treatment. Experiments were repeated 3–5 times and a representative one is shown. Statistical

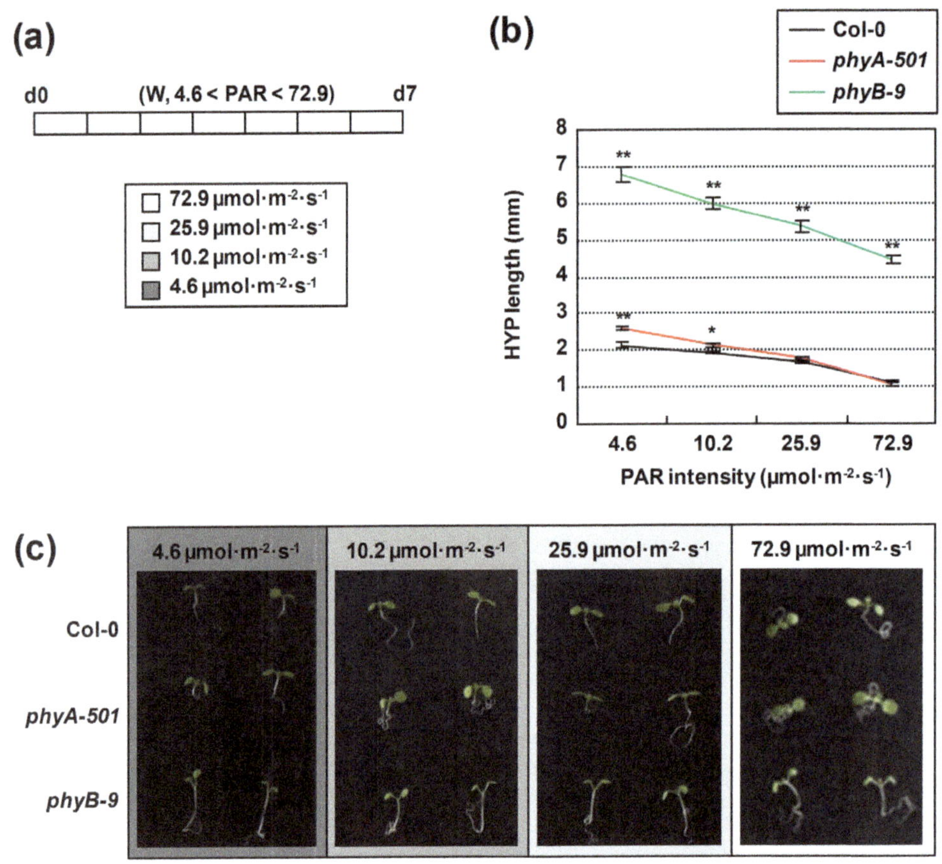

Figure 1. Effect of increasing intensities of white light on hypocotyl elongation of wild-type, *phyA-501* and *phyB-9* seedlings. (**a**) Seedlings of the indicated genotypes were grown from the day of germination until day 7 under W of increasing intensities (photosynthetic active radiation, PAR, between 4.6 and 72.9 µmol·m^{-2}·s^{-1}; R:FR>2.0). (**b**) Hypocotyl length of seedlings grown as indicated in **a**. Values are means ± SE of at least 25 hypocotyls for each light treatment. Asterisks indicate significant differences (*P<0.05, **P<0.01) relative to the control grown under the same light intensity. (**c**) Representative seedlings, grown as indicated in **a**, are shown for the three genotypes analyzed.

analyses of the data (t-test) were performed using GraphPad Prism version 4.00 for Windows (www.graphpad.com/) or Excel.

Protein extraction and Western blot analyses

Extracts were prepared following the direct extract protocol indicated elsewhere [43] with the modifications described below. Extracts shown in Figure S1 were prepared from Arabidopsis Col-0 and *phyA-501* seedlings germinated and grown in the dark for 5 days. Fifteen seedlings from each genotype were harvested and placed in 1.5 mL Eppendorf tubes containing 150 µL of (1×) Laemmli buffer. Extracts shown in other Figures were prepared from Arabidopsis Col-0 seedlings germinated and grown in the dark for 5 days and then exposed to 4 and 8 h of W or W+FR. On the day of harvest, 20 seedlings from each treatment were harvested and placed in 1.5 mL Eppendorf tubes containing 300 µL of Laemmli buffer supplemented with protease inhibitors (10 µg/mL Aprotinin, 1 µg/mL E-64, 10 µg/mL Leupeptin, 1 µg/mL Pepstatin A, 10 µM PMSF). Plant material was ground using disposable grinders in the Eppendorf tube at room temperature until the mixture was homogeneous (usually less than 15 s). Once all the samples were prepared, tubes were placed in boiling water for 3 minutes. Tubes were centrifuged in a microfuge at maximum speed (13000 *g*, 10 min) immediately before loading. Fifteen µL of each extract, equivalent to about 1.5 (Figure S1) or 1 seedling (Figures 5, S4), were loaded per lane in an SDS - 8% PAGE. Immunoblot analyses of phyA and TUB were performed as indicated [43] with some minor changes. Mouse monoclonal antibody (mAb) 0.73D, that recognizes phyA from both monocots and dicots [43], were used at 1:5000 dilutions. Membranes were stripped and rehybridized with a commercial mouse mAb against α-tubulin (www.sigmaaldrich.com) at a 1:10000 dilution. Anti-mouse horseradish peroxidase-conjugated antibody (www.promega.com) was used as a secondary antibody. ECL or ECL-plus chemiluminescence kits (www3.gehealthcare.com) were used for detection.

Accession numbers

Sequence data from this paper can be found in the Arabidopsis Genome Initiative or GenBank/EMBL databases under the following accession numbers: *PHYA* (At1g09570) and *PHYB* (At2g18790).

Results

The intensity of continuous white light affects seedling morphology

For this study, we have systematically analyzed the hypocotyl response of wild-type (Col-0), *phyA* and *phyB* mutant seedlings to different light conditions. As a phyB-deficient line, we have employed *phyB-9*, a well characterized line [41]. As a phyA-deficient line we

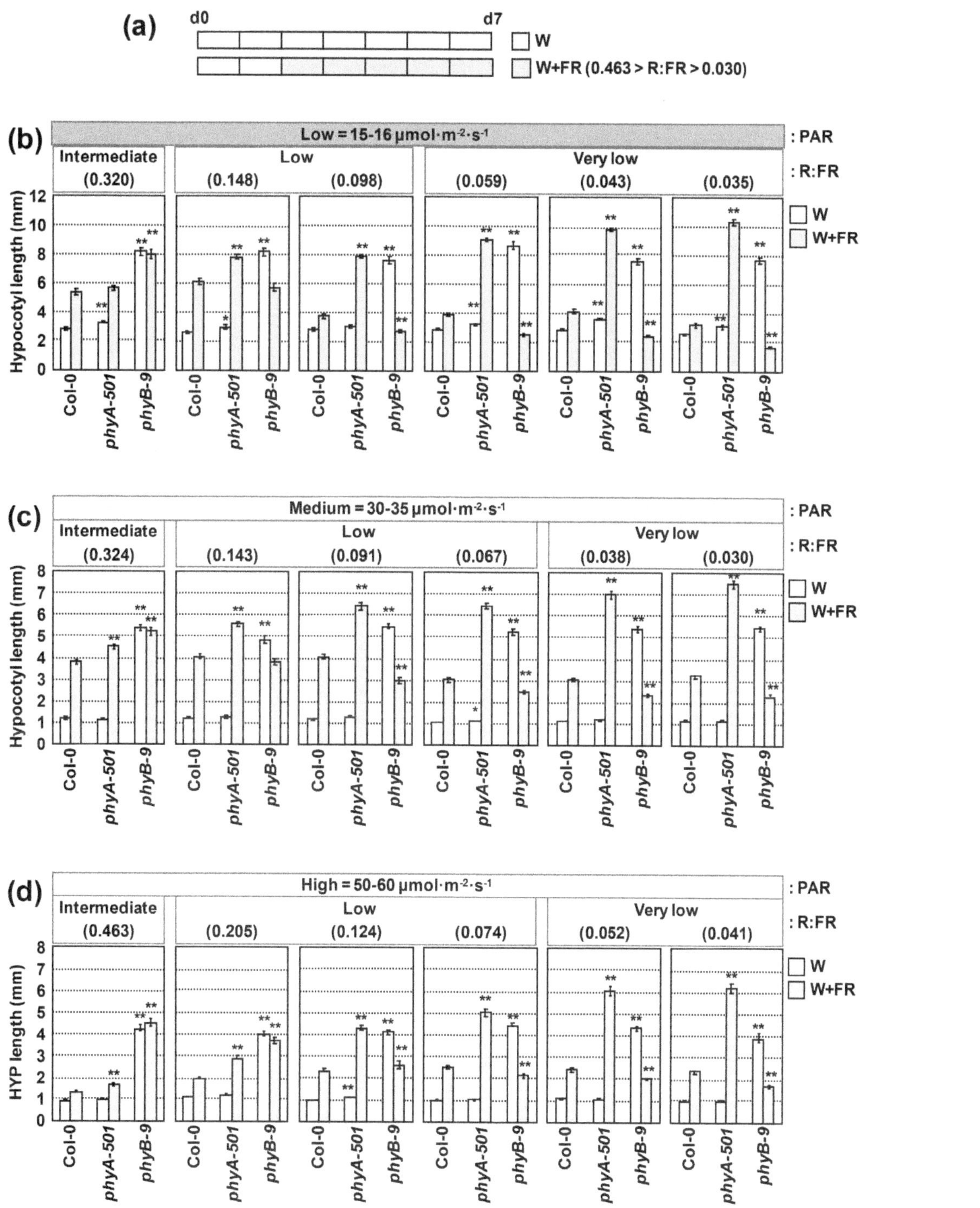

Figure 2. Effect of different R:FR on hypocotyl elongation of wild-type, *phyA-501* and *phyB-9* seedlings under low, medium or high PAR. (**a**) Seedlings were germinated and grown for 2 days under W light and then either kept in W or transferred to W supplemented with increasing amounts of FR for 5 more days. Hypocotyl length of seedlings grown as indicated in **a** under (**b**) low, (**c**) medium and high (**d**) PAR. The amount of PAR is given at the top of each section. The type of R:FR applied (nomenclature provided in Table S1) in the given W+FR treatments is indicated at the top, of the graphs; the R:FR value of each experiment is indicated at the top of each graph. In **b**, **c** and **d**, values are means ± SE of at least 25 hypocotyls for each light treatment. Asterisks indicate significant differences (*P<0.05, **P<0.01) relative to the control (Col-0) grown under the same light conditions.

have employed a SALK line (SALK_014575) that contains a T-DNA insertion in the middle of the first intron of the *PHYA* gene, about 1260 bp downstream of the ATG start codon (Figure S1a) [40]. In etiolated mutant seedlings, no levels of phyA were detected, whereas tubulin levels were similar to those of wild type (Figure S1b). This line was blind to continuous monochromatic FR, whereas it was as responsive to monochromatic R light as the wild type (Figure S1c).

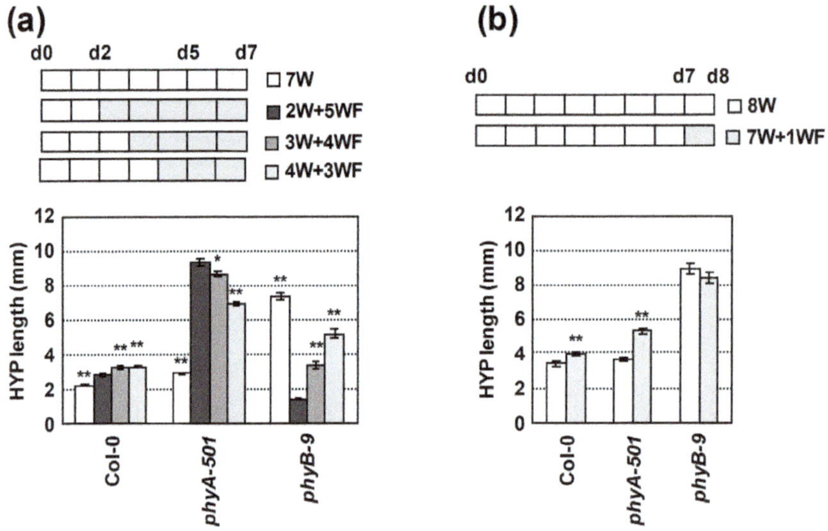

Figure 3. Effect of time of W+FR treatment on hypocotyl elongation of wild-type, *phyA-501* and *phyB-9* seedlings under low PAR. (**a**) Hypocotyl length of seedlings germinated and grown for 2, 3, 4 or 7 days under W light and then transferred to W+FR for 5, 4, 3 or 0 more days, respectively. (**b**) Hypocotyl length of seedlings germinated and grown for 7 days under W light and then either kept in W or transferred to W+FR for 1 more day. PAR was of 15–16 $\mu mol \cdot m^{-2} \cdot s^{-1}$ and R:FR of 0.059. Values are means ± SE of at least 25 hypocotyls for each light treatment. Asterisks indicate significant differences (*P<0.05, **P<0.01) relative to the same genotype (Col-0) grown for 2 days under W and 5 days under W+FR.

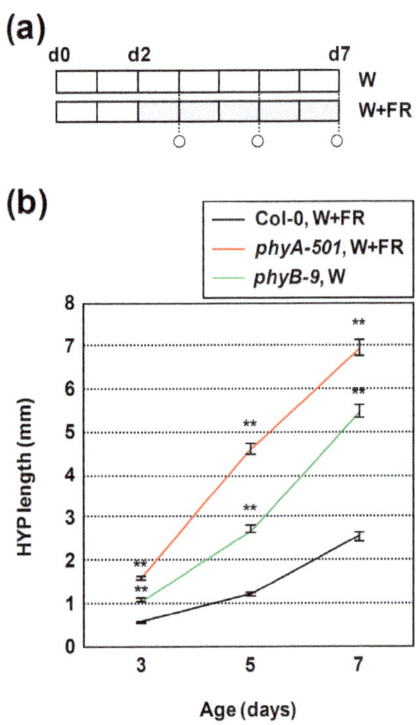

Figure 4. Effect of *phyA* and *phyB* mutations on the temporal evolution of the hypocotyl length. (**a**) Seeds were germinated and grown for 2 days under W (PAR was of 20–25 $\mu mol \cdot m^{-2} \cdot s^{-1}$) and then either kept under W (*phyB* seedlings) or transferred to W+FR (R:FR = 0.038) for 5 more days (Col-0 and *phyA* seedlings). Circles indicate the days on which hypocotyls were measured. (**b**) Hypocotyl length of seedlings grown as indicated in **a**. Values are means ± SE of at least 25 hypocotyls for each light treatment. Asterisks indicate significant differences (*P<0.05, **P<0.01) relative to the wild type seedlings grown under the corresponding light conditions.

Together, these results confirm that this T-DNA insertion line is a null mutant for *PHYA*. We named this new allele as *phyA-501*.

It is already known that light intensity affects seedling development, particularly hypocotyl length [20]. To get different light intensities, we employed neutral filters that reduced the intensity of white (W) light provided by 4 fluorescent tubes. As a result, PAR ranged from a minimum of 4.6 $\mu mol \cdot m^{-2} \cdot s^{-1}$ to a maximum of 72.9 $\mu mol \cdot m^{-2} \cdot s^{-1}$ (with no filters) (Figure 1a). As expected, hypocotyl elongation of wild-type, *phyA* and *phyB* seedlings was decreased when PAR amount was increased. Hypocotyls of *phyB* seedlings were always longer than those of wild-type and *phyA* seedlings under all different light intensities, as previously reported [41]. By contrast, *phyA* hypocotyls were generally longer than Col-0 at lower light intensities, whereas at high light intensities the differences in length were reduced or abolished (Figure 1b). Additional differences were evident in other morphological traits of the seedlings of the three genotypes analyzed: low light intensities reduced cotyledon expansion and delayed primary leaf development (Figure 1c). No higher PAR conditions were applied because under the highest light intensity employed here (i.e., about 73 $\mu mol \cdot m^{-2} \cdot s^{-1}$ from the beginning of germination) some seedlings showed signs of stress, such as a purple color and small size (data not shown).

Light of different R:FR differentially affects the hypocotyl elongation of wild-type, phyA and phyB seedlings

Next, we addressed the influence of light with varying reductions in R:FR on hypocotyl length. To manipulate the R:FR, W light of a fixed PAR was enriched with increasing fluence rates of FR light. We started our experiments with a relatively low level of W light, 15–16 $\mu mol \cdot m^{-2} \cdot s^{-1}$ in the PAR region. For simplicity, from now on we will refer to this intensity of W light as "low PAR". When supplementing with FR, the applied R:FR ranged from 0.320 to 0.035 depending on the amount of FR provided (Figure 2a). Whereas W light provided a high R:FR (> 1.5), W+FR provided light with moderate (intermediate R:FR of 0.5-0.3), substantial (low R:FR of 0.29-0.06) and very large (very

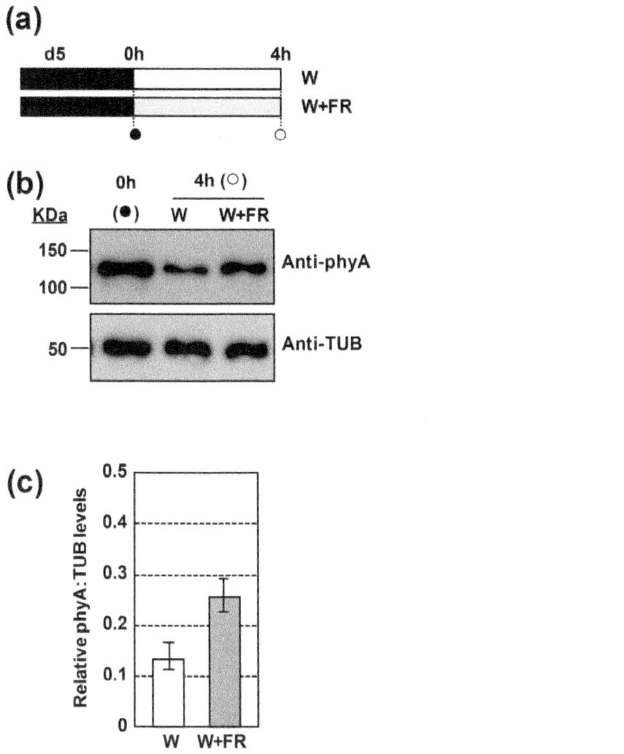

Figure 5. Phytochrome A is stabilized by white light of very low R:FR. (**a**) phyA levels were assayed in extracts from Col-0 seedlings grown in darkness for 5 days and then either transferred to W or W+FR for 4 hours. Circles indicate the harvest time of the plant material. (**b**) Representative steady-state levels of phyA (upper panel) and tubulin (TUB, lower panel) in extracts from seedlings grown as indicated in **a**,. Bands were detected by immunoblot using the phyA-specific mAb 073D or a TUB-specific mAb. TUB was used as a loading control. (**c**) Relative levels of phyA normalized to TUB in seedlings differentially grown for 4 h under W or W+FR, as indicated in **a**; n = 3 independent biological replicas; the P-value between the W and W+FR treated samples was 0.053.

low R:FR of 0.05-0.03) decrease in R:FR (see Table S1 for nomenclature used here). At the highest R:FR applied (0.320), W+FR light strongly induced hypocotyl elongation of Col-0 (about 5 mm) compared to the W-grown seedlings (about 2.5 mm). Further reductions in the R:FR were first slightly more effective (about 6 mm, R:FR of 0.148), and then rapidly less effective, in promoting hypocotyl elongation (about 4 mm for R:FR between 0.098-0.043), until shade-induced hypocotyl elongation was almost abolished (about 3.3 mm for the lowest R:FR tested of 0.035) (Figures 2b, S2a). These results indicate that quantitative variation in the R:FR provides a dual signal: when the R:FR is moderately or substantially lowered (R:FR of 0.320-0.148), it strongly induces hypocotyl elongation of Col-0, whereas when it is strongly reduced (R:FR<0.043; we define this range of R:FR as "very low R:FR") it is less effective in promoting the elongation of Col-0 hypocotyl elongation. At the highest R:FR tested (0.320), W+FR light induced the hypocotyl length of *phyA* to an extent similar to that observed for Col-0. But in striking contrast to the behavior of the Col-0 seedlings, the progressive reduction of the R:FR resulted in a gradual and strong promotion of hypocotyl length in the *phyA* seedlings (from about 7.8 mm at R:FR = 0.148 to more than 10 mm at the lowest R:FR tested of 0.035) (Figures 2b, S2a). These results agree with the reported negative role of phyA in shade-induced hypocotyl elongation [10], that was most apparent at very low R:FR. W+FR light had a contrasting effect on hypocotyl elongation of *phyB* seedlings: at the highest R:FR (0.320) it did not increase the already W-grown long hypocotyls (about 8 mm). At lower R:FR (0.148 and 0.035), W+FR inhibited (rather than promoted) hypocotyl elongation compared to W-grown seedlings. As a result, at very low R:FR (i.e., the lowest R:FR used of 0.035), W+FR-grown *phyB* hypocotyls were even shorter than those of Col-0 growing under W (Figures 2b, S2a). A mild inhibition of the *phyB* hypocotyl elongation by W+FR has also been observed previously, an effect attributed to the phyA-imposed inhibition of hypocotyl elongation in W+FR, quite apparent in the absence of phyB [17,44,45].

We next used an intensity of W of 30–35 $\mu mol \cdot m^{-2} \cdot s^{-1}$ in the PAR region (from now on, we will refer to this light as of "medium PAR"). The R:FR applied ranged from 0.324 (intermediate R:FR) to 0.030 (very low R:FR) (Figure 2c). W+FR light strongly induced hypocotyl elongation of Col-0 compared to the W-grown seedlings (about 4 mm) for R:FR between 0.324-0.091. Further reductions in the R:FR (i.e., low and very low R:FR) were less effective in promoting hypocotyl elongation (about 3 mm for R:FR between 0.067-0.030) (Figures 2c, S2b). W+FR light also strongly induced hypocotyl length of *phyA* seedlings; the progressive reduction of the R:FR resulted in a very strong promotion of hypocotyl length (from about 4.5 mm at R:FR = 0.324 to more than 7 mm at the lowest R:FR tested of 0.030). For *phyB* seedlings, W+FR light did not affect hypocotyl length at the highest R:FR used (0.324) compared to W-grown seedlings, but it was progressively more effective in inhibiting hypocotyl elongation at lower R:FR (from 0.143 to 0.038) (Figures 2c, S2b).

Finally, we used an intensity of W of 50–60 $\mu mol \cdot m^{-2} \cdot s^{-1}$ in the PAR region (we will refer to this light as of "high PAR"). The applied R:FR ranged from 0.463 to 0.041 (Figures 2d). As for the low and medium PAR experiments, W+FR light induced hypocotyl elongation of Col-0 compared to the W-grown seedlings depending on the R:FR applied: the progressive reduction of the R:FR resulted in longer hypocotyls (from about 1.5 mm at a R:FR = 0.463 to almost 2.5 mm at a R:FR of 0.074). Further reductions in the R:FR were equally effective in promoting hypocotyl elongation (for R:FR<0.074) (Figures 2d, S2c). W+FR light affected hypocotyl length of *phyA* and *phyB* seedlings essentially as described for low and medium PAR cases: the reduction of the R:FR resulted in a progressive promotion of *phyA* hypocotyl length (from almost 2 mm at R:FR = 0.463 to more than 6 mm at the lowest R:FR tested of 0.041) and a progressive inhibition of *phyB* hypocotyls (it did not affect the already long hypocotyls at R:FR of 0.463 and 0.205, but it inhibited hypocotyl elongation at low and very low R:FR of 0.124 to 0.041) (Figures 2d, S2c).

We next analyzed the hypocotyl response to very low R:FR light (very low R:FR of 0.043, low PAR) of *phyA* and *phyB* mutant seedlings in the *Ler* background. As shown in Figure S3, the hypocotyl elongation response of these *Ler* genotypes was similar to the one observed in Col-0, confirming that similar effects were observed in other genetic backgrounds. Together, these results led us to conclude that (1) the reported negative role of phyA in the shade-induced hypocotyl elongation [10,13] becomes more apparent at very low R:FR (0.05-0.03) and (2) it is qualitatively independent of the range of PAR intensity tested.

The timing of treatment with very low R:FR strongly affects the hypocotyl elongation response

We also addressed the influence of the timing and duration of the W+FR treatment on the hypocotyl elongation response of Col-0, *phyA* and *phyB* seedlings. Seedlings were grown at low PAR and submitted to a low R:FR of 0.059, conditions that allow distinction between the hypocotyl phenotypes of all three genotypes (Figure 2b). When W+FR was applied from day 2 after germination (2W+5WF), hypocotyl length of the analyzed genotypes was affected as observed before: Col-0 hypocotyls were poorly responsive, and *phyA* and *phyB* hypocotyl elongation was strongly promoted and inhibited respectively. When W+FR was applied from days 3 or 4 from germination (3W+4WF and 4W+3WF), Col-0 and *phyB* hypocotyls elongated significantly more than those grown under 2W+5WF, the promotion of elongation being more obvious for *phyB* seedlings. By contrast, *phyA* hypocotyls elongated significantly less than those grown under 2W+5WF (Figure 3a). When W+FR was applied from day 7 after germination (7W+1WF), hypocotyl length of Col-0 and *phyA* seedlings was modestly promoted although *phyA* hypocotyls still were more responsive than those of Col-0 seedlings; by contrast *phyB* seedlings were unaffected (Figure 3b). These results indicated that shade-induced phyA repression of hypocotyl elongation was operative during the entire time of exposure to W+FR (from days 2 to 8). However, based on the effect of the timing of the W+FR application on the *phyA* and *phyB* final hypocotyl elongation, it seems that the repression was stronger at the early stages of seedling development, i.e., from days 2 to 4.

We also investigated whether the elongation in the *phyA* and *phyB* mutants occurred at the same time along the course of seedling development. Wild-type and *phyA* seedlings were grown at a PAR of 20–25 $\mu mol \cdot m^{-2} \cdot s^{-1}$ (medium PAR) and exposed to W+FR of R:FR of 0.083 (low R:FR) from day 2 to day 7 after germination, and *phyB* seedlings were grown from germination under W of the same PAR. Under these conditions we expected a noticeable hypocotyl elongation of all three genotypes. As shown in Figure 4a, hypocotyl length was recorded in 3-, 5- and 7-day-old seedlings. *phyA* hypocotyls were already longer than those of *phyB* and Col-0 on day 3 (in our growth conditions, 2-day-old hypocotyls of all three genotypes are still emerging from the seeds and, therefore, their length is close to 0 mm). Although elongation in both *phyA* and *phyB* seedlings was sustained along the whole period, *phyA* hypocotyls elongated more from days 3 to 5 (d3–d5, 3.01 mm; d5–d7, 2.35 mm), whereas *phyB* hypocotyls elongated substantially more from days 5 to 7 (d3–d5, 1.65 mm; d5–d7, 2.75 mm). In addition, Col-0 hypocotyls elongated more from days 5 to 7 (d3–d5, 0.63 mm; d5–d7, 1.36 mm), suggesting that under the W+FR conditions applied its elongation was strongly inhibited from days 2 to 5, the same time window in which *phyA* hypocotyls elongated more. These results are consistent with our previous conclusion that phyA-mediated repression was stronger at the early stages of seedling development.

Treatment with very low R:FR stabilizes phyA levels

phyA is abundant in etiolated tissues but is light-labile and so is rapidly depleted in light-treated tissues. The long hypocotyl phenotype of *phyA* seedlings grown under W+FR of very low R:FR suggested that phyA may be more abundant under very low R:FR in Col-0 than under high R:FR light. To test this possibility, phyA levels were analyzed by western blot in 5-day-old etiolated Col-0 seedlings exposed to W (PAR of 20–25 $\mu mol \cdot m^{-2} \cdot s^{-1}$, R:FR = 2.5) and W+FR (same PAR, very low R:FR of 0.038) for 4 h (Figure 5a). As shown in Figure 5b, phyA levels declined after exposure to either W or W+FR light. However, in extracts from W+FR-exposed seedlings, phyA levels were higher (Figure 5c). Longer exposure times of W+FR (8 h) showed a similar tendency (Figure S4). Altogether, our data indicate that the balance between phyA synthesis and degradation in seedlings is affected by the R:FR of the incoming light, whereby light of very low R:FR has a milder destabilizing effect compared to W light of high R:FR.

Discussion

As mentioned above, nearby vegetation selectively reflects FR, thus lowering the R:FR, a signal that induces plant responses in anticipation of neighboring vegetation becoming a competitive threat [46]. If, despite these responses, neighboring vegetation directly shades the plant, light quantity becomes limiting, i.e., there is a reduction in the amount of radiation active in photosynthesis (i.e., PAR, between 400–700 nm), resulting in additional or more dramatic SAS responses [27]. Measurements by different authors of different natural light environments agree with this view [1–4,6,47,48]. In the laboratory, conditions that mimic plant proximity before actual canopy shading occurs (i.e., with lowered R:FR only, without changing PAR) have been termed *simulated shade*, and those that mimic natural situations when canopy closure occurs (which reduce both R:FR and PAR), have been termed *canopy shade* [25,28]. We have shown here that perception of these types of light-environment are genetically distinguishable by analyzing the hypocotyl elongation of *phyA* and *phyB* mutant seedlings in response to light of different R:FR, even without altering the PAR intensity: the inhibitory effect of phyA is readily observed under very low R:FR by (1) the conspicuous hypocotyl elongation of the *phyA* seedlings compared to the wild-type, and (2) the strong inhibition of the long-hypocotyl phenotype of the *phyB* seedlings. However, our data reveal a dichotomy in responsiveness across the range of R:FR tested. As mentioned, these phenotypes are observed under the lowest R:FR ratios (e.g. about 0.05-0.03) in our W+FR experiments here (Figure 2). However, these responses are essentially absent under our intermediate R:FR treatments (about 0.5-0.3), whereby *phyA* hypocotyls behave almost as wild-type ones and the long *phyB* hypocotyls are unaffected by the simulated shade signal (Figures 2, S2, 6a).

The antagonistic role of phyA in the SAS regulation has been noted previously when examining seedling deetiolation using natural shade conditions (provided by densely-grown plants of common wheat), and it was shown to be important for seedling establishment and survival under these specific conditions: *phyA* deficient mutants displayed extreme elongation growth, poor cotyledon development (phenotypes similar to those observed here under very low R:FR, i.e. R:FR = 0.05-0.03) (Figures 2, S3) and a lower survival rate than wild-type seedlings [11]. Our data show that the inhibitory role of phyA occurs independently of the range of PAR levels tested here (Figure 2), suggesting that the level of PAR in the incident light, although broadly recognized to modulate plant development (Figure 1), does not alone contain the essential differential information between these two extreme types of R:FR conditions (i.e. simulated and canopy shade). Hence, the R:FR signal alone seems sufficient to differentiate between plant proximity (mild reductions in the R:FR, without PAR decrease, defined here as "intermediate R:FR") and dense, direct canopy shade (strong reductions in the R:FR signal, defined here as "very low R:FR"). In this regard, *phyA* hypocotyls progressively elongated from intermediate to very low R:FR, and at the lowest levels tested they were much longer than those of *phyB* seedlings under W (high R:FR) (Figures 2, S2, S3), providing additional evidence for the effectiveness of the information contained in the R:FR signal. In addition, it suggests that the strong promotion of hypocotyl elongation observed in the *phyA* mutant background is due to the inactivation of the Pfr form of phyB and other photostable phys induced by the very low R:FR

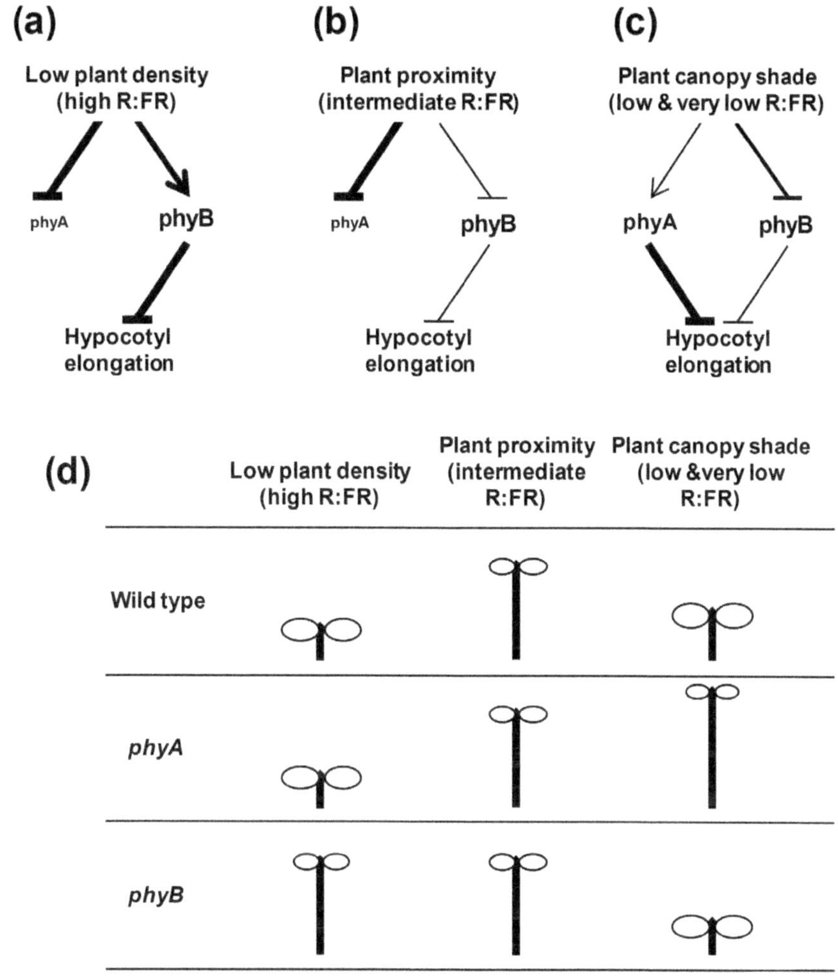

Figure 6. Model depicting the antagonistic effect of phyA and phyB on the shade-induced hypocotyl elongation. (**a**) Under low plant density, the high R:FR induces phyA degradation and stimulates the phyB active Pfr form, which strongly inhibits hypocotyl elongation. (**b**) In close proximity of vegetation, phyA degradation still occurs, but the low R:FR displaces the photoequilibrium of phyB towards the inactive Pr form, causing hypocotyls to elongate. (**c**) Under a plant canopy, the low or very low R:FR still displaces the photoequilibrium of phyB towards the inactive Pr form that stimulates hypocotyls to elongate. However, under these conditions phyA is stabilized, particularly at the beginning of the seedling emergence; as a consequence, phyA signaling is enhanced, thereby counteracting the inhibitory effect of the absence of active phyB, so that hypocotyls elongate only moderately. (**d**) Summary of phenotypes shown by the wild-type, *phyA* and *phyB* seedlings growing under the light conditions indicated in **a**, **b** and **c**.

light treatment. It might be argued that the PAR intensities employed in these experiments are low compared to what might be observed in nature (e.g., 750 $\mu mol \cdot m^{-2} \cdot s^{-1}$ in a clear midday in Buenos Aires, Argentina) [11]. Nonetheless, under natural dense canopies, PAR intensity might be reduced to less than 0.5% at the solar zenith, reaching even lower intensities than the ones employed in this work (1 $\mu mol \cdot m^{-2} \cdot s^{-1}$). Under these low PAR intensities, the R:FR results in values similar to those described here as low (0.10±0.02; Table S1) [11,49,50]. Therefore, lower R:FR values might be reached at other times of the day or latitudes. It is interesting to note that in chlorophyll rich organs, such as leaves, light absorption from the ultraviolet to the visible region by chlorophyll a leads to the emission (by fluorescence) of FR light [51]. Therefore, leaf chlorophylls can actively contribute to create low and very low R:FR signals in natural deep-shaded environments.

During the first week of seedling emergence and development, the inhibitory role of phyA is very apparent and easily observed at the beginning of this period (from days 2 to 5) when the potential to elongate is very high. Once this potential is diminished, the role of phyA also becomes less relevant (Figures 3, 4) [13]. Seedlings grown under photoperiodic conditions also respond to transient (2 h) low R:FR, which has revealed that there is also a circadian component in phyA action: whereas simulated shade given at subjective dusk increases hypocotyl length, when given at subjective dawn leads to a small inhibition of hypocotyl elongation (compared to untreated seedlings), an antagonistic effect of low R:FR shown to be dependent on phyA [18]. PhyA also (1) inhibits hypocotyl elongation under short-day conditions [10], under continuous R [52,53], or under continuous W of high R:FR but low PAR (Figure 1) and (2) promotes cotyledon expansion [52,54], likely also at the early stages of seedling development. However, the role of phyA in light-grown plants is not restricted to these early stages of the seedling development. In adult plants grown under short-day conditions, phyA has also a role in suppressing internode growth and leaf elongation [54].

The various roles of phyA in light-grown seedlings and plants are consistent with the evidence that, although phyA levels have been strongly decreased in these light-exposed tissues, they are not reduced to zero [3,55]. During seedling de-etiolation, phyA activation by FR results from the fact that continuous monochromatic FR light establishes and maintains a small fraction of the phyA population in the Pfr form, operating via the FR-HIR, over an extended period [3,23,56]. Indeed, although phyA is photolabile, FR-grown seedlings retain phyA at higher levels than R-grown seedlings as determined by western analyses blot analyses [55,57]. Also, under light/dark cycles, phyA accumulates during the night and is rapidly degraded during the day [55], which might explain the long hypocotyl phenotype of *phyA* seedlings grown under short-days conditions [10] and the absence of growth inhibition in *phyA* seedlings when a transient low R:FR treatment is given at subjective dawn [18]. We have observed that phyA is degraded under W in a R:FR-dependent manner: under very low R:FR conditions phyA is more stable than under high R:FR (Figures 5, S4). The rapid increase in the expression of *PHYA* in response to low R:FR light very likely contributes to the observed maintenance of high phyA protein levels [44]. Therefore, it seems likely that in fully de-etiolated seedlings the activation of phyA by W+FR of very low R:FR maintains a small fraction of phyA cycling in the active Pfr form that likely results in the observed suppression of the hypocotyl elongation [13,54,58].

It has been discussed in the literature that certain aspects of seedling SAS and de-etiolation affect the same traits but in opposite directions, such as accelerated hypocotyl cell elongation, retarded cotyledon expansion and reduced photosynthetic pigment accumulation. Indeed, gene-expression analyses provided initial molecular evidence for this view [17] and the continuous modulation by the phy-PIF signaling system has been implicated [21]. The analyses of seedling de-etiolation using Arabidopsis *phyA* and *phyB* mutants have established that both phytochromes have roles in seedling de-etiolation: phyA is the only photoreceptor responsible for the response of seedlings to continuous monochromatic FR, whereas phyB is mainly responsible for the responses of seedlings to continuous monochromatic R (Figure S5). The complementary actions of phyA and phyB in this process has been considered to provide optimum regulation of seedling growth after emergence from the soil [23]. We show here that during the SAS response of seedlings, phyB is deactivated by shade of intermediate, low and very low R:FR, whereas phyA is only strongly activated by shade of low, and very low R:FR, partly because of its higher levels. As a result, the phyA-Pfr produced and sustained in a cycling state strongly inhibits hypocotyl elongation via the FR-HIR activity of this phy [13,56] (Figure 6d). The differential effects of the *phyA* and *phyB* mutants on this process genetically defines the operation of two different pathways in SAS regulation [13], an additional similarity between the SAS and de-etiolation responses (Figures 6, S5). In the natural environment, continuous monitoring of the R:FR will determine the participation of phyA in the response to shade; when the R:FR is very low (such as in deep shade), phyA activation will prevent seedlings from exhibiting excessive elongation mediated principally by deactivation of phyB [58]. Whereas the overlapping actions of phyA and phyB will substantially promote de-etiolation in sparse vegetation [12], the antagonistic action of phyA and phyB will ensure the optimum elongation under deep shade, conditions in which R:FR can be strongly reduced partly due to the active emission of FR by the chlorophyll from the leaves [51].

Collectively, our data provide evidence that phyA functions in natural light environments to attenuate the SAS in response to direct canopy shading, but not to simple neighbor-proximity. This deduction refines the existing concept that phyA can "antagonize" the SAS via the FR-HIR [2,10,13], and supports the notion that plants have evolved a sophisticated intrafamily photosensory attenuation mechanism that can discriminate between the threat and imposition of competition for PAR by neighboring vegetation. This dual-track mechanism provides young seedlings with the capacity for both rapid elongation upon sensing of impending competition (intermediate R:FR) (the "neighbor-detection response mode"), or attenuation of potentially deleterious excessive elongation upon direct interception of canopy shade (low or very low R:FR) (the "direct-shade response mode").

Supporting Information

Figure S1 The *phyA-501* line is deficient in phyA. (**a**) Scheme of the *PHYA* genomic structure and the site of insertion of the T-DNA in the SALK_014575 line; arrows indicate the approximate location of primers used for genotyping (LBb1, MSO31 and MSO32). (**b**) Steady-state levels of phyA measured by protein blot. Immunoblot detection of phyA (upper panel) and tubulin (TUB, lower panel) levels in extracts from Col-0 and *phyA-501* seedlings grown in darkness for 5 days. Bands were detected as indicated in Figure 5b. (**c**) Hypocotyl length in 4-day-old Col-0 and *phyA-501* seedlings grown in darkness, continuous FR (3.7 $\mu mol \cdot m^{-2} \cdot s^{-1}$) and R (12.8 $\mu mol \cdot m^{-2} \cdot s^{-1}$), as shown in the upper part of the panel. Mean and SE values represent at least 25 seedlings from each treatment. Asterisks indicate significant differences (*$P<0.05$; **$P<0.01$) relative to control seedlings (Col-0) grown under the same conditions.

Figure S2 Effect of different R:FR on hypocotyl elongation of wild-type, *phyA-501* and *phyB-9* seedlings under low (a), medium (b) or high (c) PAR represented as the difference (left panels) or the ratio (right panels) between values under W+FR and W. Data were recalculated from the experiments generated for Figure 2. Values are the mean and SD from the 3 independent experiments.

Figure S3 Effect of very low R:FR ratios on hypocotyl elongation of wild-type (*Ler*), *phyA* and *phyB* seedlings under high light intensity. Seedlings were germinated and grown (PAR was of 15–16 $\mu mol \cdot m^{-2} \cdot s^{-1}$) as indicated in Figure 2a. Under W+FR, R:FR was 0.043. Mean and SE values represent at least 25 seedlings from each light treatment. Asterisks indicate significant differences (*$P<0.05$; **$P<0.01$) relative to control seedlings (*Ler*) grown under the same light conditions.

Figure S4 Levels of phyA are stabilized by very low R:FR treatments. (**a**) phyA levels were detected in extracts from Col-0 seedlings grown in darkness for 5 days and then either transferred to W or W+FR for 4 and 8 hours. Symbols indicate the harvest time of the plant material. (**b**) Immunoblot detection of steady-state levels of phyA (upper panel) and TUB (lower panel) in extracts from seedlings grown as indicated in (**a**). Bands were detected as indicated in Figure 5b.

Figure S5 Schematic summary of the phenotypes of wild-type, *phyA* and *phyB* seedlings grown in the dark or monochromatic R o FR light (indicated at the top).

Table S1 Terminology of the various R:FR regimes applied in this work and its proposed equivalence under natural conditions.

Acknowledgments

We are grateful to the greenhouse services of CRAG and PGEC for plant care; NASC, for providing seed stocks; and Pablo Leivar, Anne Pfeiffer and Manuel Rodríguez-Concepción for comments on the manuscript.

Author Contributions

Conceived and designed the experiments: JFMG PHQ MG. Performed the experiments: JFMG MG MJMC BL MB. Analyzed the data: JFMG MG MJMC BL MB PHQ. Contributed reagents/materials/analysis tools: JFMG PHQ. Wrote the paper: JFMG PHQ.

References

1. Casal (2012) Shade Avoidance. The Arabidopsis Book.
2. Franklin KA (2008) Shade avoidance. New Phytol 179: 930–944.
3. Smith H (1995) Physiological and ecological function within the phytochrome family. Annual Review of Plant Physiology and Plant Molecular Biology 46: 289–315.
4. Martinez-Garcia JF, Galstyan A, Salla-Martret M, Cifuentes-Esquivel N, Gallemí M, et al. (2010) Regulatory components of shade avoidance syndrome. Advances in Botanical Research 53: 65–116.
5. Ruberti I, Sessa G, Ciolfi A, Possenti M, Carabelli M, et al. (2012) Plant adaptation to dynamically changing environment: The shade avoidance response. Biotechnology Advances 30: 1047–1058.
6. Casal JJ (2013) Photoreceptor signaling networks in plant responses to shade. Annu Rev Plant Biol 64: 403–427.
7. Bae G, Choi G (2008) Decoding of light signals by plant phytochromes and their interacting proteins. Annual Review of Plant Biology 59: 281–311.
8. Franklin KA, Quail PH (2010) Phytochrome functions in Arabidopsis development. J Exp Bot 61: 11–24.
9. Hennig L, Poppe C, Sweere U, Martin A, Schafer E (2001) Negative interference of endogenous phytochrome B with phytochrome A function in Arabidopsis. Plant Physiol 125: 1036–1044.
10. Johnson E, Bradley M, Harberd NP, Whitelam GC (1994) Photoresponses of Light-Grown phyA Mutants of Arabidopsis (Phytochrome A Is Required for the Perception of Daylength Extensions). Plant Physiology 105: 141–149.
11. Yanovsky MJ, Casal JJ, Whitelam GC (1995) Phytochrome A, phytochrome B and HY4 are invovled in hypocotyl growth responses to natural radiation in Arabidopsis: Weak de-etiolation of the phyA mutant under dense canopies. Plant, Cell and Environment.
12. Smith H, Xu Y, Quail PH (1997) Antagonistic but complementary actions of phytochromes A and B allow seedling de-etiolation. Plant Physiol 114: 637–641.
13. McCormac AC, Whitelam GC, Boylan MT, Quail PH, Smith H (1992) Contrasting Responses of Etiolated and Light-Adapted Seedlings to Red: Far-Red Ratio: A Comparison of Wild Type, Mutant and Transgenic Plants has Revealed Differential Functions of Members of the Phytochrome family. Journal of Plant Physiology 140: 707–714.
14. Martinez-Garcia JF, Huq E, Quail PH (2000) Direct targeting of light signals to a promoter element-bound transcription factor. Science 288: 859–863.
15. Lorrain S, Allen T, Duek PD, Whitelam GC, Fankhauser C (2008) Phytochrome-mediated inhibition of shade avoidance involves degradation of growth-promoting bHLH transcription factors. The Plant Journal 53: 312–323.
16. Li L, Ljung K, Breton G, Schmitz RJ, Pruneda-Paz J, et al. (2012) Linking photoreceptor excitation to changes in plant architecture. Genes and Development 26: 785–790.
17. Roig-Villanova I, Bou J, Sorin C, Devlin PF, Martinez-Garcia JF (2006) Identification of primary target genes of phytochrome signaling. Early transcriptional control during shade avoidance responses in Arabidopsis. Plant Physiology 141: 85–96.
18. Salter MG, Franklin KA, Whitelam GC (2003) Gating of the rapid shade-avoidance response by the circadian clock in plants. Nature 426: 680–683.
19. Steindler C, Matteucci A, Sessa G, Weimar T, Ohgishi M, et al. (1999) Shade avoidance responses are mediated by the ATHB-2 HD-zip protein, a negative regulator of gene expression. Development 126: 4235–4245.
20. Hornitschek P, Kohnen MV, Lorrain S, Rougemont J, Ljung K, et al. (2012) Phytochrome interacting factors 4 and 5 control seedling growth in changing light conditions by directly controlling auxin signaling. The Plant Journal 71: 699–711.
21. Leivar P, Tepperman JM, Cohn MM, Monte E, Al-Sady B, et al. (2012) Dynamic Antagonism between Phytochromes and PIF Family Basic Helix-Loop-Helix Factors Induces Selective Reciprocal Responses to Light and Shade in a Rapidly Responsive Transcriptional Network in Arabidopsis. The Plant Cell 24: 1398–1419.
22. Jiao Y, Lau OS, Deng XW (2007) Light-regulated transcriptional networks in higher plants. Nat Rev Genet 8: 217–230.
23. Quail PH (2002) Phytochrome Photosensory Signalling Networks. Nature Reviews Molecular Cell Biology 3: 85–93.
24. Cifuentes-Esquivel N, Bou-Torrent J, Galstyan A, Gallemi M, Sessa G, et al. (2013) The bHLH proteins BEE and BIM positively modulate the shade avoidance syndrome in Arabidopsis seedlings. The Plant Journal 75: 989–1002.
25. Roig-Villanova I, Bou-Torrent J, Galstyan A, Carretero-Paulet L, Portoles S, et al. (2007) Interaction of shade avoidance and auxin responses: a role for two novel atypical bHLH proteins. The EMBO Journal 26: 4756–4767.
26. Sorin C, Salla-Martret M, Bou-Torrent J, Roig-Villanova I, Martinez-Garcia JF (2009) ATHB4, a regulator of shade avoidance, modulates hormone response in Arabidopsis seedlings. The Plant Journal 59: 266–277.
27. Sessa G, Carabelli M, Sassi M, Ciolfi A, Possenti M, et al. (2005) A dynamic balance between gene activation and repression regulates the shade avoidance response in Arabidopsis. Genes and Development 19: 2811–2815.
28. Crocco CD, Holm M, Yanovsky MJ, Botto JF (2010) AtBBX21 and COP1 genetically interact in the regulation of shade avoidance. The Plant Journal 64: 551–562.
29. Gangappa SN, Crocco CD, Johansson H, Datta S, Hettiarachchi C, et al. (2013) The Arabidopsis B-BOX Protein BBX25 Interacts with HY5, Negatively Regulating BBX22 Expression to Suppress Seedling Photomorphogenesis. The Plant Cell 25: 1243–1257.
30. Galstyan A, Cifuentes-Esquivel N, Bou-Torrent J, Martinez-Garcia JF (2011) The shade avoidance syndrome in Arabidopsis: a fundamental role for atypical basic helix-loop-helix proteins as transcriptional cofactors. The Plant Journal 66: 258–267.
31. Hersch M, Lorrain S, de Wit M, Trevisan M, Ljung K, et al. (2014) Light intensity modulates the regulatory network of the shade avoidance response in Arabidopsis. Proc Natl Acad Sci U S A 111: 6515–6520.
32. Bou-Torrent J, Galstyan A, Gallemi M, Cifuentes-Esquivel N, Molina-Contreras MJ, et al. (2014) Plant proximity perception dynamically modulates hormone levels and sensitivity in Arabidopsis. J Exp Bot.
33. Keuskamp DH, Pollmann S, Voesenek LA, Peeters AJ, Pierik R (2010) Auxin transport through PIN-FORMED 3 (PIN3) controls shade avoidance and fitness during competition. Proceedings of the National Academy of Sciences, USA 107: 22740–22744.
34. Tao Y, Ferrer JL, Ljung K, Pojer F, Hong F, et al. (2008) Rapid synthesis of auxin via a new tryptophan-dependent pathway is required for shade avoidance in plants. Cell 133: 164–176.
35. Galstyan A, Bou-Torrent J, Roig-Villanova I, Martinez-Garcia JF (2012) A dual mechanism controls nuclear localization in the atypical basic-helix-loop-helix protein PAR1 of Arabidopsis thaliana. Molecular Plant 5: 669–677.
36. Keuskamp DH, Keller MM, Ballare CL, Pierik R (2012) Blue light regulated shade avoidance. Plant Signal Behav 7: 514–517.
37. Keller MM, Jaillais Y, Pedmale UV, Moreno JE, Chory J, et al. (2011) Cryptochrome 1 and phytochrome B control shade-avoidance responses in Arabidopsis via partially independent hormonal cascades. The Plant Journal 67: 195–207.
38. Tsukaya H, Kozuka T, Kim GT (2002) Genetic control of petiole length in Arabidopsis thaliana. Plant Cell Physiol 43: 1221–1228.
39. Martinez-Garcia JF, Virgos-Soler A, Prat S (2002) Control of photoperiod-regulated tuberization in potato by the Arabidopsis flowering-time gene CONSTANS. Proceedings of the National Academy of Sciences, USA 99: 15211–15216.
40. Ruckle ME, DeMarco SM, Larkin RM (2007) Plastid signals remodel light signaling networks and are essential for efficient chloroplast biogenesis in Arabidopsis. The Plant Cell 19: 3944–3960.
41. Reed JW, Nagpal P, Poole DS, Furuya M, Chory J (1993) Mutations in the gene for the red/far-red light receptor phytochrome B alter cell elongation and physiological responses throughout Arabidopsis development. The Plant Cell 5: 147–157.
42. Al-Sady B, Ni W, Kircher S, Schäfer E, Quail PH (2006) Photoactivated Phytochrome Induces Rapid PIF3 Phosphorylation Prior to Proteasome-Mediated Degradation. Molecular Cell 23: 439–446.
43. Martinez-Garcia JF, Monte E, Quail PH (1999) A simple, rapid and quantitative method for preparing Arabidopsis protein extracts for immunoblot analysis. The Plant Journal 20: 251–257.
44. Devlin PF, Yanovsky MJ, Kay SA (2003) A genomic analysis of the shade avoidance response in Arabidopsis. Plant Physiology 133: 1617–1629.
45. Leivar P, Monte E, Cohn MM, Quail PH (2012) Phytochrome Signaling in Green Arabidopsis Seedlings: Impact Assessment of a Mutually Negative phyB-PIF Feedback Loop. Molecular Plant 5: 734–749.
46. Ballaré CL, Sánchez RA, Scopel AL, Casal JJ, Ghersa CM (1987) Early detection of neighbour plants by phytochrome perception of spectral changes in reflected sunlight. Plant, Cell and Environment 10: 551–557.
47. Vandenbussche F, Verbelen J-P, Van Der Straeten D (2005) Of light and length: Regulation of hypocotyl growth inArabidopsis. Bioessays 27: 275–284.
48. Smith H (1982) Light quality, photoperception, and plant strategy. Annual Review of Plant Physiology 33: 481–518.
49. Lei TT, Nilsen ET, Semones SW (2006) Light environment under Rhododendron maximum thickets and estimated carbon gain of regenerated forest tree seedlings. Plant Ecology 184: 143–156.
50. Lee DW (1987) The spectral Distribution of Radiation in Two Neotropical Rainforsts. Biotropica 19: 161–166.
51. Thornber JP (1975) Chlorophyll-proteins: light-harvesting and reaction center components of plants. Annual Review of Plant Physiology 26: 127–158.

52. Reed JW, Nagatani A, Elich TD, Fagan M, Chory J (1994) Phytochrome A and Phytochrome B Have Overlapping but Distinct Functions in Arabidopsis Development. Plant Physiol 104: 1139–1149.
53. Casal JJ, Mazzella MA (1998) Conditional synergism between cryptochrome 1 and phytochrome B is shown by the analysis of phyA, phyB, and hy4 simple, double, and triple mutants in Arabidopsis. Plant Physiol 118: 19–25.
54. Franklin KA, Praekelt U, Stoddart WM, Billingham OE, Halliday KJ, et al. (2003) Phytochromes B, D, and E Act Redundantly to Control Multiple Physiological Responses in Arabidopsis. Plant Physiology 131: 1340–1346.
55. Sharrock RA, Clack T (2002) Patterns of Expression and Normalized Levels of the Five Arabidopsis Phytochromes. Plant Physiology 130: 442–456.
56. Rausenberger J, Tscheuschler A, Nordmeier W, Wüst F, Timmer J, et al. (2011) Photoconversion and Nuclear Trafficking Cycles Determine Phytochrome A's Response Profile to Far-Red Light. Cell 146: 813–825.
57. Hudson M, Ringli C, Boylan MT, Quail PH (1999) The FAR1 locus encodes a novel nuclear protein specific to phytochrome A signaling. Genes Dev 13: 2017–2027.
58. Smith H, Whitelam G (1997) The shade avoidance syndrome: multiple responses mediated by multiple phytochromes. Plant, Cell and Environment 20: 840–844.

6

Whole Plastome Sequences from Five Ginger Species Facilitate Marker Development and Define Limits to Barcode Methodology

Justin N. Vaughn[1], Srinivasa R. Chaluvadi[1], Tushar[2], Latha Rangan[2], Jeffrey L. Bennetzen[1]*

1 Department of Genetics, University of Georgia, Athens, Georgia, United States of America, **2** Department of Biotechnology, Indian Institute of Technology Guwahati, Assam, India

Abstract

Plants from the Zingiberaceae family are a key source of spices and herbal medicines. Species identification within this group is critical in the search for known and possibly novel bioactive compounds. To facilitate precise characterization of this group, we have sequenced chloroplast genomes from species representing five major groups within Zingiberaceae. Generally, the structure of these genomes is similar to the basal angiosperm excepting an expansion of 3 kb associated with the inverted repeat A region. Portions of this expansion appear to be shared across the entire Zingiberales order, which includes gingers and bananas. We used whole plastome alignment information to develop DNA barcodes that would maximize the ability to differentiate species within the Zingiberaceae. Our computation pipeline identified regions of high variability that were flanked by highly conserved regions used for primer design. This approach yielded hitherto unexploited regions of variability. These theoretically optimal barcodes were tested on a range of species throughout the family and were found to amplify and differentiate genera and, in some cases, species. Still, though these barcodes were specifically optimized for the Zingiberaceae, our data support the emerging consensus that whole plastome sequences are needed for robust species identification and phylogenetics within this family.

Editor: Shilin Chen, Chinese Academy of Medical Sciences, Peking Union Medical College, China

Funding: This research was funded by endowment funds from the University of Georgia Giles Professorship, the Georgia Research Alliance and Department of Information Technology, and Ministry of Information Technology, Government of India (DIT Grant No. DIT No: 0526/T/IITG/014/0809/38). The funders had no role in study design, data collection and analysis, decision to publish, or preparation of the manuscript.

Competing Interests: The authors have declared that no competing interests exist.

* Email: maize@uga.edu

Introduction

The Zingiberaceae family includes ginger, turmeric, and cardamom spices, as well as many species with medicinal applications [1]. Although these plants have received little characterization at the genome level, there are many justifications for initiating such studies. Divergence in metabolite production in the Zingiberaceae can occur quite rapidly, exhibiting substantial variation even within the same species [2]. In contrast, morphological evolution is slower and often convergent [3]. Hence, DNA polymorphism analysis may be the most effective way to identify new germplasms that are sources of novel bioactive compounds. Previous work on the phylogenetic characterization of Zingiberaceae has focused on a small number of shared loci. While these loci can be used to differentiate genera, the approach fails to resolve genera into species [3]. Moreover, even for deeper phylogentic relationships, the Zingiberaceae contain many discrepancies between morphological and molecular classification of species; genera such as the *Alpinia* [4] and, to a lesser extent, *Curcuma* [3] are apparently paraphyletic.

Due to their high levels of conservation and maternal inheritance, chloroplasts can be used to develop population markers for phylogenetic and phylogeographic studies. The chloroplast DNA of most flowering plants are circular molecules that are highly conserved in terms of gene content, size and structural organization [5]. Because the size of the chloroplast genome is fairly constrained, highly conserved genes are found close to one another. These genes serve as optimal primer targets for amplifying short stretches of divergent intergenic DNA. In addition, chloroplasts outnumber nuclei in the plant cells. Thus, genes within the chloroplast are present in much greater concentrations than nuclear genes. While this may not be critical for analysis of fresh tissue, it can be a major consideration for dried specimens, gut samples, herbal stocks or other degraded DNA sources [6].

In addition to their pharmacological significance, gingers have become a model system for studying herbivore networks in tropical ecosystems through the use of DNA barcodes [7]. When the correct sequence region is chosen, DNA barcoding is an efficient way to assign an unknown specimen to a taxonomic group [6,8]. In effect, finding that a specimen matches a previously characterized barcode allows a researcher to assign that specimen to the species associated with that barcode. Difficulties arise in taxonomic groups without variation across a chosen barcode. For example,

Minami, *et al.* [9] used thirteen loci in an attempt to differentiate species within the paraphyletic *Curcuma* genus. Only one of the thirteen had haplotypes that were polymorphic across all accessions. Though these polymorphisms could be used to differentiate species, the intra-specific variation was similiar to the inter-specific variation, limiting the ability to assign unknown samples to higher taxonomic levels.

Both phylogenetic analysis and barcode classification are limited by the number of informative sites on which to base a conclusion. The entire chloroplast sequence dramatically increases the number of these sites, and, because of technological advances in sequencing, this increased information does not substantially increase total expenditures. Here, we present representative chloroplast genome sequences, or "plastomes", for *Zingiber officinale* (ginger), *Amomum cardamomum* (cardamom), *Alpinia zerumbet* (shell ginger), *Hedychium coronarium* (white ginger), and *Curcuma longa* (turmeric). These data should serve as a foundation for further characterization of this pharmacologically significant lineage. Indeed, using these sequences, we developed a set of highly variable markers. We demonstrated the utility and limits of a subset of these novel markers on a range of Zingiberaceae species.

Results and Discussion

Assembly and annotation

We performed *de novo* assembly on the *Z. officinale* chloroplast genome using DNA sequences from a shotgun library of total cell DNA. Assembly resulted in two major contigs of 87,626 and 45,356. These were aligned to the *Typha latifolia* (gi|289065068|ref|NC_013823.1) plastome and scaffolded with smaller contigs. Borders between the inverted repeat regions (IRa and IRb) and the long and short single copy regions (LSC and SSC, respectively) were manually assembled and curated by iterative mapping of reads. A final plastome with 162,598 bp was produced (Figure 1A). Excepting a gap at the IRa/SSC border, the *de novo* assembly shows perfect colinearity with the *Musa acuminata* (banana) plastome [10] (Figure 1B). Sequencing reads are fairly even in their coverage of the entire plastome (Figure 1C) - as the IR regions are effectively identical, reads were randomly placed in either region.

Annotation was performed using CpGAVAS [11]. In general, there are no major discrepancies in the gene content of the *Z. officinale* plastome compared to known plastomes, excepting an expansion of the IRa region as described below. The expansion creates a duplication of the first 3,912 bp of the *ycf1* gene. YCF1 and YCF2 are the longest proteins encoded by known plastomes and appear to be indispensable to plant survival [12]. Though the *Zingiber* plastome encodes a full-length YCF1, it is unknown if the large fragment associated with the IRa/SSC boundary is also expressed.

Repeat expansion and structural variation

The borders of inverted repeats and single copy regions are known to be hypermorphic for major structural variation. A major expansion of IRa has been noted in the *Musa* lineage [10]. As seen in Figure 1, the *Z. officinale* plastome does not completely share this expansion, although it does have a 3 kb expansion at this site relative to the otherwise highly colinear *Amborella trichopoda* (gi|34500893|ref|NC_005086.1) and *T. latifolia* plastomes (Figure S1 and S2). (Because of their deep ancestry, the colinear relationship between *Amborella* and *Typha*, among others, suggests that this is the ancestral plastome structure.) Thus, it appears that an initial 3 kb IRa expansion occurred in the *Musa/Zingiber* ancestor and was further expanded in the *Musa* lineage by another 4.7 kb. Alternatively, a 7.7 kb expansion could have occurred in the *Musa/Zingiber* ancestor followed by a deletion at the IRa/SSC boundary in the *Zingiber* lineage after divergence from the *Musa* lineage. The structure of the recently sequenced *Zingiber spectabile* plastome [13] suggests that these were separate events and that the short expansion has only occurred in *Z. officinale*. By mapping our sequencing reads from the *C. longa* plastome to the *Z. spectabile* scaffold, it is apparent that the scaffold has collapsed the 3 kb Zingiberale IRa expansion into the SSC. We see no such discrepancy in coverage when reads from *C. longa* or the more distantly related *A. cardamomum* are mapped to the *Z. officinale* plastome. Since this small expansion appears to have occurred at the root of the Zingiberaceae, the large expansion in *Musa* may have occurred in multiple steps. Whether this sequence is predisposed to expansion or such expansions are preferentially retained by natural selection remain open questions.

pan-Zingiberaceae variation

We then used the *Z. officinale* plastome as a reference on which to assemble data from four other libraries of lower coverage or shorter read length (Table 1). The final sequence reported for each plastome is the consensus sequence of these mapped reads (see Materials and Methods).

Generally, the IRs, which encode ribosomal RNAs, are highly conserved relative to both the LSC and SSC regions. Within the LSC and SSC, highly variable regions correlate with regions that either encode tRNAs or short genes and have a low gene density. Interestingly, regions of the IR with similarly low gene density exhibit much higher levels of conservation. This may be related to regulatory constraint outside of the protein coding regions or to a homology-dependent DNA repair mechanism specific to the IR region [14,15].

The tree generated from alignment of these plastomes (Figure 2) agrees with previous trees based on combined data from the nuclear ITS locus and the chloroplast *matK* gene [3]. Both trees indicate that *Hedychium* and *Zingiber* share a more recent common ancestor than *Zingiber* and *Curcuma*, although the divergence of all three lineages appears to have been nearly simultaneous on an evolutionary time scale.

Microsatellite identification and barcode development

Short sequence repeats (SSRs) or "microsatellites" are hypermorphic due to errors during DNA replication. Such sequences often serve as useful loci for differentiating closely related individuals. In contrast, DNA barcoding exploits point mutations and short indels that occur in rapidly evolving sequences flanked by more slowly evolving sequences. The lower rate of mutation in these barcodes relative to SSRs is somewhat compensated for by the presence of multiple informative sites in a single barcode sequence. Which of these strategies is most appropriate depends on the application, the taxonomic group under study, and the available instrumentation. Indeed, a very short SSR may serve as one among many informative sites within a single barcode. Using *M. accuminata* plastome sequence, researchers were able to identify 15 chloroplastic SSR markers that were polymorphic within the species [10]. By using both the *Z. officinale* sequence and the additional Zingiberaceae data produced in this study and elsewhere [13], we attempted to generate reliable chloroplastic SSR markers using a comparative approach. Briefly, alignment information from a whole plastome comparison between *Z. officinale* and *Z. spectabile* was used to find di- and tri-nucleotide repeats that were polymorphic within the *Zingiber* genus. We avoided homopolymer repeats because, generally, they are

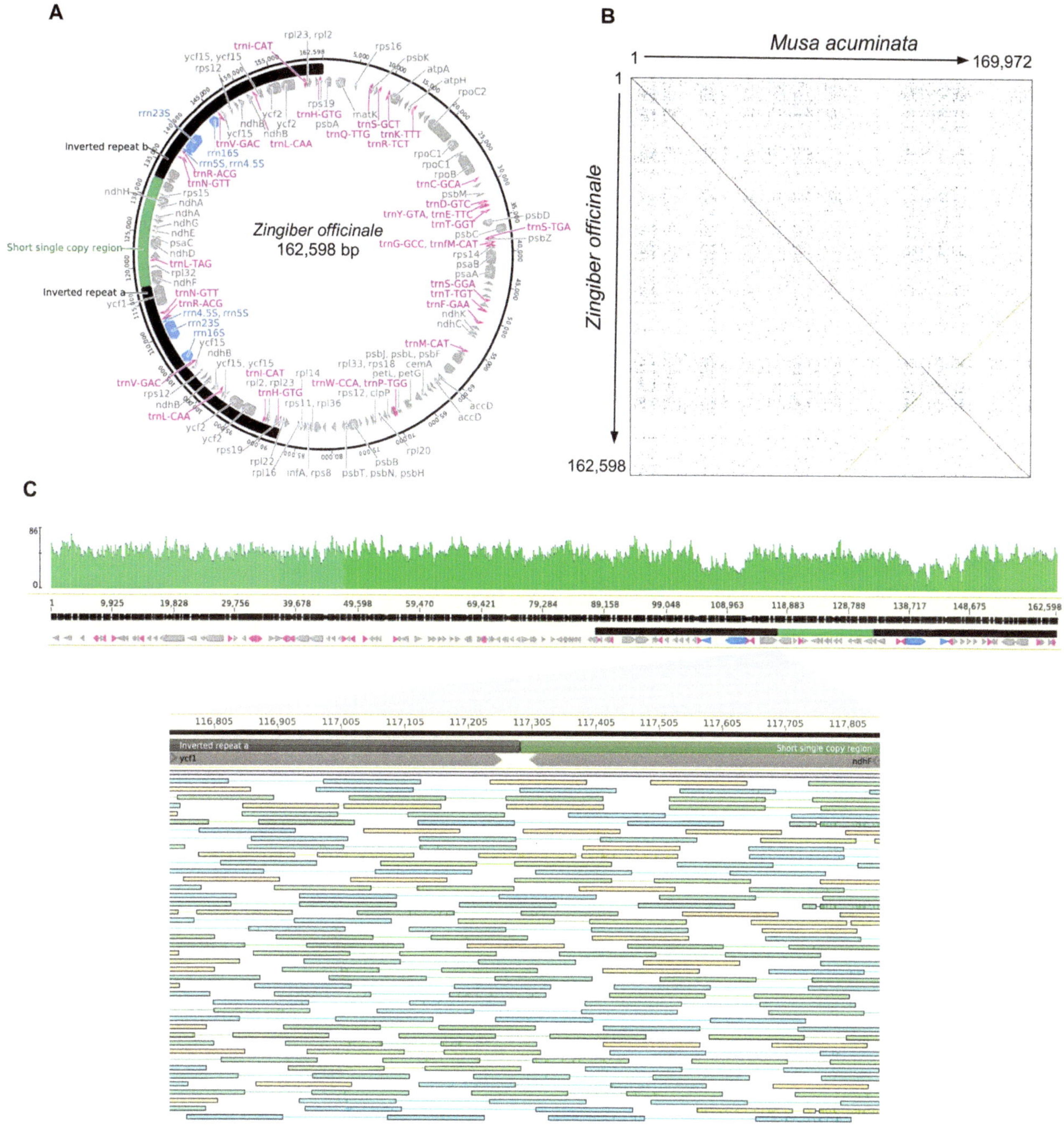

Figure 1. ***Zingiber officinale* assembly quality and annotation. A**, The *Z. officinale* plastome is shown in circular form with IRa and IRb indicated. The short single copy region between the two inverted repeats is shown in green. Protein coding genes are shown in grey, tRNA genes in pink, and rRNA genes in blue. **B**, Dotplot comparison of the *M. acuminata* and *Z. officinale* plastome sequences. The inverted repeat dotplot is transposed in gold. **C**, Total coverage across the *Z. officinale* plastome. The annotation color scheme is as depicted in A. The lower panel magnifies the border between the IRa and the SSC, where the large indel that differentiates *Musa* and *Zingiber* is seen to occur. Reads are shown individually; green is very near the expected insert size, while blue is slightly smaller and gold is slightly longer, respectively, than the expected size. The entire stack of reads is clipped due to space limitations.

technically difficult to use as markers. We also exploited alignment information in order to generate reliable primers for the amplification of these polymorphic regions. Using a similar pipeline for the alignment of all plastomes in this study (Figure 2A, excluding *Musa*), we also identified SSR markers that are likely to be most effective across the Zingiberaceae.

"Universal" plant barcodes, such as *matK* and *trnH-psdA*, have been tested across a range of Zingiberaceae. Shi *et al.* [16] found that the nuclear ITS2 region, because of its high interspecific divergence, was superior to universal choloroplastic barcodes. In contrast, Vinitha *et al.* [17] showed that the utility of both ITS and ITS2 regions is limited by low direct sequencing efficency caused

Table 1. General features of the material used for sequencing and the sequencing method.

Species	Site of Collection	Sequencing Method	Insert Size (bp)	Total Reads (10^6)	Chloroplast Reads (10^3)	Coverage
Zingiber officinale (TARS18100)	USDA Tropical Agriculture Research Station, Puerto Rico	MiSeq-PE150	400–600	3.0	52.9	50×
Curcuma longa (ZSD07)	Assam, India	MiSeq-PE150	400–600	1.3	58.1	54×
Alpinia zerumbet (TARS1718)	USDA Tropical Agriculture Research Station, Puerto Rico	HiSeq- PE100	400–600	3.0	35.4	22×
Amomum cardamomum (ZSB03)	Assam, India	HiSeq- PE100	400–600	10.8	112.4	70×
Hedychium coronarium (UGA-Hyd)	UGA Greenhouse, Athens, GA	HiSeq- PE100	400–600	20.2	337.1	211×

by intragenomic heterogeneity. These researchers determined that *matK* remained the superior barcode for gingers, but *matK* failed to differentiate species in 25% of cases. Similiar findings in other taxonomic groups [18,19] have led researchers to develop group-specific barcodes [20]. One aim of this study was to compare the discriminatory power of "universal" barcodes tested in the referenced works with barcodes that are designed specifically for gingers based on whole plastome sequences spanning the family.

To find suitable barcodes, we searched the alignments for regions that have at least 2% divergence. The regions were also filtered based on the number of single nucleotide polymorphisms (SNPs) that were bordered by 5 conserved bases on both sides of the SNP. Similar to SSR discovery, we identified markers using the intra-genus comparison for *Zingiber* as well as the comparison across all plastomes from this study. For the intra-genus and inter-genus comparison, we required at least 5 and 8 of such SNPs, respectively, across a 300 bp region. As with the SSR marker discovery, primers were only chosen from perfectly conserved sequence blocks.

The "Universal barcode" primers in Figure 3 are from previously proposed sets designed to work across all flowering plants [21]. Additionally, the "LR barcode" track plots another set of commonly used primer sets within the Zingiberaceae (Latha *et al.*, unpublished). Figure 3 also indicates the four sets of primer

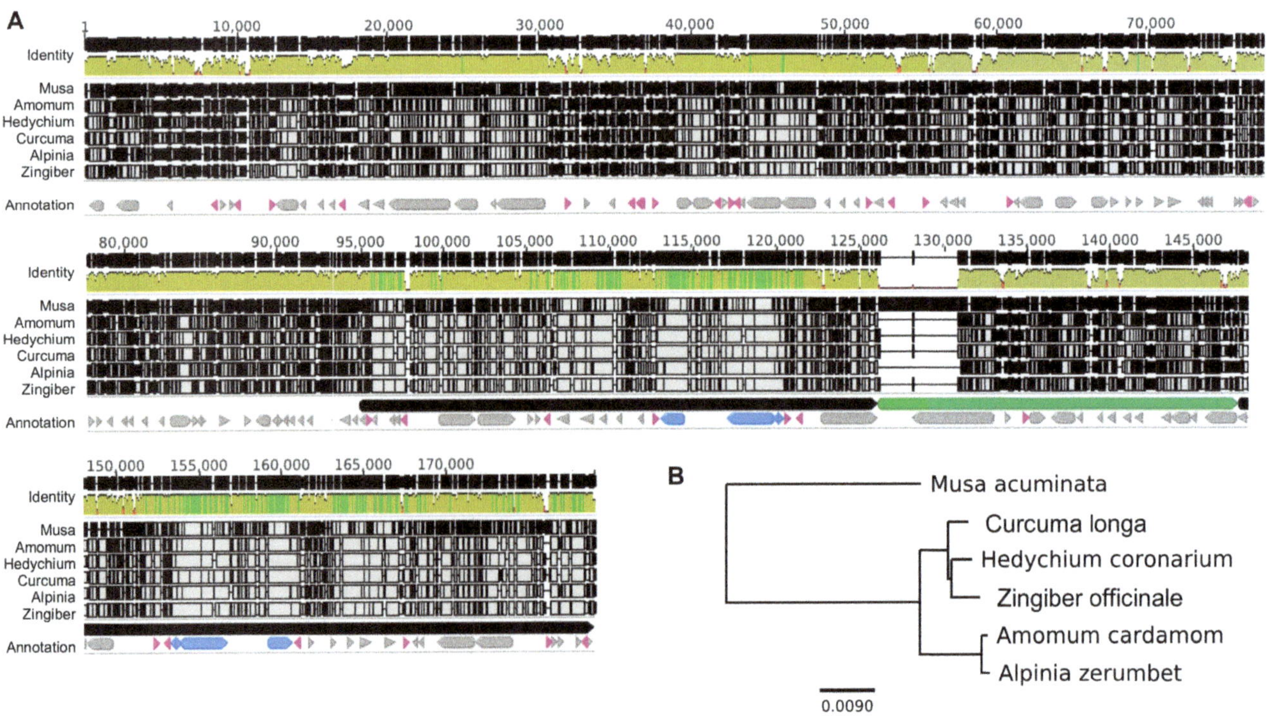

Figure 2. Plastome sequence variation across five Zingiberaceae. A, Alignment of whole plastome alignment between *M. acuminata* and Zingiberaceae species sequenced in this study. The annotation color scheme is as depicted in Figure 1A. In the coverage plot, color indicates percent identity: green, 100%; gold, >30%; red, <30%. Black bars within each species row indicate polymorphisms relative to the consensus. **B**, A phylogenetic tree based on the plastome alignments. *Musa acuminata* was used as the outgroup. Each branch point has 100% bootstrap support. Scale bar indicates the number of substitutions per site estimated to have occurred for such a length along the tree.

pairs resulting from our marker discovery pipeline. Looking across all four sets, we chose a subset of primers that exhibit a high number of polymorphisms at both the intra-genus and intra-family levels ("JNV barcode"). As shown, there is very little overlap between the "JNV barcode" markers and those used in previous studies. Though not shown in detail, only one of the "Universal barcode" primer sets perfectly matches the *Z. officinale* plastome, while the rest that are shown contain one to many mismatches. (Those with more than two mismatches are not shown.) While these mismatched primers may still produce a clear PCR product, they are not optimal with regard to heterogeneous starting material. Again, we aimed to make our primers as unbiased as possible by selecting from only highly conserved sites. Thus, in theory, "JNV barcodes" are optimal markers for the Zingiberaceae.

Marker testing on multiple Zingiberaceae lineages

To demonstrate the utility of JNV barcodes (Table S1), we amplified and Sanger sequenced these regions from a range of previously characterized and uncharacterized Zingiberaceae specimens (Table S2). Generally, all pairs amplified across the Zingiberaceae phylogeny, but JNV-A and JNV-C exhibited superior performance under the conditions tested (Table S2, Figures S4, S5). These pairs were followed in efficacy by JNV-E, JNV-D, and JNV-B. Pairs JNV-G, JNV-H, and JNV-I failed to amplify in any of the tested samples. Importantly, all pairs were tested in a single experiment as a proof-of-concept; we expect that PCR optimization on a pair-specific basis would greatly improve their amplification efficiency and phylogenetic distribution. Pairs JNV-A and JNV-C were able to efficiently amplify sequences for all genera in this study (Table S2, Figures S4, S5). They resolve a major branch point between *Alpinia/Amomum* samples and *Zingiber/Hedychium/Curcuma* samples (Figure 3B–C). The latter group could generally be differentiated by one or more polymorphic sites into three subgroups coinciding with the three different genera. As predicted by the whole plastome alignment, the branch between the *Zingiber/Hedychium* and *Curcuma* split was very short and thus difficult to resolve (Figure 2B). Still, there are clearly polymorphisms in both data sets that cluster *Zingiber* and *Hedychium* to the exclusion of *Curcuma* (Figure 3B–C). In addition, these barcodes clearly discriminate species mis-identified as belonging to the *Alpinia* genus (Figure 3B–C).

In terms of species resolution, there are cases where two species can be resolved, such as within the *Hedychium* and *Zingiber* groups, as well as within the *Alpinia* group (Figure 3B–C and File S1 and S2). *Curcuma* species are more difficult to resolve using these primers. Either this is a result of broad similarity across the whole plastomes of this group, or the markers we have chosen do not correspond to the most variable loci within this group. Additional plastome sequences within the *Curcuma* should resolve this question. As observed in other studies [3,4], the paraphyletic nature of genera such as *Alpinia* and *Curcuma* is also evident from our data, while genera such as *Zingiber*, *Kaempferia*, and *Hedychium* show monophyletic distributions (Figure 3B–C).

Conclusions

There is growing consensus that the entire chloroplast genome sequence is an optimal 'barcode' in plant identification and phlyogenetics [22–25]. Our study supports this consensus: even markers optimized to focus on the most variable (but still alignable) regions of these plastomes failed to produce enough informative sites for differentiating the most closely related species and for generating a tree with comparable bootstrap support of the same nodes shown in Figure 2B. Moreover, given the low cost of next-generation sequencing, the price of sequencing an entire plastome versus amplifying and Sanger sequencing a single barcode are approaching equivalence. This is particularly true when a reference plastome is available to facilitate assembly using lower read coverage.

Still, even a comparison of entire plastome sequences will reach resolution limits as samples are drawn from very closely related individuals [23]. In terms of the *Zingiber officinale/spectabile* comparison, plastomes exhibit hundreds of genuine polymorphisms, while universal barcodes *matK* and *rbcL* possess only 4 and 2, respectively (Figure S3). Though the ginger-specific barcodes reported in this study, JNV-C and JNV-A, have at least 5 polymorphisms between these two species, they too are ineffective with regard to *Curcuma* species (Figure 3B–C), suggesting far fewer polymorphisms exist across plastomes of that genus. As discussed above, whole-plastome sequences from the *Curcuma* genus will be an interesting test case for the differentiating power of organellar sequences.

Short-read sequencing and assembly strategies will also be ineffective when faced with a heterogeneous pool of samples from similar species, as might be associated with an ecological study or with herbal/spice contamination detection. Both the former [7] and latter applications are of particular interest within the Zingiberaceae. Therefore, exploiting the range of chloroplast sequences presented in this study, we have also developed a new set of barcode markers optimized for the Zingiberaceae. Two of those tested primer pairs, individually resolve the major and minor branches within the family. Though closely related species could be resolved in some cases, a combination of these markers is still likely to be required for confident identification of the constituent species in a mixed pool.

Methods

Sampling of Zingiberaceae accessions

Zingiberaceae accessions used in the current study were collected from US and India (Table 1). They are being maintained in greenhouses at the University of Georgia, Athens, GA and the Indian Institute of Technology in Guwahati, Assam, India.

DNA extraction and Illumina Sequencing

DNA extractions were carried out using PlantDNeasy mini kits (Qiagen, USA) following the manufacturer's instructions. The TruSeq sample preparation kit (Illumina, USA) was used to prepare DNA libraries with inserts from 400–600 bp for paired-end multiplexed sequencing. Library preparations and sequencing on an Illumina HiSeq2000 instrument were carried out at the Interdisciplinary Center for Biotechnology Research, University of Florida. Library preparations and sequencing on an Illumina MiSeq instrument were performed at the Georgia Genomics Facility, University of Georgia. Library-specific indexed adapter sequences were used to separate reads from individual samples from pooled libraries. All sequencing reads used in the assemblies have been made available in the Short Read Archive (http://www.ncbi.nlm.nih.gov/sra) under the BioProject accession PRJNA253694.

Assembly and annotation

For the *Z. officinale* plastome, shotgun reads were assembled as follows. Total reads were first assembled using *edena* with default parameters. Reads where then filtered to remove read pairs with polynucleotide runs of >30 bp and consecutive Ns of >75. Contigs resulting from *edena* assembly were used as long pseudo-

Figure 3. Assorted primer combinations from this and prior studies. A, "Universal barcode" primers are those described in [21]. "LR barcode" primers are from a recent unpublished study. "JNV barcode" primers were developed in this study and were generated based on the microsatellite and barcode primers identified by the computational pipeline described in the text. These are indicated in the figure as the last four tracts, where "Zingiber" indicates the intraspecific *Z. officinale* and *Z. spectabile* comparison and "pan-Zing" indicates a comparison across all sequences described in this study. Underlined markers indicate multiple overlapping markers at that site. The blue and orange boxes demarcate the markers used to generate sequences in Figure 3B and 3C, respectively. **B** and **C**, Sequences from a range of Zingiberaceae species were amplified using the primer pairs indicated in Figure 3A. **B** is associated with the blue box and shows products amplified by barcode set JNV-C (Table S1). **C** is associated with the orange box and shows products amplified by barcode set JNV-A (Table S1). Grey color indicates identity with the consensus of the alignment. Colored sites indicate a polymorphism, where different colors represent 4 different bases and light grey represents an ambiguous base. Appropriate regions from the fully sequenced plastomes were extracted and included in the alignment. These sequences are indicated by bold font. Names are not shown for samples that were not keyed to species-level.

reads ('-long' parameter) in a second round of assembly by *velvet* in paired-end mode with *k*-mer size set to 51 bp and the minimum k-mer coverage set to 2 ('-cov_cutoff 2'). Additionally, we required only 1 long read to scaffold two contigs ('-long_mult_cutoff 1'). Masked reads were then mapped back to the final assemblies using *bowtie-2* in local mode (–local), and scaffolded gaps were manually curated. We arrived at this hybrid strategy - a combination of overlap-consensus-layout and *k*-mer graph approaches – by observing that, while *velvet* was more robust to sequencing errors and could extend contigs further, *edena* was less prone to make errors when resolving the ends of duplicated sequences. The CpGAVAS web service [11], based on the MAKER pipline [26], was used to annotate the *Z. officinale* plastome. The sequence with annotation has been deposited in GenBank (http://www.ncbi.nlm.nih.gov/genbank/) under the accession KM213122.

For the assembly of other species in this study, total shotgun reads from a given individual were mapped to the *Z. officinale* plastome using Geneious (6.0.6) with 'Medium Sensitivity' and 5 iterations. Matches with equal scores were randomly assigned to an appropriate position. A consensus sequence was generated such that 50% of the bases across all mapped reads were required to be identical. An 'N' was called if the quality score was <20. Total shotgun reads were remapped to these consensus sequences and manually curated. All sequence assemblies are available as supporting information (File S3).

The MAUVE aligner ('mauveAligner' algorithm) [27] was used to create whole plastome alignments between the *Musa acuminata* sequence (gi|525312436|embl|HF677508.1) and all Zingiberaceae sequences generated in this study. A seed weight of 8 was used, and plastomes were assumed to be colinear. Resultant alignments and PhyML were used to estimate the topology, branch length, and substitution rate of a tree relating these species. A Kimura (K80) substitution model was used with 50 bootstrap trials. *M. acuminata* was chosen as the outgroup.

Identifying optimal markers

The method we employed to identify both microsatellite and barcode markers use an alignment to maximize both the variability of the marker and the consistency of the primer sites used to amplify the marker. To that end, two consensus sequences were generated from whole plastome alignments: the "best-base" consensus, for which the majority base in every column was used and gaps were ignored, and the "ambiguous" consensus, for which 100% of bases were required to match and gaps were considered. Thus, explicit bases (i.e., A, C, G, and T) in the ambiguous consensus are indicative of a perfectly conserved column.

For microsatellite marker development, we used the best-base consensus in conjunction with MISA software [28] to identify di- and tri-nucleotide repeats of 4 units or more. Using a custom Perl program, the resultant coordinates were then screened against the ambiguous consensus to determine if they were in fact polymorphic across the species analyzed, such that an ambiguous base was required to be present within the microsatellite or within 4 bp to the left or right of it. Primer3 [29] was then used to identify primer sites within the ambiguous consensus that amplify polymorphic positions, with the following parameters: PRIMER_PRODUCT_SIZE_RANGE = 100–280, PRIMER_MAX_END_STABILITY = 250, and PRIMER_LIBERAL_BASE = 1. Primers were not picked from ambiguous sequence, so, by using the ambiguous consensus, only highly conserved regions were considered.

A similar approach was used for barcode development, except that, instead of MISA software, a custom Perl program was written to find 300 bp blocks that 1) had >2% divergence, 2) did not contain more that 7 consecutive ambiguous bases, 3) did not have >10% or 15% divergence for genera or family comparisons, respectively, and 4) had >4 or >8 single-nucleotide polymorphisms with 5 conserved flanking bases for genera or family comparisons, respectively. Primer sequences were sought for within 300 bp upstream and 300 bp downstream of a 300 bp block passing the above criteria. The same parameters were used as for microsatellite primer identification (excepting the different size range). All programs not from external sources are available upon request.

PCR analysis with novel barcode primers designed in the current study

Genomic DNA was isolated from 26 Zingiberaceae accessions collected from Assam, India and the USA. These accessions included 20 identified species and 6 samples whose species had not been determined. Table S2 lists details of the various accessions used in this study. We tested 11 primer pairs designed in the hypervariable regions of the chloroplast genome. All the primers used are listed in Table S1. Six primer pairs showed consistent amplification across most of the species. PCR products amplified with two primer pairs, A and C (Table S1), were used for Sanger sequencing. PCR reactions were performed in 50-µL reactions: 1 × Phusion High Fidelity Buffer (Finzyme, Finland), (NEB, USA), 0.25 µM of each primer, 0.5 µM each dNTP, 1 U Phusion High-Fidelity DNA Polymerase (NEB, USA), and 100 ng DNA. PCR conditions were as follows: 98°C for 2 min; 26 cycles of 98°C for 10 s, 58°C for 20 s, 72°C for 30 s; with a final extension at 72°C for 10 min. Approximately 5 uL of the PCR reaction was tested on the gel to verify the presence of amplification product. PCR reactions were purified with the AGENCOURT AMpure kit (Beckman Coulter, USA) as per manufacturer's instructions. PCR product purification and direct sequencing were performed by standard procedures at the Georgia Genomics Facility (University of Georgia, Athens, GA) on an ABI 3730 instrument.

Supporting Information

Figure S1 Dot-plot alignment of *Zingiber officinale* and *Amborella trichopoda* plastomes.

Figure S2 Dot-plot alignment of *Zingiber officinale* and *Typha latifolia* plastomes.

Figure S3 Alignments for matK and rbcL barcode sequences from *Zingiber officinale*, *Zingiber spectabile*, and *Hedychium coronarium*. Colored letters indicate polymorphic sites.

Figure S4 Gel documenting amplification efficiency of JNV-C barcode sequence among Zingiberaceae accessions. Lanes are coded based on the matrix given below the gel. Plant accessions are as they appear in Table S2.

Figure S5 Gel documenting amplification efficiency of JNV-A barcode sequence among Zingiberaceae accessions. Lanes are coded based on the matrix given below the gel. Plant accessions are as they appear in Table S2.

Table S1 List of primers identified as optimal barcodes. Shown as 'JNV barcodes' in Figure 3.

Table S2 Species list and results for barcode A and C amplification.

File S1 Alignments of sequenced products from barcode A PCR in fasta format.

File S2 Alignments of sequenced products from barcode C PCR in fasta format.

File S3 Fasta file containing each assembled plastome sequence.

Acknowledgments

We offer our sincere thanks to the Botany Department, Gauhati University, for identification and maintenance of plant material. Tushar thanks the Department of Information Technology (DIT) and India-US Fulbright program for his student Fellowship. We thank Brian M. Irish, USDA-ARS Tropical Agriculture Research Station, Mayaguez, PR for providing Zingiberaceae accessions. Thanks to Dr. Jim Leebens-Mack at UGA for useful discussions about the phylogeny of the Zingiberales. We also thank Roger Nelson at the Georgia Genomics Facility and Savita Shankar at the University of Florida for their assistance with Illumina sequencing.

Author Contributions

Conceived and designed the experiments: JNV LR JLB. Performed the experiments: T SRC JNV. Analyzed the data: JNV SRC T. Contributed reagents/materials/analysis tools: LR T. Wrote the paper: JNV LR JLB.

References

1. Tushar, Basak S, Sarma GC, Rangan L (2010) Ethnomedical uses of Zingiberaceous plants of Northeast India. J Ethnopharmacol 132: 286–296.
2. Xiao W-L, Motley TJ, Unachukwu UJ, Lau CBS, Jiang B, et al. (2011) Chemical and genetic assessment of variability in commercial *Radix astragali* (*Astragalus* spp.) by ion trap LC-MS and nuclear ribosomal DNA barcoding sequence analyses. J Agric Food Chem 59: 1548–1556.
3. Kress WJ, Prince LM, Williams KJ (2002) The phylogeny and a new classification of the gingers (Zingiberaceae): evidence from molecular data. Am J Bot 89: 1682–1696.
4. Kress WJ, Liu A-Z, Newman M, Li Q-J (2005) The molecular phylogeny of Alpinia (Zingiberaceae): a complex and polyphyletic genus of gingers. Am J Bot 92: 167–178.
5. Wicke S, Schneeweiss GM, dePamphilis null CW, Müller KF, Quandt D (2011) The evolution of the plastid chromosome in land plants: gene content, gene order, gene function. Plant Mol Biol 76: 273–297.
6. Group CBOLPW (2009) A DNA barcode for land plants. Proc Natl Acad Sci U S A 106: 12794–12797.
7. García-Robledo C, Erickson DL, Staines CL, Erwin TL, Kress WJ (2013) Tropical plant-herbivore networks: reconstructing species interactions using DNA barcodes. PLoS One 8: e52967.
8. Fang Y, Wu H, Zhang T, Yang M, Yin Y, et al. (2012) A complete sequence and transcriptomic analyses of date palm (*Phoenix dactylifera* L.) mitochondrial genome. PLoS One 7: e37164.
9. Minami M, Nishio K, Ajioka Y, Kyushima H, Shigeki K, et al. (2009) Identification of Curcuma plants and curcumin content level by DNA polymorphisms in the trnS-trnfM intergenic spacer in chloroplast DNA. J Nat Med 63: 75–79.
10. Martin G, Baurens F-C, Cardi C, Aury J-M, D'Hont A (2013) The complete chloroplast genome of banana (*Musa acuminata*, Zingiberales): insight into plastid monocotyledon evolution. PLoS One 8: e67350.
11. Liu C, Shi L, Zhu Y, Chen H, Zhang J, et al. (2012) CpGAVAS, an integrated web server for the annotation, visualization, analysis, and GenBank submission of completely sequenced chloroplast genome sequences. BMC Genomics 13: 715.
12. Drescher A, Ruf S, Calsa T, Carrer H, Bock R (2000) The two largest chloroplast genome-encoded open reading frames of higher plants are essential genes. Plant J 22: 97–104.
13. Barrett CF, Davis JI, Leebens-Mack J, Conran JG, Stevenson DW (2013) Plastid genomes and deep relationships among the commelinid monocot angiosperms. Cladistics 29: 65–87.
14. Wolfe KH, Li WH, Sharp PM (1987) Rates of nucleotide substitution vary greatly among plant mitochondrial, chloroplast, and nuclear DNAs. Proc Natl Acad Sci U S A 84: 9054–9058.
15. Birky Jr C, Walsh JB (1992) Biased gene conversion, copy number, and apparent mutation rate differences within chloroplast and bacterial genomes. Genetics 130: 677–683.
16. SHI L-C, Zhang J, HAN J-P, SONG J-Y, Yao H, et al. (2011) Testing the potential of proposed DNA barcodes for species identification of Zingiberaceae Journal of Systematics and Evolution 49: 261–266.
17. Vinitha MR, Kumar US, Aishwarya K, Sabu M, Thomas G (2014) Prospects for discriminating Zingiberaceae species in India using DNA barcodes Journal of integrative plant biology.
18. Ran J-H, Wang P-P, Zhao H-J, Wang X-Q (2010) A test of seven candidate barcode regions from the plastome in Picea (Pinaceae) Journal of integrative plant biology 52: 1109–1126.
19. Ashfaq M, Asif M, Anjum ZI, Zafar Y (2013) Evaluating the capacity of plant DNA barcodes to discriminate species of cotton (*Gossypium*: Malvaceae) Molecular ecology resources 13: 573–582.
20. Hollingsworth ML, ANDRA CLARK A, Forrest LL, Richardson J, Pennington R, et al. (2009) Selecting barcoding loci for plants: evaluation of seven candidate loci with species-level sampling in three divergent groups of land plants. Molecular Ecology Resources 9: 439–457.
21. Kress WJ, Wurdack KJ, Zimmer EA, Weigt LA, Janzen DH (2005) Use of DNA barcodes to identify flowering plants. Proc Natl Acad Sci U S A 102: 8369–8374.
22. Parks M, Cronn R, Liston A (2009) Increasing phylogenetic resolution at low taxonomic levels using massively parallel sequencing of chloroplast genomes. BMC Biol 7: 84.
23. Nock CJ, Waters DLE, Edwards MA, Bowen SG, Rice N, et al. (2011) Chloroplast genome sequences from total DNA for plant identification. Plant Biotechnol J 9: 328–333.
24. Zhang Y-J, Ma P-F, Li D-Z (2011) High-throughput sequencing of six bamboo chloroplast genomes: phylogenetic implications for temperate woody bamboos (Poaceae: Bambusoideae). PLoS One 6: e20596.
25. Yang J-B, Tang M, Li H-T, Zhang Z-R, Li D-Z (2013) Complete chloroplast genome of the genus Cymbidium: lights into the species identification, phylogenetic implications and population genetic analyses. BMC Evol Biol 13: 84.
26. Cantarel BL, Korf I, Robb SMC, Parra G, Ross E, et al. (2008) MAKER: an easy-to-use annotation pipeline designed for emerging model organism genomes. Genome Res 18: 188–196.
27. Rissman AI, Mau B, Biehl BS, Darling AE, Glasner JD, et al. (2009) Reordering contigs of draft genomes using the Mauve aligner. Bioinformatics 25: 2071–2073.
28. Thiel T, Michalek W, Varshney RK, Graner A (2003) Exploiting EST databases for the development and characterization of gene-derived SSR-markers in barley (*Hordeum vulgare* L.). Theor Appl Genet 106: 411–422.
29. Untergasser A, Cutcutache I, Koressaar T, Ye J, Faircloth BC, et al. (2012) Primer3–new capabilities and interfaces. Nucleic Acids Res 40: e115.

7

A Model for Cell Wall Dissolution in Mating Yeast Cells: Polarized Secretion and Restricted Diffusion of Cell Wall Remodeling Enzymes Induces Local Dissolution

Lori B. Huberman[1,2], **Andrew W. Murray**[1,2]*

1 Molecular and Cellular Biology, Harvard University, Cambridge, MA, United States of America, **2** Faculty of Arts and Sciences Center for Systems Biology, Harvard University, Cambridge, MA, United States of America

Abstract

Mating of the budding yeast, *Saccharomyces cerevisiae*, occurs when two haploid cells of opposite mating types signal using reciprocal pheromones and receptors, grow towards each other, and fuse to form a single diploid cell. To fuse, both cells dissolve their cell walls at the point of contact. This event must be carefully controlled because the osmotic pressure differential between the cytoplasm and extracellular environment causes cells with unprotected plasma membranes to lyse. If the cell wall-degrading enzymes diffuse through the cell wall, their concentration would rise when two cells touched each other, such as when two pheromone-stimulated cells adhere to each other via mating agglutinins. At the surfaces that touch, the enzymes must diffuse laterally through the wall before they can escape into the medium, increasing the time the enzymes spend in the cell wall, and thus raising their concentration at the point of attachment and restricting cell wall dissolution to points where cells touch each other. We tested this hypothesis by studying pheromone treated cells confined between two solid, impermeable surfaces. This confinement increases the frequency of pheromone-induced cell death, and this effect is diminished by reducing the osmotic pressure difference across the cell wall or by deleting putative cell wall glucanases and other genes necessary for efficient cell wall fusion. Our results support the model that pheromone-induced cell death is the result of a contact-driven increase in the local concentration of cell wall remodeling enzymes and suggest that this process plays an important role in regulating cell wall dissolution and fusion in mating cells.

Editor: Yanchang Wang, Florida State University, United States of America

Funding: This work was supported by National Institute of General Medical Sciences (http://www.nigms.nih.gov/) Grant P50GM068763 of the National Centers for Systems Biology Program and a National Science Foundation (http://www.nsf.gov/) Graduate Research Fellowship and Ashford Fellowship (http://www.gsas.harvard.edu/fellowships.php) to L.B.H. The funders had no role in study design, data collection and analysis, decision to publish, or preparation of the manuscript.

Competing Interests: The authors have declared that no competing interests exist.

* Email: amurray@mcb.harvard.edu

Introduction

Cell fusion is essential for sexual reproduction and plays an important role in the development of many organisms [1]. In mammals, cell fusion is involved in the formation of myoblasts [2], osteoclasts [3], giant cells [4], and placental cells [5]. It is also important in the development of *Caenorhabditis elegans* [6] and *Drosophila melanogaster* [7]. Perhaps the simplest and most well studied form of cell fusion is the mating of the budding yeast, *Saccharomyces cerevisiae* [8].

Budding yeast can exist in both a diploid and haploid state. In either state, cells can replicate asexually by budding, producing daughters that are genetically identical to their mothers [9]. Haploid cells can be one of two mating types, **a** or α, which are defined by two alternative alleles of a single locus, *MAT***a** or *MAT*α. These mating types express reciprocal pheromones and pheromone receptors, which they use to signal to each other. Exposing a *MAT***a** cell to α-factor, the pheromone secreted by *MAT*α cells, (or vice versa) induces a pheromone response that includes transcription of pheromone response genes, cell cycle arrest in G1, and polarization in the direction of highest pheromone concentration to form a mating projection known as a shmoo [10].

After *MAT***a** and *MAT*α cells have successfully communicated and grown towards each other, they must fuse [9]. The two cells initially bind to each other at their shmoo tips using mating agglutinins [11–13], but their plasma membranes are still separated by two, approximately 100nm thick, cell walls [14]. Before the mating partners can fuse, the cell wall that lies between the two membranes must be dissolved and the boundaries of the remaining cell walls, which surround the site of cell fusion, must fuse to form a single, continuous structure that will enclose the newly formed zygote [8]. The osmotic pressure differential between the cytoplasm and the extracellular environment makes this spatially regulated cell wall dissolution and fusion a dangerous task [15,16]. If the cell wall is opened at the wrong time or place, exposing the plasma membrane directly to the environment, there will be no elastic force to resist the turgor pressure of the cell, water will rush into the cell from the extracellular environment, and the cell will lyse [15,16].

Various studies have been done on the molecular basis for cell wall dissolution. In 1996, Brizzio *et al.* showed that high pheromone concentrations are required for efficient fusion and hypothesized that vesicles found at the shmoo tip might contain cell wall remodeling enzymes [17]. Later, Cappellaro *et al.* found several proteins with homology to known cell wall glucanases, including *SCW4*, whose deletion makes mating less efficient [18]. The promoter of another putative glucanase gene identified by Cappellaro *et al.* [18], *SCW11*, has a binding site for Ste12 [19], the transcription factor that induces genes in response to pheromone stimulation [20].

Several proteins that are required for efficient cell fusion play a role in delivering secretory vesicles to the shmoo tip [21–24]. A complex containing Rvs161, an amphiphysin-like protein that binds curved membranes [25,26], and Fus2 [22,27,28], is hypothesized to direct vesicle transport to the cell fusion zone [23]. Once the vesicles reach the plasma membrane, they are anchored by Fus1 [23], a membrane spanning protein [29] that interacts with the polarisome [30], a protein complex associated with polarized actin polymerization [31], presumably ensuring tight clustering of the secretory vesicles. Although these proteins direct vesicle secretion towards the shmoo tip, their roles do not explain how cell wall dissolution is limited to the site of contact with a polarized partner.

The problem of remodeling the cell wall is not unique to mating. Even when cells are growing isotropically, there must be a balance between cell wall synthesis and destruction to allow the continual increase in cell diameter and volume, which is accomplished through spatially uniform secretion of synthesizing and remodeling enzymes (Figure 1A) [32,33]. Polarized growth, such as that associated with budding and shmooing, is achieved through polarized secretion of these enzymes [33] (Figure 1B). Most cell wall synthesizing enzymes are attached to the plasma membrane, whereas most wall-degrading enzymes are free to diffuse through the cell wall [34]. Synthesis and destruction must be carefully balanced: an excess of synthesis over degradation will lead to an increased cell wall thickness and eventually to slow growth, whereas an excess of degradation will weaken the cell wall until it is unable to resist the osmotic pressure inside the cell [35].

We propose a simple model to explain how cell walls are dissolved at the point where two polarized mating partners contact each other. When cells are not stuck to each other by mating agglutinins, the degradative enzymes diffuse through the cell wall and are then lost into the medium (Figure 1C). But when two mating partners stick to each other, using agglutinins, the enzymes must take a much longer path to escape, and because distance diffused only rises as the square root of time, their concentration at the site of fusion must rise, leading to an excess of destruction over synthesis and the eventual dissolution of the cell wall (Figure 1D).

If our model is correct, it should be possible to cause pheromone-induced cell death by tightly apposing pheromone-treated cells to impermeable surfaces, thus, mimicking the attachment of two cells to each other during mating (Figure 1E). Although previous studies [36,37] have reported that pheromone treatment can cause cell death, they neither hypothesized a mechanism through which this process is regulated, nor carefully examined the effect of holding cells against impermeable surfaces. We therefore set out to test the idea that slowing the escape of cell wall-degrading enzymes would lead to cell wall dissolution and death.

We observed that the frequency of cell death increases as the amount of cell contact with an impermeable surface increases and as the osmotic pressure differential between a cell and its environment rises, whereas decreasing the osmotic pressure differential reduces cell death. Deleting Fus1 and Fus2, proteins important for cell wall fusion [24], as well as the putative cell wall glucanases Scw4 and Scw11 [18], also decreases the frequency of cell death. Our evidence argues that the pheromone-induced cell death is due to a contact-dependent increase in the local concentration of cell wall remodeling enzymes, leading to the dissolution of the cell wall and eventual lysis of the cell. This mechanism may ensure safe and accurate cell wall fusion during mating.

Results

A model for pheromone-stimulated cell wall dissolution

We propose a simple model for cell wall dissolution: cell-cell contact increases the concentration of cell wall remodeling enzymes because they have to diffuse further within the cell wall from their site of secretion to reach the aqueous solution that surrounds the cells. We mathematically analyzed the distribution of cell wall remodeling enzymes in two situations: cells that are free in aqueous medium and those apposed to an impermeable surface. In both cases, we assume that the enzymes diffuse much more slowly through the cell wall than they do in the surrounding medium, and that this medium represents an infinite sink, allowing us to set the enzyme concentration outside the cell wall to zero. For unapposed cells, the enzymes need only diffuse through the thickness of the cell wall. At steady state, the flux through all points from the external surface of the plasma membrane to the external surface of the cell wall must be constant, implying a linear gradient in the enzymes' concentration. For apposed cells, we assume that they have a circular area of the cell wall pressed against an impermeable surface and that secreted enzymes must diffuse through the wall, parallel to the impermeable surface, before they can escape. At the center of the apposed region, the cell is secreting enzymes into the wall and the enzymes are diffusing away from the center of this region. In this region, the flux through the circumference of circles inscribed in the cell wall increases as the radius of the circles increases. Because the area of secretory activity increases with the square of the radius, whereas its circumference increases only linearly, the flux per unit length of the circumference increases, and thus the steepness of the gradient increases, moving outwards from the center of the secretory zone. Beyond this zone, no new enzyme secretion occurs, the flux through successively larger circles remains constant, and since their circumference increases, the radial concentration gradient becomes progressively shallower. If we assume that the cell wall is 0.1 μm thick, the radius of the secretory zone is 0.25 μm, and the radius of the apposed zone is 1 μm, the concentration of cell wall degrading enzymes at the center of the apposed zone is more than ten times the mean enzyme concentration in the wall of an unapposed cell. Details of this analysis are found below.

Mathematical analysis of pheromone-stimulated cell wall dissolution

First we consider an enzyme diffusing one dimensionally through a cell wall with diffusion coefficient D. If the radius of the secretion zone is substantially greater than the thickness of the cell wall, it is reasonable to treat the escape of enzymes secreted at the center of this zone as proceeding by one-dimensional diffusion through the thickness of the cell wall. As it diffuses through the wall, the enzyme's flux through a unit area of the cell wall, parallel to the surface of the cell, J, must be constant and is given by Fick's law

$$J = -D\frac{dC}{dx}$$

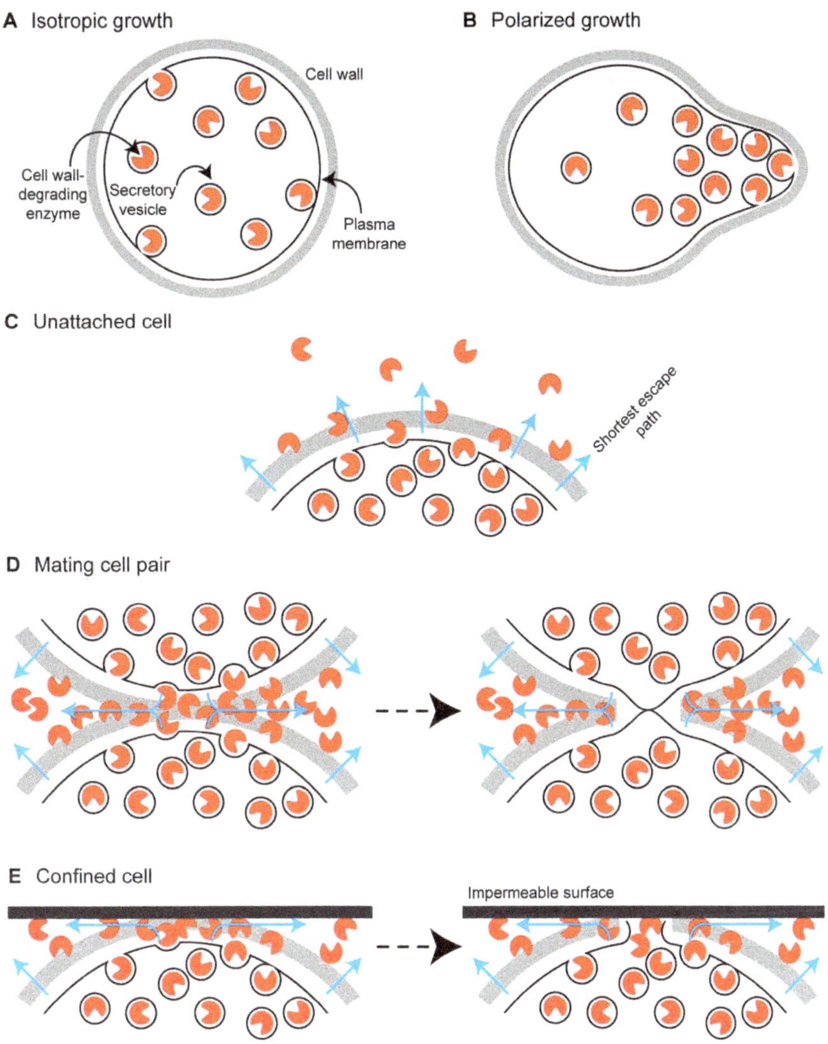

Figure 1. Model: Confining cell wall degrading enzymes in the fusion zone leads to cell wall destruction. A. Isotropically growing cells increase the size of their cell walls equally in all directions to grow larger while maintaining an ellipsoidal shape, so cell wall remodeling enzymes are secreted equally in all directions. **B.** Polarized cells grow anisotropically, so they polarize secretion of cell wall remodeling enzymes to expand their cell walls in the direction of polarization. **C.** When pheromone stimulated cells are unattached, the cell wall remodeling enzymes secreted from the shmoo tip exit the cell wall along the shortest path by traveling perpendicular to the plasma membrane. These enzymes break cell wall bonds as they diffuse through the wall to allow continual expansion of the shmoo up the pheromone gradient, but the wall is not breached. **D.** When two pheromone-stimulated cells are attached by mating agglutinins, the cell wall remodeling enzymes secreted into the future fusion zone must now travel further to exit the cell wall, traveling parallel to the plasma membrane until they reach the edge of the agglutinated zone, increasing the local concentration of cell wall remodeling enzymes in this zone. The cell wall remodeling enzymes dissolve the two cell walls at the point of contact while cell wall synthesizing enzymes simultaneously interlock them, allowing the plasma membranes of the two cells to contact one another and fuse without exposing the cell to osmotic lysis. **E.** We mimicked the attachment of two cells by tightly apposing single cells to impermeable surfaces, forcing cell wall remodeling enzymes to exit the wall by traveling parallel to the plasma membrane until they reached bulk solution and thus increasing the concentration of cell wall remodeling enzymes at the point of attachment to the impermeable surface. This causes a hole to form in the cell wall, exposing the plasma membrane to the extracellular environment and causing the cell to undergo osmotic lysis.

where x is the distance along the axis that runs perpendicular to the external surface of the plasma membrane to the outer surface of the cell wall. Since the flux is constant at all the points along this axis, the gradient must be the same at all points through the thickness of the wall, and thus the total concentration difference across the wall ΔC must increase linearly with the thickness of the wall, $\Delta C = x/D$. If we set D within the cell wall to be much lower than it is in solution, the concentration outside the wall will be close to zero, and C_0, the concentration at the site of secretion will be given by $C_0 = Jx/D$, and if we set the units of x to the thickness of the cell wall (roughly 100 nm), $C_0 = J/D$.

Now we consider an apposed cell, in which diffusion proceeds radially, in the plane of the cell wall, from the site of secretion to the edge of the apposed area. We consider two concentric regions within the opposed area, a central one where both secretion and diffusion occurs, and a peripheral one, where there is just diffusion. Remembering that we have set the unit of length equal to the thickness of the cell wall, within the region with secretion and diffusion, the flux that must leave an area of radius r is

$$\text{Flux} = J\pi r^2$$

where J is the rate of enzyme secretion per unit area. At any radius, r, within the secretory zone, the enzyme must pass through a ring whose area is $2\pi r$, specifying the radial concentration gradient, according to Fick's Law, and then, by integration, the concentration difference between the center (C_O) and any radial distance, r, within the secretion zone,

$$J\pi r^2 = -2\pi r D\frac{dC}{dr}$$

$$\frac{dC}{dr} = -\frac{J\pi r^2}{2\pi r D} = -\frac{Jr}{2D}$$

$$dC = -\frac{Jr}{2D}dr$$

$$C_r - C_0 = -\int_0^r \frac{Jr}{2D}dr = -\frac{Jr^2}{4D}$$

$$C_0 - C_r = \frac{Jr^2}{4D}$$

To get the concentration difference between the center and edge of this region, which is at r_s, the radius of the secreting region, we substitute r_s for r.

$$C_0 - C_{r_s} = \frac{Jr_s^2}{4D}$$

In the region where there is just diffusion, we can calculate the concentration difference from the inside edge to the outside edge of the region. Within the region, we must satisfy the condition that the total flux through each successive circumference is equal to the rate of enzyme production over the total producing region, which is $J\pi r_s^2$. Thus for all $r > r_s$,

$$J\pi r_s^2 = -2\pi r D\frac{dC}{dr}$$

We can rearrange and integrate to get the change in concentration between the concentration at r_s and r.

$$J\pi r_s^2 = -2\pi r D\frac{dC}{dr}$$

$$dC = -\frac{J\pi r_s^2}{2\pi r D}dr = -\frac{Jr_s^2}{2rD}dr$$

$$C_{r_s} - C_r = \int_{r_s}^{r} \frac{Jr_s^2}{2rD}dr$$

$$C_{r_s} - C_r = \frac{Jr_s^2}{2D}\log\left(\frac{r}{r_s}\right)$$

The overall concentration drop from the center of the secreting region to the edge of the apposed area is then given by adding the concentration drop from the center to the edge of the secreting region and the drop from the edge of the secreting region to the edge of the apposed region, situated at r_{max}.

$$C_0 - C_{r_{\max}} = (C_0 - C_{r_s}) + (C_{r_s} - C_{r_{\max}}) =$$

$$\frac{Jr_s^2}{4D} + \frac{Jr_s^2}{2D}\log\left(\frac{r_{\max}}{r_s}\right) = \frac{Jr_s^2}{4D}\left(1 + 2\log\left(\frac{r_{\max}}{r_s}\right)\right)$$

If we set $r_s = 2.5$, and $r_{max} = 10$, corresponding to radii of secretion and apposition of 0.25 and 1 μm, and consider the concentration drop from the center of the secreted region to its edge, C_s, we get

$$C_s = 1.56J/D$$

and for the drop from the outer edge of the secreted region to the edge of the apposed region, C_d, we get

$$C_d = 4.33J/D$$

giving a total concentration drop, $C_{tot} = 5.89J/D$, which implies a maximum hydrolase concentration that is nearly six times that attained when a cell is not apposed to another cell.

The difference between the mean concentration at the center of the apposed region and the mean concentration in the wall of an unapposed cell is even higher. In unapposed cells, the mean concentration, felt half way through the cell wall, is $J/2D$, which is the average of a concentration J/D at the cell surface and 0 at the interface between the wall and solution. In the apposed cell however, the surface that the cell is exposed to acts as a reflecting barrier so that the concentration is constant across the thickness of the wall, and thus the mean concentration, in the scenario we have described is more than ten times higher for the apposed than for the unapposed cells (Figure 2).

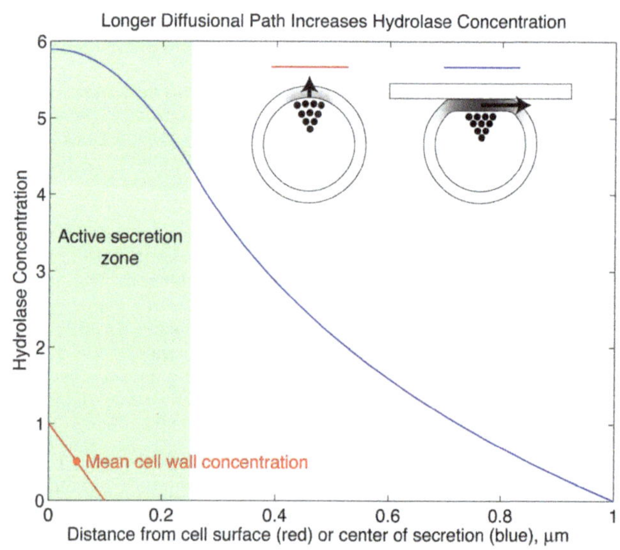

Figure 2. The role of radial diffusion through the cell wall of apposed cells in increasing the concentration of cell wall degrading enzymes. The graph shows analytical results for the relative enzyme concentration in two scenarios: *red*, diffusion through the cell wall, perpendicular to the cell surface, of a cell free in solution and not in contact with other cells or solid surfaces, and *blue*, diffusion through the cell wall, parallel to the cell surface, of a cell that is apposed to a solid surface, with a circular contact area whose radius is 1 μm.

Pheromone-stimulated cells die when attached to an impermeable surface

If contact with another cell leads to cell wall dissolution by increasing the local concentration of wall-degrading enzymes, we should be able to mimic the phenomenon by confining cells against an impermeable surface. We compared the response of pheromone-stimulated cells in environments where the cells were either free-floating, simulating cells in a mating mixture that are not attached to a fusion partner, or attached to an impermeable surface, simulating cells attached to a fusion partner via mating agglutinins. We observed cells in three different environments: bulk culture, attachment to a single, flat, impermeable surface, and confinement between two impermeable surfaces (Figure 3A). Cells expressing the α-factor protease, *BAR1* [38], are capable of decreasing the pheromone concentration at their surface [38,39], so we used *MAT***a** *bar1Δ* cells for our investigations. We incubated *bar1Δ* cells in 50 nM α-factor in bulk culture for five hours and found that roughly 10% of the cells die (Figure 3B). Although cells grown in bulk culture have no enforced contacts with the other cells or the impermeable surface of the culture tube, it is difficult to control the physical interactions of cells when they are free-floating in liquid culture and possible that cells could stick either to each other, perhaps due to incomplete separation after budding, or to the surface of the culture tube.

To mimic the attachment of two cell walls via mating agglutinins, we attached cells to the impermeable surface of a glass coverslip using the lectin, concanavalin A (ConA), which binds to carbohydrates in the cell wall [40] (Figure 3A). In order to image the yeast cells for an extended period of time, we created a chamber several hundred times the diameter of a yeast cell. Cells were adhered to the ConA-coated coverslip, and the chamber was filled with medium containing 50 nM α-factor using capillary action and then sealed and observed over a period of five hours. We found that *MAT***a** *bar1Δ* cells attached covalently to an impermeable surface were 1.6 times more likely to die than those in bulk culture, indicating that forced attachment to an impermeable surface increases the rate of cell death (Student's *t*-test, $p = 0.01$) (Figure 3B).

As cells attached to a single impermeable surface grow, they are free to expand away from the glass coverslip, resulting in a low proportion of the cell wall attached to an impermeable surface and making it likely that cells will polarize away from the impermeable surface. To address this problem we used a second technique to mimic the attachment of two cell walls via mating agglutinins. We trapped cells in a microfluidic chamber whose floor and ceiling are separated by the height of a single yeast cell and through which new medium is constantly perfused (Figure 3A). Cells are loaded into this device and then trapped between the two impermeable surfaces of a silicone ceiling and a glass floor. In addition, as the cells grow, the fraction of their surface that is pressed against the floor and ceiling rises, making them more likely to polarize towards an impermeable surface. Using an inverted microscope, it is possible to image cells over time through the glass floor as medium perfuses through the chamber. Once again we imaged *MAT***a** *bar1Δ* cells in 50 nM α-factor for five hours (Figure 3C and Movies S1–S3). In the flow chamber, the rate of death of the *MAT***a** *bar1Δ* cells was more than twice as high as in bulk culture and 1.5 times the rate of death when attached to ConA-coated coverslips, suggesting that a larger area of attachment to an impermeable surface causes increased cell death (Student's *t*-test, $p<0.003$) (Figure 3B).

Pheromone-induced cell death increases with increased cell polarization

Decreased pheromone production has been reported to cause decreased cell fusion [17]. Thus, if the cell death seen here is due to the same activities that normally promote cell fusion, we would expect to see a decrease in cell death with decreased pheromone concentration. We chose to assay the effect of decreased pheromone concentration in the flow chamber, where the highest percentage of cells died when exposed to 50 nM α-factor. As previously reported for cells in bulk culture [37], decreasing the α-factor concentration decreased the percentage of cells that died in the flow chamber. In 5 nM α-factor, 50-fold fewer cells died than when cells were exposed to 50 nM α-factor in the flow chamber (Student's *t*-test, $p = 10^{-6}$), and the fraction of dead cells increased as the concentration of α-factor was increased (Student's *t*-test, $p< 0.02$) (Figure 4A). Although shmoo formation occurs at 5 nM α-factor, as the pheromone concentration was increased, the cells became more tightly polarized, forming pointier shmoos (Figure 4B).

The flow chamber traps cells by wedging them into a space minutely smaller than a single cell in height. When cells are arrested, such as by pheromone stimulation, the cells increase in size as they continue to grow without dividing [41]. Because of this, it is possible that the increased frequency of cell death in the flow chamber, as compared to bulk culture and when cells are attached to ConA-coated coverslips, is not due to the accumulation of enzymes that would normally degrade the cell wall during cell fusion but rather because the physical strain put on the cell wall is too high, which could be increased by the modest pressure (14 kPa = 2 psi) applied to drive the flow of the perfused medium. We therefore used a different method that would arrest the cell cycle without interfering with cell growth. Like pheromone treatment, treating cells with benomyl, a drug that leads to microtubule depolymerization, causes cells to become larger without dividing, but unlike pheromone-arrest, benomyl-arrested cells are unpolarized and arrest in mitosis instead of G1 [42] (Figure 4C). If cells in the flow chamber die because they were squashed, a substantial percentage of benomyl-arrested cells should die in the flow chamber. Although it is possible to find the occasional, dead, benomyl-arrested cell, 60-fold fewer cells die during five hours of benomyl-arrest than during exposure to 50 nM α-factor, indicating that death in the flow chamber is specific to pheromone-arrest, where cells are polarized, and is not due to growth under physical confinement (Figure 4D).

Pheromone-induced cell death is due to osmotic lysis

Yeast cells require cell walls at least in part due to osmotic pressure. Since the osmolarity of the cytoplasm is higher than the typical extracellular environment, without the rigidity of a cell wall, water would rush into the cell and cause it to lyse [33], and previous studies have shown that cells that are unable to regulate the osmotic balance between the cytoplasm and the extracellular environment have a cell fusion defect [15]. One interpretation of the death of pheromone-treated cells pressed against an impermeable surface is that accumulation of cell wall-degrading enzymes causes the cells to digest part of their cell walls leading to membrane expansion through a hole in the cell wall and eventual lysis. If this interpretation is correct, it should be possible to affect the rate of death by manipulating the osmotic pressure differential between the cell and the medium [43].

We did two experiments to determine whether the pheromone-induced deaths are due to osmotic lysis: either increasing or decreasing the osmotic pressure differential between the cytoplasm and the extracellular environment. We first tested the effect of

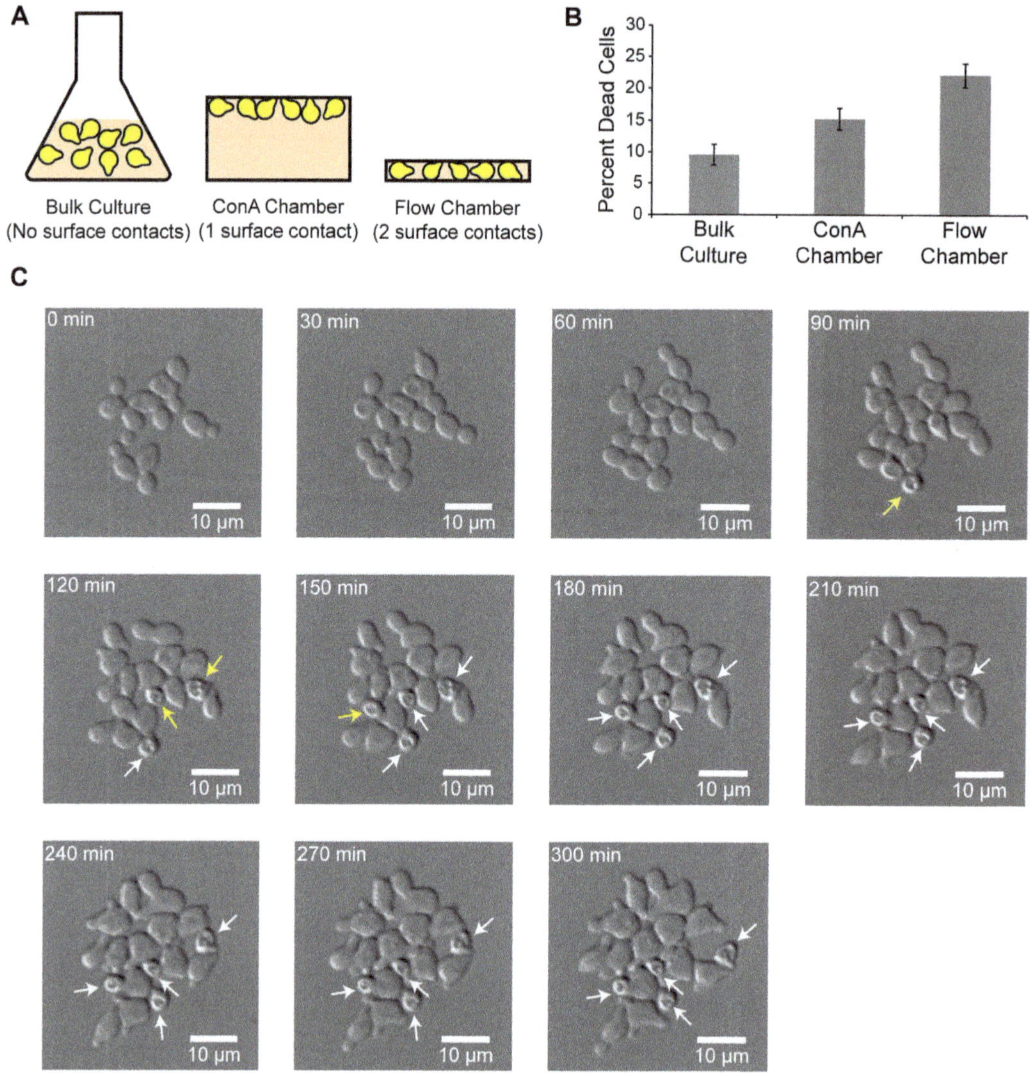

Figure 3. Pheromone-induced cell death increases with increasing attachments to an impermeable surface. A. Cells grown in bulk culture were incubated in test tubes on roller drums in liquid media without any enforced contact with impermeable surfaces. Cells grown in a concanavalin A (ConA) chamber were grown in a chamber whose depth was many times the diameter of a single yeast cell and attached to a single surface of the chamber (the ceiling provided by a glass coverslip) using the lectin, concanavalin A. For confinement, cells were loaded into a microfluidic chamber which traps cells between a ceiling and floor separated by the diameter of a single yeast cell, causing enforced contact with two surfaces. Medium is then constantly perfused through the chamber. **B.** Percent of *MAT***a** *bar1Δ* cells that died after exposure to 50nM α-factor for five hours in three different physical environments. Error bars represent the standard deviation of at least three independent experiments. **C.** Time course of *MAT***a** *bar1Δ* cells incubated in 50nM α-factor for the indicated amount of time in the flow chamber. Yellow arrows indicate cells that died since the previous time point. White arrows indicate cells that died earlier. The scale bar indicates 10 μm.

increasing the osmotic pressure differential between the cytoplasm and the extracellular environment. Cells were exposed to medium with 50 nM α-factor and 1M sorbitol, which increases the osmolarity of the medium, for five hours. When cells are exposed to high external osmolarity, they adapt to the osmotic stress by synthesizing glycerol, which can take place in a matter of minutes [44–47], thus increasing their internal osmolarity. Because of this partial restoration of the osmotic pressure gradient across the cell wall, we were not surprised to find that the fraction of cells that die when exposed to 50 nM α-factor and 1M sorbitol is similar to that of cells exposed to only 50 nM α-factor (Figure 5A). Nevertheless, we reasoned that some of the surviving cells would have holes in their cell walls that would be small enough to allow their survival until we increased the osmotic pressure difference between the inside and outside of the cells. Thus, if we replace the sorbitol-containing medium with medium lacking sorbitol, we would expect to see rapid cell death due to the large pressure differential. To test this prediction, we waited until five hours after beginning pheromone treatment and then replaced medium containing 1M sorbitol and 50 nM α-factor with medium containing only 50 nM α-factor. Immediately following the sorbitol washout, the number of dead cells in the flow chamber more than doubled, supporting the idea that the cells in the chamber are dying due to a breach in their cell walls (Figure 5A and 5B and Movie S4).

To test the effect of decreasing the osmotic pressure differential between the cells and the extracellular environment, we exposed cells to 50 nM α-factor in the absence of 1M sorbitol for 80 minutes, at which point cells are just beginning to die

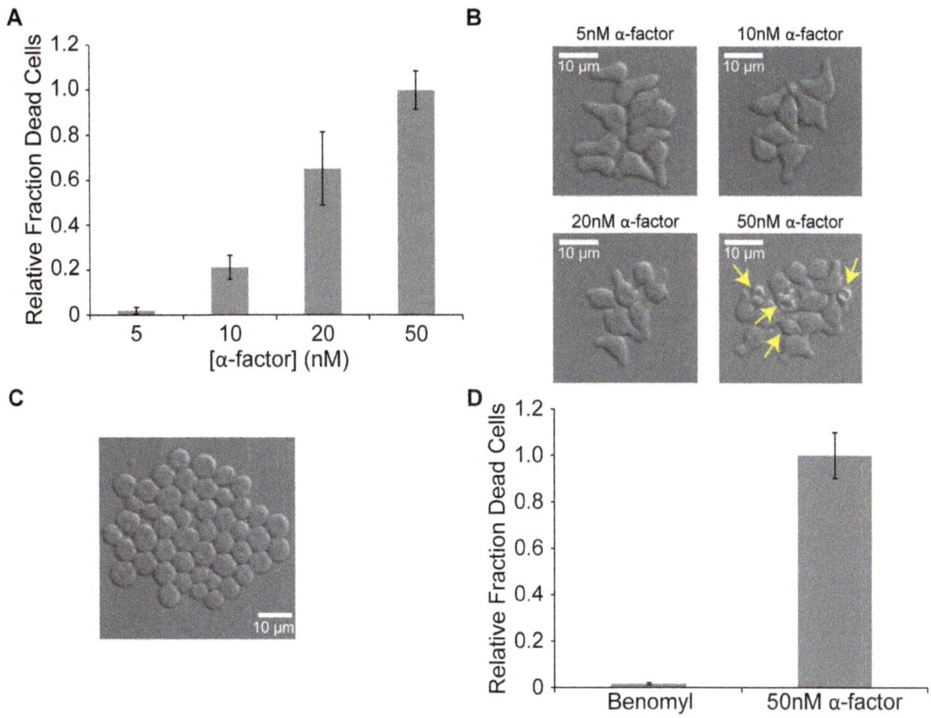

Figure 4. Pheromone-induced cell death increases with increased polarization. A. Fraction of *MAT***a** *bar1Δ* cells that died after five hours exposure to various concentrations of α-factor in the flow chamber relative to the fraction of *MAT***a** *bar1Δ* cells that died after five hours exposure to 50nM α-factor. Error bars represent the standard deviation of at least three independent experiments. **B.** *MAT***a** *bar1Δ* cells incubated in the indicated concentration of α-factor for five hours in the flow chamber. Yellow arrows indicate dead cells. The scale bar indicates 10 μm. **C.** *MAT***a** *bar1Δ* cells exposed to 0.1mM benomyl for five hours in the flow chamber. The scale bar indicates 10 μm. **D.** Fraction of *MAT***a** *bar1Δ* cells that died after five hours exposure to either 0.1mM benomyl or 50nM α-factor in the flow chamber relative to the fraction of *MAT***a** *bar1Δ* cells that died after five hours exposure to 50nM α-factor (Student's *t*-test, $p<10^{-6}$). Error bars represent the standard deviation of at least three independent experiments.

(Figure 3C and Movies S1–S3). We determined the percentage of dead cells at this point and then perfused the chamber with medium containing 1M sorbitol and 50 nM α-factor and observed the percentage of dead cells 60 minutes after the change of media. Since the sorbitol is washed in after the cells have begun to shmoo, the cells will have less time to induce the hyperosmotic response, and if the cell death is due to osmotic lysis, we should observe fewer cell deaths when 1M sorbitol is present in the medium. When we observe the fold change in cell death between 80 minutes and 140 minutes after α-factor addition in the absence of 1M sorbitol, there is an 8.9-fold increase in the fraction of dead cells (Figure 5C). However, when 1M sorbitol is added to the medium 80 minutes after α-factor addition, there is only a 1.4-fold increase in the fraction of dead cells between 80 and 140 minutes after α-factor addition, strengthening the evidence that pheromone-induced cell death is due to osmotic lysis (Figure 5C).

Proteins necessary for cell wall breakdown during mating are required for pheromone-induced cell death

We investigated the effects of deleting, *FUS1* and *FUS2*, two genes required for efficient cell fusion [24]. When *FUS1*, *FUS2*, or, both *FUS1* and *FUS2* are deleted in both mating partners, prezygotes, consisting of two shmoos bound to each other at their tips, are formed, but cells cannot dissolve their cell walls and thus fail to fuse [24,28]. Also, in *fus1* and *fus1fus2* mutants, the tightly polarized vesicles that are seen in the fusion zone of wild-type prezygotes and are hypothesized to contain cell wall remodeling enzymes are fewer and more widely dispersed than in wild-type cells [28]. If cell death in the flow chamber is due to pheromone-stimulated cell wall breakdown, mutations known to impair cell wall fusion should reduce the frequency of pheromone-induced cell death events in the flow chamber. Corroborating previous results obtained in bulk cultures [37], deleting *FUS1* and *FUS2* alone and in combination caused more than a 14-fold reduction in cell death in the flow chamber when cells were exposed to 50 nM α-factor for five hours (Student's *t*-test, $p<0.002$) (Figure 6A).

If the pheromone-induced cell death in the flow chamber is due to holes formed in the cell wall from inappropriate cell wall dissolution, the deletion of cell wall remodeling enzymes should decrease the frequency of pheromone-induced cell death. We investigated the effects of two putative cell wall glucanases that have been implicated in mating: Scw11, a target of pheromone-induced gene expression [19], and its paralog, Scw4, whose deletion interferes with mating [18]. If the observed cell death is due to accumulation of cell-wall degrading enzymes and these glucanases are major contributors to cell wall remodeling during cell wall fusion, deleting them should reduce the frequency of pheromone-induced cell death in the flow chamber. To test this prediction, we incubated *MAT***a** *bar1Δ scw11Δ* cells in a flow chamber in medium containing 50 nM α-factor for five hours. Deleting *SCW11* caused a 20% reduction in cell death compared to *MAT***a** *bar1Δ* cells (Student's *t*-test, $p=8\times10^{-4}$), and removing both Scw11 and Scw4 caused a 40% reduction in cell death compared to *MAT***a** *bar1Δ* cells (Student's *t*-test, $p=3\times10^{-5}$) (Figure 6B).

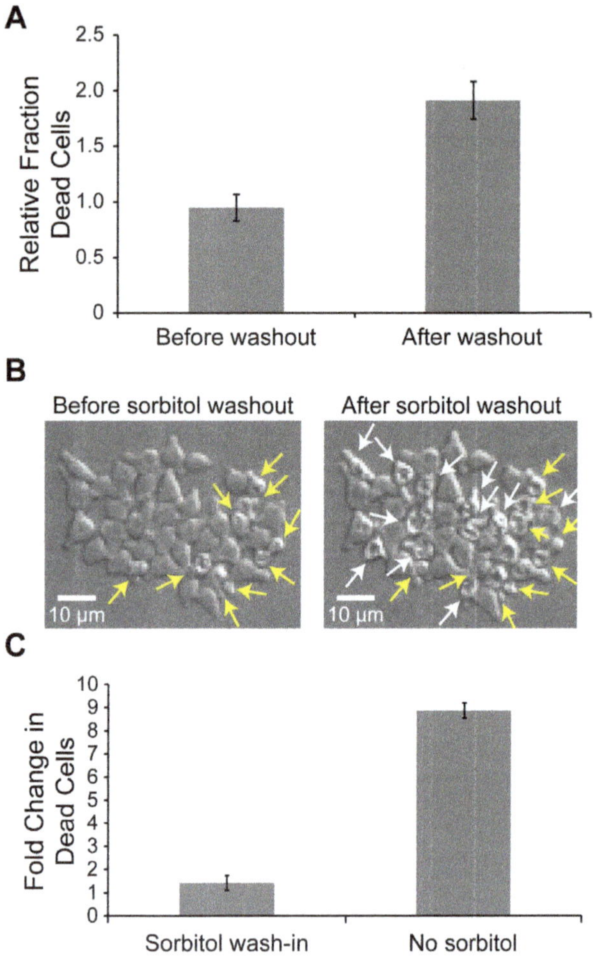

Figure 5. Pheromone-induced cell death is due to osmotic lysis. **A.** *MAT***a** *bar1Δ* cells were grown in a flow chamber for five hours in medium with 50nM α-factor and 1M sorbitol. After five hours, the sorbitol was washed out, and the cells were incubated in medium with 50nM α-factor and no sorbitol. The fraction of dead cells 10 minutes before and 10 minutes after the 1M sorbitol was washed out relative to the fraction of cells that die when exposed to 50nM α-factor for five hours without the addition of sorbitol was determined (Student's *t*-test, $p = 2\times10^{-4}$). Error bars represent the standard deviation of at least three independent experiments. **B.** Cells imaged after 290 minutes in medium with 1M sorbitol and 50nM α-factor (Before sorbitol washout) and 10 minutes after the medium was replaced with medium with 50nM α-factor and no sorbitol (After sorbitol washout). Yellow arrows indicate the cells that died during the 290 minutes of pheromone treatment prior to the sorbitol washout. White arrows in the "After sorbitol washout" picture indicate cells that died during the twenty minute period that spanned the last 10 minutes with sorbitol and the first 10 minutes after the sorbitol washout. The scale bar indicates 10 µm. **C.** Cells were grown in the flow chamber for 80 minutes in medium with 50nM α-factor. After 80 minutes, 1M sorbitol was added to the medium such that the cells were incubated in medium with 1M sorbitol and 50nM α-factor. The fold change in the number of cells that died during the 80 minutes prior to and 60 minutes after the sorbitol wash-in was determined (Sorbitol wash-in). In control chambers (No sorbitol), no sorbitol was added to the medium, and the fold change in the number of cells that died in the two corresponding periods was determined (Student's *t*-test, $p = 9\times10^{-5}$). Error bars represent the standard deviation of at least three independent experiments.

Discussion

The mating of budding yeast is risky and elaborately choreographed. When two haploid yeast cells mate, they signal through reciprocal pheromones and receptors, stimulate each other to signal ever more strongly, arrest their cell cycles, use pheromone gradients to direct their polarization towards each other, and eventually fuse their cell walls, cell membranes, and nuclei to form a single diploid cell [8,10,48,49]. Although many aspects of yeast mating have been well studied, the mechanism by which cells dissolve their cell walls to allow fusion of their plasma membranes remains mysterious. Cell wall dissolution is a particularly dangerous step in yeast mating. The plasma membranes of the two partner cells cannot touch each other and fuse until the cell walls that lie between them have been dissolved [8]. Because the osmolarity inside a cell is so much higher than outside, the elasticity of the cell wall opposes the osmotic pressure difference between the cytoplasm and the environment, thus keeping water from rushing into the cell and causing it to lyse. A cell that dissolves any part of its cell wall that does not touch a closely apposed mating partner will die [15,16].

Pheromone-induced cell death was studied previously [36,37] and hypothesized to be due to inappropriate activation of cell fusion machinery, resulting in cell wall dissolution and eventual cell lysis [37]. Although it was observed that this lysis can be reduced by increasing cell wall integrity and deleting certain proteins involved in cell fusion, a hypothesis to explain why cells were dissolving their cell walls was not given [37]. Many hypotheses can be generated to explain how cell wall dissolution is regulated in time and space to promote mating and prevent accidental deaths. Most of them posit additional signaling systems in addition to the known mechanisms of pheromone signaling, but no additional signaling molecules have been uncovered, despite a variety of searches [8,28,50,51]. The failure of these attempts led us to propose a hypothesis that requires no new components and instead appeals to the physical differences between mating cell pairs and isolated, pheromone-stimulated cells.

In an isotropically growing cell, cell wall synthesizing and remodeling enzymes are secreted uniformly around the cell, whereas the polarized growth that accompanies both budding and shmooing requires similarly polarized secretion of these enzymes [33] (Figure 1A and 1B). Thus we hypothesize that cell wall remodeling enzymes, such as Scw4 and Scw11, are preferentially released at the shmoo tip, which locally weakens the cell wall, allowing the shmoo to grow continuously up the pheromone gradient. As a shmoo approaches a suitable partner, the concentration of pheromone increases, tightening the polarization, and increasing the concentration of cell wall remodeling enzymes in the part of the cell wall that has polarized towards its partner's site of maximum pheromone secretion [49,52–55]. If the remodeling enzymes are diffusible, the maximum concentration they can reach in a shmoo that has not bound to a partner is limited: even though the secretion rate of cell wall remodeling enzymes is high, the enzymes are able to diffuse through the cell wall, keeping their concentration in the range that is high enough to allow rapid remodeling of the growing shmoo but low enough to prevent cell wall rupture (Figure 1C). But when two shmoo tips are attached to each other via mating agglutinins, it takes longer for cell wall remodeling enzymes to diffuse out of the fusion zone because they must now travel laterally through the cell wall in order to escape, thus increasing the local concentration of the remodeling enzymes and leading to the gradual dissolution of the cell wall, exposing the two plasma membranes to each other and allowing their fusion to create a single, diploid cell (Figure 1D).

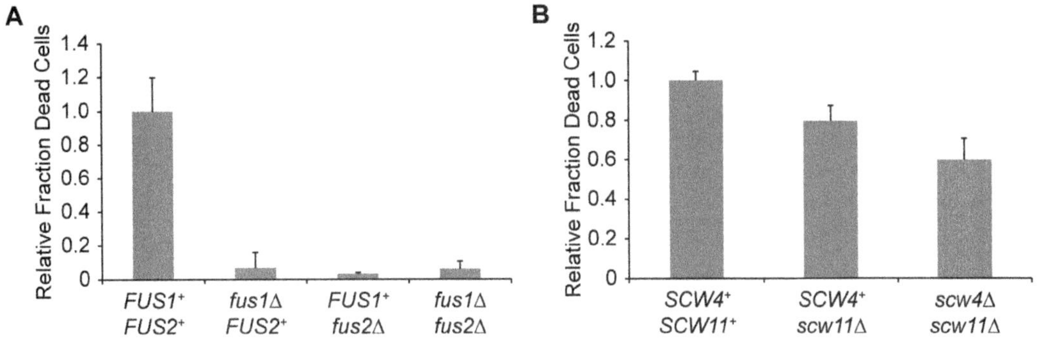

Figure 6. Pheromone-induced cell death is dependent on cell fusion proteins and putative glucanases. A. Fraction of dead *MAT***a** *bar1Δ* cells deleted for different combinations of *FUS1* and *FUS2* relative to the fraction of dead *MAT***a** *bar1Δ* cells incubated in 50nM α-factor for five hours in the flow chamber. Error bars represent the standard deviation of at least three independent experiments. **B.** Fraction of dead *MAT***a** *bar1Δ* cells deleted for different combinations of putative cell wall glucanases relative to the fraction of dead *MAT***a** *bar1Δ* cells incubated in 50nM α-factor for five hours in the flow chamber. Error bars represent the standard deviation of at least three independent experiments.

Our data supports the hypothesis that increased secretion of remodeling enzymes and the longer distances they have to diffuse when two polarization zones agglutinate to each other causes cell wall dissolution. We tightly apposed pheromone-treated cells to an impermeable surface, mimicking the effect of cell-cell attachment during mating while ensuring that the only signal these cells can receive is a uniformly high pheromone concentration (Figure 1E). The cell lysis events we observe are not due merely to the physical constraints of a flow chamber: cells also lyse when they are chemically attached to a glass coverslip with essentially infinite space to expand, and confined cells that grow larger isotropically while they are arrested in mitosis do not lyse [42]. By manipulating the presence of an osmoprotectant in the media, we show that the frequency of lysis events rises with increasing osmotic pressure difference between the osmolarity of the cell and its environment, implying that breaches in the cell wall lead to osmotically induced lysis. Lysis depends on Fus1 and Fus2, proteins that have been previously implicated in cell fusion [24], and the cell wall dissolution appears to be at least partially accomplished by two putative glucanases, Scw4 and Scw11.

Although a contact-driven increase in the concentration of cell wall remodeling enzymes is the simplest explanation for cell wall dissolution during mating, other viable hypotheses exist. One is that cells can only dissolve their cell walls in response to a high concentration of pheromone that they experience while attached to a mating partner. When cells are in the flow chamber, we do see an increase in cell lysis as we increase the pheromone concentration arguing that high pheromone concentrations promote cell wall dissolution; however, the increase in cell lysis in cells apposed to impermeable surfaces argues that factors other than pheromone concentration contribute to cell wall dissolution. Another hypothesis is that the cell is capable of detecting cell wall or cell membrane deformations that indicate that two shmoo tips are attached via mating agglutinins. Our experiments do not negate this possibility, since it is likely that contact with impermeable surfaces causes cell wall deformation, but the failure of previous attempts to identify additional signaling components [8,28,50,51] argues against this model. A third possibility is that cells respond to a direct signal from another cell, such as an additional, uncharacterized signaling mechanism similar to the G-protein coupled receptors involved in pheromone stimulation, or, perhaps, the oscillation in pheromone concentration that would occur if cells were close enough to detect the pulses of increased pheromone concentration concomitant with the fusing of individual secretory vesicles with the plasma membrane. The fact that lysis occurs without the presence of a mating partner argues against any hypothesis that requires communication aside from that of the reciprocal pheromones and pheromone receptors between the two mating cells.

Taken together with previously published studies, our data supports a model that involves pheromone-induced, polarized secretion of cell wall remodeling enzymes. When cells are pheromone stimulated, a MAP kinase cascade activates transcription of pheromone-induced genes [10]. Along with many others, these genes include the expression of mating agglutinins and cell wall remodeling enzymes, which are packaged into vesicles for secretion into the extracellular environment [11–13,17–19]. Fus2 and Rvs161, a protein that binds to curved membranes [25,26] and is involved in cell fusion [51], bind to these vesicles and travel along actin cables to the site of polarization in a Myo2-dependent fashion [56] where they are anchored to the plasma membrane by Fus1 [23], which interacts with the polarisome [30]. Fus2 and Rvs161 in conjunction with Cdc42 may then facilitate the fusion of these vesicles with the plasma membrane [57].

When cells are weakly stimulated, they form broad shmoos (Figure 4B). Although these cells are polarized, the zone of polarization is relatively large, and presumably, the vesicles containing cell wall remodeling enzymes are released into a relatively large area. The enzymes cleave carbohydrate bonds as they diffuse through the cell wall matrix, weakening the cell wall and allowing for further expansion in the direction of highest pheromone concentration [34]. As a shmoo gets closer to a cell of the opposite mating type, the pheromone concentration increases and the shmoo tip becomes more tightly polarized [53,54]. This tighter polarization focuses the secretion of cell wall remodeling enzymes into a smaller fraction of the cell surface, increasing the concentration of cell wall remodeling enzymes in this zone. Although the concentration of cell wall remodeling enzymes in this zone has increased, it is not typically high enough to cause dissolution of the cell wall unless the shmoo tip is pressed against an impermeable barrier, forcing the enzymes to travel further to reach bulk solution. Since the time taken to diffuse a given distance rises with the square of the distance, the effective speed at which the enzymes move falls, and thus their concentration in the cell wall rises (Figure 1 and 2). Similarly, when the two polarized cells attach at their shmoo tips via mating agglutinins, the presence of a second cell membrane traps the remodeling enzymes in the cell wall by requiring them to move laterally along the cell surface to exit the cell wall, increasing their concentration and breaking down the cell wall (Figure 1D). As the wall dissolves, the two plasma membranes come into contact with one another, allowing membrane fusion to begin and pushing the Fus2-bound vesicles outward [23], which allows for the rest of the intervening cell wall to be dissolved and eventually full fusion of the newly formed zygote.

Understanding more about the cell fusion of budding yeast is an important step in understanding cell fusion in more complex organisms. Although animal cells do not have a cell wall, the extracellular matrix surrounding these cells must be dissolved prior to cell fusion. A similar local increase in enzyme concentration at points where cells are very close to each other could promote the digestion of the matrix and allow the plasma membranes of the two partners to touch each other.

Materials and Methods

Yeast strains and culturing

Strains used in this study are listed in Table 1. All strains were derived from the W303 background [58] (*ade2-1 can1-100 his3-11,15 leu2-112 trp1-1 ura3-1*) using standard genetic techniques. All media was prepared as described [59] and contained 2% wt/vol of glucose. Cells were either grown in Synthetic Complete medium (SC) or Yeast Extract Peptone Dextrose (YPD) at 30°C in culture tubes on roller drums or at room temperature (25°C) for timelapse microscopy. Bovine serum albumin (BSA) was used to reduce the non-specific absorption of α-factor to glass and plastic surfaces; it was made into 10% wt/vol stocks in deionized water and then diluted into media to 0.1% wt/vol. Synthetic α-factor (Biosynthesis, Lewisville, TX) was suspended in dimethyl sulfoxide (DMSO) and then diluted into YPD+0.1% BSA or SC+0.1% BSA at the appropriate concentration. When appropriate, 1M sorbitol was added to YPD by dissolving sorbitol into YPD. YPD containing 1-(butylcarbomoyl)-2-benzimidasolecarbamate (benomyl) was prepared by heating YPD to 65°C and adding 34 mM benomyl in DMSO dropwise to a final concentration of 0.1 mM. Yeast extract was obtained from EMD Millipore (Billerica, MA). Peptone and yeast nitrogen base were obtained from BD (Franklin Lakes, NJ). Bacto-agar was obtained from US Biological (Swampscott, MA). Unless otherwise noted, all chemicals were obtained from Sigma-Aldrich (St. Louis, MO).

Microscopy

Microscopy was done at room temperature using a Nikon Ti-E inverted microscope with a 20x Plan Apo VC 0.75NA air lens, and images were acquired with a Photometrics CoolSNAP HQ camera (Roper Scientific, AZ). Timelapse photography was done using Metamorph 7.7 (Molecular Devices, CA); pictures were acquired using differential interference contrast every 10 minutes with a 10 ms exposure.

Bulk culture lysis assay

Cells were grown to log phase ($\sim 5\times10^6$ cells/mL) at 30°C in YPD and counted using a Z2 Coulter counter (Beckman-Coulter, CA). Cells were washed in YPD+0.1% BSA and resuspended at 10^6 cells/mL into plastic 14 mL culture tubes (BD Falcon, MA) in YPD+0.1% BSA with 50 nM α-factor. These cultures were then incubated on a roller drum at 30°C for five hours. Cells were then put directly onto glass slides (Corning, NY) with uncoated coverslips (VWR, PA) and imaged at 20x magnification using differential interference contrast with a 10 ms exposure. Prior to the experiment, the plastic culture tubes were coated in BSA by incubating overnight at 4°C with phosphate buffered saline (PBS) with 2% wt/vol BSA. The PBS+2% BSA was poured out immediately prior to the addition of the cell cultures. To determine the percentage of cells that lysed, more than 50 cells were counted from each trial. Statistical significance was determined using Student's *t*-Test.

Concanavalin-A coated coverslip lysis assay

Coverslips (VWR, PA) were coated in concanavalin A (MP Biomedicals, OH) in a protocol modified from Joglekar *et al.* (2008) [60]. Briefly, coverslips were soaked in 1M NaOH for one hour at room temperature (25°C), rinsed five times with deionized, filtered water, and then incubated at room temperature for one hour in a solution of 10 mM Na_2HPO_4 pH 6.0 (Fisher Biotech, MA) +1 mM $CaCl_2$+0.5 mg/mL concanavalin A. Coverslips were then rinsed five times with deionized, filtered water and air-dried over a 100°C heat block. To make a chamber, strips of parafilm (American National Can, IL) were melted at 100°C on glass slides (Corning, NY) and concanavalin A-coated coverslips were placed on top of the strips. The parafilm was allowed to cool to room temperature, creating channels with a glass slide ceiling, concanavalin A-coated coverslip floor, and parafilm walls on two, parallel sides.

Cells were grown to log phase ($\sim 5\times10^6$ cells/mL) at 30°C in YPD and then washed in SC+0.1% BSA. 50 nM α-factor was added to the cells, and the cells were immediately injected into the chamber using capillary action. The cells were allowed to adhere to the concanavalin A-coated coverslip for 10 minutes, and then 200 μL of SC+0.1% BSA with 50 nM α-factor was flowed through the chamber using capillary action to wash off excess cells.

Table 1. Strains used in this study.

Strain Name	Genotype (all cells are in the W303 background)
LBHY52	*MATa bar1Δ::KanMX6 P_{ACT1}-yCerulean-HIS3MX6 @ P_{ACT1}*
LBHY77	*MATa bar1Δ::KanMX6 fus1Δ::NatMX4 P_{ACT1}-yCerulean-HISMX3 @ P_{ACT1}*
LBHY80	*MATa bar1Δ::KanMX6 fus2Δ::HphMX4 P_{ACT1}-yCerulean-HISMX3 @ P_{ACT1}*
LBHY84	*MATa bar1Δ::KanMX6 fus1Δ::NatMX4 fus2Δ::HphMX4 P_{ACT1}-yCerulean-HISMX3 @ P_{ACT1}*
LBHY136	*MATa bar1Δ::ADE2 SPA2-YFP:HIS3 scw11Δ::HphMX4 ade2-1 can1-100 his3-11,15 leu2-3,112 trp1-1 ura3-1*
LBHY153	*MATa bar1Δ::ADE2 SPA2-YFP:HIS3 scw4Δ::KanMX6 scw11Δ::HphMX4 ade2-1 can1-100 his3-11,15 leu2-3,112 trp1-1 ura3-1*
MP 384	*MATa bar1Δ::ADE2 SPA2-YFP:HIS3 ade2-1 can1-100 his3-11,15 leu2-3,112 trp1-1 ura3-1*

All strains are from this study except for MP 384, which is from M. Piel.

The chamber was sealed with candle wax and imaged at 20x magnification, using differential interference contrast with a 10 ms exposure, every 10 minutes for five hours from the time when the cells were exposed to α-factor-containing medium. To determine the percentage of cells that lysed, more than 400 cells were counted from each trial. Statistical significance was determined using Student's *t*-Test.

Flow chamber lysis assay

Cells were grown to log phase (~5×10^6 cells/mL) at 30°C in YPD and then washed in YPD+0.1% BSA. For experiments involving α-factor, the microfluidic chambers (CellAsic, Hayward, CA) [61] were pretreated by perfusing PBS+2% BSA through the chamber at 34 kPa (5 psi) for 10 minutes and then YPD+0.1% BSA through the chamber at 34 kPa (5 psi) for 10 minutes. After cells were loaded, YPD+0.1% BSA with the appropriate concentration of α-factor was perfused through the chamber at 14 kPa (2 psi), and pictures were taken at 20x magnification every 10 minutes for five hours using differential interference contrast with a 10 ms exposure.

For experiments involving benomyl, the microfluidic chambers were pretreated by perfusing YPD through the chamber at 34 kPa (5 psi) for 10 minutes. After the cells were loaded, YPD+0.1 mM benomyl was perfused through the chamber at 14 kPa (2 psi), and pictures were taken at 20x magnification every 10 minutes for five hours using differential interference contrast with a 10 ms exposure. To determine the percentage of cells that lysed, more than 250 cells were counted from each trial. Statistical significance was determined using Student's *t*-Test.

Sorbitol wash-out assay

Cells were grown to log phase (~5×10^6 cells/mL) at 30°C in YPD and then washed in YPD+0.1% BSA. The microfluidic chambers (CellAsic, Hayward, CA) [61] were pretreated by perfusing PBS+2% BSA through the chamber at 34 kPa (5 psi) for 10 minutes and then YPD+1M sorbitol +0.1% BSA through the chamber at 34 kPa (5 psi) for 10 minutes. After cells were loaded, YPD+1M sorbitol +0.1% BSA+50 nM α-factor was perfused through the chamber at 14 kPa (2 psi) for five hours. After five hours, the medium containing 1M sorbitol was washed out, and YPD+0.1% BSA+50 nM α-factor was perfused through the chamber at 14 kPa (2 psi) for two hours. Pictures were taken at 20x magnification every 10 minutes for seven hours using differential interference contrast with a 10 ms exposure. To determine the percentage of cells that lysed, more than 500 cells were counted from each trial. Statistical significance was determined using Student's *t*-Test.

Sorbitol wash-in assay

Cells were grown to log phase (~5×10^6 cells/mL) at 30°C in YPD and then washed in YPD+0.1% BSA. The microfluidic chambers (CellAsic, Hayward, CA) [61] were pretreated by perfusing PBS+2% BSA through the chamber at 34 kPa (5 psi) for 10 minutes and then YPD+0.1% BSA at 34 kPa (5 psi) through the chamber for 10 minutes. After cells were loaded, YPD+0.1% BSA+50 nM α-factor was perfused through the chamber at 14 kPa (2 psi) for 80 minutes. At the end of 80 minutes, the YPD+0.1% BSA+50 nM α-factor was washed out, and YPD+1M sorbitol +0.1% BSA+50 nM α-factor was perfused through the chamber at 14 kPa (2 psi) for 60 minutes. Pictures were taken at 20x magnification every 10 minutes for 140 minutes using differential interference contrast with a 10 ms exposure. To determine the percentage of cells that lysed, more than 600 cells were counted from each trial. Statistical significance was determined using Student's *t*-Test.

Supporting Information

Movie S1 Pheromone-induced cell death in the flow chamber. *MAT***a** *bar1Δ* cells were incubated in medium containing 50 nM α-factor for five hours in the flow chamber. White arrows indicate cells that die during the movie. Cells were imaged every 10 minutes. The scale bar indicates 10 μm. Each movie is from an independent experiment.

Movie S2 Pheromone-induced cell death in the flow chamber. *MAT***a** *bar1Δ* cells were incubated in medium containing 50 nM α-factor for five hours in the flow chamber. White arrows indicate cells that die during the movie. Cells were imaged every 10 minutes. The scale bar indicates 10 μm. Each movie is from an independent experiment.

Movie S3 Pheromone-induced cell death in the flow chamber. *MAT***a** *bar1Δ* cells were incubated in medium containing 50nM α-factor for five hours in the flow chamber. White arrows indicate cells that die during the movie. Cells were imaged every 10 minutes. The scale bar indicates 10 μm. Each movie is from an independent experiment.

Movie S4 Pheromone-induced cell death is due to osmotic lysis. *MAT***a** *bar1Δ* cells were incubated in the flow chamber for five hours in medium containing 50 nM α-factor and 1M sorbitol. After five hours, the sorbitol was washed out, and the cells were incubated in medium containing 50 nM α-factor and no sorbitol. Cells were imaged every 10 minutes. The scale bar indicates 10 μm.

Acknowledgments

We thank V. Denic, R. Gaudet, C. Hunter, and members of the Murray laboratory for reading and commenting on the manuscript. We also thank M. Piel for strains.

Author Contributions

Conceived and designed the experiments: LBH AWM. Performed the experiments: LBH. Analyzed the data: LBH AWM. Wrote the paper: LBH AWM.

References

1. Chen EH, editor (2008) Cell fusion overviews and methods. New York, New York: Humana Press. 421 p.
2. Abmayr SM, Pavlath GK (2012) Myoblast fusion: Lessons from flies and mice. Development 139(4): 641–656.
3. Ishii M, Saeki Y (2008) Osteoclast cell fusion: Mechanisms and molecules. Mod Rheumatol 18(3): 220–227.
4. Galindo B, Lazdins J, Castillo R (1974) Fusion of normal rabbit alveolar macrophages induced by supernatant fluids from BCG-sensitized lymph node cells after elicitation by antigen. Infect Immun 9(2): 212–216.
5. Huppertz B, Borges M (2008) Placenta trophoblast fusion. Methods Mol Biol 475: 135–147.
6. Alper S, Podbilewicz B (2008) Cell fusion in *Caenorhabditis elegans*. Methods Mol Biol 475: 53–74.

7. Abmayr SM, Zhuang S, Geisbrecht ER (2008) Myoblast fusion in *drosophila*. Methods Mol Biol 475: 75–97.
8. Ydenberg CA, Rose MD (2008) Yeast mating: A model system for studying cell and nuclear fusion. Methods Mol Biol 475: 3–20.
9. Herskowitz I (1988) Life cycle of the budding yeast *Saccharomyces cerevisiae*. Microbiol Rev 52(4): 536–553.
10. Bardwell L (2005) A walk-through of the yeast mating pheromone response pathway. Peptides 26(9): 339–350.
11. Cappellaro C, Baldermann C, Rachel R, Tanner W (1994) Mating type-specific cell-cell recognition of *Saccharomyces cerevisiae*: Cell wall attachment and active sites of a- and alpha-agglutinin. EMBO J 13(20): 4737–4744.
12. Roy A, Lu C, Marykwas D, Lipke P, Kurjan J (1991) The AGA1 product is involved in cell surface attachment of the *Saccharomyces cerevisiae* cell adhesion glycoprotein a-agglutinin. Mol Cell Biol 11(8): 4196–4206.
13. Zhao H, Shen Z, Kahn PC, Lipke PN (2001) Interaction of α-agglutinin and a-agglutinin, *Saccharomyces cerevisiae* sexual cell adhesion molecules. J Bacteriol 183(9): 2874–2880.
14. Dupres V, Dufrêne YF, Heinisch JJ (2010) Measuring cell wall thickness in living yeast cells using single molecular rulers. ACS Nano 4(9): 5498–5504.
15. Philips J, Herskowitz I (1997) Osmotic balance regulates cell fusion during mating in *Saccharomyces cerevisiae*. J Cell Biol 138(5): 961–974.
16. de Nobel H, Ruiz C, Martin H, Morris W, Brul S, et al. (2000) Cell wall perturbation in yeast results in dual phosphorylation of the Slt2/Mpk1 MAP kinase and in an Slt2-mediated increase in FKS2-lacZ expression, glucanase resistance and thermotolerance. Microbiology 146 (Pt 9)(Pt 9): 2121–2132.
17. Brizzio V, Gammie AE, Nijbroek G, Michaelis S, Rose MD (1996) Cell fusion during yeast mating requires high levels of a-factor mating pheromone. J Cell Biol 135(6): 1727–1739.
18. Cappellaro C, Mrsa V, Tanner W (1998) New potential cell wall glucanases of *Saccharomyces cerevisiae* and their involvement in mating. J Bacteriol 180(19): 5030–5037.
19. Zeitlinger J, Simon I, Harbison CT, Hannett NM, Volkert TL, et al. (2003) Program-specific distribution of a transcription factor dependent on partner transcription factor and MAPK signaling. Cell 113(3): 395–404.
20. Dolan JW, Kirkman C, Fields S (1989) The yeast STE12 protein binds to the DNA sequence mediating pheromone induction. Proceedings of the National Academy of Sciences 86(15): 5703–5707.
21. Chenevert J, Valtz N, Herskowitz I (1994) Identification of genes required for normal pheromone-induced cell polarization in *Saccharomyces cerevisiae*. Genetics 136(4): 1287–1296.
22. Brizzio V, Gammie AE, Rose MD (1998) Rvs161p interacts with Fus2p to promote cell fusion in *Saccharomyces cerevisiae*. J Cell Biol 141(3): 567–584.
23. Paterson JM, Ydenberg CA, Rose MD (2008) Dynamic localization of yeast Fus2p to an expanding ring at the cell fusion junction during mating. J Cell Biol 181(4): 697–709.
24. Trueheart J, Boeke JD, Fink GR (1987) Two genes required for cell fusion during yeast conjugation: Evidence for a pheromone-induced surface protein. Mol Cell Biol 7(7): 2316–2328.
25. Friesen H, Humphries C, Ho Y, Schub O, Colwill K, et al. (2006) Characterization of the yeast amphiphysins Rvs161p and Rvs167p reveals roles for the rvs heterodimer in vivo. Mol Biol Cell 17(3): 1306–1321.
26. Peter BJ, Kent HM, Mills IG, Vallis Y, Butler PJG, et al. (2004) BAR domains as sensors of membrane curvature: The amphiphysin BAR structure. Science 303(5657): 495–499.
27. Bon E, Recordon-Navarro P, Durrens P, Iwase M, Toh-e A, et al. (2000) A network of proteins around Rvs167p and Rvs161p, two proteins related to the yeast actin cytoskeleton. Yeast 16(13): 1229–1241.
28. Gammie AE, Brizzio V, Rose MD (1998) Distinct morphological phenotypes of cell fusion mutants. Mol Biol Cell 9(6): 1395–1410.
29. Trueheart J, Fink GR (1989) The yeast cell fusion protein FUS1 is O-glycosylated and spans the plasma membrane. Proceedings of the National Academy of Sciences 86(24): 9916–9920.
30. Nelson B, Parsons AB, Evangelista M, Schaefer K, Kennedy K, et al. (2004) Fus1p interacts with components of the Hog1p mitogen-activated protein kinase and Cdc42p morphogenesis signaling pathways to control cell fusion during yeast mating. Genetics 166(1): 67–77.
31. Sheu Y, Barral Y, Snyder M (2000) Polarized growth controls cell shape and bipolar bud site selection in *Saccharomyces cerevisiae*. Mol Cell Biol 20(14): 5235–5247.
32. Klis FM (1994) Review: Cell wall assembly in yeast. Yeast 10(7): 851–869.
33. Cid VJ, Durán A, del Rey F, Snyder MP, Nombela C, et al. (1995) Molecular basis of cell integrity and morphogenesis in *Saccharomyces cerevisiae*. Microbiol Rev 59(3): 345–386.
34. Lesage G, Bussey H (2006) Cell wall assembly in *Saccharomyces cerevisiae*. Microbiol Mol Biol Rev 70(2): 317–343.
35. Lipke PN, Ovalle R (1998) Cell wall architecture in yeast: New structure and new challenges. J Bacteriol 180(15): 3735–3740.
36. Severin F, Hyman A (2002) Pheromone induces programmed cell death in *S. cerevisiae*. Current Biology: CB 12(7): R233.
37. Zhang N, Dudgeon DD, Paliwal S, Levchenko A, Grote E, et al. (2006) Multiple signaling pathways regulate yeast cell death during the response to mating pheromones. Mol Biol Cell 17(8): 3409–3422.
38. Sprague Jr GF, Herskowitz I (1981) Control of yeast cell type by the mating type locus: I. identification and control of expression of the a-specific gene BAR1. J Mol Biol 153(2): 305–321.
39. Chan RK, Otte CA (1982) Physiological characterization of *Saccharomyces cerevisiae* mutants supersensitive to G1 arrest by a factor and α factor pheromones. Mol Cell Biol 2(1): 21–29.
40. Biely P, Kratky Z, Bauer S (1976) Interaction of concanavalin A with external mannan-proteins of *Saccharomyces cerevisiae*. Glycoprotein nature of beta-glucanases. Eur J Biochem 70(1): 75–81.
41. Johnston G, Pringle J, Hartwell L (1977) Coordination of growth with cell division in the yeast *Saccharomyces cerevisiae*. Exp Cell Res 105(1): 79–98.
42. Li R, Murray AW (1991) Feedback control of mitosis in budding yeast. Cell 66(3): 519–531.
43. Hutchison HT, Hartwell LH (1967) Macromolecule synthesis in yeast spheroplasts. J Bacteriol 94(5): 1697–1705.
44. Brown A (1978) Compatible solutes and extreme water stress in eukaryotic micro-organisms. Adv Microb Physiol 17: 181–242.
45. Blomberg A, Adler L (1992) Physiology of osmotolerance in fungi. Adv Microb Physiol 33: 145.
46. Hirayarna T, Maeda T, Saito H, Shinozaki K (1995) Cloning and characterization of seven cDNAs for hyperosmolarity-responsive (HOR) genes of *Saccharomyces cerevisiae*. Molecular and General Genetics MGG 249(2): 127–138.
47. O'Rourke SM, Herskowitz I, O'Shea EK (2002) Yeast go the whole HOG for the hyperosmotic response. Trends Genet 18(8): 405–412.
48. Chang F, Herskowitz I (1990) Identification of a gene necessary for cell cycle arrest by a negative growth factor of yeast: FAR1 is an inhibitor of a G1 cyclin, CLN2. Cell 63(5): 999–1011.
49. Barkai N, Rose MD, Wingreen NS (1998) Protease helps yeast find mating partners. Nature 396(6710): 422–423.
50. Marsh L, Rose MD (1997) The pathway of cell and nuclear fusion during mating in *Saccharomyces cerevisiae*. In: Pringle JR, Broach JR, Jones EW, editors (1997) The Molecular and Cellular Biology of the Yeast *Saccharomyces* vol. 3, Cell Cycle and Cell Biology. New York: Cold Spring Harbor Laboratory Press. pp. 827–888.
51. Kurihara LJ, Beh CT, Latterich M, Schekman R, Rose MD (1994) Nuclear congression and membrane fusion: Two distinct events in the yeast karyogamy pathway. J Cell Biol 126(4): 911.
52. Jackson CL, Hartwell LH (1990) Courtship in *S. cerevisiae*: Both cell types choose mating partners by responding to the strongest pheromone signal. Cell 63(5): 1039–1051.
53. Bagnat M, Simons K (2002) Cell surface polarization during yeast mating. Proceedings of the National Academy of Sciences 99(22): 14183–14188.
54. Ayscough KR, Drubin DG (1998) A role for the yeast actin cytoskeleton in pheromone receptor clustering and signalling. Current Biology 8(16): 927–931.
55. Segall JE (1993) Polarization of yeast cells in spatial gradients of α mating factor. Proceedings of the National Academy of Sciences 90(18): 8332–8336.
56. Sheltzer JM, Rose MD (2009) The class V myosin Myo2p is required for Fus2p transport and actin polarization during the yeast mating response. Mol Biol Cell 20(12): 2909–2919.
57. Ydenberg CA, Stein RA, Rose MD (2012) Cdc42p and Fus2p act together late in yeast cell fusion. Mol Biol Cell 23(7): 1208–1218.
58. Thomas BJ, Rothstein R (1989) Elevated recombination rates in transcriptionally active DNA. Cell 56(4): 619–630.
59. Sherman F, Fink G, Lawrence C (1974) Methods in yeast genetics. New York: Cold Spring Harbor Laboratory Press. 73 p.
60. Joglekar AP, Salmon E, Bloom KS (2008) Counting kinetochore protein numbers in budding yeast using genetically encoded fluorescent proteins. Methods Cell Biol 85: 127–151.
61. Lee PJ, Helman NC, Lim WA, Hung PJ (2008) A microfluidic system for dynamic yeast cell imaging. BioTechniques 44: 91–95.

8

Photosynthetic Diffusional Constraints Affect Yield in Drought Stressed Rice Cultivars during Flowering

Marco Lauteri[1], Matthew Haworth[2]*, Rachid Serraj[3,4], Maria Cristina Monteverdi[5], Mauro Centritto[2]

1 Institute of Agro-Environmental and Forest Biology, National Research Council, Porano, Italy, **2** Trees and Timber Institute (IVALSA - Istituto per la valorizzazione del legno e delle specie arboree), National Research Council, Florence, Italy, **3** Consultative Group on International Agricultural Research, Independent Science and Partnership Council Secretariat, Food and Agriculture Organisation of the United Nations, Rome, Italy, **4** International Rice Research Institute, Los Banos, Philippines, **5** Forestry Research Centre, National Research Council, Arezzo, Italy

Abstract

Global production of rice (*Oryza sativa*) grain is limited by water availability and the low 'leaf-level' photosynthetic capacity of many cultivars. *Oryza sativa* is extremely susceptible to water-deficits; therefore, predicted increases in the frequency and duration of drought events, combined with future rises in global temperatures and food demand, necessitate the development of more productive and drought tolerant cultivars. We investigated the underlying physiological, isotopic and morphological responses to water-deficit in seven common varieties of *O. sativa*, subjected to prolonged drought of varying intensities, for phenotyping purposes in open field conditions. Significant variation was observed in leaf-level photosynthesis rates (*A*) under both water treatments. Yield and *A* were influenced by the conductance of the mesophyll layer to CO_2 (g_m) and not by stomatal conductance (g_s). Mesophyll conductance declined during drought to differing extents among the cultivars; those varieties that maintained g_m during water-deficit sustained *A* and yield to a greater extent. However, the variety with the highest g_m and yield under well-watered conditions (IR55419-04) was distinct from the most effective cultivar under drought (Vandana). Mesophyll conductance most effectively characterises the photosynthetic capacity and yield of *O. sativa* cultivars under both well-watered and water-deficit conditions; however, the desired attributes of high g_m during optimal growth conditions and the capacity for g_m to remain constant during water-deficit may be mutually exclusive. Nonetheless, future genetic and physiological studies aimed at enhancing *O. sativa* yield and drought stress tolerance should investigate the biochemistry and morphology of the interface between the sub-stomatal pore and mesophyll layer.

Editor: Wagner L. Araujo, Universidade Federal de Vicosa, Brazil

Funding: This work was funded by: Ministero dell'Istruzione, dell'Università e della Ricerca of Italy (PRIN) grant: "Going to the root of plant productivity: how the rhizosphere interact with the aboveground armament for indirect and direct defense against abiotic and biotic stressors (PRO-ROOT)" URL: http://hubmiur.pubblica.istruzione.it/web/universita/home and Marie Curie Intra European Fellowship (2010-275626) - HAWORTH - Evoutionary adaptation to atmospheric carbon dioxide, URL: http://ec.europa.eu/research/mariecurieactions/about-mca/actions/ief/index_en.htm. The funders had no role in study design, data collection and analysis, decision to publish, or preparation of the manuscript.

Competing Interests: The authors have declared that no competing interests exist.

* Email: matthew.haworth@hotmail.com

Introduction

Global production of rice (*Oryza sativa*) is limited by the availability of freshwater [1] and the tolerance of rice cultivars to drought stress [2]. Nonetheless, *O. sativa* is cultivated in a diverse range of climates and habitats, with lowland varieties generally considered to be more productive, but less tolerant of drought, than their upland counterparts [3]. Selective breeding of *O. sativa* varieties has increased the grain production of individual plants by 70% since 1965 [4]. However, these increases in productivity have been restricted by the comparatively low 'leaf-level' photosynthetic capacity of *O. sativa* relative to other C3 crops [5,6]. Phenotypic screening of the photosynthetic and gas-exchange characteristics of *O. sativa* varieties under water deficit conditions may permit the identification of attributes that confer both drought tolerance and high productivity [7,8].

Yield in varieties of *O. sativa* are closely related to photosynthesis under both well-watered and drought conditions [7,9]. The uptake of CO_2 for photosynthesis involves two major resistance steps at the stomata and at the surface of the mesophyll layer [10,11]. Stomatal (g_s) and mesophyll conductance (g_m) to CO_2 often decrease in unison in response to environmental stresses [7,12–14]. Upland and lowland *O. sativa* cultivars exposed to a soil moisture gradient ranging from high water-availability to severe water-deficit exhibited identical reductions in both g_s and g_m that correlated with the rate of photosynthesis (*A*) and yield. Most notably, while there was no observable difference in *A*, g_s or g_m response to drought between the upland and the lowland varieties, those cultivars that exhibited the highest values of total conductance to CO_2 (g_t: where $g_t = [g_s * g_m]/[g_s + g_m]$) supported higher *A* and yield under all levels of water availability [7]. Those *O. sativa* varieties with higher capacities for CO_2 uptake are therefore the most likely to offer the greatest potential improvement of leaf-level photosynthesis towards achieving enhanced yields [9,15,16]. However, an increase in the capacity for leaf gas exchange may be accompanied by reduced drought tolerance [17,18].

These differences in the photosynthetic and leaf gas-exchange capacities of *O. sativa* varieties have been ascribed to genetic variation [9]. The quantitative trait loci responsible for leaf gas-exchange and photosynthetic parameters in *O. sativa* tend to be found in adjacent regions of the plant genome; indicative of the exertion of selective pressures favouring coordination and high heritability of photosynthetic behaviour [19,20]. Therefore, effective screening and analysis of the photosynthetic and gas exchange characteristics of *O. sativa* varieties under stress conditions has the potential to identify functional traits associated with optimisation of water use efficiency (*WUE*) and the maintenance of yield and A under drought [21]. Rates of photosynthesis are not only determined by diffusive CO_2-uptake, but also by the activity of ribulose-1,5-bisphosphate (RubisCO) [22]. However, exposure of thirteen hydroponically grown *O. sativa* varieties to physiological drought did not suggest that the capacity of RubisCO carboxylation (Vc_{max}) or the maximum rate of electron transport required for ribulose-1,5-bisphosphate regeneration (J_{max}) were associated with A or grain yield. Rather, A and yield were largely determined by g_s and g_m setting the limits for diffusional CO_2 uptake [9], consistent with field based studies [7]. Furthermore, the intrinsic water-use efficiency (A/g_s) of the *O. sativa* varieties was closely related to the ratio of g_m to g_s [9], and this attribute showed the greatest variability between cultivars, and thus potential in terms of possible improvements in yield and growth through maximising *WUE* [5,9,21]. In theory, those varieties with high g_m to g_s ratios should be most suited to maintaining photosynthesis and yield under drought stress conditions, while high values of both g_m and g_s should be favourable for varieties grown under constant high water availability [21].

The observation that A and yield of *O. sativa* closely correlate to g_m [7,9,16] has led to a focus on the interface between the leaf mesophyll cells and internal sub-stomatal air space in the improvement of rice productivity [23]. Adachi et al. [16] crossed two varieties of *O. sativa* to produce a new cultivar with higher values of g_m, which were associated with modifications of the mesophyll structure that increased the surface area available for CO_2 uptake; thus demonstrating the potential for enhanced productivity through increased g_m. In addition to morphological adaptations [15], differences in g_m between *O. sativa* varieties are also likely affected by biochemical factors [23] such as the abundance of cooporin proteins responsible for active transport of CO_2 [24,25] (Kaldenhoff, 2012). Furthermore, the abundance and activity of these proteins has been observed to decline under water-stress [26]. Exposure to drought is associated with declines in values of A, g_s and g_m [11,27]. Sustained impairment of A following the alleviation of drought stress is commonly related with continued biochemical constraints to the uptake and assimilation of CO_2 [28]. However, experimental evidence suggests that *O. sativa* does not display prolonged reductions in A following the cessation of drought, and that the observed constraints to A under water-stress are largely physical impediments to CO_2 diffusion [7,9,29,30]. This suggests that artificial selection of those *O. sativa* varieties with high g_m would enhance A while not negatively affecting drought tolerance or recovery.

Rice production is sensitive to water deficit at different developmental stages of plant growth. For example, early drought stress can affect leaf expansion [3], and later drought stress during flowering or grain-filling stages can disrupt grain formation and development [31,32]. The photosynthetic and yield responses of seven *O. sativa* varieties (five upland and two lowland varieties chosen to represent genetic variability) were analysed in response to sustained drought. As different *O. sativa* varieties have different phenologies (for example Van forms flower inflorescences relatively earlier than Moro), through the use of a drip irrigation system it was possible to apply the stress at the same developmental stage for each variety. This has the advantage over a previous study into the drought tolerance of *O. sativa* varieties conducted at the IRRI in that all varieties experienced drought at the same phenological stage, therefore eliminating a crucial source of error whereby early flowering varieties outperformed late developing cultivars [cf. 7]. Mesophyll conductance to CO_2 is of key interest to the productivity and drought tolerance of *O. sativa* [7,16], however it is not possible to directly measure g_m, and due to the uncertainties associated with its calculation [33] (particularly in the use of the variable J method under stress conditions) it is preferable to simultaneously use two independent methodologies in the estimation of g_m [34]. We therefore employed two independent methodologies in the estimation of g_m: firstly, based upon the synchronous measurement of leaf gas exchange and chlorophyll fluorescence parameters (variable J), and secondly, the analysis of carbon isotope discrimination ($\Delta^{13}C$) in recently formed metabolites to assess their effectiveness in an analysis of *O. sativa* drought tolerance in the field. Further details of the variable J and '$\Delta^{13}C$ of recently synthesised sugars' methods of g_m analysis are presented in the Material and Methods section. This study aimed to: 1) assess the effect of g_s and g_m on A and yield in well-watered and drought conditions; 2) analyse differences in g_m calculation between the 'variable J' [35] and '$\Delta^{13}C$ of recently synthesised sugars' [15] methods, and; 3) identify *O. sativa* varieties that exhibit drought tolerance and/or high productivity, and may be suitable for further genomic/biochemical development to further enhance grain production.

Materials and Methods

Plant material and growth conditions

The experiment was conducted at the International Rice Research Institute Experimental Station at Los Banos, Philippines (14° 11′N 121° 15′E, 21 m above sea level) during the dry season (January to May). Rainfall was minimal, however, temperature and leaf to air vapour pressure difference (*VPD*) increased during the experimental period. The soil at the experimental site is a neutral Andaqueptic Haplaquol (pH = 6.5). Seven *O. sativa* cultivars were studied: Apo, IR55419-04 (IR55), IR64, IR71525-19-1-1 (IR71), Moroberekan (Moro), PSBRc80 (PS80) and Vandana (Van). Moroberekan failed to set or produce grains and seemed to suffer from disease and heat sensitivity, particularly under drought conditions and it was not possible to record physiological measurements for this variety under water deficit stress conditions. Seeds were sown in dry soil at a rate of 80 kg ha^{-1} in rows spaced ~25 cm apart. Fertilisation with nitrogen was conducted on two occasions at application rates of 90 and 120 kg ha^{-1}, alongside fertilisation of 30 kg ha^{-1} of phosphate and potassium. Weeds were controlled by the application of standard agricultural herbicides and manual picking. Pesticides were used to control insect pests; in particular stem borers.

Experimental plots for each *O. sativa* variety measured 3 m×2 m, with adjacent irrigated and drought-stressed treatments. The plots were arranged in a randomized block design. A sprinkler system was initially used to provide equal irrigation levels to all plants until flower (panicle) development, when the water deficit treatment was instigated. During the drought stress period, the well-watered treatments were irrigated by use of a drip irrigation system, with soil water potential maintained above -0.04 MPa (monitored using tensiometers placed at soil depths of 15 and 30 cm) – the level at which the more sensitive rice cultivars begin to exhibit a reduction in transpiration [36].

Leaf gas exchange measurements

Leaf gas-exchange and fluorescence of plants grown under both well-watered treatment or stress conditions were simultaneously measured with a LI-6400-40 leaf chamber fluorometer (Li-Cor, Inc., Nebraska, USA) on flag leaves enclosed in a temperature, light and humidity controlled cuvette. The measurements were made *in situ* between 11.00 and 15.00, in saturating PPFD (1400 μmol $m^{-2}s^{-1}$), with relative humidity ranging between 45–55%, and a leaf temperature of 30°C. Instantaneous measurements of steady-state A, g_s, internal [CO_2] (C_i) and the quantum yield of PSII in the light (ΔF/Fm') were made on six to 14 plants per treatment. Measurements of dark respiration (R_d) were also made at ambient CO_2 concentration in the dark on the same leaves. Instantaneous transpiration efficiency was calculated as the ratio of A to g_s.

Carbon isotope discrimination

Three of the same flag leaves used for *in situ* gas exchange measurements were collected in the evening, oven dried at 70°C until their weight remained constant and were then stored until soluble sugar extraction, following the protocol of Richter et al. [37]. Leaves were finely ground in liquid nitrogen and shaken for 60 min in water at room temperature. After centrifugation (15 min at 5000 g), the supernatant was sequentially mixed with cationic (Dowex-50) and anionic (Dowex-1) exchange resins. The residual solution of purified soluble sugars was freeze-dried and carbon isotope composition ($\delta^{13}C$) determined using a mass-spectrometer. Additional isotope ratio mass spectrometry analyses were performed on the pellet samples that remained after the extraction of soluble sugars. Samples of approximately 1 mg were used for $\delta^{13}C$ analysis. Stable isotope ratios were determined using a continuous-flow triple-collector isotope ratio mass spectrometer (ISOPRIME, GV, Manchester, UK). Solid samples were quantitatively combusted in an elemental analyser (Model NA 1500, Carlo Erba, Milan, Italy) and CO_2 was transferred in helium flow to the mass spectrometer. Calculations of carbon isotope discrimination ($\Delta^{13}C$) of solid samples were undertaken following the protocol of Farquhar et al. [38], assuming the carbon isotopic composition of CO_2 in air (δ_{air}) to be −8.0‰.

Estimation of mesophyll conductance

Mesophyll conductance to CO_2 diffusion is the inverse of the total resistance encountered in the transport of CO_2 across the leaf mesophyll. Estimation of g_m is subject to a number of assumptions and sources of error, and it is preferable to employ two independent approaches in its calculation [34,39]. In this study g_m was calculated using the 'variable J' and '$\Delta^{13}C$ of recently synthesised sugars' methods. The calculation of g_m using the variable J method is based on simultaneous measurements of gas-exchange and chlorophyll fluorescence parameters as described by Harley et al. [35] and Loreto et al. [40] (equations 1 and 2):

$$g_m = \frac{A}{Ci - \frac{\Gamma * [J + 8 * (A + R_d)]}{J - 4 * (A + R_d)}} \quad (1)$$

where the electron transport rate (J_F) is calculated from fluorescence [41]:

$$J_F = PPFD * \frac{\Delta F}{F_m} * \alpha * \beta \quad (2)$$

.

Where F_m is the fluorescence maximum and the partitioning factor (β) between photosystems I and II was considered to be 0.5 and leaf absorbance (α) was measured directly [42]. The variable J method is sensitive to the estimation of the CO_2 compensation point to photorespiration (Γ^*) and R_d. For the results of this study to be directly comparable with those of Centritto et al. [7], the value of Γ^* used in the gas exchange algorithm was calculated using the Rubisco specific factor for an annual herb [43], consistent with previously published *O. sativa* Γ^* values [44]. As measurements were conducted under field conditions it was not feasible to produce robust measurements of respiration in the light (R_l); respiration in the dark was therefore used in the calculation of gm, as R_l and R_d have been observed to be broadly consistent when recycling of CO_2 is taken into consideration [45]. Measurements of R_d were performed at ambient [CO_2] concentration in the dark on the same leaves by switching off the light in the leaf cuvette; when CO_2 release from the leaf had become stable for approximately five to ten minutes this was recorded and considered to represent R_d. To calibrate the values of J calculated by fluorescence (J_F) and gas-exchange (J_A), measurements were also conducted under non-photrespiratory conditions using 1% [O_2] [35,40]. Electron transport rate and R_d values are presented in Figures S1 and S2. Total conductance to CO_2 (g_t) was calculated as: $g_t = g_s * g_m/(g_s + g_m)$. The concentration of CO_2 at the chloroplast site (C_c) was then calculated from mesophyll conductance values [46].

The method described by Lauteri et al. [47] for 'on-line' determinations was adapted to calculate g_m by using the $\Delta^{13}C$ of recently synthesized sugars [48]. Discrepancies between $\Delta^{13}C$ of leaf soluble carbohydrates (Δ_{obs}) and $\Delta^{13}C$ expected on the basis of gas-exchange measurements (Δ_{exp}), allow the estimation of g_m utilising the approach of Evans et al. [49] that was subsequently modified by Lloyd et al. [50]:

$$g_m = \frac{(b - b_s - \alpha_1) * {}^{A}/_{C_a}}{(\Delta_{exp} - \Delta_{obs}) - (f\,\Gamma * / C_a)} \quad (3)$$

Where: b is the discrimination associated with carboxylation reactions, taken to be 27.5‰; b_s is the fractionation occurring when CO_2 enters solutions (1.1‰ at 25°C); a_1 is the fractionation during diffusion in water (0.7‰); f is the fractionation associated with photorespiration, and; p_a is the partial pressure of CO_2 in air. Two g_m calculations were performed and compared, since two values were taken into account for the fractionation factor f: 7‰ [50] and 0‰ [48,51]. The closest relationship between g_m estimates ($R^2 = 0.911$) using the variable J and $\Delta^{13}C$ of recently synthesized sugars methods in all varieties was found when using a f value of 0‰.

Harvesting

Plants were harvested at physiological maturity by sampling 1 m of the 2 central rows of each plot. Data collected included biomass, plant height, anthesis date, tiller number, straw production, grain yield, and harvest index (HI).

Statistical analysis

Analyses of variance were conducted using SAS to generate least squares means for each entry. Leaf gas exchange data were tested using a simple factorial ANOVA (three-way maximum interactions) and, where appropriate, the treatment means of leaf properties and gas exchange parameters were compared using a Tukey *post-hoc* test.

Results

Drought stress induced reductions in yield, *A* and rates of gas-exchange across all the *O. sativa* varieties analysed in this study. The duration of time required for 50% of plants to develop flowers (flower days) did not correlate with yield in well-watered plants, but those varieties with a shorter flowering day period exhibited enhanced yield under drought stress (Fig. 1a). Varieties with shorter flowering days allocated a higher proportion of their total biomass to yield (harvest index: *HI*) under control and drought treatments (Fig. 1b). Positive allometric correlations between plant height and both yield (Fig. 1c) and *HI* (Fig. 1d) were observed, both under watered or drought conditions.

The carbon isotopic composition of the leaves of the *O. sativa* varieties became increasingly enriched in the heavier ^{13}C isotope following water-stress, with the exception of IR64. Changes in the carbon isotopic composition of soluble sugars induced by the drought treatment were less consistent, with three varieties exhibiting increases and four decreases in the ratio of stable carbon isotopes (Table 1). Despite the lack of *genotype x water treatment* interaction, the varieties yielded differences in the 'long-term' isotopic signal of the bulk leaf material; some genotypes showed wide variation (up to −1.5‰ in pellets of Van) while others were more negligible (e.g. 0.2‰ in pellets of IR64). Drought caused also different variation in $\Delta^{13}C$ of soluble sugars: up to −3.5‰ in PS80 but 0.2‰ in IR71. Under both well-watered and rainfed conditions, soluble sugars showed generally lower values of carbon isotopic discrimination when compared with the average $\Delta^{13}C$ values of the flag leaf pellets. Significant differences were observed in the carbon isotopic discrimination values of of leaf pellets between varieties, however this was not replicated in the analysis of soluble sugars (Table 1).

A strong positive correlation was observed between the two estimates of g_m in the *O. sativa* varieties grown under control and drought conditions ($R^2 = 0.930$; $P<0.01$) (Fig. 2), excluding IR64 and PS80. Indeed, the lowland varieties IR64 and PS80 grown under well-watered conditions exhibited high values of g_m derived from the application of the isotopic methodology. It was therefore decided to utilise g_m measurements resulting from the variable J method in the subsequent analyses [an extended discussion of the contrasting merits of the approaches is given in 7]. Rates of *A*, g_s and g_m all showed consistent declines under water-stress conditions (Fig. 3). IR55 showed the highest yield and g_m under well-watered conditions. Variety IR71 exhibited the highest general

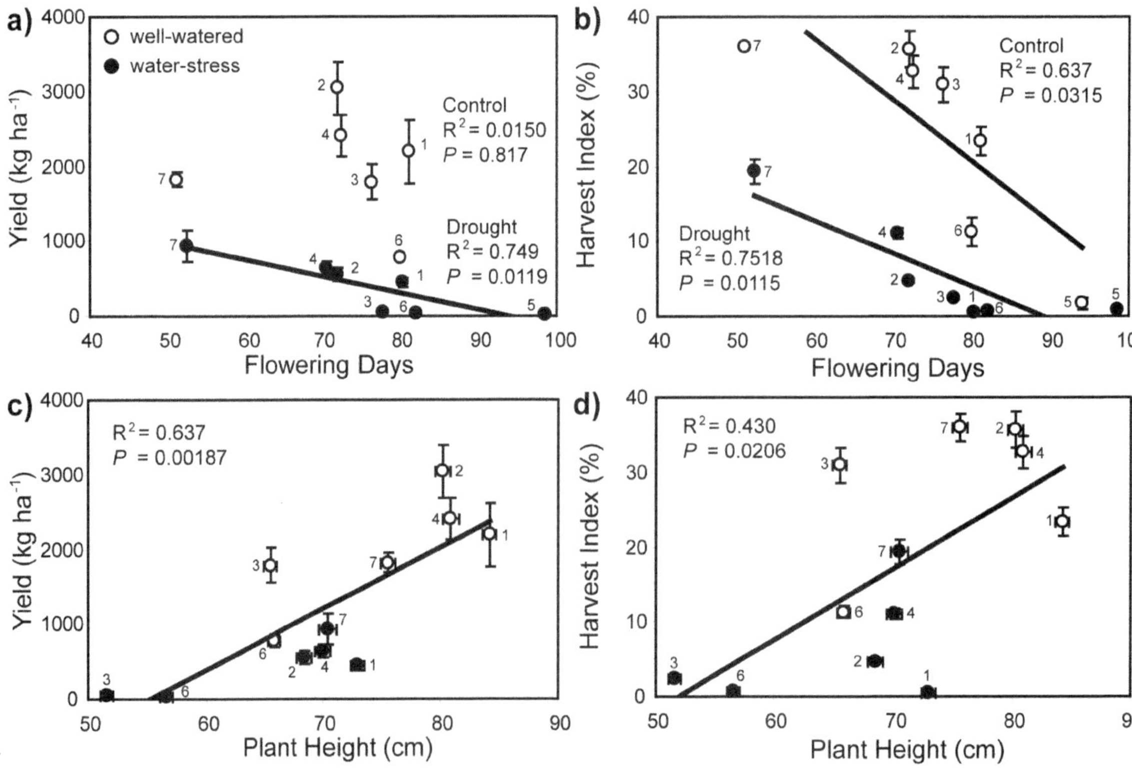

Figure 1. The performance of seven rice varieties under well-watered (open symbols) and drought conditions (closed symbols): a) relationship between yield and flowering days (period of time for 50% of plants to develop flowers) under full-water (linear regression: $R^2 = 0.0150$; $F_{1,4} = 0.0609$; $P = 0.817$) and drought (linear regression: $R^2 = 0.749$; $F_{1,5} = 14.905$; $P = 0.0119$); b) allometric relationship between yield and plant height (linear regression: $R^2 = 0.637$; $F_{1,10} = 17.536$; $P = 0.00187$); c) relationship between harvest index (*HI*: dry weight of grain relative to dry total plant biomass) and flowering days under full-water (linear regression: $R^2 = 0.637$; $F_{1,5} = 8.760$; $P = 0.0315$) and drought (linear regression: $R^2 = 0.752$; $F_{1,5} = 15.144$; $P = 0.0115$), and; d) relationship between *HI* and plant height (linear regression: $R^2 = 0.430$; $F_{1,10} = 7.548$; $P = 0.0206$). Error bars indicate one standard error either side of the mean. Numbers next to data points indicate *Oryza sativa* variety: 1 = Apo; 2 = IR55; 3 = IR64; 4 = IR71; 5 = Moro; 6 = PSBRc80; 7 = Van.

Table 1. Carbon isotope discrimination (‰) of the flag leaf pellet (bulk) and soluble sugars of seven rice cultivars grown under drought and well-watered conditions in ‰.

Genotype	Habitat	Bulk Leaf $\Delta^{13}C$ (‰)			Sugars $\Delta^{13}C$ (‰)		
		Well-watered	Drought	Mean	Well-watered	Drought	Mean
Apo	U	21.6±0.9	20.6±0.3	21.0 ab	18.3±1.3	18.7±0.6	18.5 a
IR55	U	21.4±0.5	20.8±0.0	21.1 a	18.2±0.5	16.2±0.9	17.2 a
IR64	L	20.9±0.2	21.1±0.5	21.0 ab	18.5±0.6	16.6±0.8	17.5 a
IR71	U	21.1±0.2	20.7±0.5	20.9 ab	17.3±1.3	17.5±0.7	17.4 a
Moro	U	19.9±0.4	19.7±0.2	19.8 b	18.3±0.6	19.5±0.2	18.9 a
PS80	L	21.4±0.4	21.2±0.4	21.2 a	19.3±0.3	15.8±0.2	17.2 a
Van	U	21.5±0.3	20.0±0.5	20.7 b	18.1±1.5	15.8±0.4	17.0 a
Mean		21.1 a	20.6 b		18.2 a	17.2 b	

Values indicate the mean of six plants; ± indicates the standard error of mean; different letters indicate significant differences (P≤0.05) among means according to an ANOVA.

values of A, g_s and g_m under both well-watered and drought conditions, while Apo and PS80 showed the lowest values of these photosynthesis and gas-exchange parameters under control and water-stress. The A/g_s ratio of the *O. sativa* varieties either increased (IR71; Van; IR55) or remained constant under drought stress in all varieties with the exception of PS80, where a 42.6% reduction was observed (Fig. 3d). The ratio of [CO_2] within the substomatal pore (C_i) to the external atmospheric [CO_2] (C_a) was largely unaffected by water-stress (Fig. 4a). In contrast, the ratio of the [CO_2] within the chloroplast envelope (C_c) to C_a was diminished in all *O. sativa* varieties during drought treatment (Fig. 4b), indicative of transport limitations of CO_2 in the mesophyll for photosynthesis. *Oryza sativa* variety IR71 showed the smallest proportional reduction of 16.8% in the C_c/C_a ratio, while Apo exhibited the largest reduction of 33.8%.

Yield and A exhibited significant positive correlations with g_s (Fig. 5a and 5b). However, the positive correlations between yield and A with g_m (Fig. 5c and 5d) were stronger than those observed with g_s or total conductance (g_{tot}); suggesting that g_m is the major determinant of yield and A under both well-watered and water-stressed conditions. The yield of the *O. sativa* varieties was strongly related to A under both control and drought treatments (Fig. 5g). In order to gauge the response of these diffusive limitations to CO_2 transport and their effect on yield and A, the proportional change (Δ) in response to drought was calculated. The proportional reduction in yield (Δyield) was not related to Δg_s, Δg_m or Δg_{tot} (Fig. 6a; 6b; 6c). However, ΔA was significantly correlated to Δg_m and Δg_{tot}. IR71 exhibited the highest levels of A and g_m (Fig. 3a and 3c), but also the lowest proportional reduction in g_m and A (Fig. 6d). The highest yields observed under control conditions were found in IR55 (3039.6±359.1 kg ha^{-1}) and IR71 (2407.4±284.4 kg ha^{-1}); under water-stress the yield of IR55 fell to 552.7±80.3 kg ha^{-1} and IR71 to 640.3±93.0 kg ha^{-1}. The lowest Δyield reduction under drought conditions was observed in variety Van (Fig. 6), which also produced the highest yield under

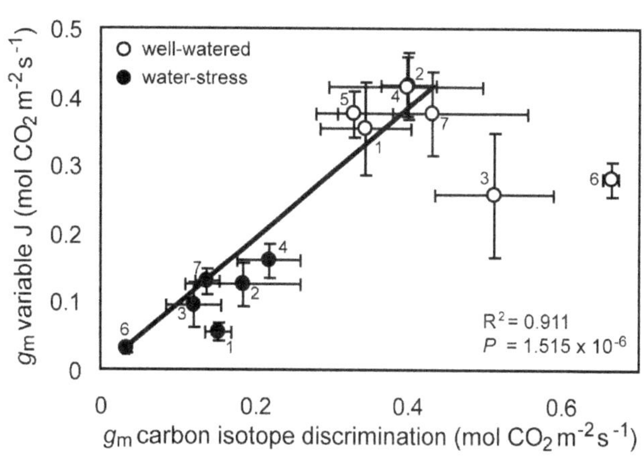

Figure 2. Comparison between the estimates of mesophyll conductance of CO_2 (g_m) obtained by applying two independent methods: the variable *J* method and the 'δ13C of recently synthesised sugars' method (linear regression: R^2=0.932; $F_{1,9}$=122.750; P=1.515×10^{-6}). Each data point represents the average value of three observations based upon the Δ13C of recently synthesised sugars and six to fourteen gas-exchange measurements utilising the variable J method. Error bars as in Figure 1. The regression line excludes the two data points on the right of the graph (IR64 and PS80) with anomalously high g_m derived from the δ13C of recently synthesised sugars. Numbers next to data points indicate *Oryza sativa* variety as in Figure 1.

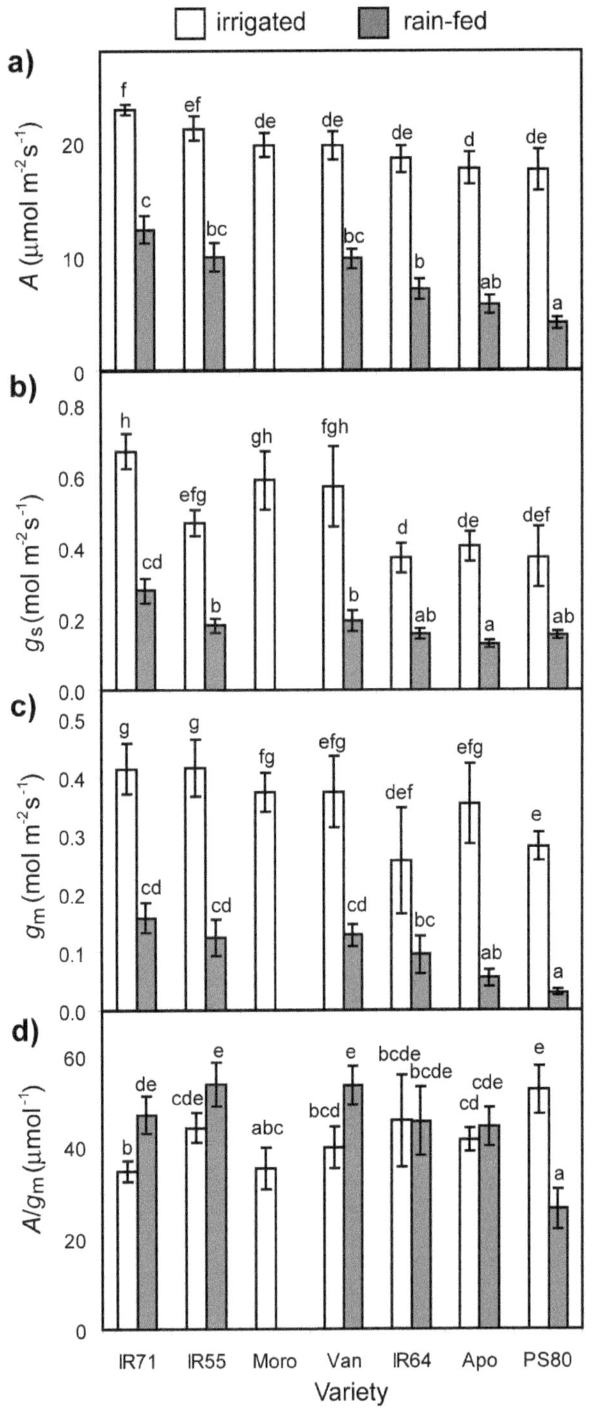

Figure 3. Measurements of (a) photosynthesis rate (A), (b) stomatal conductance (g_s), (c) mesophyll conductance (g_m), and (d) intrinsic transpiration efficiency (A/g_s) in control and water-stressed leaves of the seven *Oryza sativa* genotypes. The measurements were made on the flag leaf in saturating PPFD (1400 µmol $m^{-2}s^{-1}$), with relative humidity ranging between 45–55%, and a leaf temperature of 30°C. Data are means of 4 to 7 plants per treatment. Error bars as in Figure 1. Different letters denote significant differences among means derived using a factorial ANOVA and Tukey *post-hoc* test.

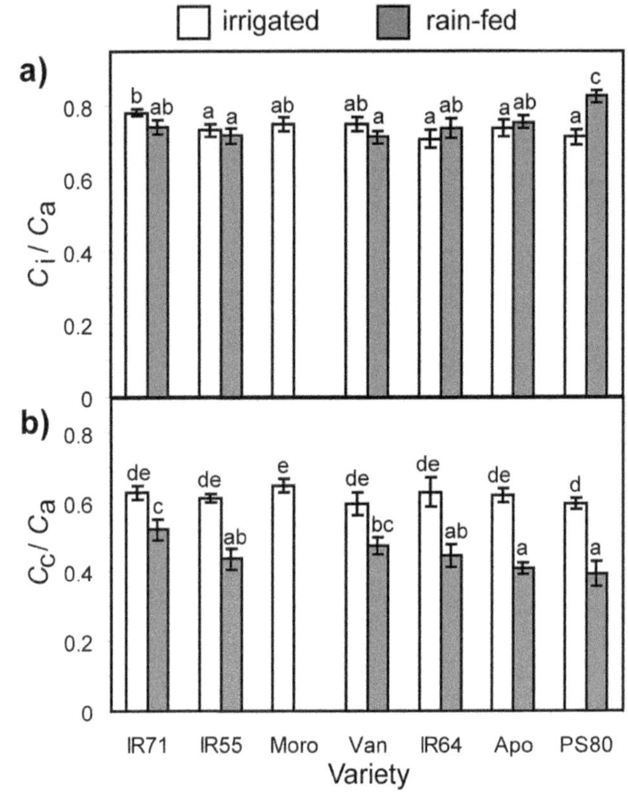

Figure 4. Measurements of (a) the intercellular [CO_2] (C_i) to the ambient [CO_2] (C_a) ratio (C_i/C_a), and (b) the chloroplastic [CO_2] (C_c) to the ambient [CO_2] ratio (C_c/C_a) in control and water-stressed leaves of the seven *Oryza sativa* genotypes. The measurements were made on the flag leaf in saturating PPFD (1400 µmol $m^{-2}s^{-1}$), with relative humidity ranging between 45–55%, and a leaf temperature of 30°C. Data are means of 4 to 7 plants per treatment. Error bars as in Figure 1. Different letters denote significant differences among means derived using a factorial ANOVA and Tukey *post-hoc* test.

water-stress (935.3±206.0 kg ha^{-1}). IR71 exhibited the second lowest Δyield and produced second largest yield during water-stress of the varieties analysed (Fig. 6g). The A/g_s ratio was not significantly related to either yield or *HI* under both well-watered or drought conditions (Fig. 7a and 7c). However, the g_m:g_s ratio was significantly correlated with both yield and *HI* (Fig. 7b and 7d), suggesting that those *O. sativa* varieties with greater g_m/g_s ratio will exhibit enhanced productivity under both well-watered and water-stressed conditions.

Discussion

The seven *O. sativa* varieties exhibited a diverse range of morphological, isotopic, photosynthetic and leaf gas-exchange responses to drought stress. A number of factors determined the yield of the *O. sativa* cultivars analysed in this study. Despite water deficit being instigated at the same developmental stage for all varieties, phenology had an effect on yield under drought as rapidly developing plants maintained yield to a greater extent than slower developers (Fig. 1a) [52]. Despite the type and phenolgical timing of the drought stress being different to that utilised in the investigation of Centritto et al. [7], the effect on A and yield observed in the two studies are broadly consistent. Those plants that were able to accumulate biomass rapidly produced higher

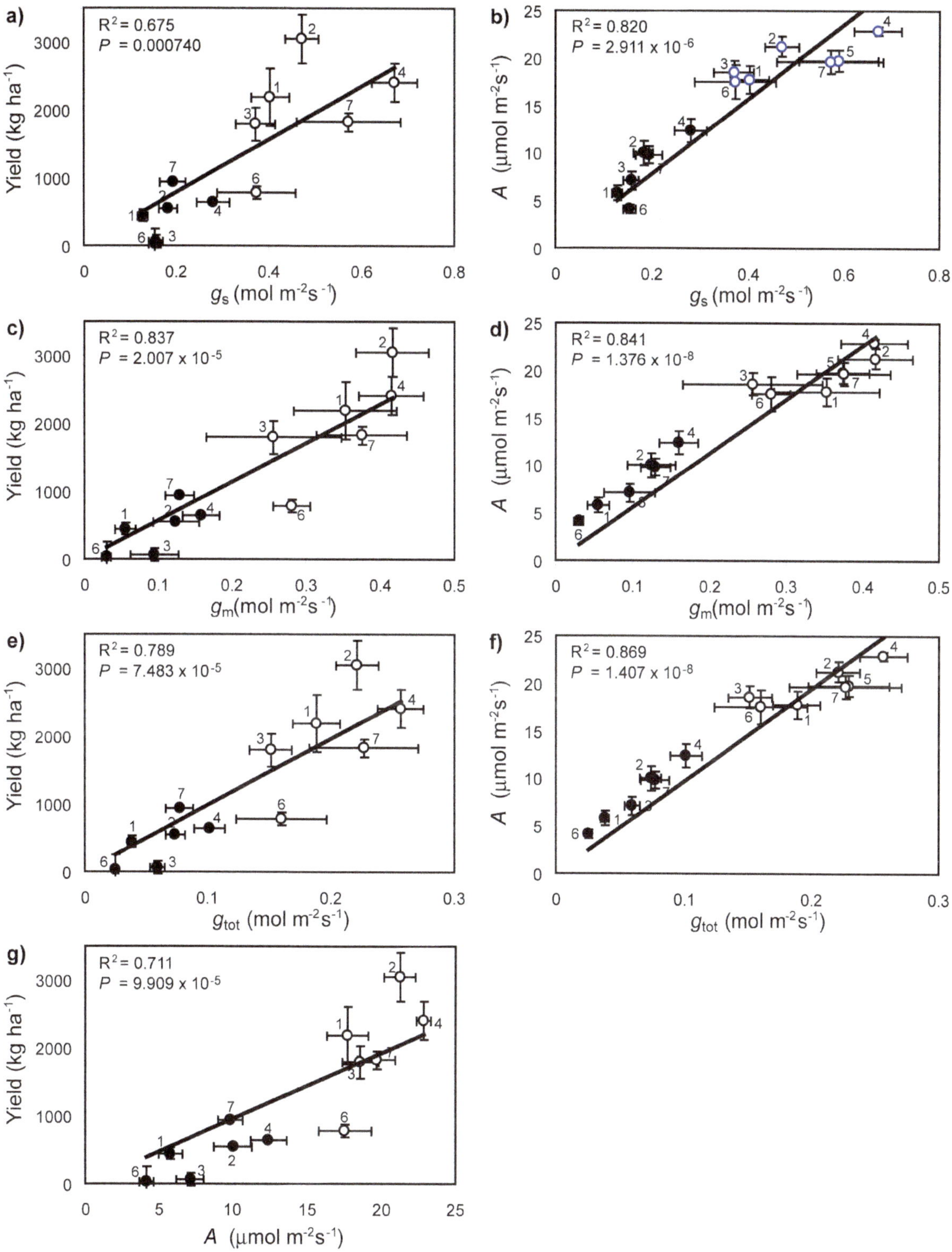

Figure 5. Interaction of diffusive conductance parameters to CO_2 uptake with yield and photosynthesis (*A*) under well-watered (open symbols) and drought conditions (closed symbols): a) relationship between yield and stomatal conductance (g_s) (linear regression: $R^2 = 0.696$; $F_{1,10} = 22.900$; $P = 0.000740$); b) relationship between *A* and g_s (linear regression: $R^2 = 0.873$; $F_{1,11} = 75.721$; $P = 2.911 \times 10^{-6}$); c) relationship between yield and mesophyll conductance (g_m) (linear regression: $R^2 = 0.850$; $F_{1,10} = 56.611$; $P = 2.911 \times 10^{-5}$); d) relationship between *A* and g_m (linear regression: $R^2 = 0.952$; $F_{1,11} = 217.071$; $P = 1.376 \times 10^{-8}$); e) relationship between yield and total conductance (g_{tot}) (linear regression: $R^2 = 0.806$; $F_{1,10} = 41.418$; $P = 7.483 \times 10^{-5}$); f) relationship between *A* and g_{tot} (linear regression: $R^2 = 0.952$; $F_{1,11} = 216.173$; $P = 1.407 \times 10^{-8}$), and; g) relationship between yield and *A* (linear regression: $R^2 = 0.795$; $F_{1,10} = 38.665$; $P = 9.909 \times 10^{-5}$). Error bars as in Figure 1. Numbers next to data points indicate *Oryza sativa* variety as in Figure 1.

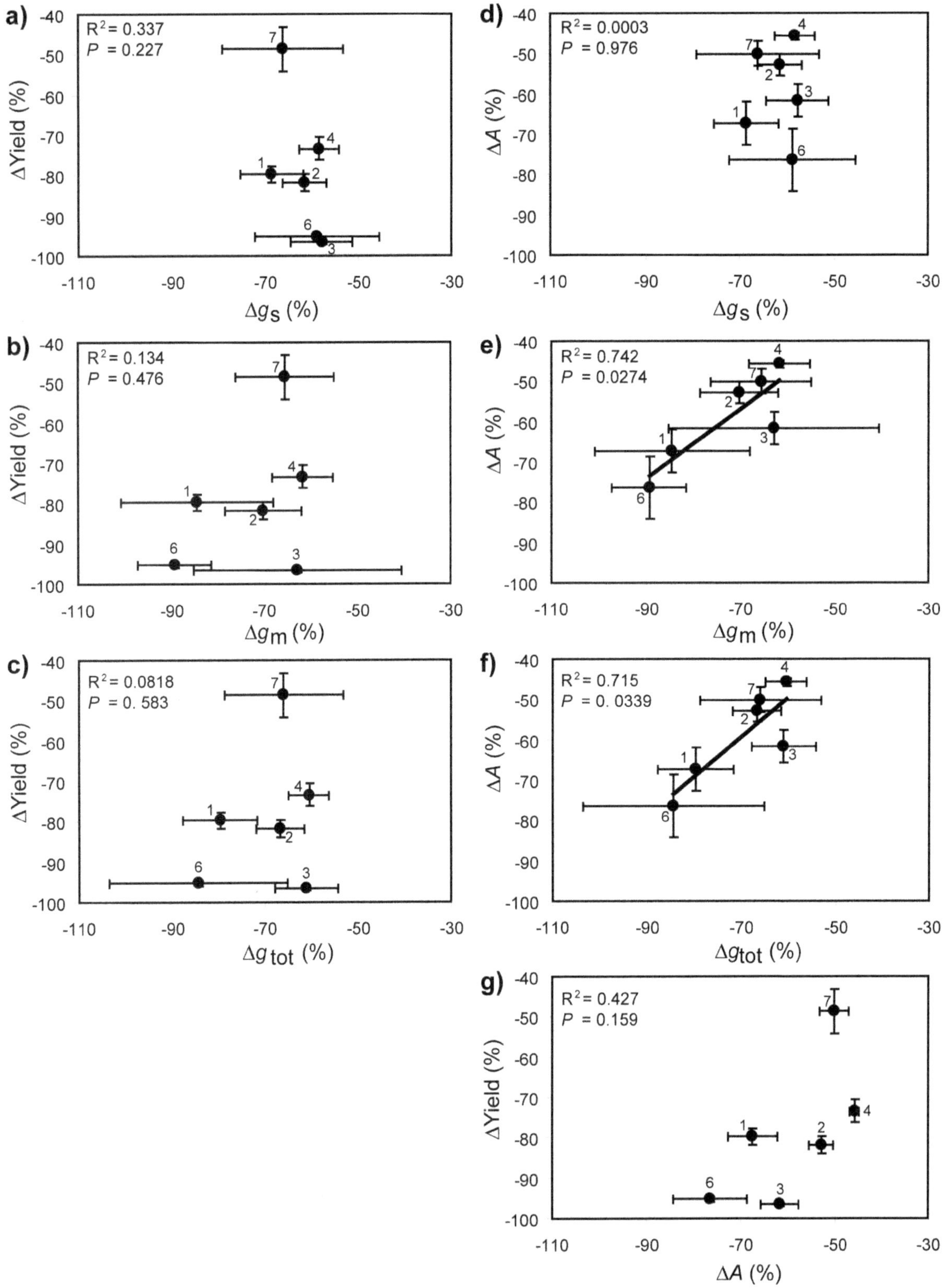

Figure 6. Changes in yield and photosynthesis in relation to modification of diffusive resistances to CO_2 uptake following water-stress. Those varieties that experienced smaller reductions in parameters were more tolerant of drought. a) relationship between Δyield and Δg_s (linear regression: $R^2=0.337$; $F_{1,4}=2.032$; $P=0.227$); b) relationship between Δyield and Δg_m (linear regression: $R^2=0.134$; $F_{1,4}=0.618$; $P=0.476$); c) relationship between Δyield and Δg_{tot} (linear regression: $R^2=0.0818$; $F_{1,4}=0.356$; $P=0.583$); d) relationship between ΔA and Δg_s (linear regression: $R^2=0.0003$; $F_{1,4}=0.00106$; $P=0.976$); e) relationship between ΔA and Δg_m (linear regression: $R^2=0.742$; $F_{1,4}=11.527$; $P=0.0274$); f) relationship between ΔA and Δg_{tot} (linear regression: $R^2=0.715$; $F_{1,4}=10.042$; $P=0.0339$), and; g) relationship between Δyield and ΔA (linear regression: $R^2=0.427$; $F_{1,4}=2.979$; $P=0.159$). Error bars as in Figure 1. Numbers next to data points indicate *Oryza sativa* variety as in Figure 1.

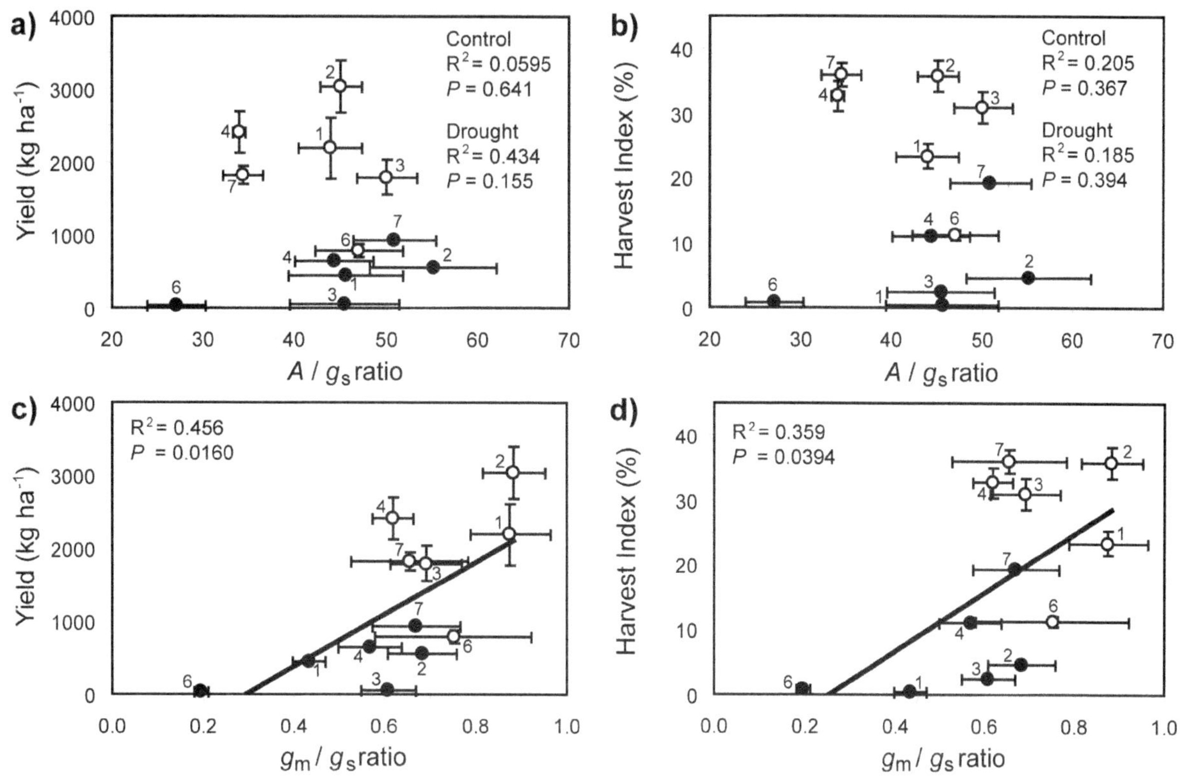

Figure 7. Interaction of yield and *A* with transpiration efficiency (A/g_s) and the ratio of g_m to g_s in well-watered (open symbols) and drought conditions (closed symbols): a) relationship between yield and A/g_s under full (linear regression: $R^2=0.0595$; $F_{1,4}=0.253$; $P=0.641$) and water-stressed (linear regression: $R^2=0.434$; $F_{1,4}=3.072$; $P=0.155$) conditions; b) relationship between harvest index (*HI*) and A/g_s under full (linear regression: $R^2=0.205$; $F_{1,4}=1.032$; $P=0.367$) and water-stressed (linear regression: $R^2=0.185$; $F_{1,4}=0.909$; $P=0.394$) conditions; c) relationship between yield and g_m:g_s (linear regression: $R^2=0.456$; $F_{1,10}=8.379$; $P=0.0160$), and; d) relationship between *HI* and g_m:g_s (linear regression: $R^2=0.359$; $F_{1,10}=5.610$; $P=0.0394$). Error bars as in Figure 1. Numbers next to data points indicate *Oryza sativa* variety as in Figure 1.

grain yields than those that accumulated biomass less efficiently under both well-watered and drought conditions. This may be due to allometric relationships between biomass and yield [53], or the positive influence of increased photosynthetic area on 'whole-plant' carbon uptake [54]. Leaf-level photosynthetic rates were also closely related to yield in all varieties under both well-watered and drought conditions (Fig. 5g). Photosynthesis rates were in turn determined by diffusive limitations to CO_2-uptake (Fig. 5b; 5d; 5f). The most significant of these resistances to CO_2-uptake in *O. sativa* occurs at the interface between the internal sub-stomatal air-space and the mesophyll layer [15,55]. Consistent with other studies, g_m in *O. sativa* correlated most closely with *A* and yield (Fig. 5d) [7,9], and represents a critical 'rate-limiting' step in the determination of grain productivity and assessment of the photosynthetic response to drought stress.

The 'bulk' carbon isotope discrimination values of the *O. sativa* varieties decreased following drought treatment; indicative of water-deficits over the life-span of the flag-leaf reducing discrimination against the heavier ^{13}C isotope [56,57]. However, this pattern of reduced carbon isotope discrimination is not apparent in the carbon isotopic composition of recently synthesised sugars (Table 1). These findings may reflect 'shorter-term' variations in *WUE* in the plants experiencing drought stress, a modification of the photosynthetic pathway affecting sugar metabolism [28] or the breakdown of previously stored carbohydrates with a different isotopic signature into more simple sugars over the course of the stress treatment [58,59]. Differences in the diffusional uptake of CO_2 likely affected carbon isotopic composition [60]; however, this was not apparent from our g_s, g_m (variable J) and g_{tot} datasets (Fig. 5), suggesting that other factors affected carbon isotopic discrimination and possibly the effectiveness of the Δ13C of recently synthesised sugars method of estimating g_m in varieties IR64 and PS80 under control conditions (Fig. 2). It is noteworthy that under water-deficit conditions, g_m estimates based on the variable J method closely correlated with those derived from the Δ13C of recently synthesised sugars (Fig. 2). These findings do not support the suggestion that reductions in C_i may affect estimation of the total photosynthetic electron transport rate associated with enhanced photorespiration, thus rendering the variable J method unreliable in the investigation of g_m in plant responses to environmental stress [34].

Selective breeding has improved the productivity of *O. sativa* varieties through improved growth form and the increased proportional allocation of biomass as grain [4]; yet these increases in yield have been constrained by the relatively low leaf-level photosynthetic rate of *O. sativa* in comparison to other C3 crops [5,61]. The results of this study indicate that under well-watered and water-deficit conditions, *A* is an important determinant of yield in *O. sativa* and is largely related to the rate of transport of CO_2 across the mesophyll (Fig. 5d). Further improvements in *O. sativa* grain yield may likely be achieved through the enhancement of g_m to CO_2 inducing enhanced *A* [6,9,16,62]. Those *O. sativa* varieties that showed the greatest photosynthetic performance and yield under control conditions were not necessarily the most effective varieties under drought conditions. Varieties that maintained transport of CO_2 during water-stress sustained A and yield to a greater extent than varieties that exhibited more pronounced diffusive limitations to photosynthetic CO_2-uptake (Fig. 5). Drought stress induced a relatively consistent mean reduction in g_s in the *O. sativa* varieties (Fig. 3), yet the C_i:C_a ratio was unaffected (Fig. 4), indicating that g_s was not limiting to A under water-deficit, or that the gas-exchange data were possibly biased by the occurrence of non-uniform stomatal closure over the leaf lamina and consequent overestimation of C_i [27,63,64]. However, the C_i/C_a ratio in the *O. sativa* varieties was reduced by 16.8 to 33.8% after drought, consistent with the impairment of A through mesophyll diffusive limitations [7,11,12,65,66]. Mesophyll conductance to CO_2 declined during the drought treatment, constraining A and growth (Fig. 5; 6). However, significant variability in the extent of reductions in g_m were observed in the *O. sativa* cultivars ranging from −61.7% (IR71) to −89.3% (PS80). Furthermore, those varieties that showed the smallest reductions in g_m also exhibited greater A; demonstrating that g_m largely determines A in *O. sativa* under differing degrees of water-deficit [7,9]. These differences in the ability to sustain g_m values under drought stress are likely related to the maintenance of biochemical function in the mesophyll such as the activity of cooporin and carbonic anhydrase [eg. 25,67].

The conductance of the mesophyll layer to the transport of CO_2 determined A and yield in the seven *O. sativa* varieties studied. Under conditions of water-stress it has been suggested that a low g_s (to reduce transpirative water-loss) and a high g_m (to maintain the uptake of CO_2) are beneficial [11,68]. However, in the case of *O. sativa*, yield and *HI* were only weakly related to the ratio of g_m:g_s (R^2<0.460) (Fig. 7). The reduction in g_s under water-deficit is consistent with active stomatal control in the optimisation of *WUE* [69,70]; nevertheless, the lack of a strong correlation to the g_m:g_s ratio may be indicative of the lack of a stomatal limitation to A under drought stress in the *O. sativa* varieties apparent in the C_i:C_a ratios (Fig. 4a). The photosynthetic and yield responses of *O. sativa* under well-watered and drought conditions were more dependent upon g_m than g_s (Fig. 5b and 5d). This is consistent with earlier studies reporting that modification and improvement of g_m is more effective in the enhancement of *O. sativa* productivity [16] and likely drought tolerance [7,9]. The strong dependence of A and yield on g_m is also evident in the lack of a relationship between A/g_s and yield or *HI* (Fig. 7a and 7c). This may suggest that stomatal control is either less effective [70,71], or plays a reduced role, in respect to mesophyll limitations to CO_2 in a drought intolerant crop such as *O. sativa* in comparison to crops adapted to more water-limited environments [72].

The disparity in A values observed between the cultivars raises the possibility of crossing different varieties to enhance leaf-level A and yield in subsequent generations [6,8,16]. Selection of drought tolerance on the basis of the ability of a particular genotype to maintain g_m function under drought stress would be viable [11,73]. However, genetic sequencing and identification of quantitative trait loci permit the possibility of extracting genetic material from a *O. sativa* cultivar with a valuable characteristic (such as high g_m and/or stable g_m under water deficit) and in the creation of a new variety with a suite of desired attributes [17,19,74–76]. For example, from the results of this study it would be desirable to combine the high g_m and yield of IR55 under well-watered conditions with the capacity to sustain g_m under water-stress exhibited by IR71 and water-stressed yield of Van (Fig. 3). In this context it may be possible to phenotypically screen a large number of *O. sativa* genotypes to identify quantitative trait loci responsible for g_m values and g_m response to drought. Those varieties developed to be resistant to drought may be suited to more marginal land-types or in scenarios of insufficient water availability. Nonetheless, given the sporadic and unpredictable nature of sustained and persistent drought events it is unlikely that large numbers of farmers would select a drought tolerant variety such as Van over a more productive cultivar such as IR55, and thus limit their harvest to accommodate the risk of interruption to water supplies [73,77]. An ideal *O. sativa* variety would therefore possess both values of g_m capable of sustaining high A and the ability to maintain g_m in the event of water-deficits [8,78]. However, the results of this study suggest that those attributes may be mutually exclusive in the species analysed. Furthermore, no single parameter analysed in this study exclusively accounted for yield under either well-watered or drought conditions; therefore attempts to improve the productivity of *O. sativa* cultivars may also involve selection for a suite of characteristics that improve productivity under specific conditions of water availability. Nonetheless, future efforts to improve *O. sativa* productivity should focus on biochemical [eg. 25] and physical modification [eg. 16] of the interface between the mesophyll layer and sub-stomatal air-space to enhance CO_2 uptake and A.

Conclusions

Physiological analysis of *O. sativa* has great potential in identifying those photosynthetic and leaf gas-exchange characteristics that confer attributes such as high yield and tolerance to drought. The analysis of the seven *O. sativa* varieties indicates that under both well-watered and drought stressed conditions yield and A are determined by diffusional constraints to CO_2 uptake and not by biochemical limitations. The varieties showed significant variation in leaf-level photosynthetic rates, symptomatic of the photosynthetic constraints to *O. sativa* production that have previously been recorded. Selection of those *O. sativa* genotypes that exhibit higher rates of leaf-level A would likely be beneficial for grain yields. Photosynthesis was most closely related to g_m, and the tolerance of the *O. sativa* varieties to drought was largely determined by the ability to maintain g_m values under water-deficits. However, the highest g_m and yield under well-watered conditions (IR55) and the most pronounced ability to sustain g_m under water-deficit (IR71) were observed in different varieties. This may suggest that the desired attributes for enhanced yield during high water availability and the ability to maintain g_m, and thus sustain grain production under water-deficits, may be mutually exclusive. Nonetheless, the results of studies such as this could make it possible through selective breeding, or the identification of a particular genetic expression associated with a desired attribute, to increase leaf-level photosynthetic rates and drought tolerance in the same variety of *O. sativa*. These findings indicate that the interface between the internal sub-stomatal air-space and the mesophyll layer offers the greatest potential in terms of possible modification to enhance the performance of *O. sativa* under optimal and stressed growth conditions in the future.

Acknowledgments

The following are thanked for scientific discussion and technical assistance: L. Spaccino for IRMS analyses (CNR-IBAF) and R. Torres G. Dimayuga and the drought physiology team at IRRI for technical support (IRRI, Philippines). The comments of two anonymous referees significantly improved this manuscript.

Author Contributions

Conceived and designed the experiments: MC ML RS. Performed the experiments: MC ML MCM. Analyzed the data: MH MC. Contributed reagents/materials/analysis tools: MC ML RS. Wrote the paper: MH MC ML.

References

1. Pandey S, Bhandari H, Ding S, Prapertchob P, Sharan R, et al. (2007) Coping with drought in rice farming in Asia: insights from a cross-country comparative study. Agricultural Economics 37: 213–224.
2. Serraj R, McNally KL, Slamet-Loedin I, Kohli A, Haefele SM, et al. (2011) Drought resistance improvement in rice: an integrated genetic and resource management strategy. Plant Production Science 14: 1–14.
3. Parent B, Suard B, Serraj R, Tardieu F (2010) Rice leaf growth and water potential are resilient to evaporative demand and soil water deficit once the effects of root system are neutralized. Plant, Cell & Environment 33: 1256–1267.
4. Peng S, Cassman KG, Virmani SS, Sheehy J, Khush GS (1999) Yield potential trends of tropical rice since the release of IR8 and the challenge of increasing rice yield potential. Crop Science 39: 1552–1559.
5. Hubbart S, Peng S, Horton P, Chen Y, Murchie EH (2007) Trends in leaf photosynthesis in historical rice varieties developed in the Philippines since 1966. Journal of Experimental Botany 58: 3429–3438.
6. Gu J, Yin X, Stomph TJ, Struik PC (2014) Can exploiting natural genetic variation in leaf photosynthesis contribute to increasing rice productivity? A simulation analysis. Plant, cell & environment 37: 22–34.
7. Centritto M, Lauteri M, Monteverdi MC, Serraj R (2009) Leaf gas exchange, carbon isotope discrimination, and grain yield in contrasting rice genotypes subjected to water deficits during the reproductive stage. Journal of Experimental Botany 60: 2325–2339.
8. Long SP (2014) We need winners in the race to increase photosynthesis in rice, whether from conventional breeding, biotechnology or both. Plant, Cell & Environment 37: 19–21.
9. Gu J, Yin X, Stomph T-J, Wang H, Struik PC (2012) Physiological basis of genetic variation in leaf photosynthesis among rice (*Oryza sativa* L.) introgression lines under drought and well-watered conditions. Journal of Experimental Botany 63: 5137–5153.
10. Niinemets Ü, Diaz-Espejo A, Flexas J, Galmes J, Warren CR (2009) Role of mesophyll diffusion conductance in constraining potential photosynthetic productivity in the field. Journal of Experimental Botany 60: 2249–2270.
11. Flexas J, Ribas-Carbó M, Diaz-Espejo A, Galmēs J, Medrano H (2008) Mesophyll conductance to CO_2: current knowledge and future prospects. Plant, Cell & Environment 31: 602–621.
12. Centritto M, Loreto F, Chartzoulakis K (2003) The use of low [CO_2] to estimate diffusional and non-diffusional limitations of photosynthetic capacity of salt-stressed olive saplings. Plant, Cell & Environment 26: 585–594.
13. Pallozzi E, Fortunati A, Marino G, Loreto F, Agati G, et al. (2013) BVOC emission from *Populus* × *canadensis* saplings in response to acute UV-A radiation. Physiologia Plantarum 148: 51–61.
14. Hu L, Wang Z, Huang B (2010) Diffusion limitations and metabolic factors associated with inhibition and recovery of photosynthesis from drought stress in a C3 perennial grass species. Physiologia Plantarum 139: 93–106.
15. Scafaro AP, Von Caemmerer S, Evans JR, Atwell BJ (2011) Temperature response of mesophyll conductance in cultivated and wild *Oryza* species with contrasting mesophyll cell wall thickness. Plant Cell and Environment 34: 1999–2008.
16. Adachi S, Nakae T, Uchida M, Soda K, Takai T, et al. (2013) The mesophyll anatomy enhancing CO_2 diffusion is a key trait for improving rice photosynthesis. Journal of Experimental Botany 64: 1061–1072.
17. Manickavelu A, Nadarajan N, Ganesh SK, Gnanamalar RP, Chandra Babu R (2006) Drought tolerance in rice: morphological and molecular genetic consideration. Plant Growth Regulation 50: 121–138.
18. Huang X-Y, Chao D-Y, Gao J-P, Zhu M-Z, Shi M, et al. (2009) A previously unknown zinc finger protein, DST, regulates drought and salt tolerance in rice via stomatal aperture control. Genes & Development 23: 1805–1817.
19. Hu SP, Zhou Y, Zhang L, Zhu XD, Li L, et al. (2009) Correlation and quantitative trait loci analyses of total chlorophyll content and photosynthetic rate of rice (*Oryza sativa*) under water stress and well-watered conditions. Journal of Integrative Plant Biology 51: 879–888.
20. Gu J, Yin X, Struik PC, Stomph TJ, Wang H (2012) Using chromosome introgression lines to map quantitative trait loci for photosynthesis parameters in rice (*Oryza sativa* L.) leaves under drought and well-watered field conditions. Journal of Experimental Botany 63: 455–469.
21. Flexas J, Niinemets Ü, Gallé A, Barbour M, Centritto M, et al. (2013) Diffusional conductances to CO_2 as a target for increasing photosynthesis and photosynthetic water-use efficiency. Photosynthesis Research 117: 45–59.
22. Farquhar GD, Caemmerer S, Berry JA (1980) A biochemical model of photosynthetic CO_2 assimilation in leaves of C3 species. Planta 149: 78–90.
23. Giuliani R, Koteyeva N, Voznesenskaya E, Evans MA, Cousins AB, et al. (2013) Coordination of leaf photosynthesis, transpiration, and structural traits in rice and wild relatives (Genus *Oryza*). Plant Physiology 162: 1632–1651.
24. Kaldenhoff R (2012) Mechanisms underlying CO_2 diffusion in leaves. Current Opinion in Plant Biology 15: 276–281.
25. Hanba YT, Shibasaka M, Hayashi Y, Hayakawa T, Kasamo K, et al. (2004) Overexpression of the barley aquaporin HvPIP2;1 increases internal CO_2 conductance and CO_2 assimilation in the leaves of transgenic rice plants. Plant and Cell Physiology 45: 521–529.
26. Tyerman SD, Niemietz CM, Bramley H (2002) Plant aquaporins: multifunctional water and solute channels with expanding roles. Plant, Cell & Environment 25: 173–194.
27. Loreto F, Centritto M (2008) Leaf carbon assimilation in a water-limited world. Plant Biosystems 142: 154–161.
28. Chaves M, Flexas J, Pinheiro C (2009) Photosynthesis under drought and salt stress: regulation mechanisms from whole plant to cell. Annals of Botany 103: 551–560.
29. Cornic G (2000) Drought stress inhibits photosynthesis by decreasing stomatal aperture – not by affecting ATP synthesis. Trends in Plant Science 5: 187–188.
30. Pinheiro C, Chaves MM (2011) Photosynthesis and drought: can we make metabolic connections from available data? Journal of Experimental Botany 62: 869–882.
31. Kamoshita A, Rodriguez R, Yamauchi A, Wade LJ (2004) Genotypic variation in response of rainfed lowland rice to prolonged drought and rewatering. Plant Production Science 7: 406–420.
32. Boonjung H, Fukai S (1996) Effects of soil water deficit at different growth stages on rice growth and yield under upland conditions. 2. Phenology, biomass production and yield. Field Crops Research 48: 47–55.
33. Pons TL, Flexas J, von Caemmerer S, Evans JR, Genty B, et al. (2009) Estimating mesophyll conductance to CO_2: methodology, potential errors, and recommendations. Journal of Experimental Botany 60: 2217–2234.
34. Gilbert ME, Pou A, Zwieniecki MA, Holbrook NM (2012) On measuring the response of mesophyll conductance to carbon dioxide with the variable J method. Journal of Experimental Botany 63: 413–425.
35. Harley PC, Loreto F, Dimarco G, Sharkey TD (1992) Theoretical considerations when estimating the mesophyll conductance to CO_2 flux by analysis of the response of photosynthesis to CO_2. Plant Physiology 98: 1429–1436.
36. Bois JF, Couchat PH, Lasceve G (1985) Relationships between transpiration and photosynthesis during a water stress. Acta Horticulturae 171: 297–314.
37. Richter A, Wanek W, Werner RA, Ghashghaie J, Jaeggi M, et al. (2009) Preparation of starch and soluble sugars of plant material for the analysis of carbon isotope composition: a comparison of methods. Rapid Communications in Mass Spectrometry 23: 2476–2488.
38. Farquhar GD, Ehleringer JR, Hubick KT (1989) Carbon isotope discrimination and photosynthesis. Annual Review of Plant Physiology and Plant Molecular Biology 40: 503–537.
39. Rodeghiero M, Niinemets Ü, Cescatti A (2007) Major diffusion leaks of clamp-on leaf cuvettes still unaccounted: how erroneous are the estimates of Farquhar et al. model parameters? Plant, cell & environment 30: 1006–1022.
40. Loreto F, Harley PC, Dimarco G, Sharkey TD (1992) Estimation of mesophyll conductance to CO_2 flux by three different methods. Plant Physiology 98: 1437–1443.
41. Genty B, Briantais J-M, Baker NR (1989) The relationship between the quantum yield of photosynthetic electron transport and quenching of chlorophyll fluorescence. Biochimica et Biophysica Acta (BBA) - General Subjects 990: 87–92.
42. Laisk A, Loreto F (1996) Determining photosynthetic parameters from leaf CO_2 exchange and chlorophyll fluorescence (ribulose-1, 5-bisphosphate carboxylase/oxygenase specificity factor, dark respiration in the light, excitation distribution between photosystems, alternative electron transport rate, and mesophyll diffusion resistance. Plant Physiology 110: 903–912.
43. Galmes J, Flexas J, Keys AJ, Cifre J, Mitchell RAC, et al. (2005) Rubisco specificity factor tends to be larger in plant species from drier habitats and in species with persistent leaves. Plant Cell and Environment 28: 571–579.
44. Furbank RT, von Caemmerer S, Sheehy J, Edwards G (2009) C4 rice: a challenge for plant phenomics. Functional Plant Biology 36: 845–856.
45. Pinelli P, Loreto F (2003) $^{12}CO_2$ emission from different metabolic pathways measured in illuminated and darkened C-3 and C-4 leaves at low, atmospheric and elevated CO_2 concentration. Journal of Experimental Botany 54: 1761–1769.

46. Loreto F, Dimarco G, Tricoli D, Sharkey TD (1994) Measurements of mesophyll conductance, photosynthetic electron transport and alternative electron sinks of field grown wheat leaves. Photosynthesis Research 41: 397–403.
47. Lauteri M, Scartazza A, Guido MC, Brugnoli E (1997) Genetic variation in photosynthetic capacity, carbon isotope discrimination and mesophyll conductance in provenances of *Castanea sativa* adapted to different environments. Functional Ecology 11: 675–683.
48. Scartazza A, Lauteri M, Guido MC, Brugnoli E (1998) Carbon isotope discrimination in leaf and stem sugars, water-use efficiency and mesophyll conductance during different developmental stages in rice subjected to drought. Australian Journal of Plant Physiology 25: 489–498.
49. Evans JR, Sharkey TD, Berry JA, Farquhar GD (1986) Carbon isotope discrimination measured concurrently with gas exchange to investigate CO_2 diffusion in leaves of higher plants. Australian Journal of Plant Physiology 13: 281–292.
50. Lloyd J, Syvertsen JP, Kriedemann PE, Farquhar GD (1992) Low conductances for CO_2 diffusion from stomata to the sites of carboxylation in leaves of woody species. Plant Cell and Environment 15: 873–899.
51. von Caemmerer S, Evans JR (1991) Determination of the average partial pressure of CO_2 in chloroplasts from leaves of several C3 plants. Australian Journal of Plant Physiology 18: 287–305.
52. Pantuwan G, Fukai S, Cooper M, Rajatasereekul S, O'Toole JC (2002) Yield response of rice (*Oryza sativa* L.) genotypes to different types of drought under rainfed lowlands: Part 1. Grain yield and yield components. Field Crops Research 73: 153–168.
53. Siddique K, Kirby E, Perry M (1989) Ear: stem ratio in old and modern wheat varieties; relationship with improvement in number of grains per ear and yield. Field Crops Research 21: 59–78.
54. Koyama K, Kikuzawa K (2009) Is whole-plant photosynthetic rate proportional to leaf area? A test of scalings and a logistic equation by leaf demography census. The American Naturalist 173: 640–649.
55. Tomás M, Flexas J, Copolovici L, Galmés J, Hallik L, et al. (2013) Importance of leaf anatomy in determining mesophyll diffusion conductance to CO_2 across species: quantitative limitations and scaling up by models. Journal of experimental botany 64: 2269–2281.
56. Zhao B, Kondo M, Maeda M, Ozaki Y, Zhang J (2004) Water-use efficiency and carbon isotope discrimination in two cultivars of upland rice during different developmental stages under three water regimes. Plant and Soil 261: 61–75.
57. Tipple BJ, Berke MA, Doman CE, Khachaturyan S, Ehleringer JR (2013) Leaf-wax n-alkanes record the plant–water environment at leaf flush. Proceedings of the National Academy of Sciences 110: 2659–2664.
58. Quick W, Chaves M, Wendler R, David M, Rodrigues M, et al. (1992) The effect of water stress on photosynthetic carbon metabolism in four species grown under field conditions. Plant, Cell and Environment 15: 25–35.
59. Yang J, Zhang J, Wang Z, Zhu Q (2001) Activities of starch hydrolytic enzymes and sucrose-phosphate synthase in the stems of rice subjected to water stress during grain filling. Journal of Experimental Botany 52: 2169–2179.
60. Scheidegger Y, Saurer M, Bahn M, Siegwolf R (2000) Linking stable oxygen and carbon isotopes with stomatal conductance and photosynthetic capacity: a conceptual model. Oecologia 125: 350–357.
61. Schmitt MR, Edwards GE (1981) Photosynthetic capacity and nitrogen use efficiency of maize, wheat, and rice - a comparison between C-3 and C-4 Photosynthesis. Journal of Experimental Botany 32: 459–466.
62. Kiran TV, Rao YV, Subrahmanyam D, Rani NS, Bhadana VP, et al. (2013) Variation in leaf photosynthetic characteristics in wild rice species. Photosynthetica 51: 350–358.
63. Terashima I (1992) Anatomy of non-uniform leaf photosynthesis. Photosynthesis research 31: 195–212.
64. Mott KA, Buckley TN (2000) Patchy stomatal conductance: emergent collective behaviour of stomata. Trends in Plant Science 5: 258–262.
65. Sun P, Wahbi S, Tsonev T, Haworth M, Liu S, et al. (2014) On the use of leaf spectral indices to assess water status and photosynthetic limitations in *Olea europaea* L. during water-stress and recovery. Plos One 9: e105165.
66. Tsonev T, Wahbi S, Sun P, Sorrentino G, Centritto M (2014) Gas exchange, water relations and their relationships with photochemical reflectance index in *Quercus ilex* plants during water stress and recovery. International Journal of Agriculture and Biology 16: 335–341.
67. Warren CR. Stand aside stomata, another actor deserves centre stage: the forgotten role of the internal conductance to CO_2 transfer; 2007 Jul 22–27; Glasgow, Scotland. Oxford University Press. pp.1475–1487.
68. Centritto M, Tognetti R, Leitgeb E, Střelcová K, Cohen S (2011) Above ground processes - anticipating climate change influences. In: Bredemeier M, Cohen S, Godbold DL, Lode E, Pichler V et al., editors. Forest Management and the Water Cycle: An Ecosystem-Based Approach. London: Springer. pp.31–64.
69. Haworth M, Elliott-Kingston C, McElwain J (2013) Co-ordination of physiological and morphological responses of stomata to elevated [CO_2] in vascular plants. Oecologia 171: 71–82.
70. Haworth M, Elliott-Kingston C, McElwain JC (2011) Stomatal control as a driver of plant evolution. Journal of Experimental Botany 62: 2419–2423.
71. Haworth M, Gallagher A, Elliott-Kingston C, Raschi A, Marandola D, et al. (2010) Stomatal index responses of *Agrostis canina* to carbon dioxide and sulphur dioxide: implications for palaeo-[CO_2] using the stomatal proxy. New Phytologist 188: 845–855.
72. Kubiskem ME, Abrams MD (1993) Stomatal and nonstomatal limitations of photosynthesis in 19 temperate tree species on contrasting sites during wet and dry years. Plant, Cell and Environment 16: 1123–1129.
73. Venuprasad R, Lafitte HR, Atlin GN (2007) Response to direct selection for grain yield under drought stress in rice. Crop Sci 47: 285–293.
74. Karaba A, Dixit S, Greco R, Aharoni A, Trijatmiko KR, et al. (2007) Improvement of water use efficiency in rice by expression of HARDY, an *Arabidopsis* drought and salt tolerance gene. Proceedings of the National Academy of Sciences of the United States of America 104: 15270–15275.
75. Lanceras JC, Pantuwan G, Jongdee B, Toojinda T (2004) Quantitative trait loci associated with drought tolerance at reproductive stage in rice. Plant Physiology 135: 384–399.
76. Xu Y, This D, Pausch RC, Vonhof WM, Coburn JR, et al. (2009) Leaf-level water use efficiency determined by carbon isotope discrimination in rice seedlings: genetic variation associated with population structure and QTL mapping. Theoretical and Applied Genetics 118: 1065–1081.
77. Wassmann R, Jagadish SVK, Heuer S, Ismail A, Redona E, et al. (2009) Climate change affecting rice production: the physiological and agronomic basis for possible adaptation strategies. In: Sparks DL, editor. Advances in Agronomy, Vol 101. pp.59–122.
78. Blum A (2005) Drought resistance, water-use efficiency, and yield potential–are they compatible, dissonant, or mutually exclusive? Australian Journal of Agricultural Research 56: 1159–1168.

Intraspecific Trait Variation Driven by Plasticity and Ontogeny in *Hypochaeris radicata*

Rachel M. Mitchell[1]*, Jonathan D. Bakker[1,2]

1 School of Environmental and Forest Sciences, University of Washington, Seattle, Washington, United States of America, **2** Smithsonian Environmental Research Center, Edgewater, Maryland, United States of America

Abstract

The importance of intraspecific variation in plant functional traits for structuring communities and driving ecosystem processes is increasingly recognized, but mechanisms governing this variation are less studied. Variation could be due to adaptation to local conditions, plasticity in observed traits, or ontogeny. We investigated 1) whether abiotic stress caused individuals, maternal lines, and populations to exhibit trait convergence, 2) whether trait variation was primarily due to ecotypic differences or trait plasticity, and 3) whether traits varied with ontogeny. We sampled three populations of *Hypochaeris radicata* that differed significantly in rosette diameter and specific leaf area (SLA). We grew nine maternal lines from each population (27 lines total) under three greenhouse conditions: ambient conditions (control), 50% drought, or 80% shade. Plant diameter and relative chlorophyll content were measured throughout the experiment, and leaf shape, root:shoot ratio, and SLA were measured after five weeks. We used hierarchical mixed-models and variance component analysis to quantify differences in treatment effects and the contributions of population of origin and maternal line to observed variation. Observed variation in plant traits was driven primarily by plasticity. Shade significantly influenced all measured traits. Plant diameter was the only trait that had a sizable proportion of trait variation (30%) explained by population of origin. There were significant ontogenetic differences for both plant diameter and relative chlorophyll content. When subjected to abiotic stress in the form of light or water limitation, *Hypochaeris radicata* exhibited significant trait variability. This variation was due primarily to trait plasticity, rather than to adaptation to local conditions, and also differed with ontogeny.

Editor: Eric Gordon Lamb, University of Saskatchewan, Canada

Funding: The authors have no support or funding to report.

Competing Interests: The authors have declared that no competing interests exist.

* Email: rmm57@duke.edu

Introduction

Recent and growing interest in plant functional traits has deepened our understanding of critical ecological mechanisms. For example, interspecific trait-based approaches have identified differing processes of community assembly across spatial scales [1] and in temperate and tropical ecosystems [2], shaped understanding of abiotic drivers such as altitude [3,4], climate [5,6], temperature [7,8], and resource gradients [9] on community assembly and diversity, and identified global patterns of plant-trait expression [10]. In addition, trait-based approaches have provided a mechanistic understanding of correlations between biodiversity and ecosystem function [11–13].

Although trait-based approaches have improved our understanding of ecological mechanisms, the role of intraspecific trait variation - variation from individual to individual within a species - is less well understood. Intraspecific variation in plant functional traits can be large, though it has often been ignored or its contribution underestimated [14–16]. Intraspecific variation can structure communities [17] and influence ecosystem function [18]. For example, intraspecific variation in plant functional traits has been shown to structure communities along environmental gradients [19], to drive diversity in forest ecosystems [20], and to enhance community resistance to plant invasion [21].

Changes in intraspecific variation in plant functional traits have been observed in response to a wide range of environmental conditions (e.g., flooding [19], fire [22], drought [23] and temperature [24]), but few studies have examined the mechanisms controlling these changes (but see [25–27]). Two primary processes are hypothesized to drive variation in trait expression: adaptation to local conditions, creating distinct genetic ecotypes [28], and phenotypic plasticity in response to prevailing abiotic conditions [29]. Those studies that have investigated sources of intraspecific variation have, at times, found contrasting results, depending on the species and functional trait being investigated [30–32]. For example, Firn et al. found significant plasticity in specific leaf area (SLA; leaf area per unit mass) for an exotic, invasive grass but not for co-occurring native grasses [33]. In *Quercus suber*, genetically distinct populations displayed significantly different mean values for SLA, suggesting local adaptation to climate regimes, but populations also demonstrated high phenotypic plasticity [34]. In contrast, Robson et al. found that differences in SLA in *Fagus sylvatica* were primarily driven by genetic provenance [35].

If observed intraspecific trait variation is largely driven by the genetics of a given population, abiotic or biotic filters can selectively eliminate individuals with traits that exceed the bounds of the filter. For example, species or populations with low variability in traits related to drought stress (e.g., leaf size) may be unable to cope with reduced water availability under shifting climate. However, if variation in critical plant traits is largely a plastic response, independent of genetic diversity, even rare species may respond to filtering by altering their trait expression and thus persist in the community. However, plasticity is not without cost, as very plastic species may express phenotypes unsuited to abiotic filters, or experience costs associated with the trait of plasticity [36,37].

In addition to being influenced by adaptation and plasticity, trait expression also changes due to ontogeny. Determining how traits vary ontogenetically is important because traits influence ecosystem function [11], and temporal differences in plant trait expression may facilitate species coexistence and biodiversity maintenance [38]. Smith *et al.* found that SLA decreased with age in *Populus tremuloides* [39], while Cornelissen et al. found poor correlation between greenhouse seedlings and field-measured adults for 90 woody plant species, indicating strong ontogenetic differences [40]. The effects of ontogeny can differ depending on the trait being measured. For example, wood density, but not leaf mass per unit area, was strongly affected by ontogeny in tropical *Nothofagus pumilio* [24]. Despite evidence of correlation between trait variability and ontogenetic state, this source of variability is rarely accounted for, and often explicitly avoided by use of standard collection protocols [41].

Examination of the mechanisms controlling intraspecific variation is confounded by the fact that variation is both adaptive and with cost [37,42,43]. One species, population within a species, or genetic lineage may be inherently more plastic than others for some traits. In a widely distributed species, some populations may exhibit greater plasticity in response to abiotic stress due to high environmental heterogeneity selecting for phenotypically plastic genotypes, or through genetic correlation to other traits that are under selection [44]. For example, Balaguer *et al.* [45] found that populations of *Quercus coccifera* growing in homogenous conditions displayed significantly less phenotypic plasticity in light-responsive traits than those growing in more heterogeneous conditions. Pratt & Mooney found a similar pattern for growth and flower production in *Artemisia californica* in response to precipitation variability [46].

We used *Hypochaeris radicata*, a globally widespread weed, as a model species for examining trait variation in terms of both trait means (average trait values) and trait dispersion (variability of trait values). *H. radicata* occurs in a number of ecosystems, from undisturbed grasslands to highly disturbed urban environments. Because it persists across such a wide variety of conditions, it likely exhibits considerable trait plasticity [47]. However, persistence under differing abiotic conditions may lead to distinct population-level trait differences through ecotypic adaptation or transgenerational plastic responses [14,28,38,48].

We collected and germinated seeds from three populations, exposed the seedlings to abiotic stress (shade, drought) in the greenhouse, and measured a range of traits on each individual. We used these data to experimentally test three hypotheses. First, we hypothesized that abiotic stress would significantly impact trait expression for all populations, leading to a convergence in trait values for plants experiencing abiotic stress. Second, we hypothesized that population of origin and maternal line of origin, and not random variation between individuals, would explain the majority of observed trait variation that was not explained by abiotic conditions. Third, we hypothesized that over the lifetime of a plant, individuals, maternal lines, and populations within a treatment would become more similar in terms of both trait means and variances in order to accommodate stressful abiotic conditions.

Materials and Methods

Ethics Statement

No special permits were required for this research. Permission to conduct research at Glacial Heritage Preserve was granted by Thurston County. Permission to conduct research at Smith Prairie was granted by the non-profit owners, Pacific Rim Institute for Environmental Stewardship. Permission to collect samples at Union Bay Natural Area was granted by the University of Washington. This research did not include any populations of endangered or threatened species. Data used in this manuscript are available through the TRY database and can be accessed by submitting a data request at http://www.try-db.org.

Study Species

Hypochaeris radicata is a short-lived perennial herb species common in intact and degraded grassland ecosystems around the globe, including Europe, North America, Australia, New Zealand and Japan. It forms basal rosettes with oblong-lanceolate leaves and produces several upright, branching flower stalks during a long flowering period – June through October in the Pacific Northwest of the USA. *H. radicata* is self-incompatible [49,50] and its seeds are wind dispersed up to a few hundred meters [51].

Experimental Design

Three populations (Glacial Heritage (GL), Smith Prairie (SP), and Union Bay Natural Area (UB)) spanning a 200 km range in Washington state, USA, were sampled for this experiment. Populations occurred in exotic-dominated grasslands, but experienced different abiotic conditions in terms of annual precipitation, soil type, and land-use history (Table S1). Specific leaf area and rosette diameter were measured on 20–25 plants (maternal lines) per population, and one mature seed-head was collected from each plant. Measured plants were at least 3 m apart to prevent sampling from closely related individuals. Previous analyses detected significant differences in trait variability for SLA and rosette diameter [16]; follow-up analyses using PermDISP and PERMANOVA (see "Statistical Analysis", below) indicated significant differences in trait dispersion and mean values for diameter (Fig. 1A, $P<0.0001$) and significantly different dispersion for SLA (Fig. 1B, $P=0.05$) but no significant differences in seed mass (Fig. 1C).

The single seed head collected per maternal line did not contain enough seeds to allow individual maternal lines to be replicated in each abiotic treatment. Instead, maternal lines were subjected to Principal Components Analysis (PCA), and the first principal component (PC1) was used to select maternal lines spanning the range of traits expressed in each population. Thus, for each population, the three highest, three most central, and three lowest maternal lines (hereafter referred to as "high", "central", and "low") along PC1, which explained 49% of the variation, were identified (Fig. 1D). One maternal line from each group was then randomly assigned to an abiotic treatment (Figure S1).

Seeds from the selected maternal lines (3 populations ×9 maternal lines/population) were cold-stratified for 10 days and germinated at 25°C. Twenty germinants per maternal line (540 total) were planted individually into 164 ml "conetainers" (Stuewe & Sons, Oregon, USA) filled with Sunshine Mix #3 potting soil (Sun Gro Horticulture, Massachusetts, USA). Germinants were allowed to acclimate for seven days before treatments were applied.

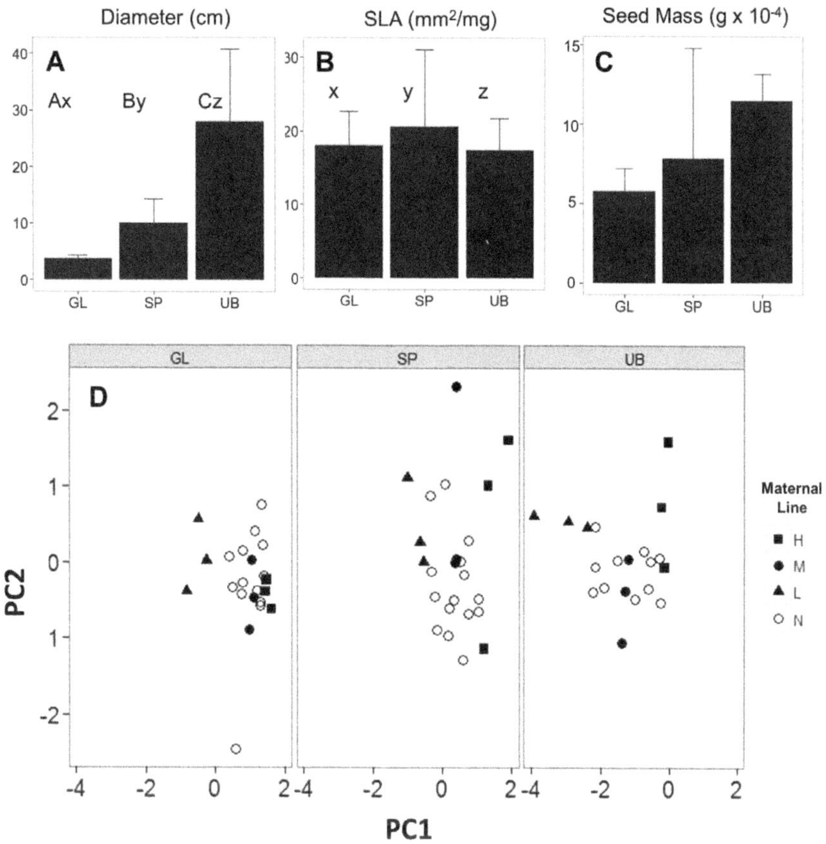

Figure 1. Mean (+ SD) of field-based measurements of three traits for the populations of origin (GL: Glacial Heritage; SP: Smith Prairie; UB: Union Bay). Significant PERMANOVA results ($\alpha = 0.05$) are indicated by different uppercase letters and significant PERMDISP results are indicated by different lowercase letters. D) Results of a Principal Components Analysis of the three traits measured on individuals from each population. Populations were analyzed together, but are shown in separate graphs for clarity. PC1 explained 49% of observed variation. Filled symbols indicate maternal "high" (H), "medium" (M) and "low" (L) maternal lines used in the greenhouse experiment, while open symbols (N) indicate maternal lines measured in the field but not included in the experiment.

Plants were grown in the greenhouse under one of three treatments: ambient conditions, 80% shade (80% of light is blocked), and drought conditions (50% of winter precipitation). The shade and drought treatment levels were selected to mimic marginal conditions where *H. radicata* is known to survive. Beginning on day seven, the high, low and central maternal lines were randomly assigned to the three treatments, and individuals were randomly arranged on the bench to account for edge effects. The ambient and shade treatments received weekly top-watering equivalent to average October-November rainfall across the three populations (53 ml per week per conetainer). The drought treatment received half as much (26 ml per week per conetainer) and visibly induced plant stress; individuals wilted between waterings. Each treatment received weekly 10-10-10 NPK fertilizer plus micronutrients per manufacturer specifications.

Mortality and rosette diameter (a measure of competitive ability similar to height [16]) were measured weekly, beginning on day seven. Relative chlorophyll content (correlated with photosynthesis and leaf life-span [52]) was measured weekly beginning on day 14 (plant leaves were too small on day seven for data collection) using a SPAD 502 plus chlorophyll meter (Spectrum Technologies, Illinois, USA). The length and width of the largest leaf on each individual were measured after five weeks to quantify whether leaves were more lanceolate or ovate, which can have consequences for a number of functions, including light interception [53]. Plants were destructively sampled after five weeks. One leaf per individual was harvested, scanned, dried at 60°C, and weighed using a Mettler AE 163 analytical scale (Mettler Toledo, Ohio, USA). One-sided leaf area was calculated from the scanned images using ImageJ version 1.45 (http://rsb.nih.gov/ij). SLA, which is correlated with resource acquisition and leaf life-span [54,55], was calculated as leaf area (mm^2) divided by leaf dry mass (mg). Above and belowground biomass were harvested separately, dried at 60°C, weighed, and used to calculate root:shoot ratio, a measure of resource allocation.

Statistical Analysis

Differences between trait means for the populations of origin measured in the field were quantified using PERMANOVA, a non-parametric permutational form of ANOVA [56]. Differences in trait dispersion were quantified using PermDISP, a permutational analog to the Levene's test [57]. These analyses were conducted using the PERMANOVA+ add-on to PRIMER-E [58], with Euclidean distances, type III sums of squares, and 9,999 permutations. Populations were considered to differ if $P<0.05$.

For plants grown under greenhouse conditions, measured traits were checked for normality, and corrective measures were taken when traits were non-normal. Correlation between dependent variables was checked and, when response variables were strongly correlated, only one representative variable was tested (the weakest

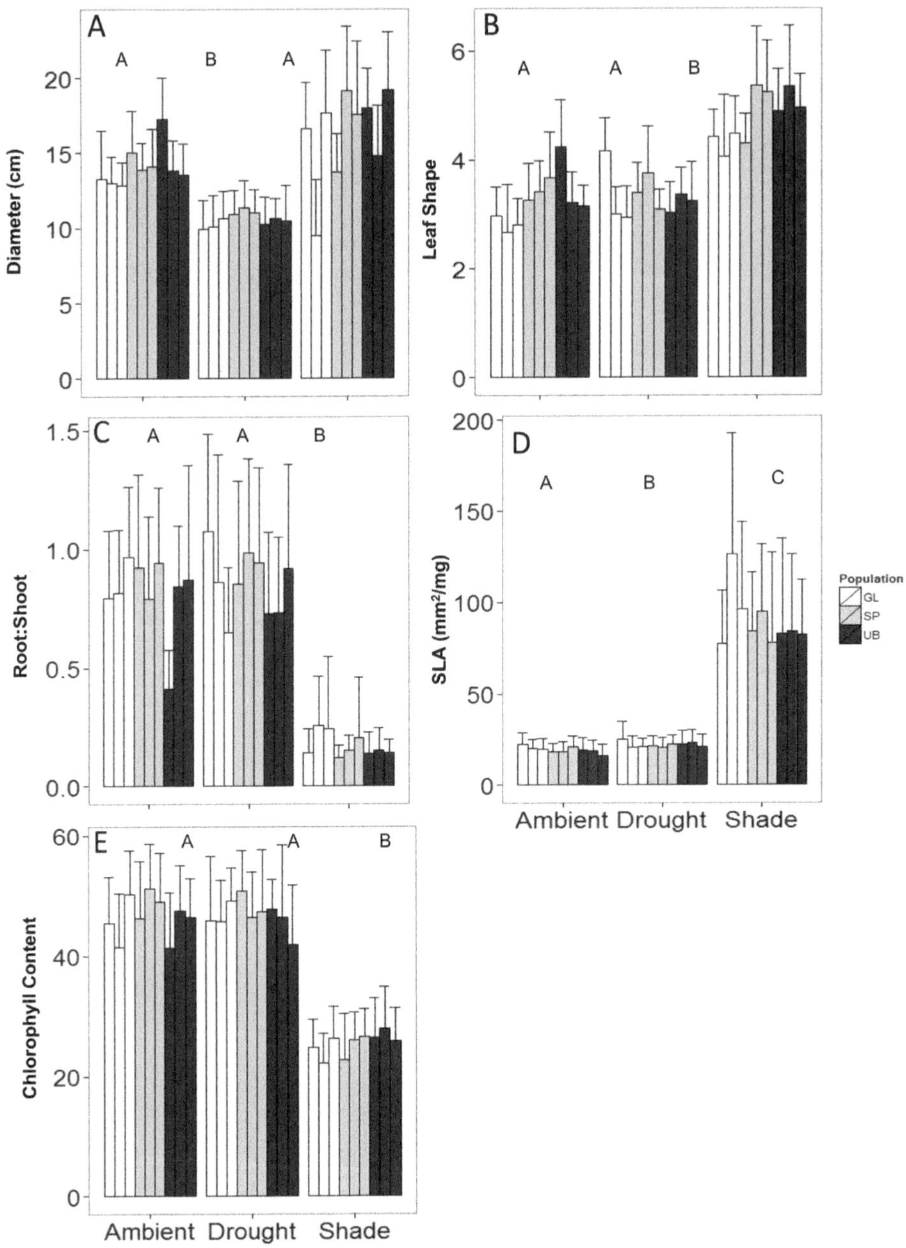

Figure 2. Mean (+ SD) values of A) final diameter, B) leaf shape (length:width ratio), C) root:shoot ratio, D) SLA, and E) final relative chlorophyll content for each maternal line (individual bar, arranged from low to high based on scores from PC1) in each population (unique color; defined in Table S1) for each treatment. Significant treatment differences ($\alpha = 0.05$) in trait means are indicated by different letters.

correlation among tested variables was $r^2 > 0.65$, see Fig S2). Five traits were selected for analysis: diameter, leaf shape (leaf length:width ratio; larger numbers indicate more lanceolate leaves and lower numbers indicate more ovate leaves), root:shoot ratio, SLA, and final relative chlorophyll content. SLA and diameter were log transformed to correct for non-normality and heteroscedacity.

We used hierarchical linear mixed effects models and variance component analysis to quantify the influence of abiotic treatment and ecotypic factors on observed traits. Population and maternal line were designated as random effects, with maternal line nested within populations. The variance attributable to population of origin, maternal line, and individual were expressed as proportions of the total random-effects variance. Abiotic treatment (ambient, shade, or drought) was treated as a fixed effect. Models were fit with the "lme4" package in R (version 2.14.2) and *P*-values were generated using the "languageR' package. Significant treatment effect was followed by pairwise comparisons using TukeyHSD through the "multcomp" package (version 1.2–19), with attention focused on the differences between the ambient treatment and each of the shade and drought treatments.

Ontogenetic changes were examined by analyzing changes over time in diameter and relative chlorophyll content. These traits were selected because they did not require repeated, destructive

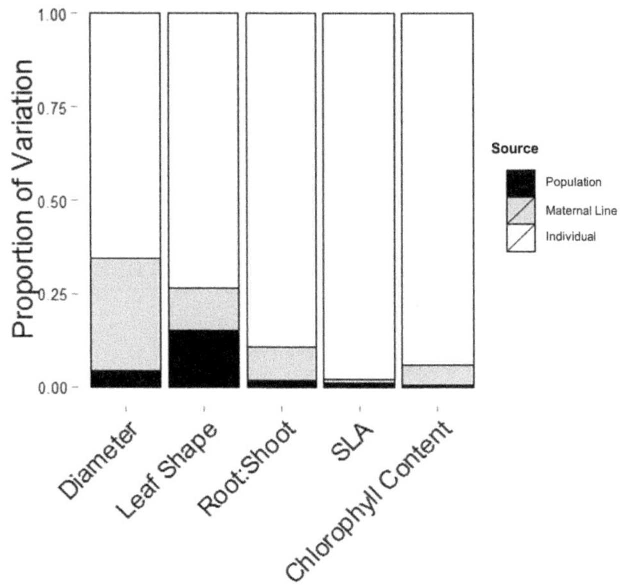

Figure 3. Proportion of total random effects variance explained by population of origin, maternal line, and individual plant (residual) for five measured traits.

measurements. Individual plants were nested within maternal lines within populations in these models, and time was included as a fixed effect. The fixed effects (abiotic treatment and time) and their interaction were iteratively added, beginning with the null model (including only random effects) and selecting the best model using Akaike's Information Criterion (AIC). Significant treatment by time effects were followed by separate analyses of each date to quantify the variance attributable to population of origin, maternal line, and individuals.

Results

Response to Abiotic Stress

Mortality was low, even for plants grown under abiotic stress: only 8 of 540 individuals died. However, the abiotic treatments affected plant trait expression (Fig. 2, See Table S2 for GLMM results). Mean final diameter was larger for plants in the shade treatment (16.4 cm) than the ambient treatment (14.1 cm), but smaller in the drought treatment (10.6 cm) (Fig. 2A). Plants in the shade treatment had significantly more lanceolate leaves (4.8) than those in the ambient (3.3) and drought treatments (3.3) (Fig. 2B), and also had significantly lower root:shoot ratios (0.17 vs 0.82 and 0.86, respectively) (Fig. 2C). Both the shade and drought treatments increased mean SLA (88.2 and 22.1 mm^2/mg, respectively) compared to those in the ambient treatment (19.4 mm^2/mg) (Fig. 3D). Relative chlorophyll content was significantly lower for plants in the shade treatment (25.6 SPAD units) compared with those in the drought (46.8 units) or ambient treatments (46.7 SPAD units) (Fig. 2E).

Contribution of Population and Maternal Line to Trait Variation

A surprisingly small amount of variation was explained by population of origin or maternal line for most measured traits (Fig. 3). Diameter was the only trait that showed an appreciable maternal contribution to observed variation; 30% of the total random effects variation was attributable to maternal lines, while 5% was attributable to populations (Fig. 3). Variation in leaf shape showed nearly equal contributions from population (15%) and maternal sources (12%). Visual inspection of the mean values for each maternal line showed no clear relationship with the maternal plant's score on PC1 (Fig. 1).

Ontogenetic Effects

There were significant interactions between treatment and time for both plant diameter and relative chlorophyll content (Fig. 4A and C). Treatment groups were indistinguishable on days 7 and 14. On day 21, plants in the shade treatment had significantly smaller diameter than those in the drought and ambient treatments ($P = 0.03$). On days 28 and 35, diameter in the drought treatment was significantly lower than the shade or ambient treatments ($P \leq 0.0001$; Fig 4A). Individuals accounted for an increasing amount of the variation in plant diameter, while variation between populations decreased over time (Fig. 4B). Variation between maternal lines remained fairly consistent.

Shading rapidly and strongly affected relative chlorophyll content, as indicated by significant treatment effects already on day 14, while drought did not ($P<0.0001$; Fig 4C). Variation between individuals increased between days 14 and 21, and remained high throughout the experiment (Fig. 4D). Variation between maternal lines and populations decreased over the course of the experiment.

Discussion

The roles of plasticity and adaptation in shaping trait expression

We set out to determine whether observed intraspecific variation was driven by plasticity or by adaptation to local abiotic conditions. Our results indicate that, for *H. radicata*, the observed trait variation is largely attributable to plasticity. Neither shading nor drought stress were strong enough to cause appreciable mortality, confirming that *H. radicata* can tolerate a very wide range of abiotic conditions. However, plants grown under these conditions exhibited markedly different functional traits. For example, SLA values previously reported for *H. radicata* individuals have ranged from 1.7–46.1 mm^2/mg [16,59,60]; in this study, SLA ranged from 6.6–239.1 mm^2/mg. Much of the increased range of SLA in this study was due to increases in response to shading (Fig. 2D), a response which has been demonstrated in the past [61]. The inclusion of shade-grown individuals, which have significantly higher SLA than sun-grown individuals, would dramatically shift the trait distribution for this species. Inclusion of such plants would also likely influence the overall trait distribution of the plant community, and may in turn impact estimations of critical ecosystem processes like litter production and nutrient cycling [13] through differing nutrient contributions and breakdown times in the decomposition pool [62]. Standard trait collection procedures [41,63] focus trait measurements exclusively on sun-exposed plants and therefore do not account for the full trait variation of species that can tolerate varying levels of shade. The omission of individuals in partial or full-shade from trait measurements may hinder our understanding of trait mediated ecosystem processes and responses. We suggest it would be more appropriate to sample the full range of conditions experienced by a species.

The variance explained by population of origin and maternal line was surprisingly small for every measured trait. Only leaf shape and diameter had significant variation attributable to differences across populations and maternal lines respectively, and even this amount of variation (30%) was dwarfed by treatment

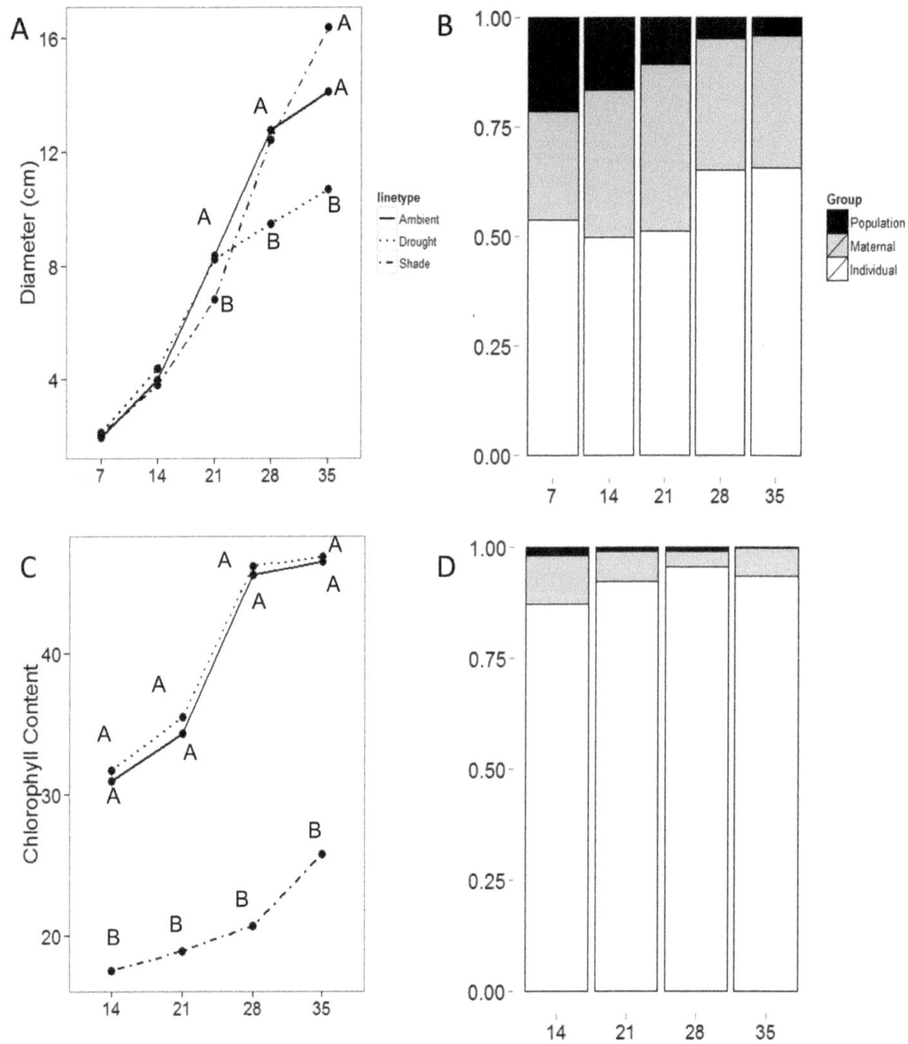

Figure 4. Change in average trait values in response to treatments (A and C; in each time period, letters indicate significantly different groups) and (B and D) the amount of variation attributed to population of origin, maternal line, and individual plant for diameter (top) and relative chlorophyll content (bottom).

effects. Virtually none of the observed variation could be attributed to population or maternal line for SLA (Fig. 3), confirming that this trait is highly plastic, a finding which supports previous analyses of this trait [61]. Furthermore, the position of each maternal line on PC1 had little correlation with the ranking of each maternal line in the greenhouse, suggesting that individuals within maternal lines exhibited significant variability and plasticity, and are not necessarily constrained by maternal genetics. The very small amount of variation attributable to population or maternal sources indicates that, for the traits measured in this experiment, plastic responses to prevailing conditions are the primary driver of observed variation.

The fact that all of the measured traits displayed significant plasticity has important implications for how traits are measured, and how they are incorporated into larger, trait-based models. Although some species are shade-intolerant, many species experience variable light conditions through interactions with neighboring plants in the community. Trait values based on fully sun-exposed leaves may inaccurately represent the trait contributions of such species, and thus may misrepresent community-level trait values, and the ecosystem functions dependent on those trait values. In addition, attempts to model species presence or response to abiotic change may be hindered by a reliance on trait means derived from sun-exposed individuals. If species, or populations within species, are variable and highly plastic for traits critical to survival, such taxa may be present in a wider range of habitat types, and be better able to cope with a wider range of conditions [64,65] than expected from trait values measured on individuals growing in a subset of the taxa's possible growing conditions.

The role of ontogeny in shaping trait expression

Variation in trait values over the lifespan of a plant is an additional, and often unaccounted for, source of intraspecific trait variability. This source of variation may be underappreciated as collection techniques typically focus on "mature" plants and tissues [41] yet maturity can be difficult to distinguish in the field. Because plants of differing ontogenies may appear "mature" in the field, intraspecific trait variation measured in field populations could be partially attributed to differing ontogeny between individuals, rather than to genetic or environmental conditions.

Furthermore, ontogenetic differences in trait variability may be important in the structure and function of communities. For example, ontogenetic trait differences may lead to temporal niche partitioning, facilitating biodiversity maintenance through limiting similarity [66]. Ontogenetic differences leading to higher palatability can also alter plant fitness by influencing interactions with herbivores and predators [67]. Furthermore, temporal variation in plant traits may exacerbate or mediate the impacts of abiotic filters. Young plants face similar pressures during establishment, and tend to exhibit similar trait values, in both this study, and others (e.g., [68]). This lack of trait variability in young plants may increase susceptibility to stress, while increased variability in older plants may buffer the population against abiotic change. Incorporating trait values of plants at different ontogenetic stages could provide a clearer understanding of community and population dynamics, and more accurate extrapolation to ecosystem processes and functions.

Conclusion

Taken together, these results indicate that plasticity in response to abiotic conditions and ontogeny, and not adaptation and genetically distinct ecotypes, drive trait expression in *H. radicata*. However, the generality of this conclusion needs to be explored. *Hypochaeris radicata* is globally widespread, and thus may be inherently more plastic than species with restricted distributions [69,70]. To determine whether plasticity is a characteristic trait of widespread or dominant species, this research should be replicated with other widespread species and with species from restricted ranges. In addition, further research should examine how sensitive these conclusions are to the identity of the trait being examined.

The treatments applied during this experiment represented strong abiotic filters for this species, but are within the range of conditions experienced by *H. radicata*. Studies involving light or moisture gradients could assess whether the importance of population of origin or plasticity changes depending on the strength of the abiotic filter, a question which was not addressed in this study. Studies could also examine other abiotic filters, such as soil fertility. Finally, the additive and interactive effects of multiple abiotic filters on trait expression could be tested in both field and greenhouse settings.

Despite growing interest in understanding and incorporating intraspecific variation in plant functional traits into larger community- and ecosystem-level studies, the factors driving observed variation remain poorly understood. Our results suggest that observed intraspecific variation is largely driven by trait plasticity rather than genetic factors. We also found strong evidence of variation in plant traits due to ontogeny, a source of trait variation that is often ignored.

Acknowledgments

We thank Ty Robinson for assistance in the field, and Loretta Fisher for assistance in the lab. We also thank Janneke Hille Ris Lambers, Soo-Hyung Kim, Sarah Reichard, Regina Rochefort, and Martha Groom for their advice and guidance.

Author Contributions

Conceived and designed the experiments: RMM JDB. Performed the experiments: RMM. Analyzed the data: RMM. Contributed reagents/materials/analysis tools: RMM JDB. Wrote the paper: RMM JDB.

References

1. Messier J, McGill BJ, Lechowicz MJ (2010) How do traits vary across ecological scales? A case for trait-based ecology. Ecol Lett 13: 838–848. doi:10.1111/j.1461-0248.2010.01476.x.
2. Myers JA, Chase JM, Jiménez I, Jørgensen PM, Araujo-Murakami A, et al. (2013) Beta-diversity in temperate and tropical forests reflects dissimilar mechanisms of community assembly. Ecol Lett 16: 151–157. doi:10.1111/ele.12021.
3. Long W, Zang R, Schamp BS, Ding Y (2011) Within- and among-species variation in specific leaf area drive community assembly in a tropical cloud forest. Oecologia 167: 1103–1113. doi:10.1007/s00442-011-2050-9.
4. Swenson NG, Anglada-Cordero P, Barone Ja (2011) Deterministic tropical tree community turnover: evidence from patterns of functional beta diversity along an elevational gradient. Proc R Soc B 278: 877–884. doi:10.1098/rspb.2010.1369.
5. Swenson NG, Weiser MD (2010) Plant geography upon the basis of functional traits: an example from eastern North American trees. Ecology 91: 2234–2241.
6. Thuiller W, Lavorel S, Midgley G, Lavergne S (2004) Relating plant traits and species distributions along bioclimatic gradients for 88 Leucadendron taxa. Ecology 85: 1688–1699.
7. Laughlin DC, Fulé PZ, Huffman DW, Crouse J, Laliberté E (2011) Climatic constraints on trait-based forest assembly. J Ecol 99: 1489–1499. doi:10.1111/j.1365-2745.2011.01885.x.
8. Hudson JMG, Henry GHR, Cornwell WK (2011) Taller and larger: shifts in Arctic tundra leaf traits after 16 years of experimental warming. Glob Chang Biol 17: 1013–1021. doi:10.1111/j.1365-2486.2010.02294.x.
9. Fonseca C, McC Overton J, Collins B, Westoby M (2000) Shifts in trait-combinations along rainfall and phosphorus gradients. J Ecol 88: 964–977.
10. Díaz S, Cabido M, Casanoves F (1998) Plant functional traits and environmental filters at a regional scale. J Veg Sci 9: 113–122.
11. Diaz S, Hodgson JG, Thompson K, Cabido M, Cornelissen JHC, et al. (2004) The plant traits that drive ecosystems: Evidence from three continents. J Veg Sci 15: 295–304. doi:10.1111/j.1654-1103.2004.tb02266.x.
12. Díaz S, Cabido M (2001) Vive la différence: plant functional diversity matters to ecosystem processes. 16: 646–655.
13. William KC, Johannes HCC, Kathryn A, Ellen D, Valerie TE, et al. (2008) Plant species traits are the predominant control on litter decomposition rates within biomes worldwide. Ecol Lett 11: 1065–1071.
14. Albert CH, Thuiller W, Yoccoz NG, Douzet R, Aubert S, et al. (2010) A multi-trait approach reveals the structure and the relative importance of intra- vs. interspecific variability in plant traits. Funct Ecol 24: 1192–1201. doi:10.1111/j.1365-2435.2010.01727.x.
15. Albert CH, Thuiller W, Yoccoz NG, Soudant A, Boucher F, et al. (2010) Intraspecific functional variability: extent, structure and sources of variation. J Ecol 98: 604–613. doi:10.1111/j.1365-2745.2010.01651.x.
16. Mitchell RM, Bakker JD (2014) Quantifying and comparing intraspecific functional trait variability: a case study with Hypochaeris radicata. Funct Ecol 28: 258–269. doi:10.1111/1365-2435.12167.
17. Siefert A (2012) Incorporating intraspecific variation in tests of trait-based community assembly. Oecologia 170: 767–775. doi:10.1007/s00442-012-2351-7.
18. Lecerf A, Chauvet E (2008) Intraspecific variability in leaf traits strongly affects alder leaf decomposition in a stream. Basic Appl Ecol 9: 598–605. doi:10.1016/j.baae.2007.11.003.
19. Jung V, Violle C, Mondy C, Hoffmann L, Muller S (2010) Intraspecific variability and trait-based community assembly. J Ecol 98: 1134–1140. doi:10.1111/j.1365-2745.2010.01687.x.
20. Clark JS (2010) Individuals and the variation needed for high species diversity in forest trees. Science (80-) 327: 1129–1132. doi:10.1126/science.1183506.

21. Crutsinger GM, Souza L, Sanders NJ (2008) Intraspecific diversity and dominant genotypes resist plant invasions. Ecol Lett 11: 16–23. doi:10.1111/j.1461-0248.2007.01118.x.
22. Moreira B, Tavsanoglu C, Pausas JG (2012) Local versus regional intraspecific variability in regeneration traits. Oecologia 168: 671–677. doi:10.1007/s00442-011-2127-5.
23. Jung V, Albert CH, Violle C, Kunstler G, Loucougaray G, et al. (2014) Intraspecific trait variability mediates the response of subalpine grassland communities to extreme drought events. J Ecol 102: 45–53. doi:10.1111/1365-2745.12177.
24. Fajardo A, Piper FI (2011) Intraspecific trait variation and covariation in a widespread tree species (Nothofagus pumilio) in southern Chile. New Phytol 189: 259–271. doi:10.1111/j.1469-8137.2010.03468.x.
25. Grassein F, Till-Bottraud I, Lavorel S (2010) Plant resource-use strategies: the importance of phenotypic plasticity in response to a productivity gradient for two subalpine species. Ann Bot 106: 637–645. doi:10.1093/aob/mcq154.
26. Wellstein C, Chelli S, Campetella G, Bartha S, Galiè M, et al. (2013) Intraspecific phenotypic variability of plant functional traits in contrasting mountain grasslands habitats. Biodivers Conserv 22: 2353–2374. doi:10.1007/s10531-013-0484-6.
27. Freschet GT, Bellingham PJ, Lyver PO, Bonner KI, Wardle Da. (2013) Plasticity in above- and belowground resource acquisition traits in response to single and multiple environmental factors in three tree species. Ecol Evol: n/a–n/a. doi:10.1002/ece3.520.
28. Joshi J, Schmid B, Caldeira MC, Dimitrakopoulos PG, Good J, et al. (2001) Local adaptation enhances performance of common plant species. Ecol Lett 4: 536–544. doi:10.1046/j.1461-0248.2001.00262.x.
29. Schlichting C (1986) The evolution of phenotypic plasticity in plants. Annu Rev Ecol Syst 17: 667–693.
30. Hansen C, Garcia MB, Ehlers BK (2013) Water availability and population origin affect the expression of the tradeoff between reproduction and growth in Plantago coronopus. J Evol Biol 26: 1–10. doi:10.1111/jeb.12114.
31. Grassein F, Till-Bottraud I, Lavorel S (2010) Plant resource-use strategies: the importance of phenotypic plasticity in response to a productivity gradient for two subalpine species. Ann Bot 106: 637–645. doi:10.1093/aob/mcq154.
32. Bonser SP, Ladd B, Monro K, Hall MD, Forster M (2010) The adaptive value of functional and life-history traits across fertility treatments in an annual plant. Ann Bot 106: 979–988. doi:10.1093/aob/mcq195.
33. Firn J, Prober SM, Buckley YM (2012) Plastic traits of an exotic grass contribute to its abundance but are not always favourable. PLoS One 7: e35870. doi:10.1371/journal.pone.0035870.
34. Ramírez-Valiente JA, Sánchez-Gómez D, Aranda I, Valladares F (2010) Phenotypic plasticity and local adaptation in leaf ecophysiological traits of 13 contrasting cork oak populations under different water availabilities. Tree Physiol 30: 618–627. doi:10.1093/treephys/tpq013.
35. Robson TM, Sánchez-Gómez D, Cano FJ, Aranda I (2012) Variation in functional leaf traits among beech provenances during a Spanish summer reflects the differences in their origin. Tree Genet Genomes 8: 1111–1121. doi:10.1007/s11295-012-0496-5.
36. Dewitt TJ, Sih A, Wilson DS (1998) Costs and limits of phenotypic plasticity. Trends Ecol Evol 13: 77–81.
37. Auld JR, Agrawal A, Relyea R (2010) Re-evaluating the costs and limits of adaptive phenotypic plasticity. Proc R Soc B 277: 503–511. doi:10.1098/rspb.2009.1355.
38. Violle C, Enquist BJ, McGill BJ, Jiang L, Albert CH, et al. (2012) The return of the variance: intraspecific variability in community ecology. Trends Ecol Evol 27: 244–252. doi:10.1016/j.tree.2011.11.014.
39. Smith EA, Collette SB, Boynton TA, Lillrose T, Stevens MR, et al. (2011) Developmental contributions to phenotypic variation in functional leaf traits within quaking aspen clones. Tree Physiol 31: 68–77. doi:10.1093/treephys/tpq100.
40. Cornelissen JHC, Cerabolini B, Puyravaud JP, Maestro M, Werger MJA, et al. (2003) Functional traits of woody plants: correspondence of species rankings between field adults and laboratory-grown seedlings? J Veg Sci 14: 311–322.
41. Cornelissen J, Lavorel S, Garnier E, Diaz S, Buchmann N, et al. (2003) A handbook of protocols for standardised and easy measurement of plant functional traits worldwide. Aust J Bot 51: 335–380.
42. Bradshaw AD (2006) Unravelling phenotypic plasticity-Why should we bother? New Phytol 170: 639–641. doi:10.1111/j.1469-8137.2006.01758.x.
43. Nicotra AB, Atkin OK, Bonser SP, Davidson AM, Finnegan EJ, et al. (2010) Plant phenotypic plasticity in a changing climate. Trends Plant Sci 15: 684–692. doi:10.1016/j.tplants.2010.09.008.
44. Valladares F, Gianoli E, Gómez JM (2007) Ecological limits to plant phenotypic plasticity. New Phytol 176: 749–763. doi:10.1111/j.1469-8137.2007.02275.x.
45. Balaguer L, Martinez-Ferri E, Valladares F, Perez-Corona ME, Baquedano FJ, et al. (2001) Population divergence in the plasticity of the response of Quercus coccifera to the light environment. Funct Ecol 15: 124–135. doi:10.1046/j.1365-2435.2001.00505.x.
46. Pratt JD, Mooney KA (2013) Clinal adaptation and adaptive plasticity in Artemisia californica: implications for the response of a foundation species to predicted climate change. Glob Chang Biol 19: 2454–2466. doi:10.1111/gcb.12199.
47. Davidson AM, Jennions M, Nicotra AB (2011) Do invasive species show higher phenotypic plasticity than native species and, if so, is it adaptive? A meta-analysis. Ecol Lett 14: 419–431. doi:10.1111/j.1461-0248.2011.01596.x.
48. Herman JJ, Sultan SE (2011) Adaptive transgenerational plasticity in plants: case studies, mechanisms, and implications for natural populations. Front Plant Sci 2: 102. doi:10.3389/fpls.2011.00102.
49. Pico F, Ouborg NJ, Van Groenendael JM (2004) Influence of selfing and maternal effects on life-cycle traits and dispersal ability in the herb *Hypochaeris radicata* (Asteraceae). Bot J Linn Soc 146: 163–170.
50. Ortiz MA, Talavera S, Garcia-Castano JL, Tremetsberger K, Stuessy T, et al. (2006) Self-Incompatibility and Floral Parameters in Hypochaers sec. Hypocaheris (Asteraceae). Am J Bot 93: 234–244.
51. Mix C, Arens PFP, Rengelink R, Smulders MJM, Van Groenendael JM, et al. (2006) Regional gene flow and population structure of the wind-dispersed plant species *Hypochaeris radicata* (Asteraceae) in an agricultural landscape. Mol Ecol 15: 1749–1758. doi:10.1111/j.1365-294X.2006.02887.x.
52. Poorter L, Bongers F (2006) Leaf traits are good predictors of plant performance across 53 rain forest species. Ecology 87: 1733–1743.
53. Nicotra A, Leigh A, Boyce C (2011) The evolution and functional significance of leaf shape in the angiosperms. Funct Plant … 38: 535–552.
54. Reich PB, Walters MB, Ellsworth D (1992) Leaf life-span in relation to leaf, plant and stand characteristics among diverse ecosystems. Ecol Monogr 62: 365–392.
55. Westoby M (1998) A leaf-height-seed (LHS) plant ecology strategy scheme: 213–227.
56. Anderson MJ (2001) A new method for non-parametric multivariate analysis of variance. Austral Ecol 26: 32–46. doi:10.1046/j.1442-9993.2001.01070.x.
57. Anderson MJ (2004) PERMDISP: a FORTRAN computer program for permutational analysis of multivariate dispersions (for any two-factor ANOVA design) using permutation. Department of Statistics, University of Auckland, Auckland, New Zealand.
58. Clarke K, Gorley R (2006) Primer v6: User Manual/Tutorial. Plymouth: PRIMER-E.
59. Mokany K, Ash J (2008) Are traits measured on pot grown plants representative of those in natural communities? J Veg Sci 19: 119–126. doi:10.3170/2007-8-18340.
60. Kleyer M, Bekker RM, Knevel IC, Bakker JP, Thompson K, et al. (2008) The LEDA Traitbase: a database of life-history traits of the Northwest European flora. J Ecol 96: 1266–1274. doi:10.1111/j.1365-2745.2008.01430.x.
61. Poorter H, Niinemets Ü, Poorter L, Wright IJ, Villar R (2009) Causes and consequences of variation in leaf mass per area (LMA): a meta-analysis. New Phytol 182: 565–588. doi:10.1111/j.1469-8137.2009.02830.x.
62. Cornwell WK, Cornelissen JHC, Amatangelo K, Dorrepaal E, Eviner VT, et al. (2008) Plant species traits are the predominant control on litter decomposition rates within biomes worldwide. Ecol Lett 11: 1065–1071.
63. Pérez-Harguindeguy N, Díaz S, Garnier E, Lavorel S, Poorter H, et al. (2013) New handbook for standardised measurement of plant functional traits worldwide. Aust J Bot: 167–234.
64. Sultan SE (2000) Phenotypic plasticity for plant development, function and life history. Trends Plant Sci 5: 537–542.
65. Richards CL, Bossdorf O, Muth NZ, Gurevitch J, Pigliucci M (2006) Jack of all trades, master of some? On the role of phenotypic plasticity in plant invasions. Ecol Lett 9: 981–993. doi:10.1111/j.1461-0248.2006.00950.x.
66. Diaz S, Cabido M (2001) Vive la difference: plant functional diversity matters to ecosystem processes. Trends Ecol Evol 16: 646–655.
67. Boege K (2010) Induced responses to competition and herbivory: natural selection on multi-trait phenotypic plasticity. Ecology 91: 2628–2637.
68. Mediavilla S, Escudero A (2004) Stomatal responses to drought of mature trees and seedlings of two co-occurring Mediterranean oaks. For Ecol Manage 187: 281–294. doi:10.1016/j.foreco.2003.07.006.
69. Sultan SE, Barton K, Wilczek AM (2009) Contrasting patterns of transgenerational plasticity in ecologically distinct congeners. Ecology 90: 1831–1839. doi:doi:10.1890/08-1064.1.
70. Gitzendanner M, Soltis P (2000) Patterns of genetic variation in rare and widespread plant congeners. Am J Bot 87: 783–792.

10

Proteomic Analysis of MG132-Treated Germinating Pollen Reveals Expression Signatures Associated with Proteasome Inhibition

Candida Vannini[1]*, Marcella Bracale[1], Rita Crinelli[2], Valerio Marconi[2], Paola Campomenosi[1], Milena Marsoni[1], Valeria Scoccianti[3]

1 Dipartimento di Biotecnologie e Scienze della Vita, Università degli Studi dell'Insubria, Varese, Italy, **2** Dipartimento di Scienze Biomolecolari, Università di Urbino "Carlo Bo", Urbino, Italy, **3** Dipartimento di Scienze della Terra, della Vita e dell'Ambiente, Sezione di Biologia Vegetale, Università di Urbino "Carlo Bo", Urbino, Italy

Abstract

Chemical inhibition of the proteasome has been previously found to effectively impair pollen germination and tube growth *in vitro*. However, the mediators of these effects at the molecular level are unknown. By performing 2DE proteomic analysis, 24 differentially expressed protein spots, representing 14 unique candidate proteins, were identified in the pollen of kiwifruit (*Actinidia deliciosa*) germinated in the presence of the MG132 proteasome inhibitor. qPCR analysis revealed that 11 of these proteins are not up-regulated at the mRNA level, but are most likely stabilized by proteasome inhibition. These differentially expressed proteins are predicted to function in various pathways including energy and lipid metabolism, cell wall synthesis, protein synthesis/degradation and stress responses. In line with this evidence, the MG132-induced changes in the proteome were accompanied by an increase in ATP and ROS content and by an alteration in fatty acid composition.

Editor: Sue Cotterill, St. Georges University of London, United Kingdom

Funding: This work was funded by the University of Insubria RFO funds to MB and by University of Urbino RFO funds to VS. The funders had no role in study design, data collection and analysis, decision to publish, or preparation of the manuscript.

Competing Interests: The authors have declared that no competing interests exist.

* Email: candida.vannini@uninsubria.it

Introduction

Proteolysis is a key mechanism regulating protein homeostasis in living cells. The ubiquitin/proteasome pathway (UPP) is the most important proteolytic system in eukaryotic cells and is essential for protein quality control because it selectively eliminates misfolded, damaged and oxidized proteins. Since the accumulation of altered polypeptides may become cytotoxic, UPP failure may have severe consequences for cellular homeostasis and survival. Even more crucial, the UPP is an important regulatory tool of the cell that controls the concentrations of numerous signaling mediators and transcription factors [1,2]. In plants, selective UPP-mediated protein degradation is critical for many key processes such as hormone perception and signaling, light response, flower development, self-incompatibility, epigenetic regulation, stress response and disease control [2–7].

Pollen germination and tube growth are essential for higher plant sexual reproduction. These processes are highly dynamic and coordinated, integrating many internal and external environmental signals in both space and time [8,9]. Transcriptomic and proteomic data suggest that quick pollen germination and early tube growth largely depend on presynthesized mRNA and protein [10]. On the other hand, protein turnover seems to be essential during tube organization and elongation and several observations have provided evidence that in this context the UPP system is the major regulator. In particular, the upregulation of six subunit components of the 20S core complex of the 26S proteasome has been reported to occur during pollen tube growth [10]. Furthermore, Arabidopsis mutants in regulatory particles of proteasome display alterations in pollen development [11,12]. Finally, chemical proteasome inhibitors impair pollen tube emergence and growth severely and induce aberrant morphology in *Picea wilsonii* and *Actinidia deliciosa* [13–16]. In particular, treatment with benzyloxycarbonyl-leucinyl-leucinyl-leucinal (MG132) caused cytoskeleton disorganization, accompanied by cytoplasmic streaming inhibition and loss of polarized distribution of organelles and nuclei [13,16]. These findings suggest that the ubiquitin proteolytic pathway may control the levels of regulatory proteins essential for cytoskeleton organization. In addition, following proteasome inhibition, a marked decline in cell wall components (pectins and cellulose) was observed especially in tube apical and subapical regions [13]. It has been hypothesized that inhibitor-induced cytoplasmic vacuolization and weakening of cell walls could be responsible for tip swelling and irregularly broadened tube diameters. More recently, it has also been reported that MG132 strongly disturbs mitochondrial remodeling and reduces mitochondrial membrane potential during pollen tube development in *Picea wilsonii* [14]. Finally, Qiao et al. [17] revealed that MG132 specifically blocks compatible pollination in *Antirrhinum deliciosa* by inhibiting degradation of non-self S-RNases.

While these studies have focused mainly on the morphological and cytological effects of UPP malfunction during pollen germination, little is known about molecular mechanisms affected by proteasome inhibition. In the present study, the effect of proteasome inhibitor MG132 on *Actinidia deliciosa* pollen germination was investigated by combining two-dimensional electrophoresis (2-DE) and liquid chromatography-electro spray tandem mass spectrometry (LC-ESI-MS/MS) analysis with the aim of identifying proteins whose abundance and/or activity are directly/indirectly affected by the ubiquitin-proteasome proteolytic pathway. ATP content, lipid profile and ROS production were also analyzed, based on the predicted function of the differentially expressed proteins. The results of these studies are expected to contribute to the shedding of light on the nature and regulation of molecular players crucial for pollen germination and tube growth. The factors that control these processes are, in fact, of considerable interest to the understanding of plant fertility and reproduction mechanisms, as well as in the development of molecular tools aimed at manipulating pollen tube growth for practical purposes.

Materials and Methods

Plant material

Pollen of male kiwifruit *Actinidia deliciosa* var.deliciosa (A. Chev.), C.F. Liang et A.R. Ferguson] was obtained from plants of cv Tomuri growing in experimental plots at the Azienda Tarozzi near Faenza (Italy). Anthers, collected from central flowers of the three-six-flowered inflorescence, were allowed to dehisce under controlled conditions as described by Calzoni et al. [18]. Pollen was separated by sieving and stored at −20°C until use.

The percentage of viable pollen grains before freezing was 94±1%, as determined using the fluorochromatic reaction with fluoresceine diacetate which assesses the integrity of the plasmalemma of the vegetative cell [19]. Compared to fresh pollen of the same year no significant decline in viability was registered during the lasting of the experiments described in this paper.

In vitro pollen germination

Pollen was rehydrated for 30 min at 30°C under 100% relative humidity. Germination was performed by suspending the rehydrated pollen grains (1 mg mL^{-1}) in a liquid medium containing 0.29 M sucrose and 0.4 mM boric acid according to Scoccianti et al. [16]. Proteasome inhibitor MG132 (Biomol Research Laboratory, Plymouth Meeting, PA) was added into the medium at the beginning of incubation to a final concentration of 40 μM (MG132-treated). Since the inhibitor was dissolved in dimethylsulphoxide (DMSO), parallel incubations containing the solvent alone, at the same concentration (0.08%) that was present in MG132-treated samples, were set up (DMSO-controls). In some cases, pollen cultures receiving neither MG132 nor DMSO (controls) were run in parallel. Cultures were incubated for 90 min, unless otherwise specified, at 30°C in the dark. After incubation, pollen cultures were centrifuged at 1400 *g* for 2 min, the supernatants were discarded and the cells were washed with fresh medium containing 0.29 M sucrose. Cells were then collected by Millipore vacuum filtration (Millipore Corporation, Billerica, MA), detached from the filter membrane (8.0 μm pore size) and immediately used for the subsequent analyses or frozen in liquid nitrogen and stored at −80°C until use.

The percentage of tube emergence (% germination) was determined with an image analyzer-Axioplan microscope (Zeiss, Jena, Germany) combination by scoring at least 1000 randomly chosen pollen grains per sample, resulting from the sum of several non overlapping fields. Pollen was considered germinated only when the tube length was greater than the diameter of the grain.

Protein sample preparation

Samples were homogenized by using mortar and pestle in liquid nitrogen with an addition of sand quartz. Total proteins were extracted as described by Marsoni et al. [20]. After sample clarification at 13000 *g* for 10 min, the protein concentration was measured by Bio-Rad protein assay (Hercules, CA, USA), using bovine serum albumin as a standard. The samples were directly loaded for isoelectrofocusing (IEF) or stored in aliquots at −80°C until use. The extractions were performed in triplicate and the results were highly reproducible.

Two-dimensional IEF/SDS-PAGE

IEF was carried out with 700 μg of total protein extract by using an immobilized 4 to 7 pH gradient (Immobiline DryStrip, 18 cm; Amersham Biosciences, Uppsala, Sweden). The strips were rehydrated in the IPGphor system (Amersham Biosciences, Bucks, UK) for 1 h at 0 V, 20°C and 10 h at 30 V, 16°C with the solubilization buffer containing 7 M urea, 2 M thiourea, 4% 3-[(3-Cholamidopropyl) dimethylammonio]-1-propanesulfonate (CHAPS), 50 mg/mL dithiothreitol (DTT), 0.5% of carrier ampholyte (3–10 NL IPG buffer; Amersham Biosciences), bromophenol blue 0.005% and the protein extracts. IEF was performed at 16°C in the IPGphor system (Amersham Biosciences) for 4 h at 200 V, from 200 to 3500 V in gradient during 30 min, 3 h at 3500 V, from 3500 to 8000 V in gradient during 30 min, after which the run was continued at 8000 V to give a total of 70 kVh.

Each focused strip was equilibrated for 30 min against 6 M urea, 30% glycerol, 2% SDS, 50 mM Tris-HCl pH 8.8, 2% DTT and then a further 30 min with the substitution of the DTT with 2.5% iodoacetamide in the equilibration buffer. The equilibrated strips were placed on top of vertical 12.5% polyacrylamide gels (acrylamide/PDA 12.5% T, 2.6% C, 0.375 M Tris-HCl pH 8.8, TEMED 0.05% v/v, APS 0.1% w/v). The molecular weight markers, covering a 10 to 250 kDa range, were run on the acidic side of each gel. Electrophoresis was performed at 4°C in a Laemmli running buffer (25 mM Tris-HCl pH 8.3, 192 mM glycine, 0.1% SDS) for 30 min at 15 mA/gel then at 45 mA/gel until the dye front reached the bottom of the gel. Each extraction was analyzed by three gel replicates.

Staining and analysis of 2DE gels

Proteins were detected with Colloidal Coomassie Brilliant Blue (CCBB) modified as described in Marsoni et al. [21]. Thereafter, gels were digitalized by ImageScanner (Amersham Bioscience). Image and data analysis of the scanned (300 dpi, 16-bit greyscale pixel depth) gels as TIFF files were performed by using the Image Master 2D Platinum imaging software version 5.0, which allows spot detection, spot matching among multiple gels, background subtraction and spot area and optical intensity quantization. For this purpose, data were normalized by expressing protein abundance as percent spot volume relative to volume of total protein in the gel (%Vol). Three replicate gels in each experimental group were averaged and the resulting gels (average-gels) contained only spot presents in all the replicates. The average-gels were compared and only proteins with a fold change of ±1.5, significant in Student's *t*-test at a level of 95%, were accepted as differentially expressed. These spots were selected for MS/MS analysis.

Mass Spectrometry analysis and protein identification

Spots digestion was performed as described in Marsoni et al. [21]. The extracted tryptic fragments were analysed by MS/MS after reverse phase separation of peptides (liquid chromatography-electro spray tandem mass spectrometry, LC-ESI-MS/MS). For all experiments, a Finnigan LXQ mass spectrometer, equipped with a Finnigan Surveyor MS Pump Plus HPLC system (Thermo Electron Corporation, CA, and USA), was used. Chromatography separations were conducted on a BioBasic C18 column (150 μm I.D.×150 mm length and 5 μm particle size; Thermo Electron Corporation, USA), using a linear gradient from 5 to 75% acetonitril (ACN), containing 0.1% formic acid with a flow of 2 μL min^{-1}. Including the regeneration step, one run lasted 80 min. Acquisitions were performed in the data-dependent MS/MS scanning mode (full MS scan range of 400–1400 m z^{-1} followed by Zoom scan for the most intense ion from the MS scan and full MS/MS for the most intense ion from the zoom scan), thus enabling a dynamic exclusion window of 3 min.

Protein identifications were conducted by correlation of uninterpreted tandem mass spectra to the entries of a non-redundant protein and/or EST-viridiplantae database downloaded from the National Center for Biotechnology Information (NCBI NR 04/29/2011: 13841106 sequences, 4750259772 residues. Viridiplantae: 881708 sequences. Plants_EST: 144355992 sequences, 25218209504 residues) using the MASCOT open source (http://www.matrixscience.com). The parameters were set to allow one missed cleavage and considering as variable modification cysteine carbamido-methylation and methionine oxidation. The precursor ion tolerance was set at 1.2 Da and fragment ion tolerance 0.6 Da. Only proteins with a minimum of two matching peptides were considered and Mowse scoring system was used to assign correct identification.

RNA extraction and qPCR analysis

Total RNA from pollen was extracted with the Spectrum Plant Total RNA Kit (Sigma, Milan, Italy), according to the manufacturer's instructions and treated with DNase 1 (AMPD, Sigma, Milan, Italy). For real-time quantitative PCR (qPCR), cDNA was generated from 1.5 μg of RNA by using the iScript cDNA synthesis kit (Biorad, Milan, Italy). qPCR was performed on an AB7000 thermal cycler (AB, Milan, Italy) using the iTAQ Universal Sybr Green Supermix (Biorad, Milan, Italy). The total RNA was extracted twice (in two independent experiments). From each extraction two independent cDNA were prepared to minimize the impact of random fluctuations. Three amplifications were subsequently performed for each relevant gene and all amplifications were repeated on a second independent plate. Relative mRNA quantification was obtained using the ΔΔCt method, using Actin and 18S rRNA as reference. Tubulin was not used as reference as its expression was found affected by the treatment. Melting curve analysis was performed to ensure that single amplicons were obtained for each target. Primers for the genes under investigation were designed to encompass at least one intron, whenever possible. To do this, for each protein identified by Mascot, the full length or partial cDNA clone from Actinidia (for which no information on genomic DNA is available) was retrieved and subjected to Blast analysis to obtain the sequence of the nearest ortholog in a species for which genomic information was available. The mRNA-to-genomic alignment program "Spidey", available at NCBI (http://www.ncbi.nlm.nih.gov/spidey/), was used to obtain information on gene organization, assuming that intron-exon boundaries would be conserved among the species under consideration. This information was used to set the parameters for designing the primers using the Primer-Blast tool at NCBI (http://www.ncbi.nlm.nih.gov/tools/primer-blast/). Data on Blast analyses are provided on Table S1; primer sequences are reported in Table S2.

Phosphoglycerate kinase enzymatic assay

Samples were lysed in liquid nitrogen by four freeze-thaw cycles in the presence of 50 mM Tris-HCl, pH 7.8 containing a cocktail of proteasome inhibitors (Roche Applied Science, Indianapolis, IN, USA). The extract was tip sonicated on ice for 15 sec at 50 W, three times. Sonication was performed with a Labsonic 1510 Sonicator (Braun, Melsun- gen, Germany) equipped with a 4 mm probe. Samples were then centrifuged at 12000 rpm in an Eppendorf centrifuge for 10 min at +4°C to remove insoluble debris. Enzyme activity was assayed spectrophotometrically by monitoring the oxidation of NADH at 340 nm, according to Beutler [22].

ATP content determination

Samples were immediately extracted in cold 7% perchloric acid using a Potter–Elvehjem apparatus. The extracts were centrifuged at 14000 *g* for 10 min at 4°C. Aliquots of the supernatants (500 μL) were neutralized with 3 M K_2CO_3 solution and centrifuged at 5000 *g* for 10 min at 4°C. The ATP content was determined spectrophotometrically following the reduction of $NADP^+$ to NADPH at 340 nm, as reported by Beutler [22].

Dosage of lipids and fatty acid profile analysis

Total lipid content was determined in kiwifruit pollen after 0, 30, 60 or 90 min of incubation, in the absence (DMSO-control) or in the presence of 40 μM MG132, according to the method of Izard and Limberger [23]. Pollen cultures were centrifuged at 1400 *g* for 2 min, the supernatants were discarded and the cells were suspended with MilliQ water and lysed in liquid nitrogenby five freeze-thaw cycles. One-hundred microliter aliquots of the lysed cell suspensions, or of MilliQ water (reference tube), were added to a stoppered glass tube. Two milliliters of 96% sulphuric acid were then added to each tube. The tubes were incubated in a boiling water bath for 30 min and then cooled for 5 min in a water bath at room temperature. Five milliliters of phosphoric acid–vanillin reagent [0.6 g of vanillin (4-hydroxy-3-methoxybenzaldehyde; Sigma) in 100 mL distilled water, with the volume adjusted to 500 mL with 85% phosphoric acid] were added to the tubes and incubated at room temperature for 15 min, and then absorbance was read at 530 nm. The reference curve was composed of Triolein (1,2,3-tri[*cis*-9-octadecenoyl] glycerol; Sigma) ranging from 10 to 100 μg [23].

Fatty acids (FAs), extraction was carried out according to the method of Folch et al. [24] with few modifications. Samples were extracted with a chloroform/methanol solution (2:1, v:v) under stirring for 1 min, and then equilibrated with one fourth its volume of distilled water. Extracts were centrifuged for 20 min at 3500 *g*. The upper phase was discarded and the lower was filtered through a 0.2 μm pore PVDF membrane. The filtrate was collected and taken to dryness in a rotary evaporator. Methyl esters of fatty acids were obtained suspending the samples in 100 μL of 2 M KOH-CH_3OH with shaking for 10 min and then extracted with 100 μL of n-hexane by stirring with a vortex for 30 sec at maximum speed. The hexane phase, containing fatty acid methyl esters, was transferred into a clean tube, and the residual phase was extracted again with 100 μL of n-hexane. The hexane phases were combined and concentrated with a stream of liquid nitrogento a final volume of 50 μL. For gas chromatography/mass spectrometry analysis of fatty acid methyl esters, a gas chromatograph (Agilent 6890 N, U.S.A.), coupled with double focalization high resolution Jeol JMS-GC-Mate II (Japan) in full scan mode and equipped with Jeol Software, was used. Chromatographic separation was performed using a Supelcowax 10 GC column (0,25 mm i.d.×30 m length, 0,25 μm film thickness). The oven temperature was programmed as follows: 100°C, hold for 3 min, 20°C min^{-1} to 165°C no hold, 12°C min^{-1} to 270°C, hold for 5 min. The carrier gas was helium with a constant flow of 1 mL min^{-1}. The injector temperature was 260°C and the interface temperature was 270°C. Injection volume was 1 μL with 3 min of delay time in splitless mode.

ROS production

The nitroblue tetrazolium (NBT) assay described by Wang et al. [25] was used, with slight modifications concerning incubation time and temperature. After 30 and 90 min of incubation in the absence (control) or in the presence of 0.08% DMSO (DMSO-control) or 40 μM MG132, pollen cultures were centrifuged at 1400 *g* for 2 min. and the pelletted cells were added with 2 mM NBT. After 30 min of the reaction at 30°C, the suspensions were centrifuged and washed twice with 1 mL water. The formazan produced by NBT reduction was dissolved in 1 mL of methanol and incubated for 30 min at 30°C. After centrifugation, the absorbance of supernatants was read at 530 nm.

Detection of carbonylated protein

Protein carbonylation assay was assessed by using the commercially available OxyBlot protein oxidation detection kit (Millipore). Protein samples containing 40 μg of proteins were added to an equal volume of 12% SDS. Then protein carbonyl groups were derivatized to 2, 4-dinitrophenylhydrazone (DNP) by incubation with one additional volume of 2,4-dinitrophenylhydrazine for 30 min at room temperature.

The derivatization reaction was stopped by the addition of 20 μL neutralization solution. Proteins were separated by 12% SDS-PAGE and transferred to PVDF membrane (Serva). The oxidatively modified proteins were detected using anti-DNP antibodies (anti-dinitrophenyl-group antibodies, Sigma) and visualized by a chemiluminescence detection kit (SuperSignal, Pierce Biotechnology, Rockford, IL, USA). To monitor the equal loading of samples, CCBB was used to stain the proteins in a duplicate gel.

Statistical analysis

Statistical analysis of data was performed by ANOVA for repeated measurement, followed by the Tukey-Kramer multiple comparison test (for n groups>2) or with the paired t-test (for n groups = 2), using the GraphPad Prism package, version 5.0 for Macintosh (GraphPad Software, San Diego, CA, USA). Differences between values were assumed statistically significant at $p<0.05$ (*), very significant at $p<0.01$ (**).

Results and Discussion

Proteasome inhibition alters the expression of soluble proteome

To investigate the protein changes induced by proteasome inhibition, kiwifruit pollen was cultured in the presence of 40 μM MG132 for 90 min. Doses of MG132 ranging from 10 to 100 μM have been used to analyze the involvement of the proteasome in various aspects of plant physiology [26–28]. These differences probably reflect a different sensitivity of the cells/organisms and/or of the proteasome inhibition pathway studied; depending on the experimental model investigated, different uptake and toxicity of inhibitors could be also suggested. Concerning pollen germination, concentrations of MG132 between 20 and 80 μM have been shown to be effective in altering pollen germination and tube length in *Picea wilsonii* with a dose-dependent effect particularly evident between 10 and 40 μM, at least as shown for tube elongation [13]. Similarly, we have previously demonstrated that kiwifruit pollen germination is affected by treatment with both 40 and 80 μM MG132. In particular, the tube mass produced and the rate of tube elongation were altered in a dose-dependent manner [15,16]. Notably, at 40 μM MG132, kiwifruit pollen germination and tube growth were markedly affected, but were not almost completely blocked, as at the higher dose of 80 μM. Thus, in light of these findings and with the aim of limiting indirect versus MG132-induced direct effects, all the analyses performed in this study were conducted using MG132 at the final concentration of 40 μM.

Under these experimental conditions, the percentage of germination was 87.3±3.8%, 84.6±2.5% and 54.0±4.9% in control, DMSO-control and MG132-treated pollen, respectively. Therefore, MG132 induces a reduction in pollen germination of more than 30% compared to DMSO-control or untreated pollen. It is worth noting that we previously demonstrated that the effect of MG132 was reversible, indicating that the inhibition of pollen germination was not due to cell death [15]. Moreover, in the presence of MG132, the majority of germinated pollen grains produced markedly shorter tubes with altered morphology (Figure 1A). In particular, tubes showed diffuse tip swelling and branching. No solvent effect due to 0.08% DMSO was evident comparing both tube length as well as morphology of DMSO-controls (Figure 1B) and controls (Figure 1C) in agreement with our previous results [15].

To detect changes in the proteome, total soluble proteins from MG132-treated and DMSO-control pollen were analyzed by using both 2DE and a mass spectroscopy approach. We extracted 35.5±3.5 μg and 34.3±5.5 μg of proteins from 1 mg of DMSO-control and MG132-treated pollen respectively. Following CCBB staining, approximately 600 reproducible protein spots were detected per sample (Figure 1D). All spots were matched by gel-to-gel comparisons and the differences in the relative abundance (% vol) of each spot were evaluated by software-assisted analysis. The ANOVA test ($p<0.05$), coupled with a threshold of 1.5-fold change in levels, clearly revealed 24 differentially expressed protein spots in samples treated with MG132 compared with respective control, as shown in Figure 1D, E. These differentially expressed proteins mainly accumulated under conditions of proteasome inhibition, while only one protein (spot 19) was downregulated in MG132-treated vs. DMSO-control pollen.

The excised spots were in-gel digested and analyzed by LC-ESI-MS/MS; 22 protein spots were successfully identified, corresponding to 14 unique proteins (Table S3). The experimental molecular masses (MW) and pIs of the majority of the identified proteins were consistent with the theoretical values, as judged by the locations of the spots on 2-D gels. However, there were some exceptions: spots 10, 11, 12, 6 and 7 displayed more acidic experimental pI values than the theoretical values, while spot 1 had a higher experimental MW than the MW of the corresponding identified protein (Table S3). The uniquely identified proteins were categorized as follows into five functional groups based on their predicted function: energetic metabolism (4), lipid metabolism (3), cell wall synthesis (1), protein synthesis/degradation (4) and stress response (2).

To investigate altered protein level mechanisms, we analyzed the RNA expression of genes corresponding to the differentially accumulated proteins by qPCR. The results showed that the observed accumulation of these proteins in pollen treated with MG132 was primarily due to altered protein turnover rather than

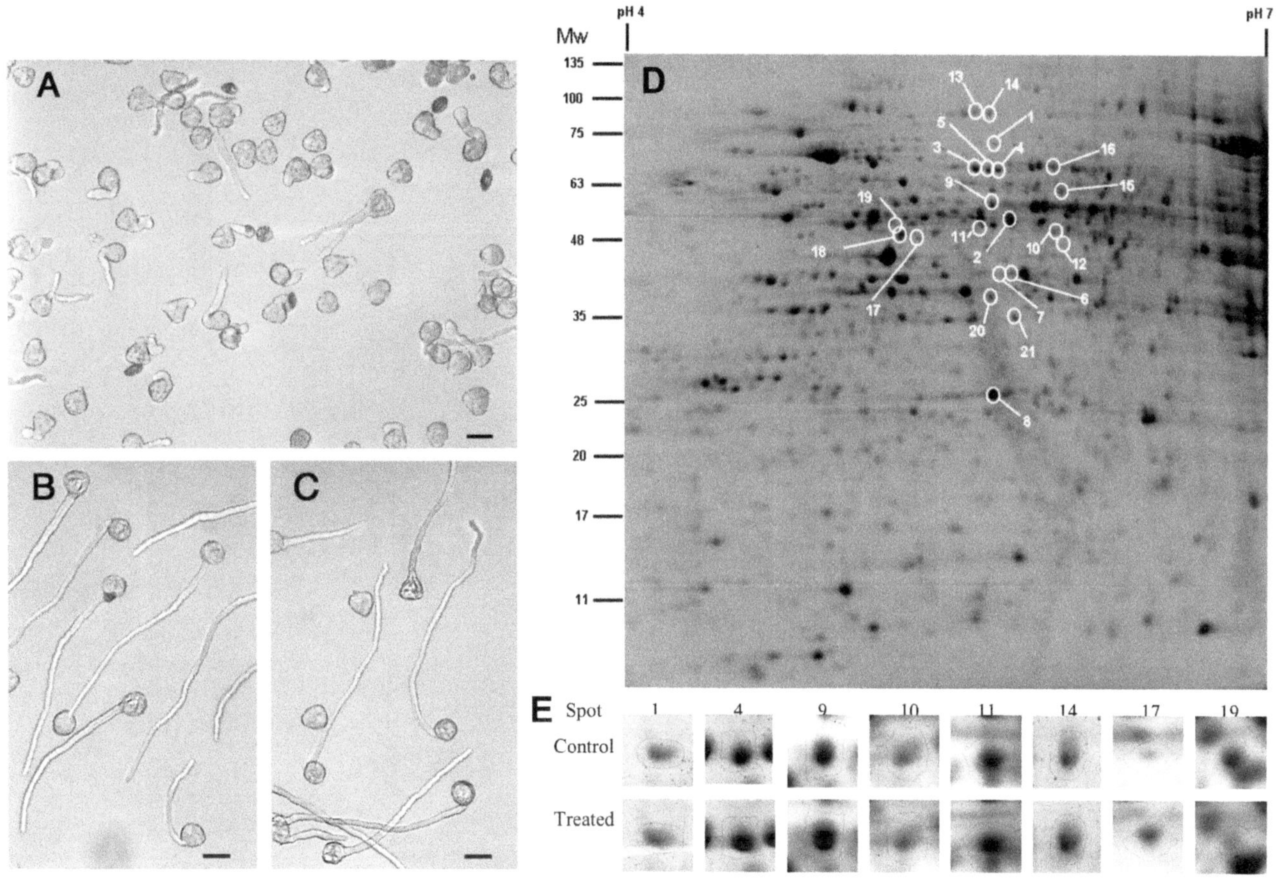

Figure 1. Effects of MG132 on kiwifruit pollen. Pollen after 90 min of incubation in the presence of 40 μM MG132 (A), 0.08% DMSO (B), or without any treatment (C). Bar = 30 μm. D) Representative gel of MG132-treated pollen showing significant ($p<0.05$) changes in protein abundance. (E) Selected differentially accumulated protein spots in pollen exposed to 40 μM MG132.

altered regulation of mRNA levels (Figure 2). Only the levels of phosphoglycerate mutase, lysyl-t-RNA synthase and isoflavone reductase transcripts showed a statistically significant increase in pollen treated with MG132 compared to the DMSO-control (Figure 2).

Energy metabolism

Treatment with MG132 induced the accumulation of four unique proteins involved in energy metabolism. Three of these, namely phosphoglycerate kinase (PK), phosphoglycerate mutase (PGM) and enolase (ENO), act consecutively in the glycolytic pathway to produce phosphoenolpyruvate (PEP), which is indispensable for energy metabolism. The qPCR analysis indicated that the accumulation of PGM was associated to increased transcript levels (Figure 2), suggesting that proteasome inhibition could indirectly affect the intracellular levels of the enzyme.

Since the proteomic data indicate that there was an accumulation of enzymes involved in energy metabolism, we evaluated the effect of MG132 on ATP content in kiwifruit pollen after 90 min of incubation. As shown in Figure 3A, a 43% increase in ATP content was observed in MG132-treated pollen compared to DMSO-control pollen. However, when PK activity was assayed no significant differences were detected between MG132- and DMSO-control pollen (1.38±0.16 vs 1.42±0.14 U/mg of total proteins, respectively). This evidence together with the fact that the pI of spots corresponding to PK showed a strong shift toward more acidic values, suggests that the protein accumulates in an oxidized inactive form. Finally, as far as ENO is concerned, the observed MW of the spots was higher than the theoretical value, suggesting possible ubiquitination of this protein. In light of this evidence, it could be hypothesized that the accumulation of ATP in MG132-treated pollen might depend on lower usage rather than on increased production. This is because, pollen tube growth is critically dependent on ATP pools which are used to power energy-consuming processes such as actin polymerization and exo- and endocytosis [29] and following MG132 treatment these processes are inhibited [13].

Lipid metabolism

In MG132-treated pollen, we detected the accumulation of three unique proteins involved in fatty acid metabolism: an acetyl-CoA biotin carboxylases (BC), a ketoacyl-ACP synthase (KAS I) and a phospholipaseDα (PLDα) (Table S3). BC is a subunit of heteromeric acetyl-coenzyme carboxylase (htACCase), which produces malonyl-CoA in plastids and is a major regulatory point for plant fatty acid synthesis. Madoka et al. [30] showed that post-transcriptional regulation is involved in the formation of a functional htACCase. Moreover, KAS I catalyzes the elongation of *de novo* synthesised fatty acids and is therefore crucial for fatty acid synthesis in Arabidopsis [31]. In the presentexperimental system, the accumulation of BC and KAS I proteins in MG132-

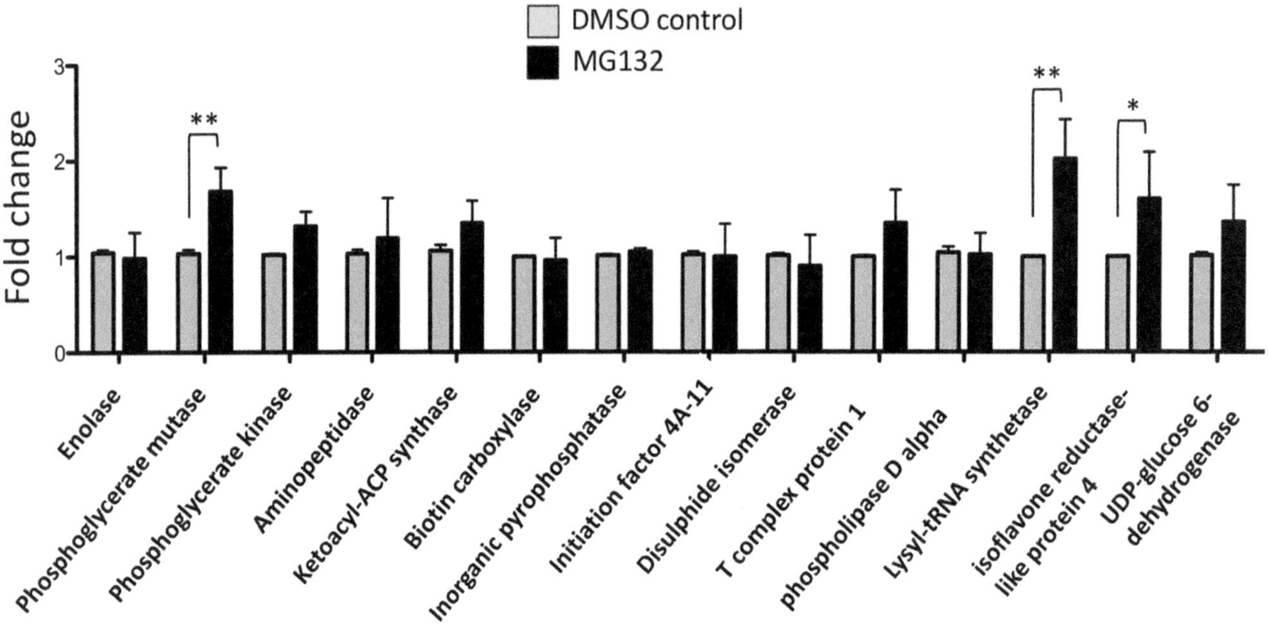

Figure 2. qPCR analysis of DMSO-control or MG132-treated kiwifruit pollen after 90 min of incubation. The values reported are means ± SD of two independent experiments and several technical replicates. Asterisks indicate statistically significant differences in MG132-treated samples relative to DMSO-controls (* $p<0.05$; ** $p<0.01$).

treated pollen may affect the biosynthesis of fatty acids required for normal pollen tube development.

To test whether proteasome inhibition resulted in an altered lipid metabolism, we investigated total lipid content and fatty acid composition. In both MG132-treated and DMSO-control pollen, we observed a statistically significant increase in lipid content after 90 min of germination. However, no differences were found between DMSO-control and proteasome inhibitor-treated cells at all time points considered (Figure 3B).

The following fatty acids were separated and identified in kiwifruit pollen: palmitic acid (C16:0), stearic acid (C18:0), oleic acid (C18:1, n-9), linoleic acid (C18:2, n-6), linolenic acid (C18:3) (Figure 4). Similar results were obtained in pollen from other species [32,33]. Small quantities of eicosanoic acid (C20:0) (2.98%), eicosadienoic acid (C20:2) (0.6%), eicosatrienoic acid (C20:3) (0.68%) and docosanoic acid (C22:0) (2.05) were also detected.

After 90 min of incubation in the presence of MG-132 the content of palmitic acid was lower than that found in the

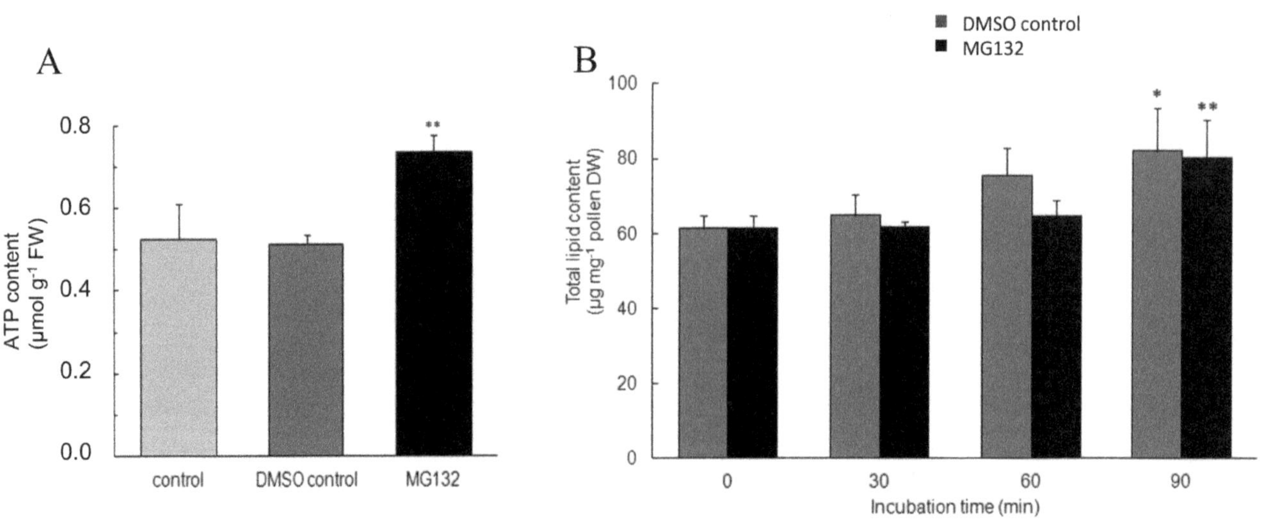

Figure 3. ATP and total lipid content. A) ATP content in control, 0.08% DMSO-control or 40 µM MG132-treated kiwifruit pollen after 90 min of incubation. Asterisks indicate statistically significant differences relative to controls. FW, fresh weight. B) Lipid content at different times of germination in the presence of MG132 or 0.08% DMSO. Asterisks indicate statistically significant differences compared with the respective time zero. DW, dry weight. All data are the means ± SD of two independent experiments each with three biological replicates.

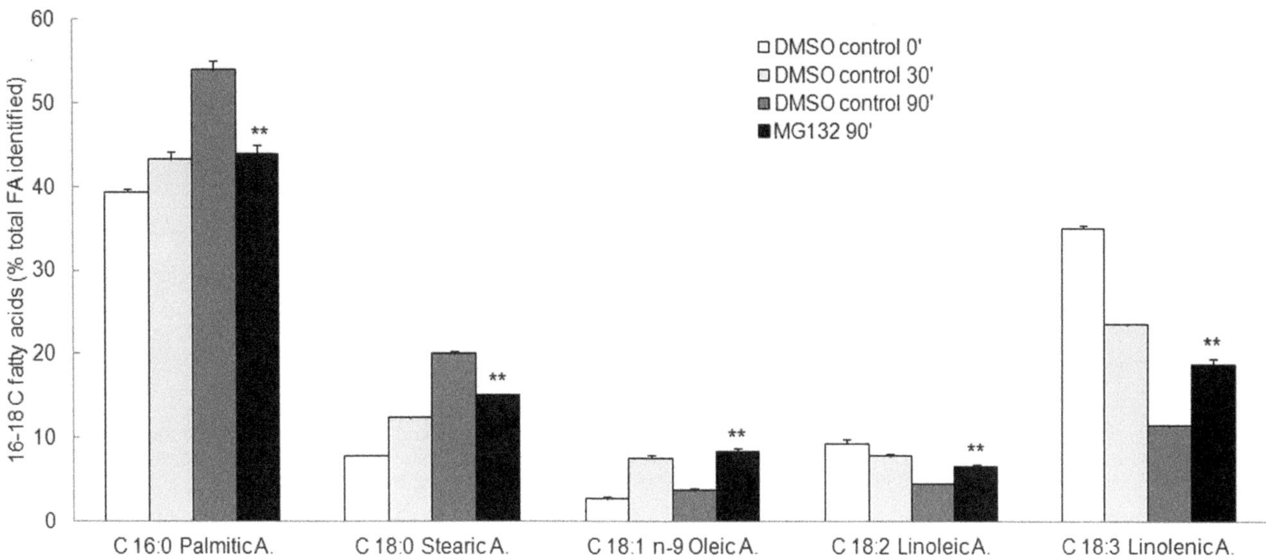

Figure 4. Fatty acid profile in kiwifruit pollen at different times of incubation in the presence of 0.08% DMSO (DMSO-control) or 40 μM MG132. Each fatty acid was identified by comparing its fragmentation pattern with the mass spectrometer database. Results of triplicate analysis were averaged and the content of individual fatty acids in pollen was expressed as percentages of the total fatty acids identified ± S.D. Asterisks indicate significant differences compared with the respective DMSO-control. In some cases bars are too close to be visible. FA, fatty acids.

respective DMSO-control, suggesting that KAS I, used for making 6–16 C fatty acids, accumulates in an inactive form. Significantly, the isoelectric point of the accumulated protein was more acidic than that of the native form, possibly as a consequence of oxidative damage (Table S3). From the fatty acids profile it was also evident that upon MG132 treatment, a higher amount of the unsaturated fatty acids oleic, linoleic and linolenic acids was detected (Figure 4). From these data it is possible to speculate that proteasome modulates the activity of desaturating enzymes. It is worth noting that during germination, a time-dependent increase in saturated palmitic and stearic fatty acids occurs together with a concomitant decrease in unsaturated linoleic and linolenic acids (Figure 4). Thus, the alteration in fatty acid composition associated with proteasome inbition may contribute in explaining the impairement of pollen germination and tube growth.

The secretion of cell wall polysaccharides in growing pollen tips also requires spatial and temporal coordination of secretory vesicle trafficking. In MG132-treated pollen, we detected the accumulation of a PLDα (Table S3). PLD hydrolyzes common structural phospholipids to produce phosphatidic acid (PA). PLD and PA produced by PLD are involved in the production of exocytotic vesicles, in their delivery to the pollen tube apex and in their fusion with the plasma membrane [34,35]. In pollen tubes, the majority of PA is produced via the PLD pathway, and the expansion of the apical region of the pollen tube is associated with a several-fold increase in PA content [36]. Moreover, Pleskot et al. [35] demonstrated that actin polymerization is specifically regulated by PLD-derived PA in tobacco pollen tubes. These data suggest the proteasome exerts a regulatory effect on pollen tube development through the modulation of phospholipid signaling.

Cell wall biosynthesis

Pollen tubes are tip-growing cells and their growth requires the continuous deposition of cell wall material in a highly controlled manner to ensure the morphogenesis of a perfectly cylindrical structure. MG132-treated pollen exhibits dramatic modifications in tube-tip morphology, tip swelling and irregularly broadened diameters [16,13].

In pollen treated with MG132, we identified the accumulation of UDP-glucose dehydrogenase (UGD) (Table S3), an enzyme that oxidizes the conversion of UDP-D-glucose to UDP-D-glucuronate, a key sugar nucleotide involved in the biosynthesis of plant cell wall polysaccharides [37]. UGD is a good candidate for a control point in the metabolic pathway of cell wall synthesis because it is present at low concentrations with respect to other enzymes in the pathway and it operates far from equilibrium [38]. UGD homeostasis is highly important for cell wall integrity. In fact, both over-expression and suppression of this enzyme reduce the polysaccharide concentrations of the cell wall [39–42]. The altered homeostasis of UGD in the presence of MG132 correlates very well with the observed inhibition of pollen tube growth and with the morphological disorder of the pollen tube. Our data confirm the hypothesis of Sheng et al. [13] that MG132 inhibits the turnover of enzymes associated with cell wall organization.

Stress response

MG132 caused a significant increase in ROS content with respect to the controls at both 30 min of incubation, when pollen tube emergence was initiated, (Figure 5A) and at 90 min, when a large number of tubes had already emerged and elongated in the control pollen (Figure 5B). This increase in intracellular ROS levels in response to MG132 treatment has been previously observed both in plants and animal cells [43–45]. ROS are well known to play a dual role as both deleterious and beneficial species depending on their concentration in plants. At high concentrations ROS causes damage to biomolecules, whereas at low/moderate concentrations it acts as a second messenger in intracellular signaling cascades that mediate several responses in plant cells. ROS levels play an important role in pollen tube emergence and elongation and must be strictly regulated [46,47]. Thus, ROS over-accumulation induced by proteasome inhibition may contribute to the observed inhibition of pollen germination and tube elongation.

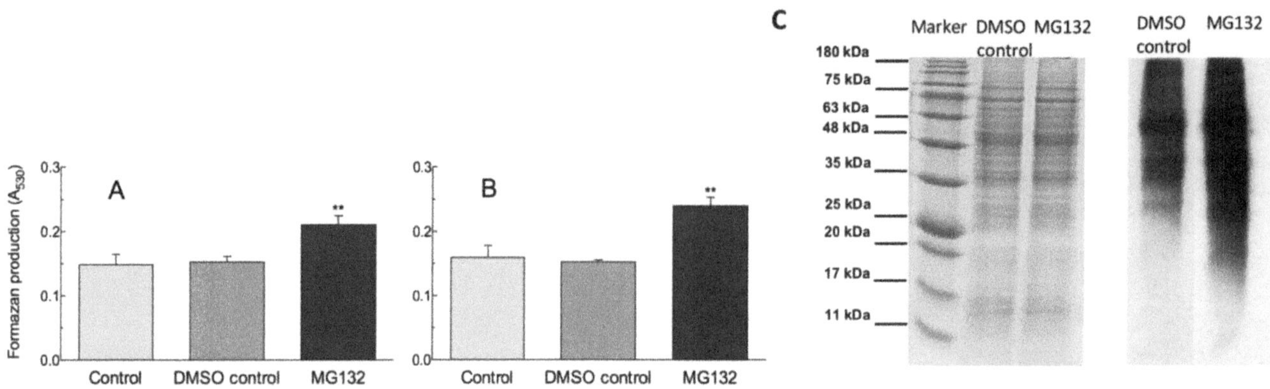

Figure 5. ROS levels and oxidatively modified proteins. ROS generated by germinating kiwifruit pollen after 30 (A) or 90 min of incubation (B) in the absence (control) or in the presence of DMSO (DMSO-control) or MG132. Data are mean values ± S.D. of three replicates. Asterisks indicate significant differences relative to controls. C) Protein stain (left) and anti-DNP immunoassay (right) are shown.

Interestingly, the inhibition of the proteasome resulted in the accumulation of oxidatively-modified polypeptides, as revealed by an increase in carbonylated proteins (Figure 5C). This could be the consequence of the increased intracellular ROS levels as well as the impaired degradation of oxidatively damaged proteins, due to proteasome inactivation. Moreover, MG132 led to the accumulation of an isoflavone reductase-like protein (ISFR), however qPCR data indicate that this accumulation is associated

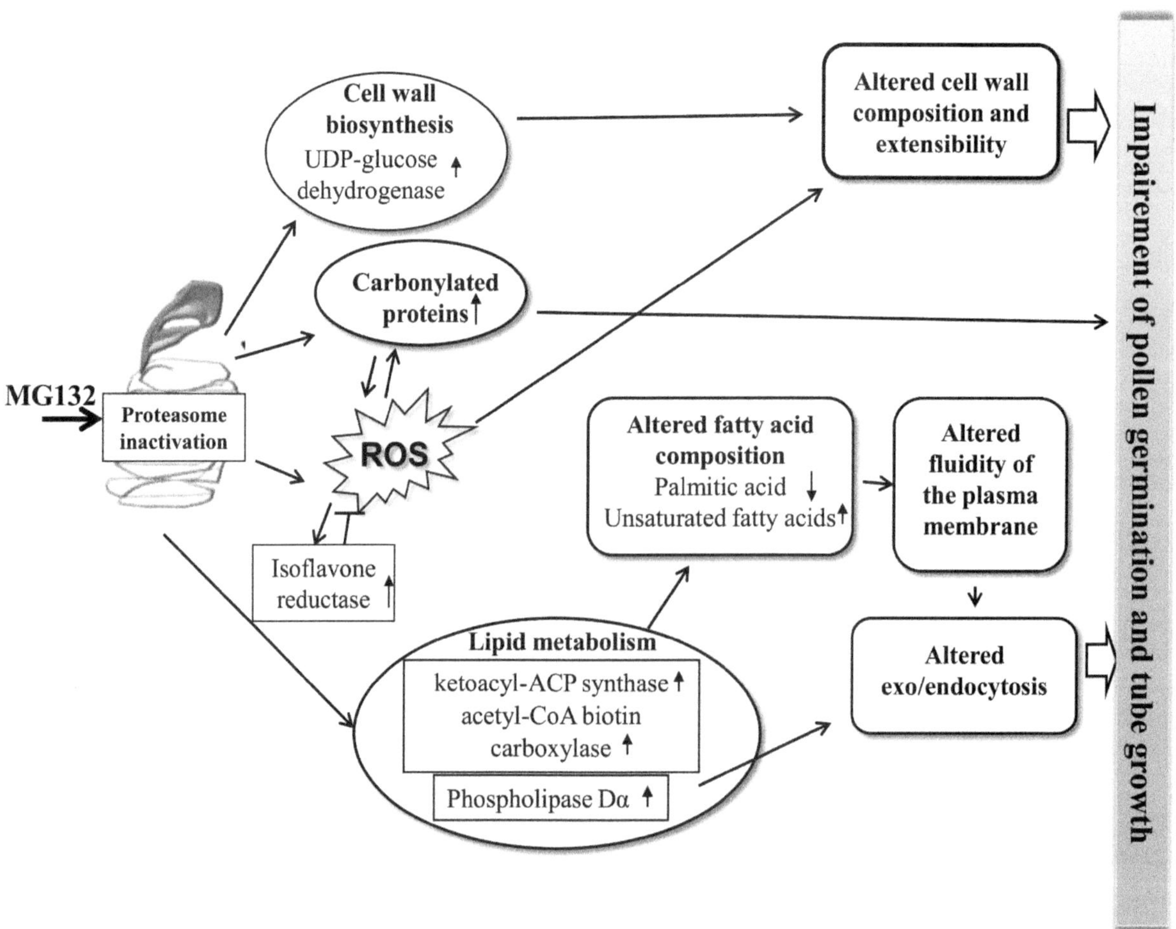

Figure 6. Model of the response of kiwifruit pollen to MG132 treatment. The proposed mechanisms suggest that ROS-induction, lipid metabolism and cell wall biosynthesis are important cellular events in pollen tube development under MG132 treatment. Only some of the responsive proteins are indicated. Arrows "↑" and "↓" indicate an increasing or a decreasing amount of proteins and metabolites.

with an increased gene expression (Figure 2), suggesting that the proteasome could affect protein levels indirectly. Indeed, it has been demonstrated that ISFR transcript levels are upregulated by oxidants [48,49], further confirming a role of proteasome inhibition in modulating ROS homeostasis.

In the current analysis, MG132 treatment led also to the accumulation of a protein disulfide isomerase-like (PDI). PDI facilitates the retrotranslocation of some misfolded proteins from the ER to the proteasome in yeast and animal cells [50,51]. Immunogold labeling with anti-ubiquitin antibody has provided direct proof of UbP accumulation in ER membranes upon MG132 treatment, leading to significant ER stress [13]. In addition, PDI is recognized and selectively degraded by the proteasome [52]. Thus, the inhibition of the proteasome may contribute to alter PDIcellular homeostasis, which has been shown to be important for pollen tube growth in *Arabidopsis thaliana* [53].

In conclusion, the accumulation of ATP in pollen treated with MG132 indicates that the inhibition of tube growth is not due to a deficit in energy power. Our data indicate that the increase in the content of ATP is due to a lower usage rather than an increased production.

As shown in Figure 6, proteasome inhibition affects the accumulation of some proteins involved in lipid metabolism (ketoacyl-ACP synthase, acetyl-CoA biotin carboxylase and phospholipase Dα) and induces the alteration of the composition of fatty acids. This could lead to a change in the fluidity of the plasma membrane and to the impairment of the exo/endocytosis during the elongation of the pollen tube. The ROS increase in MG132-treated pollen disrupts the redox homeostasis as indicated also by the induction of the isoflavone reductase. This fact could increase the level of oxidatively damaged proteins as well as the level of the impaired degradation due to proteasome inhibition. Moreover, the intracellular ROS accumulation and the altered UDP-glucose dehydrogenase homeostasis could modify the composition of the cellular wall and extensibility, contributing in this way to the altered growth and morphology of the pollen tube.

Acknowledgments

We are grateful to Prof. Anna Speranza for her considerable help, advice and critical reading of the manuscript. We also thank Nicodemo Mele (University of Bologna) for technical assistance in preparing pollen photographs.

Author Contributions

Conceived and designed the experiments: VS RC MB CV. Performed the experiments: VS RC CV VM PC. Analyzed the data: MM CV RC VS MB PC. Contributed reagents/materials/analysis tools: VS CV. Wrote the paper: CV MB VS RC PC.

References

1. Hershko A, Ciechanover A (1998) The ubiquitin system. Annu Rev Biochem 67: 425–479.
2. Vierstra RD (2007) The ubiquitin-specific protease subfamily UBP3/UBP4 is essential for pollen development and transmission in Arabidopsis. Plant Physiol 145: 801–813.
3. Wang F, Deng XW (2011) Plant ubiquitin-proteasome pathway and its role in gibberellin signaling. Cell Res 21: 1286–1294.
4. Santner A, Estelle M (2010) The UPS regulates plant hormone signaling. Plant J 61:1029–1040.
5. Monte E, Al-Sady B, Leivar P, Quail PH (2007). Out of the dark: How the PIFs are unmasking a dual temporal mechanism of phytochrome signalling. J Exp Bot 58: 3125–3133.
6. Moon J, Parry G, Estelle M (2004) The ubiquitin-proteasome pathway and plant development. Plant Cell 16: 3181–3195.
7. Sullivan JA, Shirasu K, Deng XW (2003) The diverse roles of ubiquitin and the 26S proteasome in the life of plants. Nat Rev Genet 4: 948–958.
8. Taylor LP, Hepler PK (1997) Pollen germination and tube growth. Annu Rev Plant Physiol Plant Mol Biol 48: 461–491.
9. Moscatelli A, Cresti M (2001) Pollen germination and pollen tube growth. In: Bhoiwani SS, Soh WY, editors. Current Trends in the Embryology of Angiosperms. Dordrecht: Kluwer Academic Publishers. pp. 33–65.
10. Dai SJ, Chen TT, Chong K, Xue YB, Liu SQ et al. (2007) Proteomics identification of differentially expressed proteins associated with pollen germination and tube growth reveals characteristics of germinated Oryza sativa pollen. Mol Cell Proteomics 6: 207–230.
11. Gallois JL, Guyon-Debast A, Lecureuil A, Vezon D, Carpentier V et al. (2009) The *Arabidopsis* proteasome RPT5 subunits are essential for gametophyte development and show accession-dependent redundancy. Plant Cell 21: 442–459.
12. Lee KH, Minami A, Marshall RS, Book AJ, Farmer LM et al. (2011) The RPT2 subunit of the 26S proteasome directs complex assembly, histone dynamics, and gametophyte and sporophyte development in Arabidopsis. Plant Cell 23: 4298–4317.
13. Sheng X, Hu Z, Lu H, Wang X, Baluska F et al. (2006) Roles of the ubiquitin/proteasome pathway in pollen tube growth with emphasis on MG132-induced alterations in ultrastructure, cytoskeleton, and cell wall components. Plant Physiol 141: 1578–1590.
14. Sheng XY, Dong XL, Zhang SS, Jiang LP, Zhu J et al. (2010) Mitochondrial dynamics and its responds to proteasome defection during *Picea wilsonii* pollen tube development. Cell Biochem Funct 28: 420–425.
15. Speranza A, Scoccianti V, Crinelli R, Calzoni GL, Magnani M (2001) Inhibition of proteasome activity strongly affects kiwifruit pollen germination: involvement of the ubiquitin/proteasome pathway as a major regulator. Plant Physiol 126: 1150–1161.
16. Scoccianti V, Ovidi E, Taddei AR, Tiezzi A, Crinelli R et al. (2003) Involvement of the ubiquitin/proteasome pathway in the organisation and polarised growth of kiwifruit pollen tubes. Sex Plant Reprod 16: 123–133.
17. Qiao H, Wang H, Zhao L, Zhou J, Huang J et al. (2004) The F-box protein AhSLF-S2 physically interacts with S-RNases that may be inhibited by the ubiquitin/26S proteasome pathway of protein degradation during compatible pollination in *Antirrhinum*. Plant Cell 16: 582–595.
18. Calzoni GL, Speranza A, Bagni N (1979) In vitro germination of apple pollen. Sci Hortic 10:49–55.
19. Speranza A, Crinelli R, Scoccianti V, Geitmann A (2012) Reactive oxygen species are involved in pollen tube initiation in kiwifruit. Plant Biol 14: 64–76.
20. Marsoni M, Bracale M, Espen L, Prinsi B, Negri AS et al. (2008) Proteomic analysis of somatic embryogenesis in Vitis vinifera. Plant Cell Reports 27: 347–356.
21. Marsoni M, Cantara C, De Pinto MC, Gadaleta C, De Gara L et al. (2010) Exploring the soluble proteome of Tobacco Bright Yellow-2 cells at the switch towards different cell fates in response to heat shocks. Plant Cell Environ 33: 1161–1175.
22. Beutler E (1984) Red Cell Metabolism. A Manual of Biochemical Methods. New York: Grune and Stratton editors.
23. Izard J, Limberger RJ (2003) Rapid screening method for quantitation of bacterial cell lipids from whole cell. Journal of Microbiological Methods 55: 411–418.
24. Folch J, Lees M, Sloane Stanley GH (1957) A simple method for the isolation and purification of total lipides from animal tissues. J Biol Chem 226: 497–509.
25. Wang CL, Xu GH, Jiang XT, Chen G, Wu J et al. (2009) S-RNase triggers mitochondrial alteration and DNA degradation in the incompatible pollen tube of Pyrus pyrifolia in vitro. Plant J 57: 220–229.
26. Jang IC, Yang JY, Seo HS, Chua NH (2005) HFR1 is targeted by COP1 E3 ligase for post-translational proteolysis during phytochrome A signaling. Genes Dev 19: 593–602.
27. Nishizawa-Yokoi A, Tainaka H, Yoshida E, Tamoi M, Yabuta Y et al. (2010) The 26S proteasome function and Hsp90 activity involved in the regulation of HsfA2 expression in response to oxidative stress. Plant Cell Physiol 51: 486–96.

28. Kim DY, Scalf M, Smith LM, Vierstra RD (2013) Advanced proteomic analyses yield a deep catalog of ubiquitylation targets in Arabidopsis. Plant Cell 25: 1523–40.
29. Chebli Y, Geitmann A (2007) Mechanical principles governing pollen tube growth. Funct Plant Sci Biotechnol 1: 232–245.
30. Madoka Y, Tomizawa K, Mizoi J, Nishida I, Nagano Y et al. (2002) Chloroplast transformation with modified *accD* operon increases acetyl-CoA carboxylase and causes extension of leaf longevity and increase in seed yield in tobacco. Plant Cell Physiol 43: 1518–1525.
31. Wu G Z, Xue HW (2010) *Arabidopsis* β-ketoacyl-[acyl carrier protein] synthase I is crucial for fatty acid synthesis and plays a role in chloroplast division and embryo development. Plant Cell 22: 3726–3744.
32. Thompson TE, Senter SD, Grauche LJ (1993) Lipid content and fatty acids of Pecan pollen. Hort Sci 28: 1191–1193.
33. Obermeyer L, Fragner A, Lang V, Weckwerth W (2013) Dynamic Adaption of Metabolic Pathways during Germination and Growth of Lily Pollen Tubes after Inhibition of the Electron Transport Chain. Plant Physiol 162: 1822–1833.
34. Potocký M, Eliáš M, Profotová B, Novotná Z, Valentová O et al. (2003) Phosphatidic acid produced by phospholipase D is required for tobacco pollen tube growth. Planta 217: 122–130.
35. Pleskot R, Pejchar P, Bezvoda R, Lichtscheidl IK, Wolters-arts M et al. (2012) Turnover of phosphatidic acid through distinct signalling pathways affects multiple aspects of tobacco pollen tube tip growth. Front Physiol 3: 54–59.
36. Zonia L, Munnik T (2004) Osmotically induced cell swelling versus cell shrinking elicits specific changes in phospholipid signals in tobacco pollen tubes. Plant Physiology 134: 813–823.
37. Hempel J, Perozich J, Romovacek H, Hinich A, Kuo I et al. (1994) UDP-glucose dehydrogenase from bovine liver primary structure and relationship to other dehydrogenases. Protein Sci 3:1074–1080.
38. Gibeaut DM, Cramer GR, Seemann JR (2001) Growth, cell walls, and UDP-Glc dehydrogenase activity of *Arabidopsis thaliana* grown in elevated carbon dioxide. J Plant Physiol 158: 569–576.
39. Samac DA, Litterer L, Temple G, Jung HJG, Somers DA (2004) Expression of UDP-glucose dehydrogenase reduces cell wall polysaccharide concentration and increases xylose content in alfalfa stems. Appl Biochem Biotechnol A: Enzyme Eng Biotechnol 116: 1167–1182.
40. Roman E, Roberts I, Lidholt K, Kusche-Gullberg M (2003) Overexpression of UDP-glucose dehydrogenase in *Escherichia coli* results in decreased biosynthesis of K5 polysaccharide. Biochem J 374: 767–772.
41. Reboul R, Geserick C, Pabst M, Frey B, Wittmann D et al. (2011) Down-regulation of UDP-glucuronic acid biosynthesis leads to swollen plant cell walls and severe developmental defects associated with changes in pectic polysaccharides. J Biol Chem 286: 39982–39992.
42. Reboul R, Tenhaken R (2012) An emerging role of pectic rhamnogalacturonanII for cell wall integrity. Plant Signal Behav 7: 298–299.
43. Kim M, Ahn JW, Jin UH, Chai D, Poek KH et al. (2003) Activation of the programmed cell death pathway by inhibition of proteasome function in plant. Journal of Biological Chemistry 278: 19406–19415.
44. Lee DH, Goldberg AL (1998) Proteasome inhibitors: Valuable new tools for cell biologists. Trends Cell Biol 8: 397–403.
45. Bieler S, Meiners S, Stangl V, Pohl T, Stangl K (2009) Comprehensive proteomic and transcriptomic analysis reveals early induction of a protective anti-oxidative stress response by low-dose proteasome inhibition. Proteomics 9: 3257–3267.
46. Smirnova AV, Matveyeva NP, Polesskaya OG, Yermakov IP (2009) Generation of reactive oxygen species during pollen grain germination. Russian J Devel Biol 40: 345–353.
47. Speranza A, Scoccianti V (2012) New insights into an old story: pollen ROS also play a role in hay fever. Plant Signal Behav 7: 994–998.
48. Petrucco S, Bolchi A, Foroni C, Percudani R, Rossi GL et al. (1996) A maize gene encoding an NADPH binding enzyme highly homologous to isoflavone reductases is activated in response to sulfur starvation. Plant Cell 8: 69–80.
49. Kim SG, Kim ST, Wang Y, Kim S-K, Lee CH et al. (2010) Overexpression of rice isoflavone reductase-like gene (OsIRL) confers tolerance to reactive oxygen species. Physiologia Plantarum 138: 1–9.
50. Gillece P, Luz JM, Lennarz WJ, de La Cruz FJ, Romisch K (1999) Export of a cysteine-free misfolded secretory protein from the endoplasmic reticulum for degradation requires interaction with protein disulfide isomerase. J Cell Biol 147:1443–1456.
51. Forster ML, Sivick K, Park YN, Arvan P, Lencer WI et al. (2006) Protein disulfide isomerase-like proteins play opposing roles during retrotranslocation. J Cell Biol 173: 853–859.
52. Grune T, Reinheckel T, Li R, North JA, Davies KJ (2002) Proteasome-dependent turnover of protein disulfide isomerase in oxidatively stressed cells. Arch Biochem Biophys 397: 407–413.
53. Boavida LC, Ron M, McCormick S (2008) Truncation of a protein disulfide isomerase, PDIL2-1, delays embryo sac maturation and disrupts pollen tube guidance in Arabidopsis thaliana. Plant Cell 20: 3300–3311.

Evaluation of the Agronomic Performance of Atrazine-Tolerant Transgenic *japonica* Rice Parental Lines for Utilization in Hybrid Seed Production

Luhua Zhang[1,§], Haiwei Chen[1,§,¤], Yanlan Li[1], Yanan Li[1], Shengjun Wang[2], Jinping Su[2], Xuejun Liu[2], Defu Chen[1*], Xiwen Chen[1*]

1 Laboratory of Molecular Genetics, College of Life Sciences, Nankai University, Tianjin, China, **2** Tianjin Crop Research Institute, Tianjin, China

Abstract

Currently, the purity of hybrid seed is a crucial limiting factor when developing hybrid *japonica* rice (*Oryza sativa* L.). To chemically control hybrid seed purity, we transferred an improved atrazine chlorohydrolase gene (*atzA*) from *Pseudomonas* ADP into hybrid *japonica* parental lines (two maintainers, one restorer), and Nipponbare, by using *Agrobacterium*-mediated transformation. We subsequently selected several transgenic lines from each genotype by using PCR, RT-PCR, and germination analysis. In the presence of the investigated atrazine concentrations, particularly 150 μM atrazine, almost all of the transgenic lines produced significantly larger seedlings, with similar or higher germination percentages, than did the respective controls. Although the seedlings of transgenic lines were taller and gained more root biomass compared to the respective control plants, their growth was nevertheless inhibited by atrazine treatment compared to that without treatment. When grown in soil containing 2 mg/kg or 5 mg/kg atrazine, the transgenic lines were taller, and had higher total chlorophyll contents than did the respective controls; moreover, three of the strongest transgenic lines completely recovered after 45 days of growth. After treatment with 2 mg/kg or 5 mg/kg of atrazine, the atrazine residue remaining in the soil was 2.9–7.0% or 0.8–8.7% respectively, for transgenic lines, and 44.0–59.2% or 28.1–30.8%, respectively, for control plants. Spraying plants at the vegetative growth stage with 0.15% atrazine effectively killed control plants, but not transgenic lines. Our results indicate that transgenic *atzA* rice plants show tolerance to atrazine, and may be used as parental lines in future hybrid seed production.

Editor: Jin-Song Zhang, Institute of Genetics and Developmental Biology, Chinese Academy of Sciences, China

Funding: This work was supported by the Key Project of Tianjin Science and Technology Support Program (11ZCGYNC01000), the Key Program of the Natural Science Foundation of Tianjin (12YFJZJC01700, 14JCZDJC34100), the grant of Natural Science Foundation of China (No. 31070273, 31070717) and the 111 Project (No. B08011). The funders had no role in study design, data collection and analysis, decision to publish, or preparation of the manuscript.

Competing Interests: The authors have declared that no competing interests exist.

* Email: chendefu@nankai.edu.cn (DC); xiwenchen@nankai.edu.cn (XC)

§ These authors contributed equally to this work.

¤ Current address: College of Life Sciences, Chifeng College, Chifeng, China

Introduction

Rice (*Oryza sativa*) is one of the most important staple food crops globally. According to the National Grain and Oil Information Center, the area of China planted with rice in 2012 was 3.0×10^7 hm^2, including 9.0×10^6 hm^2 of *japonica* rice [1]. *Japonica* rice is mainly planted in the northern region of the Qinling Mountains–Huai River, and its planted area has increased in recent years because of its high quality and good taste. *Japonica* rice production is currently dominated by conventional varieties, with hybrid rice accounting for only 3% of the cultivated area. On the other hand, *indica* hybrid rice represents 70–80% of the total planted area of *indica* rice [2]. Therefore, there is considerable potential for the development of *japonica* hybrid rice. An increase in the annual planted area of *japonica* hybrid rice from 3% to 50%, i.e., to reach 4.0×10^6 hm^2, is estimated to lead to the production of 3.5×10^9 kg of high-quality grain (www.cngrain.com/Publish/qita/200503/207290), thereby contributing considerably to meet consumer's demand for high-quality food both in China and globally.

The three-line system is a traditional and effective production method for hybrid *japonica* rice seed [3]. The most widely used male sterile line in the system is BT-type cytoplasmic male sterile (CMS). However, the panicle of this line is loosely enclosed when heading, and this appearance closely resembles that of the maintainer. This makes it difficult for farmers to distinguish the BT-CMS line when eliminating off-type plants [2]. Furthermore, the BT CMS line has good restorability, and may therefore be easily pollinated with exotic pollens that contaminated during mechanical harvesting and storage of seeds, and also with exotic pollens from other plants [4]. The use of contaminated CMS lines in seed production results in decreased hybrid seed purity. Therefore, off-type contamination must be eliminated as early as possible. This is largely a manual process and requires considerable labor input, particularly in Asia. On the one hand, the need for increased labor will increase the price of hybrid seeds, while on the other hand, the increase in manual procedures may lead to the production of false hybrids. Furthermore, as the Chinese economy develops, increasing numbers of young men are leaving their home towns to seek work in the cities, leaving the elderly and women to work on the farms. The transformation of heavy and complex farming to light and simple farming is therefore becoming increasingly important. Thus, ensuring hybrid seed purity and reducing labor costs are two key issues in hybrid *japonica* rice seed production.

Genetic engineering, especially herbicide resistance engineering, provides an efficient means of controlling purity in hybrid seed production. Yan first proposed a strategy of utilizing herbicide resistance genes to chemically control purity in hybrid seed production [5]. Since then, two-line hybrid rice production has been extensively investigated [6–12] and progress has recently been reviewed [13]. Additionally, some transgenic hybrid rice combinations have been used in field trials [7,8,10,12]. However, the research has mainly focused on the *bar* gene isolated from *Streptomyces hygroscopicus* [6,8,10,12]; other genes, such as the EPSPS (5-enolpyruvylshikimate-3-phosphate synthase) gene from *Agrobacterium* strain CP4, and the protoporphyrinogen oxidase gene from *Bacillus subtilis*, have rarely been investigated [11]. If the strategy is proven to be effective, chemical control of hybrid seed purity will be mainly dependent on an herbicide with a single mode of action, and this will hinder sustainable weed management.

Atrazine (6-chloro-N^2-ethyl-N^4-isopropyl-1,3,5-triazine-2,4-diamine) is a triazine herbicide, and is commonly used in maize, sorghum, and sugarcane fields [14]. By inhibiting electron transport to plastoquinone in the photosystem PSII, atrazine terminates photosynthesis and kills weeds [15]. Atrazine was once the most widely used herbicide worldwide, because of its low cost and high effectiveness [14]. We previously isolated an atrazine chlorohydrolase gene (*atzA*) from a soil bacterium *Pseudomonas* ADP [16], and modified this gene by using directed evolution, to improve the enzymatic activity [17]. In the present study, we transferred the improved *atzA* gene into breeding hybrid *japonica* parental lines. Our results indicate that the transgenic *atzA* rice lines show tolerance to atrazine, and may be used as parental lines to chemically improve seed purity in hybrid seed production.

Materials and Methods

Construction of plant expression vector

Ubiquitin promoter and an improved atrazine chlorohydrolase gene *atzA*-22-4 [17] were respectively amplified from pSTARLING, an RNAi intermediate vector for monocots (a kind gift from the Commonwealth Scientific and Industrial Research Organization, Australia), and *AtzA*-22-4 using primers SL-Ubi-F/SL-Ubi-R and atzA-TJ-F/atzA-TJ-R (Table 1). After respectively inserted into TaKaRa pMD19T-simple vector and confirmed by sequencing, the pAtzA-22-4-19T-simple was restricted with *Eco*RV to collect the 1.4 kb fragment, and then lignated with pUbi-19T-simple. The resulting plasmid was restricted with *Sac*I/*Spe*I to collect the 3.4 kb fragment, and then ligated with pCAMBIA1301. The recombinant plasmid p1301-ubi-22-14 was then introduced into *Agrobacterium tumefaciens* EHA105 by the freeze-thaw method [18].

Genetic transformation and plant regeneration

Oryza sativa L. Nipponbare and *japonica* hybrid rice parental lines in the three-line system, Jindao7 (maintainer), Jindao8 (maintainer) and Jinhui3 (restorer) were used for transformation. Mature seeds were dehulled, surface-sterilized and placed on NB medium (N6 macro elements, B5 micro elements and vitamins) supplemented with 2 g/L proline, 3 mg/L 2, 4-D and 300 mg/L casein hydrolysate in dark at 28°C. After 2–3 weeks, the scutellum-derived calli were excised and subcultured every four weeks on the same medium but with 0.5 g/L proline, 2 mg/L 2, 4-D in dark at 28°C. The highly embryogenic compact calli (3–5 mm in diameter) that subcultured for less than five generations, were selected and co-cultivated with *A. tumefaciens* EHA105 harboring p1301C-ubi-22-14 on the co-cultivation medium (subculture medium but with 100 μM acetosyringone) for 3 days in dark at 28°C. Following that, the explants were then transferred into selection medium (subculture medium but with 50 mg/L hygromycin and 500 mg/L cefotaxime) in dark at 28°C for selection. After two cycles of selection, hygromycin-resistant calli were transferred onto pre-regeneration medium (NB medium with 0.5 g/L proline, 2 mg/L 6-BA, 1 mg/L NAA, 5 mg/L ABA, 300 mg/L casein hydrolysate and 50 mg/L hygromycin) for 14 to 21 days in dark at 28°C, then to regeneration medium (pre-regeneration medium but without 5 mg/L ABA) for 30 days under 54 $\mu mol/m^2/s$ light at 28°C and finally to the rooting medium (MS medium with 1 mg/L IBA) for 15 days under light at 28°C.

Molecular analysis of transgenic lines

Genomic DNA was isolated from young leaves using a modified CTAB method [19]. PCR was performed to preliminarily select the transformed plants using primers atzA7/atzA8 [20] and SL-Ubi-C-F/SL-Ubi-C-R (Table 1). RT-PCR was performed to further confirm the expression of *atzA* in the PCR-positive transformed plants. Total RNA was extracted from young leaves using an RNAultra Extraction Kit (Qiagen). cDNA was synthesized using atzA8 and oligo(dT) as primers and SuperScript™ II RNase H− (Invitrogen) as reverse transcriptase. RT-PCR was amplified using the cDNA as template and atzA7/atzA2 [20] as primers. *β*-actin (AB047313) was also amplified as the internal control using actin-R/actin-F as primers [21].

Germination and seedling growth in the presence of atrazine

Fifty seeds from each transgenic line and the respective control plant were directly sown on the surface of filter paper in plates containing 0, 75 or 150 μM atrazine. Seeds were placed in a growth chamber at 28°C with 16 h of 54 $\mu mol/m^2/s$ light per day. Germination (based on radicles >2 mm) was recorded daily and the cumulative values at day 3 and day 7 were calculated to represent as germination potential and germination percentage. The shoot length, root length and their biomass were measured after 7 days. For seedling growth test, seven-day old seedlings germinated in absence of atrazine were placed on filter paper in pots, and incubated in Kimura B nutrition solution [22] containing 0, 75 or 150 μM atrazine in growth chamber described above. During the period, the nutrition solution containing the respective concentration of atrazine was added to keep the filter paper wet. The growth parameters as described above were measured after 10 days.

Soil-grown transgenic T_2 lines in the presence of atrazine

Two-week old seedlings of similar size that germinated in absence of atrazine were transplanted into pots containing 1.12 kg of soil with 0, 2, or 5 mg/kg of atrazine. Nine plants were planted in each pot, and irrigated with the same amount of water every day and with Kimura B nutrition solution [22] twice a week. Plants were incubated in the greenhouse in a 14-h light/10-h dark cycle (28/25°C) at 300 $\mu mol\ m^{-2}s^{-1}$ light and 75% relative humidity. Plant growth and chlorophyll content were measured at

Table 1. Primers used in this study.

Primer	Sequence (5′→3′)
SL-Ubi-F	tgagctcctgcagtgcagcgtgacccggtcgt, with *Sac*I site
SL-Ubi-R	tgatatcctgcagaagtaacaccaaacaacag, with *Eco*RV site
atzA-TJ-F	tgatatcatgcaaacgctcagcatccag, with *Eco*RV site
atzA-TJ-R	tgatatcactagtctagaggctgcgccaagctg, with *Eco*RV and *Spe*I site
SL-Ubi-C-F	cccgccgtaataaatagacac
SL-Ubi-C-R	accacaccacatcatcacaac

15 -day intervals. For chlorophyll content analysis, approximately 10 mg of leaves was extracted with 5 ml acetone and then quantified the absorbance at 663.6 nm and 646.6 nm, as described by Porra et al. [23]. Chlorophyll a = $12.25\times A_{663.6}-2.55\times A_{646.6}$, Chlorophyll b = $20.31\times A_{646.6}-4.91\times A_{663.6}$.

HPLC analysis of atrazine and its metabolite hydroxyatrazine in the plant and soil

Atrazine and hydroxyatrazine in the rice leaves and soil were also determined in the soil-grown experiment. One gram of leaves, or five grams of soil samples that collected from the soil layer mixture as thick as possible, were extracted with 5 mL dichloromethane. After 30 min of incubation at room temperature, the mixture was centrifuged at 15,294 *g* for 10 min. The supernatant was transferred to a new tube and filtered through a 0.2-μm filter and then subjected to CoM 6000 HPLC system analysis on an analytical C18 column (5 μm, 250 mm×4.6 mm) at 30°C with linear gradients phase as follows: 0 to 6 min, 10% to 25% acetonitrile; 6 min to 21 min, 25% to 65% acetonitrile; 21 min to 23 min, 65% to 100% acetonitrile; and 23 min to 25 min, 100% acetonitrile [24] at the wavelength of 228 nm.

Spraying atrazine to transgenic T2 lines

Two-week old seedlings from the germination experiment were also transplanted into pots with soil. Each pot contained 16 plants. After 30 days' growth, the last second leaves were taken and cut into 2–3 cm section, soaked in 0, 75 and 150 μM atrazine, and incubated at 25°C under light for 2 days. The color of the leaves was observed every day. After 40 days' growth, each pot of plants was sprayed with 20 ml of 0.15% atrazine solution.

Statistical analysis

All the experiments were performed for three times. Significant difference between the specific transgenic line and the respective control was performed using independent-samples *t*-test at 95% or 99% confidence with IBM SPSS Statistics 11.0. Values indicated by * or ** represented significantly difference at $P<0.05$ or $P<0.01$.

Results

Selection of transgenic *atzA* plants

PCR analysis revealed that 18 (100%) Nipponbare transformants, 15 out of 18 (83%) Jindao7 transformants, 14 out of 16 (88%) Jinhui1 transformants, and 17 (100%) Jindao8 transformants were positive. RT-PCR analysis further confirmed the expression of *atzA* in the PCR-positive lines (data not shown). Subsequently, the self-pollinated seeds (T_1 progenies) from each plant were germinated in the presence of hygromycin. Transgenic lines that showed a segregation pattern of 3:1 resistant/sensitive in the germination test were selected to produce the T_2 generations. Individual plants, whose herbicide tolerance did not segregate in the germination test, were considered as homogenous lines and selected. For simplicity, in further research, we used only two or three independent transgenic T_2 lines for each genotype.

Germination of transgenic *atzA* lines in the presence of atrazine

To investigate the atrazine tolerance of transgenic lines during germination, we germinated seeds of three lines of each genotype and the respective controls in the presence of atrazine (Fig. 1A, Table 2). The germination potential of the wild types decreased as the atrazine concentration increased (58.3–98%, 22.4–64.2%, and 0%, respectively, in 0 μM, 75 μM, and 150 μM atrazine); on the other hand, the germination percentage significantly decreased only in presence of 150 μM atrazine (95.2–100%, 72.1–96.7%, and 0–42.2%, respectively, in 0 μM, 75 μM, and 150 μM atrazine). In the presence of the investigated atrazine concentrations, the highest germination potential was determined for Nipponbare, followed by Jinhui1, Jindao8, and Jindao7. In the presence of 0 μM or 75 μM atrazine, the germination potentials and percentages of almost all of the transgenic lines did not differ significantly from those of the respective controls. The exceptions were the germination potential for Jindao8 in the presence of 75 μM atrazine and the germination percentage for Jinhui1 in the presence of 75 μM atrazine. On the other hand, in the presence of 150 μM atrazine, all of the transgenic lines showed significantly higher germination potentials and germination percentages than did the respective controls. Moreover, two of the wild types (Nipponbare and Jinhui1) failed to germinate or only rarely germinated at day 3 or day 7. Interestingly, in the presence of 150 μM atrazine, the highest germination percentage was determined for Jindao7, followed by Jindao8, Jinhui1, and Nipponbare; on the other hand, Jindao7 showed the lowest germination potential. When germinated in the presence of atrazine, all of the transgenic lines produced larger seedlings (with taller shoots and longer roots) than did the respective control plants (Fig. 1A). However, the presence of atrazine significantly inhibited the growth of transgenic lines and wild types (Fig. 1A).

Seedling growth of transgenic *atzA* lines in the presence of atrazine

To investigate the tolerance of transgenic lines to atrazine, we transplanted 7-day old seedlings germinated in the absence of atrazine, to pots of soil containing different concentration of atrazine (Fig. 1B and Table 3). In the absence of atrazine, we determined no differences in shoot or root growth between any of the transgenic lines and the respective controls. However, in the

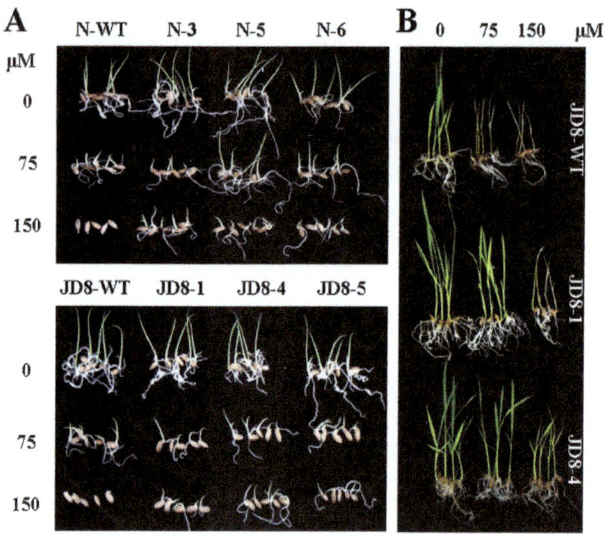

Figure 1. Germination and seedling growth of transgenic *atzA* rice lines in the presence of atrazine. (**A**) Representative images of seeds from the transgenic lines and the respective controls sown on plates containing 0, 75 or 150 μM atrazine for 7 days. (**B**) Representative images from seven-day old seedlings germinated in absence of atrazine were transplanted into pots containing 0, 75 or 150 μM atrazine for 10 days. For simplicity, only images of transgenic Jindao8 (JD8) and/or Nipponbare (N) were shown here.

presence of 75 μM atrazine, all of the transgenic lines (except for Jindao8) were significantly taller than were the respective controls; further, all of the transgenic lines (except for Jinhui1) produced significantly more root biomass than did the respective controls. On the other hand, in the presence of 150 μM atrazine, shoot and root growth of almost all of the transgenic lines and the respective controls were significantly inhibited. The exceptions were the shoot growth of Nipponbare and the root growth of Jinhui1–5. Plants became yellowish, and subsequently rotted and died after 10 days (Fig. 1B).

Tolerance of soil-grown transgenic *atzA* lines to atrazine

To further investigate the tolerance of transgenic lines to atrazine, we transplanted 15-day-old seedlings germinated in the absence of atrazine, to pots of soil containing 0 mg/kg, 2 mg/kg, or 5 mg/kg of atrazine. We measured the plant height (as a non-destructive assay of plant growth) and chlorophyll content at 15-day intervals.

In the absence of atrazine, we determined no significant differences in plant height between the transgenic lines and the respective controls (Fig. 2A). In soil containing 2 mg/kg or 5 mg/kg of atrazine, all the transgenic lines were significantly taller than were the respective control plants. However, the presence of atrazine significantly inhibited the growth of transgenic lines and control plants (by 68.1–111.9% and 46.4–68.2%, respectively). In contrast to the control plants, all of the transgenic lines (except for the Jindao8) were slightly taller in soil containing 2 mg/kg of atrazine than in soil containing 5 mg/kg of atrazine. The growth of transgenic lines gradually recovered with an increase in the growth time. Further, three of the strongest transgenic lines (Jindao7–4, Jinhui1–5, and Jindao8–4) completely recovered after 45 days of growth (data not shown).

Atrazine may disrupt the electron flow of photosystem II and destroy photosynthetic pigments in plant leaves [15]. In the absence of atrazine, we observed no difference in chlorophyll content between transgenic lines and control plants (Fig. 2B). However, in soil containing 2 mg/kg or 5 mg/kg of atrazine, the transgenic lines retained more chlorophyll than did the respective control plants. With a few exceptions, the chlorophyll content of the transgenic lines decreased significantly in soil containing 5 mg/kg of atrazine, but not in soil containing 2 mg/kg of atrazine; on the other hand, the chlorophyll content of the control plants decreased significantly under both conditions. For each genotype grown in soil containing the same atrazine concentration, we observed no difference in chlorophyll content at different growth stages. After 45 days, the chlorophyll contents of the three

Table 2. Germination of transgenic *atzA* rice lines in the presence of atrazine.

Line	0 μM		75 μM		150 μM	
	3 d	7 d	3 d	7 d	3 d	7 d
N[a]-WT	98.0±2.0	100±0.0	54.0±1.6	94.5±1.4	0.0±0.0	0.0±0.0
N-3	98.0±1.6	94.2±0.0	60.0±2.2	98.2±1.1	64.3±1.6**	96.2±1.8**
N-5	96.0±2.0	100±0.0	64.0±1.3	92.1±1.5	62.3±2.6**	92.6±1.5**
JD7-WT	58.3±1.6	100±0.0	24.3±2.1	90.2±2.3	0.0±0.0	42.2±1.1
JD7-2	56.0±1.1	98.1±1.4	22.3±1.7	88.2±2.1	2.3±0.1**	88.5±1.6**
JD7-4	62.3±2.2	98.6±1.6	26.3±1.3	90.2±0.8	2.4±0.2**	86.5±1.3**
JH3-WT	86.3±2.0	95.2±2.8	64.2±2.2	72.1±1.1	0.0±0.0	4.5±0.0
JH-3	82.3±1.2	94.2±1.8	58.7±1.6	90.2±1.2*	32.3±0.9**	78.6±1.2**
JH3-5	88.3±3.2	92.2±1.3	66.3±2.7	94.1±1.0*	38.7±1.1**	80.5±1.2**
JD8-WT	82.0±4.2	98.7±0.8	22.4±1.6	96.7±1.0	0.0±0.0	26.5±1.3
JD8-1	78.3±2.5	99.2±0.8	48.3±1.1**	97.2±0.2	34.3±0.7**	94.5±2.8**
JD8-4	78.0±2.6	100±0.0	54.0±2.6**	98.3±1.0	35.2±1.6**	96.8±1.3**

Note: In each independent treatment 50 seeds were used for the experiment. Values are shown as mean ±SEM ($n = 3$). * or ** indicates significant difference ($P<0.05$) or highly significant difference ($P<0.01$) between the treatment and the control.
[a]N, JD7, JH3 and JD8 were the abbreviations of Nipponbare, Jindao7, Jindao8 and Jinhui3, respectively. For simplicity, only two transgenic lines for each genotype were shown here.

Table 3. Seedling growth performance of transgenic *atzA* rice lines in the presence of atrazine.

Line	0 μM		75 μM		150 μM	
	Shoot length (cm)	Root biomass (mg)	Shoot length (cm)	Root biomass (mg)	Shoot length (cm)	Root biomass (mg)
N [a]-WT	9.90±1.37	32.00±3.16	6.54±0.86 (66.0)[a]	21.00±2.55 (65.6)	5.07±1.17 (51.2)	21.75±3.77 (68.0)
N-3	10.55±1.49	30.25±3.50	9.56±2.08* (90.7)	29.50±3.14** (97.5)	7.55±1.20* (71.6)	21.50±6.40 (71.1)
N-5	8.78±2.42	32.75±4.50	8.94±1.33* (101.8)	29.85±1.83** (91.2)	9.39±0.51** (106.9)	21.00±2.71 (64.1)
JD7-WT	8.47±0.37	28.50±2.56	5.28±0.57 (62.4)	20.00±3.10 (70.2)	6.54±0.61 (77.2)	17.75±5.19 (62.3)
JD7-2	9.40±2.61	31.50±3.00	7.08±0.78* (75.3)	32.50±2.65** (103.2)	5.81±0.56 (61.8)	17.00±3.46 (54.0)
JD7-4	7.99±1.65	27.50±3.35	8.16±1.27* (102.2)	27.50±2.65** (100.0)	6.78±0.53 (84.8)	17.75±4.19 (64.5)
JH3-WT	9.83±1.25	24.00±3.07	5.80±0.43 (59.0)	21.75±3.36 (90.6)	4.62±0.42 (47.0)	13.25±6.24 (55.2)
JH-3	10.44±1.47	26.75±2.75	8.27±0.59* (79.2)	24.25±2.22* (90.7)	5.12±0.61 (49.0)	15.25±2.63 (57.0)
JH3-5	8.89±1.58	25.50±3.57	8.25±0.78* (92.8)	23.25±2.87* (83.8)	6.04±0.19* (67.9)	22.00±3.37** (86.3)
JD8-WT	7.88±0.93	26.75±3.55	7.20±1.85 (91.4)	21.75±2.87 (81.3)	6.43±0.35 (81.5)	12.25±3.46 (45.8)
JD8-1	8.11±1.51	27.50±2.15	7.18±0.40 (88.6)	27.50±1.52** (100.0)	5.65±1.12 (69.6)	11.50±1.91 (41.8)
JD8-4	7.90±0.72	27.75±2.53	7.64±0.69 (96.7)	27.75±1.79** (100.0)	5.64±0.78 (71.3)	11.75±2.99 (42.3)

Note: The parameters were recorded after 10 days. Values are shown as mean ± SEM ($n=3$). * or ** indicates significant difference ($P<0.05$) or highly significant difference ($P<0.01$) between the treatment and the control.
[a]N, JD7, JH3 and JD8 were the abbreviations of Nipponbare, Jindao7, Jindao8 and Jinhui3, respectively.
[b]Data in the bracket are the percentage of the treated values to the mock values.

strongest lines (Jindao7–4, Jinhui1–5, and Jindao8–4) were almost identical when grown in soil containing 5 mg/kg of atrazine and when grown in the absence of atrazine (data not shown).

Ability of soil-grown T_2 transgenic lines to degrade atrazine

To investigate the ability of the soil-grown transgenic plants to degrade atrazine, we determined the atrazine residue in the soil at 15-day intervals, after growth of seedlings in pots of soil containing 0 mg/kg, 2 mg/kg, or 5 mg/kg of atrazine. We observed that the atrazine residue in the soil decreased with an increase in growth time, for transgenic and control plants (Fig. 2C). However, the decrease was greater for transgenic lines than for control plants. After treatment with 2 mg/kg of atrazine, the atrazine residue remaining in the soil was 2.9–7.0% for transgenic lines, and 44.0–59.2% for control plants; after treatment with 5 mg/kg of atrazine, the atrazine residue remaining in the soil was 0.8–8.7% for transgenic lines and 28.1–30.8% for control plant. For each genotype at the same growth stage, the atrazine residue in the soil was higher after growth in 2 mg/kg atrazine than after growth in 5 mg/kg atrazine. We also observed that hydroxyatrazine, not atrazine (only trace) appeared in the leaves of the transgenic plants (Fig. S1), indicating that atrazine was metabolized rather than accumulated in the transgenic plants.

Utilization of transgenic plants in hybrid seed production

To investigate the tolerance of transgenic plants to the atrazine concentration used in weed control, we first examined the tolerance of leaf sections to atrazine solution. In the presence of 75 μM atrazine, the leaf sections of transgenic lines remained green, whereas those of control plants became bleached (Fig. 3A). In the presence of 150 μM atrazine, the leaf sections of transgenic lines became slightly bleached, but remained greener than those of control plants. We subsequently sprayed plants with 0.15% atrazine, and observed that wild-type plants became curled and withered after 6 days, whereas transgenic lines continued to grow well (Fig. 3B). Our results indicate that the transgenic rice plants showed tolerance to the atrazine concentration used in weed control.

Discussion

Current research on the utilization of herbicide resistance genes in hybrid rice is focused on the two-line hybrid production system. Sterile lines are not stable in the two-line system, and therefore most *japonica* hybrid combinations are still produced by using three-line system. In the present study, we have developed atrazine-tolerant transgenic *japonica* rice parental lines with the potential to be used in future hybrid seed production.

Herbicide tolerance during germination is very important when utilizing transgenic *atzA* lines in hybrid seed production, because off-type plants should be eliminated as early as possible. Atrazine is a slightly water-soluble herbicide, with a saturated concentration of 153 μM. Kawahigashi [25] previously showed that an atrazine concentration of 100 μM did not affect the germination of rice plants. Therefore, we used two atrazine concentrations (75 μM and 150 μM) in the germination test, to evaluate the tolerance of transgenic rice lines. In contrast to Kawahigashi [25], we observed that the transgenic lines germinated well in the presence of atrazine whereas the respective controls did not. However, as the seedling growth of transgenic lines were also inhibited by atrazine, chemical control of contamination during germination may not be appropriate.

To evaluate the tolerance of seedlings to atrazine, we used 7-day-old seedlings germinated in the absence of atrazine as a starting material, to avoid cumulative effects derived from germination. We observed that almost all of the transgenic lines grew well in the presence of 75 μM atrazine with no difference from that in absence of atrazine. Therefore, 75 μM atrazine may be used for effective control of off-type plants during the early seedling growth stage. Our data also suggested spraying plants with 0.15% atrazine (equivalent to the standard dosage used in the field for weed control, i.e., 200–250 g 40% suspension concentrate in 30~50 kg of water) at the subsequently vegetative growth stage could serve as an alternative means of chemically controlling

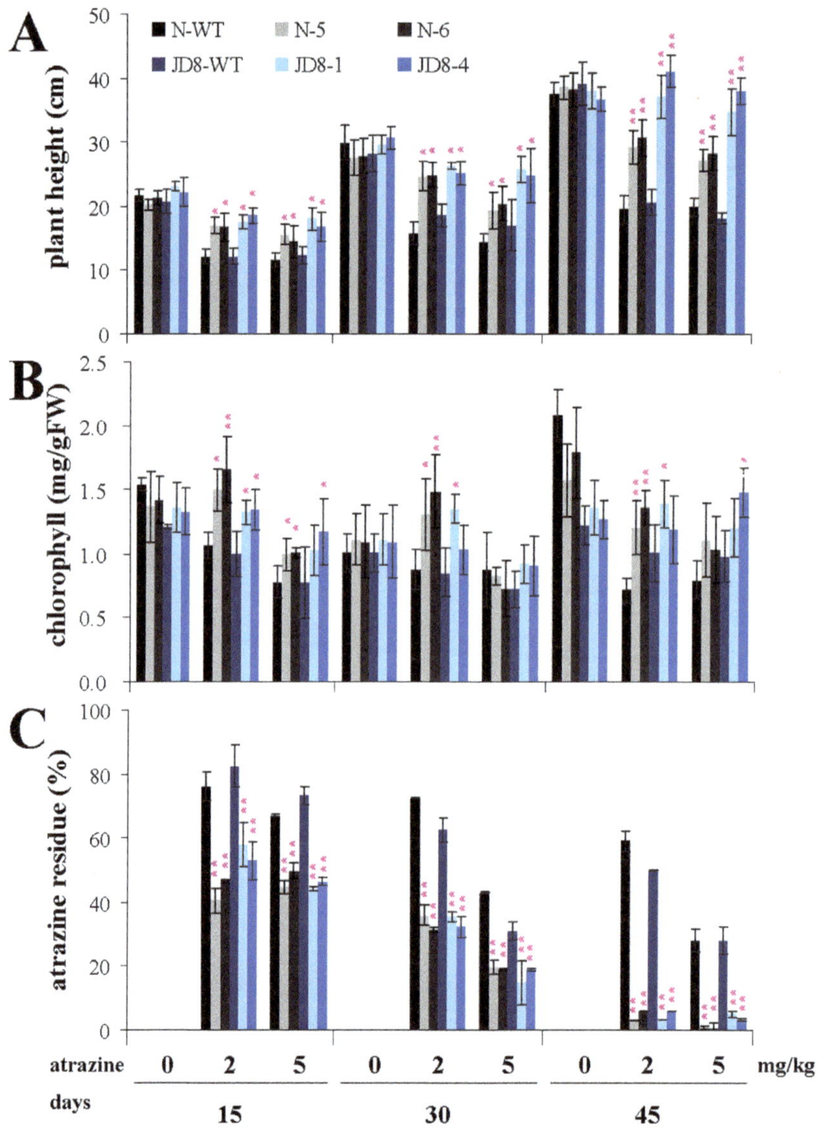

Figure 2. Growth and physiology of transgenic *atzA* rice lines in the presence of atrazine. Fifteen-day old seedlings germinated in absence of atrazine were transplanted into pots with 1.12 kg soil containing 0, 2 or 5 mg/kg of atrazine. (**A**) plant height; (**B**) chlorophyll content; and (**C**) atrazine residue were measured at 15-day intervals. Values are shown as mean ±SEM ($n = 3$). * or ** over the bar indicates significant difference ($P < 0.05$) or highly significant difference ($P < 0.01$) between the treatment and the control. For simplicity, only data of transgenic Nipponbare (N) and Jindao8 (JD8) were shown here.

off-type plants. However, we did not test the tolerance of transgenic *atzA* lines at the reproductive growth stage, which is important for the utilization in mechanical harvesting of hybrid seed production. Further studies are required to investigate whether the lethal dosage for untransformed parental lines is tolerated at the reproductive stage.

Atrazine degrades relatively slowly, with an average half-life of 4–57 weeks [26]. Therefore, application of this herbicide will inevitably lead to accumulation in the field, which may affect current crop growth or damage sensitive succession crops [27]. The transgenic rice lines were taller, and had higher chlorophyll contents than did the respective control plants when growing in the soil of 2 mg/kg or 5 mg/kg atrazine. Three of the transgenic lines (Jindao7–4, Jinhui1–5, and Jindao8–4) completely recovered after 45 days of growth in the presence of atrazine, suggesting that an atrazine concentration up to 5 mg/kg does not affect sustained growth of transgenic lines. Our data also suggested that almost all of the atrazine was degraded after growth of the transgenic rice plants for 45 days. This was also verified by the appearance of hydroxyatrazine in the leaves of transgenic plants. Besides, atrazine metabolite hydroxyatrazine could be directly determined in transgenic plants. However, as the leave, stem and root of the plants all possibly metabolize atrazine [28], it would be more complicated if we want to quantitatively evaluate the atrazine metabolized by plants. That is why we chose to determine the atrazine residue in the soil after growth of the transgenic plants in this study.

Utilization of herbicide resistance genes in hybrid seed production may be achieved by introducing the genes into a maintainer or restorer line in the three-line system. When an herbicide-resistant maintainer is obtained, its corresponding CMS line, which possesses herbicide resistance, may be created by using

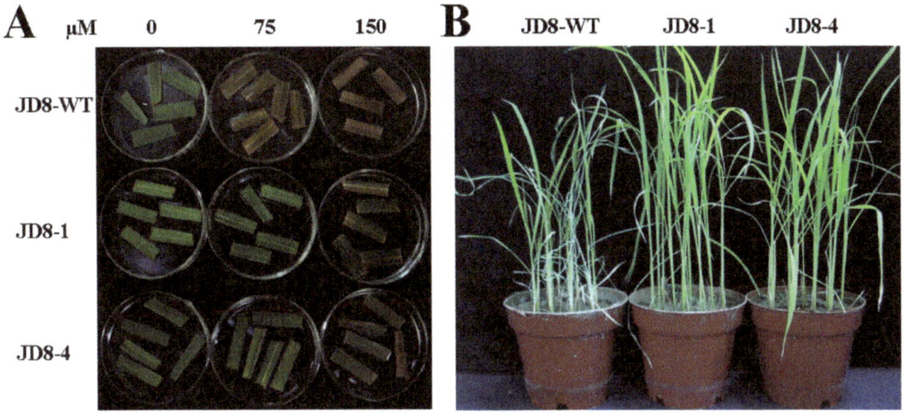

Figure 3. Tolerance of mature transgenic *atzA* plants to atrazine. (**A**) Representative images of leave sections in the presence of atrazine. Fifteen-day old seedlings were transplanted into pots containing soil. After 30 days' growth, the last second leaves were cut into 2–3 cm section, and soaked in 0, 75 and 150 μM atrazine, and incubated at 25°C under light for 2 days. (**B**) Representative images of mature plants after sprayed with atrazine. Fifteen-day old seedlings were transplanted into pots containing soil. After 40 days' growth, each pot of transgenic lines (9 plants) was sprayed with 20 mL of 0.15% atrazine solution. For simplicity, only data of transgenic Jindao8 (JD8) was shown here.

a backcross. The resulting herbicide-resistant CMS breeder seeds may be used to cross with the original maintainer (herbicide sensitive), to multiply the herbicide-resistant foundation seeds. During this process, the selfed maintainer and off-type plants may easily be eliminated by herbicide spraying, thus ensuring the purity of CMS seeds. The herbicide-resistant CMS line may also be used in mechanical harvesting, by spraying herbicide to kill the pollen plants after pollination, and harvesting seeds from the surviving CMS plants. A similar result may be achieved when crossing the herbicide-resistant CMS line with the herbicide-sensitive restorer line. Alternatively, the herbicide-resistant restorer line may be crossed with the herbicide-sensitive sterile line during the hybrid seed production process, by spraying herbicide to kill the off-type plants. Further, the herbicide-resistant maintainer and restorer lines may be useful for eliminating off-type plants, and for weed management when growing hybrid rice.

Field trials with genetically modified rice are strictly controlled in China, and therefore we were unable to evaluate our herbicide-resistant hybrid rice combinations on a field scale. Further studies are required to verify the validity of utilizing atrazine-tolerant transgenic lines in the field, and to determine whether herbicide-resistant hybrid rice combinations retain their original superiority in terms of yield. Nevertheless, the development of atrazine-tolerant transgenic *japonica* rice parental lines represents a valuable tool for future application in hybrid seed production.

Supporting Information

Figure S1 HPLC analysis for determination of atrazine and its metabolite hydroxyatrazine in leaves of transgenic and WT rice plants after grown in 5 mg/kg atrazine soil for 45 days. (**A**) Standard of hydroxyatrazine (I) and atrazine (II). (**B**) Wild type (WT). (**C–E**) Transgenic Nipponbare (N-6), Jindao7 (JD7-4) or Jindao8 (JD8-4), respectively.

Author Contributions

Conceived and designed the experiments: DFC XWC. Performed the experiments: LHZ HWC YLL YNL SJW JPS. Analyzed the data: LHZ DFC XWC. Contributed reagents/materials/analysis tools: SJW XJL DFC XWC. Wrote the paper: DFC XWC.

References

1. Peng C, Zhang H (2013) Analysis of paddy and rice markets at home and abroad in the first quarter of 2013 and its prospect. Agric Outlook (4): 4–9 (in Chinese, with English abstract).
2. Shi KB, Deng HS (2004) Chinese *japonica* hybrid rice technology innovation seminar held in Sanya. Hybrid Rice 19: 76 (in Chinese).
3. Li S, Yang D, Zhu Y (2007) Characterization and use of male sterility in hybrid rice breeding. J Integ Plant Biol 49: 791–804.
4. Tang SZ, Zhang HG, Zhu ZB, Liu C, Li P, et al. (2010) Application of HL type male sterile cytoplasm in *japonica* hybrid rice breeding. Chin J Rice Sci 24: 116–124 (in Chinese, with English abstract).
5. Yan W (2000) Crop heterosis and herbicide. US Patent 6066779.
6. Hu G, Xiao H, Yu Y, Zhu Z, Si H, et al. (2000) *Agrobacterium*-mediated transformation of the restorer lines of two-line hybrid rice with *bar* gene. Chin J Appl Environ Biol 6: 511–515 (in Chinese, with English abstract).
7. Li Y, Xu Q, Duan F, Liu G, Yan W (2000) Breeding of herbicide-resistant hybrid rice combinations. Hybrid Rice 15(6): 9–11.
8. Xue S, Zhang W, Gong X, Shen L, Huang D, et al. (2001) Breeding of isotype restorer line with *bar* gene from Miyang 46 and its combinations. J Zhejiang Agric Sci (4): 181–183 (in Chinese).
9. Ramesh S, Nagadhara D, Pasalu IC, Kumari AP, Sarma NP, et al. (2004) Development of stem borer resistant transgenic parental lines involved in the production of hybrid rice. J Biotechnol 111: 131–141.
10. Xiong X, Tang L, Deng X, Xiao G (2004) A preliminary report on the experiments of herbicide-resistant two-line hybrid rice Xiang 125S/Bar 68-1. Hybrid Rice 19(5): 41–43.
11. Wu F, Wang S, Li S, Zhang K, Li P (2006) Research progress on herbicide resistant transgenic rice and its safety issues. Mol Plant Breed 4: 846–852.
12. Xiao G, Yuan L, Sun S (2007) Strategy and utilization of a herbicide resistance gene in two-line hybrid rice. Mol Breed 20: 287–292.
13. Xiao G (2009) Recent advances in development of herbicide resistance transgenic hybrid rice in China. Rice Sci 16: 235–239.
14. Udiković-Kolić N, Scott C, Martin-Laurent F (2012) Evolution of atrazine-degrading capabilities in the environment. Appl Microbiol Biotechnol 96: 1175–1189.
15. Rutherford AW, Krieger-Liszkay A (2001) Herbicide-induced oxidative stress in photosystem II. Trends Biochem Sci 26: 648–653.
16. Cai B, Han Y, Liu B, Ren Y, Jiang S (2003) Isolation and characterization of an atrazine-degrading bacterium from industrial wastewater in China. Lett Appl Microbiol 36: 272–276.
17. Wang Y, Li X, Chen X, Chen D (2013) Directed evolution and characterization of atrazine chlorohydrolase variants with enhanced activity. Biochemistry (Moscow) 78: 1104–1111.
18. Höfgen R, Willmitzer L (1988) Storage of competent cells for *Agrobacterium* transformation. Nucleic Acids Res 16: 9877.

19. Loh JP, Kiew R, Kee A, Gan LH, Gan YY (1999) Amplified fragment length polymorphism (AFLP) provides molecular markers for the identification of *Caladium bicolor* cultivars. Ann Bot 84: 155–161.
20. Wang H, Chen X. Xing X, Hao X, Chen D (2010) Transgenic tobacco plants expressing *atzA* exhibit resistance and strong ability to degrade atrazine. Plant Cell Rep 29: 1391–1399.
21. Chen D, Chen H, Zhang L, Shi X, Chen X (2014) Tocopherol-deficient rice plants display increased sensitivity to photooxidative stress. Planta 239: 1351–1362.
22. Ma J, Takahashi E (1990) Effect of silicon on the growth and phosphorus uptake of rice. Plant Soil 126: 115–119.
23. Porra RJ, Thompson WA, Kriedemann PE (1989) Determination of accurate extinction coefficients and simultaneous equations for assaying chlorophylls *a* and *b* extracted with four different solvents: verification of the concentration of chlorophyll standards by atomic absorption spectroscopy. Biochim Biophys Acta - Bioenergetics 975: 384–394.
24. de Souza ML, Sadowsky MJ, Wackett LP (1996). Atrazine chlorohydrolase from *Pseudomonas* sp. strain ADP: gene sequence, enzyme purification, and protein characterization. J Bacteriol 178: 4894–4900.
25. Kawahigashi H, Hirose S, Ohkawa H, Ohkawa Y (2007) Herbicide resistance of transgenic rice plants expressing human CYP1A1. Biotechnol Adv 25: 75–84.
26. Erickson LE, Lee KH, Sumner DD (1989) Degradation of atrazine and related *s*-triazines. Crit Rev Environ Control 19: 1–14.
27. Rhine ED, Fuhrmann JJ, Radosevich M (2003) Microbial community responses to atrazine exposure and nutrient availability: linking degradation capacity to community structure. Microbiol Ecol 46: 145–160.
28. Wang L, Samac DA, Shapir N, Wackett LP, Vance CP, et al. (2005) Biodegradation of atrazine in transgenic plants expressing a modified bacterial atrazine chlorohydrolase (*atzA*) gene. Plant Biotech J 3: 475–486.

12

Computational Prediction of Candidate Proteins for S-Nitrosylation in *Arabidopsis thaliana*

Mounira Chaki[1], Izabella Kovacs[1], Manuel Spannagl[2], Christian Lindermayr[1]*

1 Institute of Biochemical Plant Pathology, Helmholtz Zentrum München-German Research Center for Environmental Health, Neuherberg, Germany, **2** Institute of Bioinformatics and Systems Biology, Helmholtz Zentrum München-German Research Center for Environmental Health, Neuherberg, Germany

Abstract

Nitric oxide (NO) is an important signaling molecule that regulates many physiological processes in plants. One of the most important regulatory mechanisms of NO is S-nitrosylation—the covalent attachment of NO to cysteine residues. Although the involvement of cysteine S-nitrosylation in the regulation of protein functions is well established, its substrate specificity remains unknown. Identification of candidates for S-nitrosylation and their target cysteine residues is fundamental for studying the molecular mechanisms and regulatory roles of S-nitrosylation in plants. Several experimental methods that are based on the biotin switch have been developed to identify target proteins for S-nitrosylation. However, these methods have their limits. Thus, computational methods are attracting considerable attention for the identification of modification sites in proteins. Using GPS-SNO version 1.0, a recently developed S-nitrosylation site-prediction program, a set of 16,610 candidate proteins for S-nitrosylation containing 31,900 S-nitrosylation sites was isolated from the entire *Arabidopsis* proteome using the medium threshold. In the compartments "chloroplast," "CUL4-RING ubiquitin ligase complex," and "membrane" more than 70% of the proteins were identified as candidates for S-nitrosylation. The high number of identified candidates in the proteome reflects the importance of redox signaling in these compartments. An analysis of the functional distribution of the predicted candidates showed that proteins involved in signaling processes exhibited the highest prediction rate. In a set of 46 proteins, where 53 putative S-nitrosylation sites were already experimentally determined, the GPS-SNO program predicted 60 S-nitrosylation sites, but only 11 overlap with the results of the experimental approach. In general, a computer-assisted method for the prediction of targets for S-nitrosylation is a very good tool; however, further development, such as including the three dimensional structure of proteins in such analyses, would improve the identification of S-nitrosylation sites.

Editor: Yu Xue, Huazhong University of Science and Technology, China

Funding: This work was supported by Marie Curie Intra-European Fellowship within the 7th European Community Framework Programme (Call: FP7-PEOPLE-2011-IEF) under grant agreement n°300176 and the Bundesministerium für Forschung und Bildung. The funders had no role in study design, data collection and analysis, decision to publish, or preparation of the manuscript.

Competing Interests: The authors have declared that no competing interests exist.

* Email: lindermayr@helmholtz-muenchen.de

Introduction

NO is a membrane-permeable free radical that plays a central role in a broad spectrum of physiological processes in plants, including germination, flowering, root development, hormonal signaling, senescence, and the establishment of adaptive responses against biotic and abiotic stress [1–9]. NO and related nitrogen species that are considered reactive can mediate various post-translational modifications (PTMs), such as metal nitrosylation, tyrosine nitration, and cysteine S-nitrosylation. Cysteine S-nitrosylation is the term used to describe the covalent binding of an NO group to a protein cysteine (Cys) residue. This PTM is considered one of the most important molecular mechanisms by which NO regulates protein functions and cell signaling and has been shown to alter protein activities, protein-protein interactions, and subcellular localization under both normal and pathological conditions [10–13].

A number of indirect MS-based proteomics approaches have been developed for the identification of S-nitrosylated proteins and their modification sites from complex biological samples [14,15]. The biotin switch technique (BST) is the most widely used method and is based on the conversion of S-nitrosylated Cys to biotinylated Cys. Such labeling allows the detection of S-nitrosylated proteins using specific anti-biotin antibodies and their isolation by affinity chromatography using neutravidin matrices. The proteins can then be identified using mass spectrometry. S-nitrosoglutathione (GSNO) is the most abundant low-molecular-weight S-nitrosothiol in plant cells and is a physiological NO reservoir and NO donor. This molecule can transfer its NO moiety to protein cysteine residues via trans-nitrosylation. GSNO has often been used to generate S-nitrosylated proteins in extracts for the subsequent isolation and identification of S-nitrosylated proteins [16–20].

The identification of redox-sensitive cysteine residues is important for understanding the regulatory functions of NO. Cysteine residues exhibiting a low-pKa sulfhydryl group are particularly susceptible to certain types of redox modification [21]. Several research groups have attempted to define consensus motifs for S-nitrosylation by comparing the amino acid sequences around identified target cysteine residues. Such analyses have revealed that the target cysteine residues often lie within an acid-base or hydrophobic motif [22]. In contrast, other studies have revealed that the primary sequence of the surrounding amino acid residues has no significant effect on the reactivity of cysteines towards S-nitrosylation at the peptide level [23]. Greco et al. (2006) supported the idea of extending the motif beyond the primary sequence to include hydrophobic motifs surrounding the identified cysteine residues [24]. Recently, 70 known S-nitrosylated sites were used to identify general structures associated with S-nitrosylation. The results obtained revealed that proximal acid–base motif, Cys pKa, sulfur atom exposure, and Cys conservation or hydrophobicity in the vicinity of the modified cysteine do not predict S-nitrosylation specificity. Instead, this analysis identified a revised acid-base motif that is located farther from the cysteine and in which the charged groups are exposed [25].

Many studies have been performed to identify and characterize S-nitrosylated proteins in plants [26]. The pioneer analysis of S-nitrosylated proteins was conducted in 2005 [16]. In this work, 63 proteins from GSNO-treated *Arabidopsis* cell culture extracts and 52 proteins from NO-treated leaves were identified as possible NO targets. In addition, Romero-Puertas and colleagues found 16 *Arabidopsis* proteins that were differentially S-nitrosylated under hypersensitive responses [27]. Moreover, endogenous S-nitrosylated proteins have been identified in an *Arabidopsis* cell culture under salt stress [28]. To date, more than two hundred proteins have been identified as putative targets for S-nitrosylation in *Arabidopsis* using proteomics approaches based on the biotin switch assay or related techniques, however only in the minority of them the exact S-nitrosylation sites have been identified. Moreover, such analyses have also been performed in other plant species such as in citrus plants exposed to salinity [29], a rice mutant overproducing NO [30], pea-leaf peroxisomes under abiotic stress [31], and a tobacco cell suspension treated with cryptogein [32]. The S-nitrosylated proteins identified from plant proteome studies have been shown to participate in major cellular activities, notably primary and secondary metabolism, protein folding and genetic information processing, photosynthesis, cellular architecture, and responses to biotic and abiotic stresses [33]. Although the number of plant proteins that have been identified as putative targets for S-nitrosylation has drastically increased during recent years, studies identifying the NO-sensitive cysteine residues involved remain rare. These analyses are essential for a better understanding of the function of protein S-nitrosylation in plants [33].

In contrast to the technical difficulties associated with experimental methods, the computational analysis of PTMs is an attractive alternative. The use of computational predictors can identify a number of potential candidates and rapidly generate useful information. Currently, approximately 170 databases and computational tools have been developed for PTM analysis [34]. The algorithms used in this field include iGPS 1.0, which is used to predict phosphorylation [35], CSS-Palm 4.0, which is used to predict S-palmitoylation [36], GPS-SUMO 1.0, which is used to predict sumoylation [37], and GPS-YNO2, which is used to predict protein nitration [38]. Moreover, several programs and algorithms have been developed to predict cysteine residues that are susceptible to S-nitrosylation, including SNOSite, iSNO-PseAAC, iSNO-AAPair, and GPS-SNO 1.0 [39–42].

In this study, we used GPS-SNO 1.0 to identify candidate proteins for S-nitrosylation within the *Arabidopsis* proteome (27,416 proteins). In total, 31,907 S-nitrosylated sites were predicted in 16,610 (approximately 61%) candidate proteins using the medium threshold. Potential target proteins were detected in all cellular compartments and ranged from 37% to 86% of the total number of proteins per compartment. More than 70% of the S-nitrosylated candidates identified were in the "chloroplast", "CUL4-RING ubiquitin ligase complex", and "membrane" compartments. In most compartments, the proportion of S-nitrosylation candidates was approximately 60%. Moreover, the 10% of S-nitrosylation sites with the highest prediction confidence were extracted for further study. This group comprised 3,190 sites in 3,005 target proteins. These candidates were detected in all compartments and ranged from 5% to 17% of the total number of proteins per compartment. These targets were enriched in the "chloroplast" (17%), "intracellular" (15%), and "plasmodesmata" (14%) compartments. In most compartments, the percentage of proteins predicted as S-nitrosylation candidates was approximately 10%. The high proportion of proteins identified as S-nitrosylation candidates reflects the importance of redox signaling in these compartments. An analysis of the functional distribution of the predicted candidates showed that the group with the highest prediction rate was the process "signaling". Moreover, a set of 46 *Arabidopsis* proteins, where 53 putative S-nitrosylation sites were previously determined using a BST-based approach, was analysed with the GPS-SNO program. The computational method predicted 60 S-nitrosylation sites within these proteins, but only 11 overlap with the results of the BST-based approach. In general, the currently available algorithm appears to be a useful tool for characterizing the S-nitrosylome but requires further improvement regarding its accuracy in identifying S-nitrosylation sites.

Materials and Methods

Data collection

First, 27,416 amino acid sequences were downloaded from the most recent version of the *Arabidopsis* information resource TAIR (TAIR10, www.arabidopsis.org). For all subsequent analyses, only one representative gene model was used per locus.

Comparison of prediction performance

To evaluate and compare the prediction performance of four S-nitrosylation prediction programs, we calculated 3 parameters (accuracy, sensitivity, and specificity) according to the definitions described previously [42]. The 4 prediction programs tested were GPS-SNO 1.0 (using the medium threshold condition), iSNO-PseAAC, iSNO-AAPair, and SNOSite [39-41].

Prediction of SNO sites using GPS-SNO software

Group-based Prediction System (GPS-SNO 1.0) software was used to predict S-nitrosylation sites [42]; this program can be executed online or downloaded at http://sno.biocuckoo.org/. In all analyses, 27,416 *Arabidopsis* amino acid sequences in FASTA format were submitted for use in predicting S-nitrosylation sites under the medium threshold condition using the batch prediction tool of the GPS-SNO 1.0 software. The predicted S-nitrosylation sites were extracted into an Excel file for further analysis.

Subcellular compartmentalization of *Arabidopsis* proteins

To determine the cellular localization of all gene predictions in *Arabidopsis*, we utilized gene ontology terms (GO) obtained from the TAIR10 annotation release (ftp://ftp.arabidopsis.org/home/tair/Ontologies/Gene_Ontology/) and filtered these terms for terms categorized as "cellular component". The distribution of proteins among the individual localization categories was plotted for all categories comprising more than 100 assignments.

MapMan analysis of the predicted candidate proteins

Protein functional classification was performed according to the MapMan Ontology of *Arabidopsis* proteins, version 3.5.1R2 (http://mapman.gabipd.org/web/guest/mapman).

Results and Discussion

In recent years, many experimental methods have been developed for the identification of S-nitrosylated proteins and the mapping of SNO-sites. The BST and related methods have enabled the high-throughput identification of hundreds of novel targets for S-nitrosylation [16,18,43-45]. However, these methods have several limitations, especially regarding the detection of low-abundance or unstable proteins or of proteins that are present only in specific tissues/organs that are difficult to handle, e.g., meristems or epidermis. Therefore, more sensitive approaches are required. ProteoMiner is a technology allowing the enrichment of low-abundance proteins [46]. However, the extracted proteins are denatured by the harsh conditions required for protein elution. Therefore, this method cannot be used in combination with the BST until a method for enriching low-abundance proteins under native conditions is established. Computational methods can overcome such technical difficulties because the analyses can be performed using the complete protein datasets that are available in databases. Thus, a nearly complete map of candidates for S-nitrosylation can be generated, providing a good starting point for more detailed, experimental approaches.

1. A comparison of programs used to predict S-nitrosylation sites

Previously, we compared three programs that are used to predict S-nitrosylation sites in proteins [26]. Here, we extended this study by including a fourth program and including all plant proteins in which modified cysteine residues have been verified using mass spectrometry and for which the physiological functions are known (Table 1). The programs GPS-SNO 1.0, iSNO-PseAAC, iSNO-AAPair, and SNOSite were tested. The performances of the 4 programs in predicting S-nitrosylation were evaluated (Table S1) as previously defined [42], using the 12 characterized S-nitrosylated proteins listed in Table 1. GPS-SNO performed best according to the three criteria chosen (accuracy, sensitivity, and specificity; 82.2%, 50%, and 87.9%, respectively, Table S1). The SNOSite software predicted almost all cysteine residues present as targets for S-nitrosylation, with accuracy and specificity of 25% and 13%, respectively, which implies that S-nitrosylation is very unspecific. The programs iSNO-PseAAC and iSNO-AAPair presented higher accuracy and specificity than SNOSite (Table S1), but their correlation with actual sites remained low. Significantly better predictions appeared possible when using the GPS-SNO 1.0 software, which exhibited a much lower rate of false positives. Approximately 60% of the proteins that were found to be S-nitrosylated using mass spectrometry were predicted using the GPS-SNO 1.0 software (which was developed by Xue and colleagues [42]). The authors of this program have improved their previous algorithm, GPS 2.0 (Group-based Prediction System), which was used for the prediction of kinase-specific phosphorylation sites, and have released GPS 3.0 [47]. Based on this algorithm, they developed the computational software GPS-SNO 1.0 for the prediction of S-nitrosylation sites. The performance of the GPS 3.0 algorithm at predicting S-nitrosylation was much better than that obtained using several other approaches, providing an accuracy of 75.70%, a sensitivity of 53.32% and a specificity of 80.11% under the low threshold condition. GPS-SNO 1.0 was applied to a test set of 485 potentially S-nitrosylated proteins collected from PubMed. These proteins were identified in large- or small-scale studies, and the actual S-nitrosylation sites have not been experimentally determined. Of the analyzed proteins, 371 (approximately 76%) were predicted to be S-nitrosylated at one or more potential S-nitrosylation sites.

2. Prediction of S-nitrosylation candidate proteins using the GPS-SNO 1.0 program

For the computer-based prediction of the S-nitrosylation of *Arabidopsis* target proteins, 27,416 amino acid sequences were extracted from the TAIR 10 database (www.arabidopsis.org) (Table S2). Of these proteins, 25,785 (94%) contain at least one cysteine residue; in total, 207,473 cysteine residues were found. All of the *Arabidopsis* amino acid sequences were analyzed with GPS-SNO 1.0 using the medium threshold, as recommended by Xue and colleagues [42]. In total, 31,907 (approximately 15% of all Cys residues) S-nitrosylation sites were predicted in 16,610 proteins (60%) (Table 2 and Table S2 and S3), suggesting that redox-related processes are closely regulated by a small number of redox-sensitive cysteine residues. The high number of putative candidate proteins reflects the importance of redox-signaling in general. Redox homeostasis during development is an evolutionary conserved strategy and the common origin of redox sensing indicate that organisms evolved similar strategies for utilizing redox-signaling during development [48]. In plant with impaired NO/S-nitrosothiol (SNO) homeostasis the importance of balancing NO/SNO levels for plant growth and development become apparent. For instance, S-nitrosoglutathione reductase knock-out plants have higher SNO levels in comparison to wild type plants and display a lot of different developmental defects, such as delayed seed germination, reduced growth, reduced trichome density, increased number of branched shoots, and generation of more flowers, which are smaller and develop to smaller siliques containing smaller seeds [49]. Moreover, leaf shape, 2,4-D sensitivity, and hypocotyl elongation is affected [50]. But S-nitrosylation of proteins might have not only a signaling function. A protection of cysteine residues against irreversible oxidation is also described [51,52]. In this way proteins can be protected against oxidative damage and after reduction they can fulfil their physiological function again.

On the other side, the high number of putative candidate proteins might indicate a high rate of false-positive predictions. Therefore, we extracted the 10% of predicted sites with the highest prediction confidence by ranking the prediction results according to the raw score divided by the threshold (Cutoff) for a particular cluster. These sites (3,190) were localized to 3005 different proteins, which comprise 18% of all predicted S-nitrosylation candidates (Table 2 and Table S2 and S3). Similarly, computational prediction has also been used for other post-translational modifications of target proteins. In the *Arabidopsis* proteome, the phosphorylation hotspot prediction algorithm has predicted 13,677 P-hotspots in 9,599 proteins corresponding to 7,847 unique genes [53]. The cited study provides a new bioinformatic method to identify phosphorylation hotspots and provides the basis for further investigation of novel candidate P-hotspots. Moreover, in the human proteome, nitration-sensitive tyrosine residues have been predicted using GPS-YNO2, a recently described 3-nitrotyrosine prediction algorithm [54]. In total, 9.27% (27,977) of all tyrosine residues (301,091) were predicted to be nitration targets. Collectively, these studies demonstrate the feasibility of using predicted datasets for whole-proteome analyses.

Table 1. Computational prediction of *S*-nitrosylation sites from experimentally identified *S*-nitrosylated proteins in plants using GPS-SNO 1.0, iSNO-PseAAC, iSNO-AAPair, and SNOSite software.

Protein name	Accession number	Total number of Cys	Physiological function demonstrated	Cys-NO sites identified by LC-MS/MS	Cys-NO sites predicted by GPS-SNO 1.0	Cys-NO sites predicted by iSNO-PseAAC	Cys-NO sites predicted by iSNO-AAPair	Cys-NO sites predicted by SNOSite	Reference
Methionine adenosyltransferase 1	At1g02500	8	Inhibited	C_{114}	C_{114}	C_{161}	C_{31}, C_{90}, C_{161}	C_{20}, C_{31}, C_{42}, C_{73}, C_{90}, C_{114}, C_{161}	[85]
Metacaspase 9	At5g04200	7	Inhibited	C_{147}	C_{17}, C_{147}	C_{17}, C_{29}	C_{117}	C_{17}, C_{29}, C_{117}, C_{147}, C_{309}	[81]
Peroxiredoxin II E	At3g52960	2	Inhibited	C_{121}	C_{121}	C_{121}, C_{146}	C_{121}	C_{121}, C_{146}	[86]
NPR1	At1g64280	17	Inhibited	C_{156}	C_{156}, C_{385}	C_{212}, C_{306}	C_{223}, C_{306}, C_{394},C_{457}	C_{82}, C_{150}, C_{155}, C_{156}, C_{160}, C_{212}, C_{223}, C_{297}, C_{306}, C_{378}, C_{385}, C_{394}, C_{457}, C_{511}, C_{529}	[83]
GAPDH	At1g13440	2	Inhibited	C_{156}, C_{160}	C_{156}, C_{160}	_	_	C_{156}, C_{160}	[87]
SABP3	At3g01500	7	Inhibited	C_{280}	C_{34}, C_{173}, C_{280}	C_{230}, C_{257}	C_{34}	C_{34}, C_{167}, C_{173}, C_{230}, C_{257}, C_{277}, C_{280}	[88]
Transcription factor-TGA1	At5g65210	4	Activated	C_{172}, C_{287}	C_{172}	_	_	C_{172}, C_{260}, C_{266}, C_{287}	[19]
NADPH oxidase	At5g47910	10	Inhibited	C_{890}	_	C_{208}, C_{387}, C_{433}, C_{480}, C_{695}	C_{412}, C_{480}, C_{695}, C_{890}	C_{208}, C_{410}, C_{412}, C_{433}, C_{480}, C_{651}, C_{695}, C_{825}, C_{890}	[89]
cALD2	At2g36460	6	Inhibited	C_{173}	C_{68}, C_{326}	C_{326}	C_{208}	C_{68}, C_{173}, C_{197}, C_{208}, C_{326}	[90]
TIR1	At3g62980	23	Activated	C_{140}	C_{516}, C_{551}	C_{34}, C_{53}, C_{121}, C_{140}, C_{155}, C_{210}, C_{269}, C_{288}, C_{311}, C_{405}, C_{480}, C_{491}	C_{121}, C_{140}, C_{405}, C_{551}	C_{34}, C_{44}, C_{53}, C_{121}, C_{140}, C_{155}, C_{193}, C_{210}, C_{264}, C_{269}, C_{288}, C_{311}, C_{337}, C_{371}, C_{405}, C_{480}, C_{491}, C_{516}, C_{523}, C_{551}	[91]
CDC48	Q1G0Z1	14	Inhibited	C_{110}, C_{526}, C_{664}	C_{426}, C_{576}	C_{74}, C_{82}, C_{110}, C_{526}, C_{539}, C_{576}, C_{664}, C_{699}	C_{74}, C_{426}, C_{539}, C_{576}	C_{74}, C_{82}, C_{110}, C_{179}, C_{189}, C_{272}, C_{419}, C_{426}, C_{539}, C_{576}, C_{664}, C_{695}, C_{699}	[32]
AtMYB30	At3g28910	7	Inhibited	C_{53}	C_{6}	C_{6}, C_{7}, C_{49}, C_{53}, C_{257}, C_{289}	C_{6}, C_{7}	C_{49}, C_{53}, C_{257}, C_{289}, C_{290}	[92]

Amino acid sequences were downloaded from the most recent version of the *Arabidopsis* information resource TAIR (TAIR10, www.arabidopsis.org) and subjected to the different programs for prediction of S-nitrosylation sites. NPR1, non-expresser of pathogenesis related genes 1; GAPDH, glyceraldehyde 3-phosphate dehydrogenase; SABP3, salicylic acid binding protein 3; TGA1, TGACG motif binding factor; cALD2, cytosolic fructose 1,6-bisphosphate aldolase; TIR1, transport inhibitor response 1; CDC48, cell division cycle 48; AtMYB30, *Arabidopsis thaliana* MYB transcription factor.
C in bold, matched cysteine residues, "_" not predicted

Table 2. Prediction of *Arabidopsis* candidate proteins for S-nitrosylation using the GPS-SNO 1.0 software.

	Arabidopsis proteome	Candidate proteins for *S*-nitrosylation	The highest 10% high-confident predicted candidates
Total number of proteins	27,416	16,610 (60%)	3,005 (18%)
Total number of Cys-NO	207,473	31,907 (15%)	3,190 (10%)

Arabidopsis amino acid sequences were extracted from TAIR 10 database (www.arabidopsis.org) and analysed by GPS-SNO 1.0 software using medium threshold condition. The 10% of predicted sites with the highest prediction confidence were determined by ranking the prediction results according to the raw score divided by the threshold (Cutoff) for a particular cluster.

3. Subcellular compartment classification of *Arabidopsis* proteins

To determine whether the identified candidates for S-nitrosylation are enriched in distinct subcellular compartments (Text S1), all *Arabidopsis* proteins and the predicted candidates were assigned to subcellular locations according to gene ontology (GO) terms using cellular component classifications (Table S4). In Table 3, only compartments with more than 100 representatives are listed. An analysis of the subcellular localization of all *Arabidopsis* proteins revealed that most were assigned to the "nucleus" (9,214 proteins) or to "membranes" (4,389 proteins). The predicted S-nitrosylation candidate proteins were also located in other compartments, comprising 37% to 86% of the total protein content in each compartment (Table 3). Similar results have been found experimentally in *Arabidopsis* suspension cell cultures: S-nitrosylated proteins were found in almost all cell

Table 3. Subcellular compartment classification of *Arabidopsis* proteins.

Compartments	Total number of proteins	Candidate proteins for *S*-nitrosylation	Candidate proteins for *S*-nitrosylation harboring the highest 10% high-confident predicted sites
Chloroplast	3795	3259 (86%)	659 (17%)
CUL4-RING ubiquitin ligase complex	121	91 (75%)	13 (11%)
Membrane	4389	3257 (74%)	493 (11%)
Plasmodesmata	848	596 (70%)	116 (14%)
Vacuole	799	556 (70%)	79 (10%)
Cell wall	469	314 (67%)	45 (10%)
Plant-type cell wall	264	176 (67%)	26 (10%)
Endosome	232	153 (66%)	13 (6%)
Trans-Golgi network	219	144 (66%)	13 (6%)
Cytoplasm	3461	2222 (64%)	364 (11%)
Nucleus	9214	5924 (64%)	1118 (12%)
Extracellular region	2390	1512 (63%)	232 (10%)
Intracellular	1015	630 (62%)	148 (15%)
Cytosol	1468	903 (62%)	151 (10%)
Integral to membrane	808	503 (62%)	67 (8%)
Golgi apparatus	877	539 (61%)	65 (7%)
Plastid	289	172 (60%)	37 (13%)
Peroxisome	170	99 (58%)	17 (10%)
Mitochondrion	3048	1744 (57%)	323 (11%)
Cytosolic ribosome	304	164 (54%)	30 (10%)
Apoplast	390	208 (53%)	35 (9%)
Endoplasmic reticulum	517	270 (52%)	26 (5%)
Anchored to membrane	237	120 (51%)	16 (7%)
Ribosome	384	143 (37%)	33 (9%)
Cellular component	1917	705 (37%)	149 (8%)

Total number of proteins, number of predicted candidates for *S*-nitrosylation, and the number of candidates with the highest 10% prediction confidence were assigned to their subcellular localization according to gene ontology cellular component classification. The prediction confidence was calculated by dividing the raw score value by the cutoff value of a particular cluster.

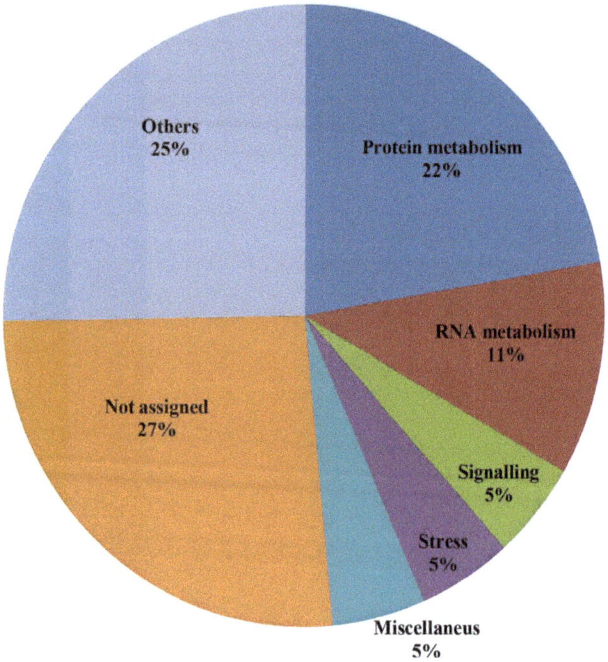

Figure 1. Functional distribution of predicted candidate proteins for *S*-nitrosylation has been determined using the MapMan Ontology tool (http://mapman.gabipd.org/). Others; include all functional classes which have less than 5% of predicted candidates.

compartments [28]. Moreover, a similar distribution was also observed in animal cells [55]. Interestingly, the predicted candidates are most enriched in the "chloroplast", "CUL4-RING ubiquitin ligase complex", and "membrane" compartments (86%, 75%, and 74%, respectively), suggesting that redox-related processes play important roles in these locations.

The nucleus is an important sub-cellular organelle that contains almost all of the genetic information required for the regulation of cellular processes. Interestingly, a high number of S-nitrosylation candidates was predicted for the "nucleus" compartment (5,924 proteins, 64% of the total), which also contained a high proportion of the proteins that harbored the 10% of sites that were predicted with the highest confidence (1,118 proteins, 12% of the total).

The 10% of S-nitrosylation sites that were predicted with the highest confidence were also found in all compartments at levels of 5% to 17% (Table 3). In particular, the compartments "chloroplast" (17%), "intracellular" (15%), and "plasmodesmata" (14%) appeared to be enriched in the sites predicted with high confidence. Interestingly, chloroplast proteins exhibited the highest percentage of S-nitrosylation candidates in both analyses. Chloroplasts are sources of redox intermediates and chloroplast signaling pathways are triggered by the redox state of the plastochinone pool, the thioredoxin system, and the acceptor availability at photosystem I [56]. Moreover, discrete redox signaling pathways regulate photosynthetic light-harvesting and chloroplast gene transcription [57]. Production of NO in plant cells arise from several different pathways and in different organelles, including chloroplasts [58,59] and target sites of NO in chloroplasts have been found in photosystem I and II, in the cytochrome *b6f* complex and in carbon dioxide reduction processes [60]. Although the chloroplast S-nitrosylome has not been analyzed yet, alterations in ribulose-1,5-bisphosphate carboxylase/oxygenase S-nitrosylation inactivated its carboxylase activity in *Brassica juncea* [61]. Furthermore, chloroplastic triosephosphate isomerase (TPI) was already identified as target for S-nitrosylation in rice, citrus, and *Chlamydomonas reinhardtii*, suggesting that this type of modification might be involved in the regulation of chloroplastic TPI activity [29,30,62,63]. Moreover, chloroplasts have been discussed as a source and a target of cellular redox regulation [56] and therefore might represent a favorable microenvironment for S-nitrosylation in *Arabidopsis*.

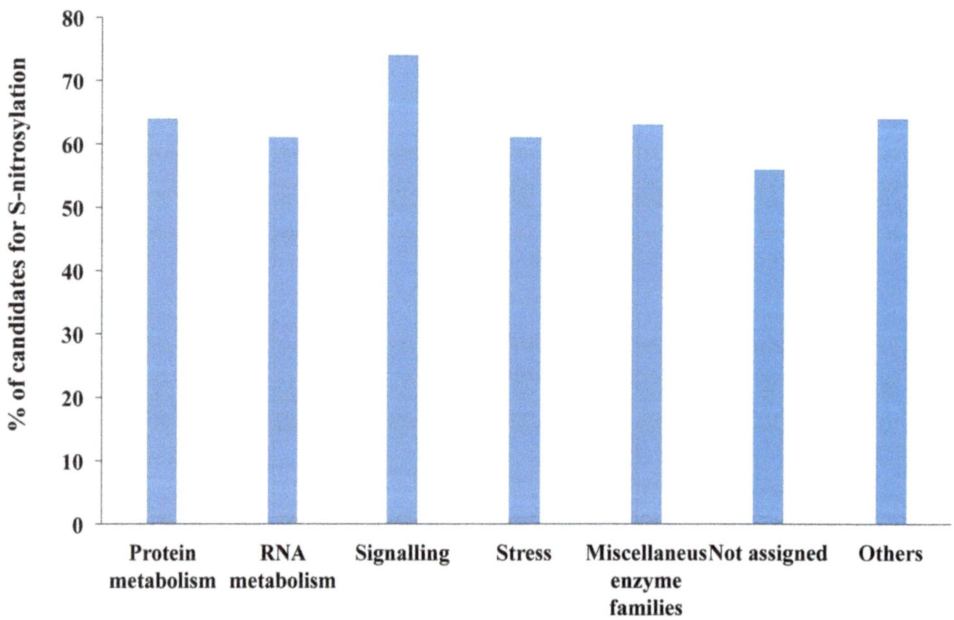

Figure 2. Percentage of candidate proteins for *S*-nitrosylation in different functional categories. Functional assignment has been done using the MapMan Ontology tool (http://mapman.gabipd.org/web/guest/mapman).

Table 4. Percentage of predicted candidate proteins for *S*-nitrosylation in signaling subclasses.

Signaling subclasses	Total proteins	Candidate proteins for *S*-nitrosylation
14-3-3 proteins	15	15 (100%)
Light	117	98 (84%)
Lipids	6	5 (83%)
MAP kinases	50	40 (80%)
Receptor kinases	1067	843 (79%)
Phosphinositides	98	76 (78%)
Sugar and nutrient physiology	82	58 (71%)
G-proteins	243	157 (65%)
Unspecified	8	5 (62%)
Calcium	230	141 (61%)
Miscellaneus enzyme families	41	21 (51%)
Phosphorelay	5	1 (20%)

Functional classification of the predicted candidates has been done using the MapMan Ontology software (http://mapman.gabipd.org/web/guest/mapman).

In most compartments, the percentage of proteins predicted as S-nitrosylation candidates using the medium threshold ranged from 51% to 70%. The smallest proportion of S-nitrosylation candidates was located in the "ribosome" compartment (37%). Ribosomes comprise the basic machinery that decodes genetic information into proteins. Increasing numbers of studies on ribosome biogenesis have been performed on *Arabidopsis*. Ribosomal protein functions have been demonstrated in embryo biogenesis, leaf and flower development, vacuolar trafficking, and the UV response [64–70]. However, the lowest percentages of predicted targets among the 10% of sites predicted with the highest confidence were found in the "endoplasmic reticulum" (5%), "trans-Golgi network" (6%), and "endosome" (6%) compartments. Several previous studies have demonstrated that S-nitrosylated proteins were localized in various organelles, including the cell membrane [71], mitochondria [72,73], the nucleus [74], the endoplasmic reticulum, the Golgi and cytosol [75], the peroxisome [31], and the apoplast [76]. This suggests that protein S-nitrosylation can occur in all subcellular compartments [77].

4. Functional distribution of *Arabidopsis* S-nitrosylation candidate proteins

To analyze the functional classification of the predicted candidates, 16,610 predicted proteins were subjected to analysis using the MapMan Ontology of *Arabidopsis* proteins (http://mapman.gabipd.org/web/guest/mapman). Most of the candidates belong to unknown categories (not assigned) or others, including categories containing less than 5% of candidates (Figure 1). Most of the candidates assigned to known categories are involved in protein and RNA metabolism (22% and 11% of all candidates, respectively), signaling (5%) and stress-related processes (5%). The proportion of predicted candidates in known functional categories was calculated in relation to the total number of proteins of each category; the results showed that approximately 60% of the proteins in each category were S-nitrosylation candidates (Figure 2). The signaling category presented the highest proportion of S-nitrosylation candidates (70%). A more detailed analysis of this group revealed that 70% to 100% of the subclasses "14-3-3 family proteins", "light", "lipids", "MAP and receptor kinases", "phosphoinositides", and "sugar and nutrient physiology", are S-nitrosylation candidates (Table 4). 14-3-3 proteins have previously been identified as S-nitrosylation targets in *Arabidopsis* [16,28] and in mesangial cells [78]. 14-3-3 proteins represent an emerging family of proteins and protein domains that bind to serine/threonine-phosphorylated residues. These proteins regulate key proteins that are involved in several physiological processes, including intracellular signaling, apoptosis, cell cycling, and transcriptional regulation. 14-3-3 proteins also act as adaptor molecules that stimulate protein-protein interactions and regulate the subcellular localization of proteins [79]. Interestingly, the 10% of sites predicted with the highest confidence in the large-scale prediction study showed the same functional classification pattern as that for all S-nitrosylated proteins (Figure S1). The functional distribution of the predicted S-nitrosylation candidates is similar to that of the major classes of S-nitrosylated proteins that have been identified experimentally in *Arabidopsis* [16,27,28,80].

5. A comparison of experimentally identified candidates with the candidates predicted using GPS-SNO 1.0 software

Two-hundred sixty-three proteins have previously been identified experimentally in *Arabidopsis thaliana* as S-nitrosylation candidates based on the BST [16,18,19,27,28,80–83]. These proteins were detected using large- and small-scale studies, most of which did not determine the exact S-nitrosylation sites experimentally. To compare the results of the computational predictions with experimental data, we analyzed these datasets using GPS-SNO. Interestingly, 160 proteins (approximately 61%) that were identified using the biotin switch approach were also predicted by the GPS-SNO software (using the medium threshold) as S-nitrosylation candidates.

In a more detailed analysis, Fares et al. experimentally identified 53 S-nitrosylation sites on 46 proteins in an *Arabidopsis* cell suspension using BS-ICAT technology [28]. However, these identified S-nitrosylation sites were not further verified on the biochemical and physiological level meaning that these S-nitrosylation sites/proteins are still candidates. This set of proteins was also analyzed using the GPS-SNO 1.0 software under the medium threshold condition (Table 5). This analysis revealed that approximately 74% of proteins (34 proteins) that were identified as S-nitrosylated using BS-ICAT were also predicted as S-nitrosylation candidates using GPS-SNO. To compare the candidate

Table 5. Prediction of *S*-nitrosylated sites from experimentally identified *S*-nitrosylated proteins by GPS-SNO software.

Accession number	Cys-NO site identified by BS-ICAT	Cys-NO site predicted by GPS-SNO	NO-peptide sequence predicted by GPS-SNO
AT1G04710	C_{130}	C_{184}	KFEQAHNCLLPMGIT
AT1G04710	-	C_{363}	FASQFVYCRNKLGLD
AT1G04710	-	C_{394}	LGATGARCVATLLHE
AT1G04710	-	C_{417}	RFGVVSMCIGSGMGA
AT1G07890	C_{32}	C_{19}	YKKAVEKCRRKLRGL
AT1G07930	C_{87}	C_{111}	TGTSQADCAVLIIDS
AT1G09780	C_{100}	C_{355}	NGVSTFACSETVKFG
AT1G19570	C_{20}	C_{6}	**MALEICVKAAVGA
AT1G22300	C_{98}	$\mathbf{C}_{98}$	EDELAKVCNDILSVI
AT1G35720	C_{111}	$\mathbf{C}_{111}$	QVLMEVACTRTSTQL
AT1G47128	C_{233}	C_{161}	DQGGCGSCWAFSTIG
AT1G47128	C_{342}	$\mathbf{C}_{342}$	IASSSGKCGIAIEPS
AT1G47128	-	C_{200}	DTSYNEGCNGGLMDY
AT1G56070	C_{370}	C_{131}	GALVVVDCIEGVCVQ
AT1G56070	-	C_{448}	ETVEDVPCGNTVAMV
AT1G60710	C_{198}	C_{5}	***MAEACGVRRMKL
AT1G60710	-	C_{254}	KIVYEKVCAISEKKG
AT1G63000	C_{162}	-	-
AT1G65930	C_{75}	C_{297}	LMTSVLVCPDGKTIE
AT1G65930	C_{363}	$\mathbf{C}_{363}$	TEKLEAACVGTVESG
AT1G65930	C_{269}	-	-
AT1G73010	C_{98}	C_{165}	GTCPPNMCKGLIIER
AT1G77120	C_{243}	C_{10}	TTGQIIRCKAAVAWE
AT1G77120	-	C_{271}	GVDRSVECTGSVQAM
AT1G78830	C_{374}	-	-
AT2G31390	C_{298}	-	-
AT2G39730	C_{175}	C_{451}	NLPVPEGCTDPVAEN
AT2G44350	C_{108}	C_{210}	WEPTYEDCLNLIARV
AT2G45290	C_{440}	$\mathbf{C}_{440}$	TRNLSQQCLNALAKA
AT2G45290	-	C_{245}	EGISNEVCSLAGHWG
AT3G08580	C_{130}	$\mathbf{C}_{130}$	PYKGIGDCFGRTIKD
AT3G09820	C_{323}	-	-
AT3G09840	C_{109}	C_{425}	CTEAALQCIREKMDV
AT3G09840	-	C_{575}	KARQSAPCVLFFDEL
AT3G11940	C_{175}	C_{69}	KRFRKAQCPIVERLT
AT3G17240	C_{372}	-	-
AT3G47370	C_{39}	-	-
AT3G51800	C_{178}	-	-
AT3G53870	C_{134}	C_{97}	KVNNRGLCAIAQAES
AT3G55440	C_{218}	C_{13}	FVGGNWKCNGTAEEV
AT3G55440	C_{127}	$\mathbf{C}_{127}$	QGLKVIACVGETLEE
AT3G56310	C_{311}	C_{117}	IHVNIDDCWSNLLRD
AT3G56310	-	C_{422}	AQVDAHDCHMYVLTP
AT3G61440	C_{72}	C_{16}	LRRETIPCFSHTVRK
AT3G61440	-	C_{87}	QEHFQPTCSIKDRPA
AT4G09320	C_{43}	C_{2}	******MCGLYINLF
AT4G09320	C_{268}	-	-
AT4G11150	C_{201}	C_{121}	LKDLIVQCLLRLKEP
AT4G11150	-	C_{134}	EPSVLLRCREEDLGL

Table 5. Cont.

Accession number	Cys-NO site identified by BS-ICAT	Cys-NO site predicted by GPS-SNO	NO-peptide sequence predicted by GPS-SNO
AT4G11650	C_{72}	-	-
AT4G13430	C_{376}	C_{12}	ISSSPFLCKSSSKSD
AT4G13940	C_{244}	C_{42}	EMPGLMACRTEFGPS
AT4G13940	C_{268}	-	-
AT4G33030	C_{357}	$\mathbf{C}_{357}$	DIRDTVQCVEIAIAN
AT4G33030	-	C_{9}	AHLLSASCPSVISLS
AT5G02500	C_{319}	$\mathbf{C}_{319}$	NMDLFRKCMEPVEKC
AT5G02500	-	C_{326}	CMEPVEKCLRDAKMD
AT5G02500	-	C_{609}	MKELESICNPIIAKM
AT5G14040	C_{104}	C_{194}	IIADIALCPFEAVKV
AT5G15490	C_{350}	-	-
AT5G25100	C_{104}	C_{363}	YVGTGVQCLGMVLVT
AT5G44340	C_{354}	$\mathbf{C}_{354}$	NNVKSSVCDIAPKGL
AT5G44340	-	C_{12}	LHIQGGQCGNQIGAK
AT5G44340	-	C_{238}	ATMSGVTCCLRFPGQ
AT5G61790	C_{108}	-	-
AT5G62690	C_{56}	C_{12}	LHIQGGQCGNQIGAK
AT5G62690	C_{301}	C_{238}	ATMSGVTCCLRFPGQ
AT5G62690	-	C_{354}	NNVKSTVCDIPPTGL
AT5G66760	-	C_{4}	****MWRCVSRGFRA
AT5G66760	-	C_{77}	EHGFNTACITKLFPT
AT5G66760	-	C_{294}	TGIYGAGCLITEGSR
AT5G66760	-	C_{457}	IVVFGRACANRVAEI
AT5G66760	C_{526}	$\mathbf{C}_{526}$	QETLEEGCQLIDKAW
ATCG00340	C_{559}	-	-
ATCG00490	C_{192}	-	-
ATCG00490	C_{427}	-	-

S-nitrosylated *Arabidopsis* candidate proteins published by Fares *et al.* (2011) were analysed by GPS-SNO software using the medium threshold condition.
C in bold, matched cysteine residues.

cysteine sites, the GPS-SNO program was used to predict 60 putative S-nitrosylation sites within these 34 proteins; however, only 11 of the predicted S-nitrosylation sites corresponded to sites identified using BS-ICAT (Table 6). These data indicate that the GPS-SNO software predicts a different set of S-nitrosylation sites in comparison to the BST-based approach.

Conclusions

Protein S-nitrosylation has emerged as an important field of the study of post-translational modification and is increasingly studied in plants. However, the proteomic approaches used to identify proteins that are targets of S-nitrosylation are associated with a variety of technical difficulties, such as the existence of side reactions in multi-step procedures, the low abundance or instability of proteins, and instrumental inaccuracy. Computational methods can help to overcome these problems. Computational analyses can be performed easily on complex protein datasets obtained from databases, regardless of protein abundance or instability or the existence of complex chemical reactions. However, computational approaches also present disadvantages. Protein S-nitrosylation is an enzyme-independent chemical

Table 6. Computational analysis of proteins, which S-nitrosylation sites were identified by BS-ICAT technology [28].

	BS-ICAT	GPS-SNO 1.0 medium threshold
Protein number	46	34
Total number of Cys-NO	53	60
Matched Cys-NO with BS-ICAT	-	11

reaction that depends on many factors, all of which define whether a given cysteine residue will be sensitive to this modification. Although GPS-SNO 1.0 appears to predict S-nitrosylation sites with better accuracy, sensitivity, and specificity than other algorithms (Table S1), further research is required to improve the accuracy of the identification of S-nitrosylated sites. In this context, a set of non-SNO proteins would be helpful to calculate the sensitivity and specificity of the predictor.

Of greatest importance, all developed programs, including GPS-SNO 1.0, are based on the primary sequence of the studied proteins. However, the 3-dimensional (3D) structure of a protein also greatly affects its sensitivity to S-nitrosylation. The 3D structure defines which cysteine residues are accessible, and the amino acids surrounding a cysteine residue in the 3D structure determine the sensitivity of this residue to S-nitrosylation. Knowledge of the tertiary and quaternary structure of the protein may identify additional cysteines that might not be identified based on the primary sequence. Conversely, cysteine residues that are predicted to be S-nitrosylation targets might be excluded because they are inaccessible based on the spatial conformation. Therefore, knowledge of the high-resolution structure of the microenvironment around each cysteine residue is essential for defining the physicochemical features that determine S-nitrosylation specificity. Protein 3D structures have been already used to identify protein phosphorylation sites [84]. In that study linear motifs and spatial amino acid composition within a specific radial distance from the phosphorylated amino acid residue have been included [84]. But in general, computer-based prediction of S-nitrosylation candidates from *Arabidopsis* can offer a starting point for experimental verification and for further studies of S-nitrosylation in plants. The combination of computational prediction and experimental verification represents a good approach to better understand the molecular mechanisms and the regulatory functions of S-nitrosylation in plants. Nevertheless, both methods must be developed further to improve the precision with which S-nitrosylation targets are identified. Finally, the identified or predicted candidates must be confirmed using recombinant proteins, cysteine mutants and *in-vivo* approaches.

Supporting Information

Figure S1 Functional distribution of the 10% of candidates that were predicted with the highest confidence levels based on the MapMan Ontology of *Arabidopsis* proteins (http://mapman.gabipd.org/web/guest/mapman). Others: functional classes with less than 5% of S-nitrosylated candidates.

Table S1 Comparison of the performance of four software tools in predicting S-nitrosylation sites. Accuracy, sensitivity and specificity were used to evaluate the performance.

Table S2 The *Arabidopsis* proteome was extracted from the TAIR 10 database, and proteins were assigned to cellular localizations according to the gene ontology cellular component classification.

Table S3 Amino acid sequences were downloaded for *Arabidopsis* from TAIR (www.arabidopsis.org) and analyzed using the GPS-SNO 1.0 program and the medium threshold. The 10% of candidates that were predicted with the highest confidence were ranked by the raw score divided by the cutoff of a particular cluster.

Table S4 The *Arabidopsis* proteome was extracted from the TAIR 10 database, and proteins were assigned to cellular localizations according to the gene ontology cellular component classification.

Text S1 Subcellular compartments assigned according to the gene ontology cellular component classification (http://amigo1.geneontology.org/cgi-bin/amigo/go.cgi).

Author Contributions

Conceived and designed the experiments: MC IK MS CL. Performed the experiments: MC IK MS. Analyzed the data: MC IK MS CL. Contributed reagents/materials/analysis tools: MS. Wrote the paper: MC IK MS CL.

References

1. Durner J, Wendehenne D, Klessig DF (1998) Defense gene induction in tobacco by nitric oxide, cyclic GMP, and cyclic ADP-ribose. Proc Natl Acad Sci U S A 95: 10328–10333.
2. Delledonne M, Xia Y, Dixon RA, Lamb C (1998) Nitric oxide functions as a signal in plant disease resistance. Nature 394: 585–588.
3. Garcia-Mata C, Lamattina L (2002) Nitric oxide and abscisic acid cross talk in guard cells. Plant Physiol 128: 790–792.
4. Lombardo MC, Graziano M, Polacco JC, Lamattina L (2006) Nitric oxide functions as a positive regulator of root hair development. Plant Signal Behav 1: 28–33.
5. Corpas FJ, Chaki M, Fernandez-Ocana A, Valderrama R, Palma JM, et al. (2008) Metabolism of reactive nitrogen species in pea plants under abiotic stress conditions. Plant Cell Physiol 49: 1711–1722.
6. Chaki M, Fernandez-Ocana AM, Valderrama R, Carreras A, Esteban FJ, et al. (2009) Involvement of reactive nitrogen and oxygen species (RNS and ROS) in sunflower-mildew interaction. Plant Cell Physiol 50: 265–279.
7. Sirova J, Sedlarova M, Piterkova J, Luhova L, Petrivalsky M (2011) The role of nitric oxide in the germination of plant seeds and pollen. Plant Sci 181: 560–572.
8. Chaki M, Valderrama R, Fernandez-Ocana AM, Carreras A, Gomez-Rodriguez MV, et al. (2011) Mechanical wounding induces a nitrosative stress by down-regulation of GSNO reductase and an increase in S-nitrosothiols in sunflower (Helianthus annuus) seedlings. J Exp Bot 62: 1803–1813.
9. Begara-Morales JC, Chaki M, Sanchez-Calvo B, Mata-Perez C, Leterrier M, et al. (2013) Protein tyrosine nitration in pea roots during development and senescence. J Exp Bot 64: 1121–1134.
10. Hess DT, Matsumoto A, Kim SO, Marshall HE, Stamler JS (2005) Protein S-nitrosylation: purview and parameters. Nat Rev Mol Cell Biol 6: 150–166.
11. Cho DH, Nakamura T, Fang J, Cieplak P, Godzik A, et al. (2009) S-nitrosylation of Drp1 mediates beta-amyloid-related mitochondrial fission and neuronal injury. Science 324: 102–105.
12. Guo CJ, Atochina-Vasserman EN, Abramova E, Foley JP, Zaman A, et al. (2008) S-nitrosylation of surfactant protein-D controls inflammatory function. PLoS Biol 6: e266.
13. Hao G, Xie L, Gross SS (2004) Argininosuccinate synthetase is reversibly inactivated by S-nitrosylation in vitro and in vivo. J Biol Chem 279: 36192–36200.
14. Jaffrey SR, Snyder SH (2001) The biotin switch method for the detection of S-nitrosylated proteins. Sci STKE 2001: pl1.
15. Hao G, Derakhshan B, Shi L, Campagne F, Gross SS (2006) SNOSID, a proteomic method for identification of cysteine S-nitrosylation sites in complex protein mixtures. Proc Natl Acad Sci U S A 103: 1012–1017.
16. Lindermayr C, Saalbach G, Durner J (2005) Proteomic identification of S-nitrosylated proteins in *Arabidopsis*. Plant Physiol 137: 921–930.
17. Abat JK, Mattoo AK, Deswal R (2008) S-nitrosylated proteins of a medicinal CAM plant Kalanchoe pinnata- ribulose-1,5-bisphosphate carboxylase/oxygenase activity targeted for inhibition. FEBS J 275: 2862–2872.
18. Palmieri MC, Lindermayr C, Bauwe H, Steinhauser C, Durner J (2010) Regulation of plant glycine decarboxylase by s-nitrosylation and glutathionylation. Plant Physiol 152: 1514–1528.

19. Lindermayr C, Sell S, Muller B, Leister D, Durner J (2010) Redox regulation of the NPR1-TGA1 system of Arabidopsis thaliana by nitric oxide. Plant Cell 22: 2894–2907.
20. Begara-Morales JC, Sanchez-Calvo B, Chaki M, Valderrama R, Mata-Perez C, et al. (2013) Dual regulation of cytosolic ascorbate peroxidase (APX) by tyrosine nitration and S-nitrosylation. J Exp Bot.
21. Spadaro D, Yun BW, Spoel SH, Chu C, Wang YQ, et al. (2010) The redox switch: dynamic regulation of protein function by cysteine modifications. Physiol Plant 138: 360–371.
22. Stamler JS, Lamas S, Fang FC (2001) Nitrosylation. the prototypic redox-based signaling mechanism. Cell 106: 675–683.
23. Taldone FS, Tummala M, Goldstein EJ, Ryzhov V, Ravi K, et al. (2005) Studying the S-nitrosylation of model peptides and eNOS protein by mass spectrometry. Nitric Oxide 13: 176–187.
24. Greco TM, Hodara R, Parastatidis I, Heijnen HF, Dennehy MK, et al. (2006) Identification of S-nitrosylation motifs by site-specific mapping of the S-nitrosocysteine proteome in human vascular smooth muscle cells. Proc Natl Acad Sci U S A 103: 7420–7425.
25. Marino SM, Gladyshev VN (2010) Structural analysis of cysteine S-nitrosylation: a modified acid-based motif and the emerging role of trans-nitrosylation. J Mol Biol 395: 844–859.
26. Kovacs I, Lindermayr C (2013) Nitric oxide-based protein modification: formation and site-specificity of protein S-nitrosylation. Front Plant Sci 4: 137.
27. Romero-Puertas MC, Campostrini N, Matte A, Righetti PG, Perazzolli M, et al. (2008) Proteomic analysis of S-nitrosylated proteins in Arabidopsis thaliana undergoing hypersensitive response. Proteomics 8: 1459–1469.
28. Fares A, Rossignol M, Peltier JB (2011) Proteomics investigation of endogenous S-nitrosylation in Arabidopsis. Biochem Biophys Res Commun 416: 331–336.
29. Tanou G, Job C, Rajjou L, Arc E, Belghazi M, et al. (2009) Proteomics reveals the overlapping roles of hydrogen peroxide and nitric oxide in the acclimation of citrus plants to salinity. Plant J 60: 795–804.
30. Lin A, Wang Y, Tang J, Xue P, Li C, et al. (2012) Nitric oxide and protein S-nitrosylation are integral to hydrogen peroxide-induced leaf cell death in rice. Plant Physiol 158: 451–464.
31. Ortega-Galisteo AP, Rodriguez-Serrano M, Pazmino DM, Gupta DK, Sandalio LM, et al. (2012) S-Nitrosylated proteins in pea (Pisum sativum L.) leaf peroxisomes: changes under abiotic stress. J Exp Bot 63: 2089–2103.
32. Astier J, Besson-Bard A, Lamotte O, Bertoldo J, Bourque S, et al. (2012) Nitric oxide inhibits the ATPase activity of the chaperone-like AAA+ ATPase CDC48, a target for S-nitrosylation in cryptogein signalling in tobacco cells. Biochem J 447: 249–260.
33. Astier J, Kulik A, Koen E, Besson-Bard A, Bourque S, et al. (2012) Protein S-nitrosylation: what's going on in plants? Free Radic Biol Med 53: 1101–1110.
34. Xue Y, Liu Z, Cao J, Ren J (2011) Computational Prediction of Post-Translational Modification Sites in Proteins. book edited by Ning-Sun Yang: 105–124.
35. Song C, Ye M, Liu Z, Cheng H, Jiang X, et al. (2012) Systematic analysis of protein phosphorylation networks from phosphoproteomic data. Mol Cell Proteomics 11: 1070–1083.
36. Ren J, Wen L, Gao X, Jin C, Xue Y, et al. (2008) CSS-Palm 2.0: an updated software for palmitoylation sites prediction. Protein Eng Des Sel 21: 639–644.
37. Ren J, Gao X, Jin C, Zhu M, Wang X, et al. (2009) Systematic study of protein sumoylation: Development of a site-specific predictor of SUMOsp 2.0. Proteomics 9: 3409–3412.
38. Liu Z, Cao J, Ma Q, Gao X, Ren J, et al. (2011) GPS-YNO2: computational prediction of tyrosine nitration sites in proteins. Mol Biosyst 7: 1197–1204.
39. Lee TY, Chen YJ, Lu TC, Huang HD, Chen YJ (2011) SNOSite: exploiting maximal dependence decomposition to identify cysteine S-nitrosylation with substrate site specificity. PLoS One 6: e21849.
40. Xu Y, Ding J, Wu LY, Chou KC (2013) iSNO-PseAAC: predict cysteine S-nitrosylation sites in proteins by incorporating position specific amino acid propensity into pseudo amino acid composition. PLoS One 8: e55844.
41. Xu Y, Shao XJ, Wu LY, Deng NY, Chou KC (2013) iSNO-AAPair: incorporating amino acid pairwise coupling into PseAAC for predicting cysteine S-nitrosylation sites in proteins. PeerJ 1: e171.
42. Xue Y, Liu Z, Gao X, Jin C, Wen L, et al. (2010) GPS-SNO: computational prediction of protein S-nitrosylation sites with a modified GPS algorithm. PLoS One 5: e11290.
43. Sun J, Morgan M, Shen RF, Steenbergen C, Murphy E (2007) Preconditioning results in S-nitrosylation of proteins involved in regulation of mitochondrial energetics and calcium transport. Circ Res 101: 1155–1163.
44. Doulias PT, Greene JL, Greco TM, Tenopoulou M, Seeholzer SH, et al. (2010) Structural profiling of endogenous S-nitrosocysteine residues reveals unique features that accommodate diverse mechanisms for protein S-nitrosylation. Proc Natl Acad Sci U S A 107: 16958–16963.
45. Kato H, Takemoto D, Kawakita K (2013) Proteomic analysis of S-nitrosylated proteins in potato plant. Physiol Plant 148: 371–386.
46. Frohlich A, Gaupels F, Sarioglu H, Holzmeister C, Spannagl M, et al. (2012) Looking deep inside: detection of low-abundance proteins in leaf extracts of Arabidopsis and phloem exudates of pumpkin. Plant Physiol 159: 902–914.
47. Xue Y, Ren J, Gao X, Jin C, Wen L, et al. (2008) GPS 2.0, a tool to predict kinase-specific phosphorylation sites in hierarchy. Mol Cell Proteomics 7: 1598–1608.
48. Schippers JH, Nguyen HM, Lu D, Schmidt R, Mueller-Roeber B (2012) ROS homeostasis during development: an evolutionary conserved strategy. Cell Mol Life Sci 69: 3245–3257.
49. Holzmeister C, Frohlich A, Sarioglu H, Bauer N, Durner J, et al. (2011) Proteomic analysis of defense response of wildtype Arabidopsis thaliana and plants with impaired NO- homeostasis. Proteomics 11: 1664–1683.
50. Kwon E, Feechan A, Yun BW, Hwang BH, Pallas JA, et al. (2012) AtGSNOR1 function is required for multiple developmental programs in Arabidopsis. Planta 236: 887–900.
51. Kohr MJ, Evangelista AM, Ferlito M, Steenbergen C, Murphy E (2014) S-nitrosylation of TRIM72 at cysteine 144 is critical for protection against oxidation-induced protein degradation and cell death. J Mol Cell Cardiol 69: 67–74.
52. Chen YY, Chu HM, Pan KT, Teng CH, Wang DL, et al. (2008) Cysteine S-nitrosylation protects protein-tyrosine phosphatase 1B against oxidation-induced permanent inactivation. J Biol Chem 283: 35265–35272.
53. Christian JO, Braginets R, Schulze WX, Walther D (2012) Characterization and Prediction of Protein Phosphorylation Hotspots in Arabidopsis thaliana. Front Plant Sci 3: 207.
54. Ng JY, Boelen L, Wong JW (2013) Bioinformatics analysis reveals biophysical and evolutionary insights into the 3-nitrotyrosine post-translational modification in the human proteome. Open Biol 3: 120148.
55. Chen YJ, Ku WC, Lin PY, Chou HC, Khoo KH, et al. (2010) S-alkylating labeling strategy for site-specific identification of the s-nitrosoproteome. J Proteome Res 9: 6417–6439.
56. Baier M, Dietz KJ (2005) Chloroplasts as source and target of cellular redox regulation: a discussion on chloroplast redox signals in the context of plant physiology. Journal of Experimental Botany 56: 1449–1462.
57. Allen JF, Santabarbara S, Allen CA, Puthiyaveetil S (2011) Discrete Redox Signaling Pathways Regulate Photosynthetic Light-Harvesting and Chloroplast Gene Transcription. Plos One 6.
58. Galatro A, Puntarulo S, Guiamet JJ, Simontacchi M (2013) Chloroplast functionality has a positive effect on nitric oxide level in soybean cotyledons. Plant Physiology and Biochemistry 66: 26–33.
59. Tewari RK, Prommer J, Watanabe M (2013) Endogenous nitric oxide generation in protoplast chloroplasts. Plant cell reports 32: 31–44.
60. Misra AN, Vladkova R, Singh R, Misra M, Dobrikova AG, et al. (2014) Action and target sites of nitric oxide in chloroplasts. Nitric Oxide-Biology and Chemistry 39: 35–45.
61. Abat JK, Deswal R (2009) Differential modulation of S-nitrosoproteome of *Brassica juncea* by low temperature: change in S-nitrosylation of Rubisco is responsible for the inactivation of its carboxylase activity. Proteomics 9: 4368–4380.
62. Tanou G, Filippou P, Belghazi M, Job D, Diamantidis G, et al. (2012) Oxidative and nitrosative-based signaling and associated post-translational modifications orchestrate the acclimation of citrus plants to salinity stress. Plant J 72: 585–599.
63. Zaffagnini M, Michelet L, Sciabolini C, Di Giacinto N, Morisse S, et al. (2014) High-resolution crystal structure and redox properties of chloroplastic triosephosphate isomerase from Chlamydomonas reinhardtii. Mol Plant 7: 101–120.
64. Tzafrir I, Pena-Muralla R, Dickerman A, Berg M, Rogers R, et al. (2004) Identification of genes required for embryo development in Arabidopsis. Plant Physiology 135: 1206–1220.
65. Pinon V, Etchells JP, Rossignol P, Collier SA, Arroyo JM, et al. (2008) Three PIGGYBACK genes that specifically influence leaf patterning encode ribosomal proteins. Development 135: 1315–1324.
66. Yao Y, Ling QH, Wang H, Huang H (2008) Ribosomal proteins promote leaf adaxial identity. Development 135: 1325–1334.
67. Fujikura U, Horiguchi G, Ponce MR, Micol JL, Tsukaya H (2009) Coordination of cell proliferation and cell expansion mediated by ribosome-related processes in the leaves of Arabidopsis thaliana. Plant Journal 59: 499–508.
68. Ferreyra MLF, Pezza A, Biarc J, Burlingame AL, Casati P (2010) Plant L10 Ribosomal Proteins Have Different Roles during Development and Translation under Ultraviolet-B Stress. Plant Physiology 153: 1878–1894.
69. Rosado A, Sohn EJ, Drakakaki G, Pan SQ, Swidergal A, et al. (2010) Auxin-Mediated Ribosomal Biogenesis Regulates Vacuolar Trafficking in Arabidopsis. Plant Cell 22: 143–158.
70. Szakonyi D, Byrne ME (2011) Ribosomal protein L27a is required for growth and patterning in Arabidopsis thaliana. Plant Journal 65: 269–281.
71. Pawloski JR, Hess DT, Stamler JS (2001) Export by red blood cells of nitric oxide bioactivity. Nature 409: 622–626.
72. Mannick JB, Schonhoff C, Papeta N, Ghafourifar P, Szibor M, et al. (2001) S-Nitrosylation of mitochondrial caspases. J Cell Biol 154: 1111–1116.
73. Camejo D, Romero-Puertas Mdel C, Rodriguez-Serrano M, Sandalio LM, Lazaro JJ, et al. (2013) Salinity-induced changes in S-nitrosylation of pea mitochondrial proteins. J Proteomics 79: 87–99.
74. Ckless K, Reynaert NL, Taatjes DJ, Lounsbury KM, van der Vliet A, et al. (2004) In situ detection and visualization of S-nitrosylated proteins following chemical derivatization: identification of Ran GTPase as a target for S-nitrosylation. Nitric Oxide 11: 216–227.
75. Zaman K, Carraro S, Doherty J, Henderson EM, Lendermon E, et al. (2006) S-nitrosylating agents: a novel class of compounds that increase cystic fibrosis transmembrane conductance regulator expression and maturation in epithelial cells. Mol Pharmacol 70: 1435–1442.

76. Sehrawat A, Deswal R (2014) S-Nitrosylation Analysis in Brassica juncea Apoplast Highlights the Importance of Nitric Oxide in Cold-Stress Signaling. J Proteome Res.
77. Liu M, Hou J, Huang L, Huang X, Heibeck TH, et al. (2010) Site-specific proteomics approach for study protein S-nitrosylation. Anal Chem 82: 7160–7168.
78. Kuncewicz T, Sheta EA, Goldknopf IL, Kone BC (2003) Proteomic analysis of s-nitrosylated proteins in mesangial cells. Mol Cell Proteomics 2: 156–163.
79. Ferl RJ, Manak MS, Reyes MF (2002) The 14-3-3s. Genome Biol 3: REVIEWS3010.
80. Maldonado-Alconada AM, Echevarría-Zomeño S, Lindermayr C, Redondo-López I, Durner J, et al. (2011) Proteomic analysis of Arabidopsis protein S-nitrosylation in response to inoculation with Pseudomonas syringae. Acta Physiol Plant 33: 1493–1514.
81. Belenghi B, Romero-Puertas MC, Vercammen D, Brackenier A, Inze D, et al. (2007) Metacaspase activity of *Arabidopsis thaliana* is regulated by S-nitrosylation of a critical cysteine residue. J Biol Chem 282: 1352–1358.
82. Perazzolli M, Dominici P, Romero-Puertas MC, Zago E, Zeier J, et al. (2004) *Arabidopsis* nonsymbiotic hemoglobin AHb1 modulates nitric oxide bioactivity. Plant Cell 16: 2785–2794.
83. Tada Y, Spoel SH, Pajerowska-Mukhtar K, Mou Z, Song J, et al. (2008) Plant immunity requires conformational changes [corrected] of NPR1 via S-nitrosylation and thioredoxins. Science 321: 952–956.
84. Su MG, Lee TY (2013) Incorporating substrate sequence motifs and spatial amino acid composition to identify kinase-specific phosphorylation sites on protein three-dimensional structures. BMC Bioinformatics 14 Suppl 16: S2.
85. Lindermayr C, Saalbach G, Bahnweg G, Durner J (2006) Differential inhibition of *Arabidopsis* methionine adenosyltransferases by protein S-nitrosylation. J Biol Chem 281: 4285–4291.
86. Romero-Puertas MC, Laxa M, Matte A, Zaninotto F, Finkemeier I, et al. (2007) S-nitrosylation of peroxiredoxin II E promotes peroxynitrite-mediated tyrosine nitration. Plant Cell 19: 4120–4130.
87. Holtgrefe S, Gohlke J, Starmann J, Druce S, Klocke S, et al. (2008) Regulation of plant cytosolic glyceraldehyde 3-phosphate dehydrogenase isoforms by thiol modifications. Physiol Plant 133: 211–228.
88. Wang YQ, Feechan A, Yun BW, Shafiei R, Hofmann A, et al. (2009) S-nitrosylation of AtSABP3 antagonizes the expression of plant immunity. J Biol Chem 284: 2131–2137.
89. Yun BW, Feechan A, Yin M, Saidi NB, Le Bihan T, et al. (2011) S-nitrosylation of NADPH oxidase regulates cell death in plant immunity. Nature 478: 264–268.
90. van der Linde K, Gutsche N, Leffers HM, Lindermayr C, Muller B, et al. (2011) Regulation of plant cytosolic aldolase functions by redox-modifications. Plant Physiol Biochem 49: 946–957.
91. Terrile MC, Paris R, Calderon-Villalobos LI, Iglesias MJ, Lamattina L, et al. (2012) Nitric oxide influences auxin signaling through S-nitrosylation of the Arabidopsis TRANSPORT INHIBITOR RESPONSE 1 auxin receptor. Plant J 70: 492–500.
92. Tavares CP, Vernal J, Delena RA, Lamattina L, Cassia R, et al. (2014) S-nitrosylation influences the structure and DNA binding activity of AtMYB30 transcription factor from Arabidopsis thaliana. Biochim Biophys Acta 1844: 810–817.

13

Molecular Characterization of Three PRORP Proteins in the Moss *Physcomitrella patens*: Nuclear PRORP Protein Is Not Essential for Moss Viability

Chieko Sugita[1§], Yoshihiro Komura[1§], Korechika Tanaka[1], Kazuki Kometani[1], Hiroyuki Satoh[2], Mamoru Sugita[1]*

1 Center for Gene Research, Nagoya University, Nagoya, Japan, **2** Department of Biomolecular Science, Toho University, Funabashi, Japan

Abstract

RNase P is a ubiquitous endonuclease that removes the 5′ leader sequence from pre-tRNAs in all organisms. In *Arabidopsis thaliana*, RNA-free proteinaceous RNase Ps (PRORPs) seem to be enzyme(s) for pre-tRNA 5′-end processing in organelles and the nucleus and are thought to have replaced the ribonucleoprotein RNase P variant. However, the evolution and function of plant PRORPs are not fully understood. Here, we identified and characterized three PRORP-like proteins, PpPPR_63, 67, and 104, in the basal land plant, the moss *Physcomitrella patens*. PpPPR_63 localizes to the nucleus, while PpPPR_67 and PpPPR_104 are found in both the mitochondria and chloroplasts. The three proteins displayed pre-tRNA 5′-end processing activity *in vitro*. Mutants with knockout (KO) of the *PpPPR_63* gene displayed growth retardation of protonemal colonies, indicating that, unlike Arabidopsis nuclear RPORPs, the moss nuclear PpPPR_63 is not essential for viability. In the KO mutant, nuclear-encoded $tRNA^{Asp}$ (GUC) levels were slightly decreased, whereas most nuclear-encoded tRNA levels were not altered. This indicated that most of the cytosolic mature tRNAs were produced normally without proteinaceous RNase P-like PpPPR_63. Single *PpPPR_67* or *104* gene KO mutants displayed different phenotypes of protonemal growth and chloroplast $tRNA^{Arg}$ (ACG) accumulation. However, the levels of all other tRNAs were not altered in the KO mutants. In addition, *in vitro* RNase P assays showed that PpPPR_67 and PpPPR_104 efficiently cleaved chloroplast pre-$tRNA^{Arg}$ (CCG) and pre-$tRNA^{Arg}$ (UCU) but they cleaved pre-$tRNA^{Arg}$ (ACG) with different efficiency. This suggests that the two proteins have overlapping function but their substrate specificity is not identical.

Editor: Lennart Randau, Max-Planck-Institute for Terrestrial Microbiology, Germany

Funding: This work was supported by JSPS KAKENHI Grant Numbers 2057003, 23657003, 25291059, 25660292 (MS). The funders had no role in study design, data collection and analysis, decision to publish, or preparation of the manuscript.

Competing Interests: The authors have declared that no competing interests exist.

* Email: sugita@gene.nagoya-u.ac.jp

§ These authors contributed equally to this work.

Introduction

RNase P is an endonuclease that removes the 5′ leader sequence from precursor tRNAs (pre-tRNAs). This endonucleolytic cleavage is an essential step in the production of mature tRNAs in all organisms, as well as in mitochondria and chloroplasts [1,2]. Bacterial, archaeal, and eukaryotic nuclear RNase P enzymes are ribonucleoprotein complexes composed of a catalytic RNA component and one or several proteins [3]. Therefore, it was believed that 5′-maturation of tRNAs in all organisms is catalyzed by ubiquitous ribonucleoprotein (RNP) RNase P enzymes. However, human mitochondrial RNase P was identified as an RNA-free enzyme composed of three proteins called mitochondrial RNase P protein 1 (MRPP1), MRPP2, and MRPP3 [4]. MRPP3 contains two RNA-binding pentatricopeptide repeat (PPR) motifs and a conserved NYN metallonuclease domain [5], which are involved in the catalytic activity of mitochondrial RNase P. In a model plant, *Arabidopsis thaliana*, three MRPP3 homologs have been identified as RNase P enzymes; they are termed proteinaceous RNase P 1 (PRORP1), PRORP2, and PRORP3 [6]. *In vitro* cleavage assays using recombinant PRORP proteins have demonstrated that three PRORP proteins display RNase P activity [6,7]. PRORPs are single-protein RNase P enzymes, whereas human mitochondrial RNase P is a multi-subunit enzyme. PRORP1 localizes in both chloroplasts and mitochondria, whereas PRORP2 and 3 localize in the nucleus. *PRORP1* is an essential gene, because *prorp1* mutants are embryonic lethal [6]. Single *PRORP2* or *PRORP3* knockout mutant lines show wild-type phenotypes, whereas the homozygous double mutation in *prorp2 prorp3* results in embryonic lethality [7]. This indicates that PRORP2 and 3 have redundant functions and are essential for embryogenesis and plant growth.

The Arabidopsis and rice genomes do not encode RNase P RNA, whereas they do code for several homologs of RNP RNase P protein subunits found in mammals and yeasts [8]. The presence of ribonucleoprotein RNase P enzyme in plants remains obscure. Recently, Gutmann et al. (2012) suggested that Arabidopsis had entirely replaced the ribonucleoprotein RNase P enzyme by

PRORP enzymes for tRNA maturation during plant evolution [7]. However, functional characterization of PRORP proteins is limited in Arabidopsis.

Unlike in land plant organelles, an RNA subunit of bacterial-type RNase P is encoded in the organellar genomes of the Glaucophyta (*Cyanophora paradoxa*) [9], the red alga *Poryphyra purpurea* [10], and green algae (*Nephroselmis olivacea*, *Ostreococcus tauri* [11,12]. These algae also possess a single nuclear gene for PRORP [6]. In *C. paradoxa*, RNase P activity in an extract of the photosynthesis organelle (cyanelle) is sensitive to micrococcal nuclease [13,14], indicating the existence of bacterial-type RNase P. The nuclear-encoded PRORP in the green alga *O. tauri* has been shown to cleave the 5′ leader of pre-tRNA *in vitro* [12,15]. Thus, algae seem to possess both a ribonuceoprotein RNase P and a PRORP. However, the latter is not known to localize in either the organelles or the nucleus.

From these differences in the status of PRORP enzymes in algae and plants, the following questions have arisen: When were plant PRORP genes duplicated during plant evolution, which PRORP protein was targeted to the organelles and which to the nucleus, and was ribonucleoprotein RNase P enzyme replaced by PRORP in plants? To answer these questions, we focused on PRORPs of the basal land plant bryophyte *Physcomitrella patens*. Here, we report the molecular characterization of three *P. patens* PRORP-like proteins (PpPPR_63, 67, and 104) that are members of the PPR protein family [16,17]. We also report the disruption mutants of each gene and show that nuclear-localized PpPPR_63 is dispensable for moss viability and that organelle-localized PpPPR_67 and PpPPR_104 have functions that are redundant, but not completely so.

Results

The moss *P. patens* has three PRORP homologs

Among 105 *Physcomitrella* PPR proteins [17], PpPPR_63 (protein ID, Phypa_174001), PpPPR_67 (Phypa_177191) and PpPPR_104 (Phypa_152956) appeared to be PRORP homologs (Fig. 1A), because these three PPR proteins contained at least two PPR motifs and an NYN metallonuclease domain and had 42%–53% amino acid identity with Arabidopsis PRORP proteins. Although algae and *Chara* have a single PRORP homolog, mosses and vascular plants have two or three PRORP homologs (Fig. 1B and Table S1). This suggests that PRORP genes were duplicated after the emergence of land plants.

PpPPR_63 is localized in the nucleus, whereas PpPPR_67 and 104 are found in the organelles

PpPPR_67 and 104 possess an N-terminal transit peptide-like sequence targeting the organelles, whereas PpPPR_63 does not. This suggests that PpPPR_67 and 104 are likely organelle-localized PRORP1 homologs, whereas PpPPR_63 seems to be a nuclear-localized PRORP2/PRORP3 homolog. To confirm this prediction we generated PpPPR_63-green fluorescent protein (GFP) knockin (KI) mosses. The four independent KI lines obtained showed normal growth (Fig. S1), and GFP fluorescence was observed in the nuclei (Fig. 2A, b and c). GFP fluorescence was observed in the protonemal cells, buds, and young leaves (Fig. 2A, d–i) but not in mature leaves (Fig. 2A, j and k).

To verify the organellar localization of PpPPR_67 and 104, their N-terminal sequences fused to GFP (N67-GFP and N104-GFP) were transiently expressed in the protonemata (Fig. 2B). GFP fluorescence of N67-GFP and N104-GFP was observed in both mitochondria and chloroplasts, coinciding with mitochondria-localized red fluorescent protein (Mt-RFP) signals and Nomarski images of chloroplasts (DIC).

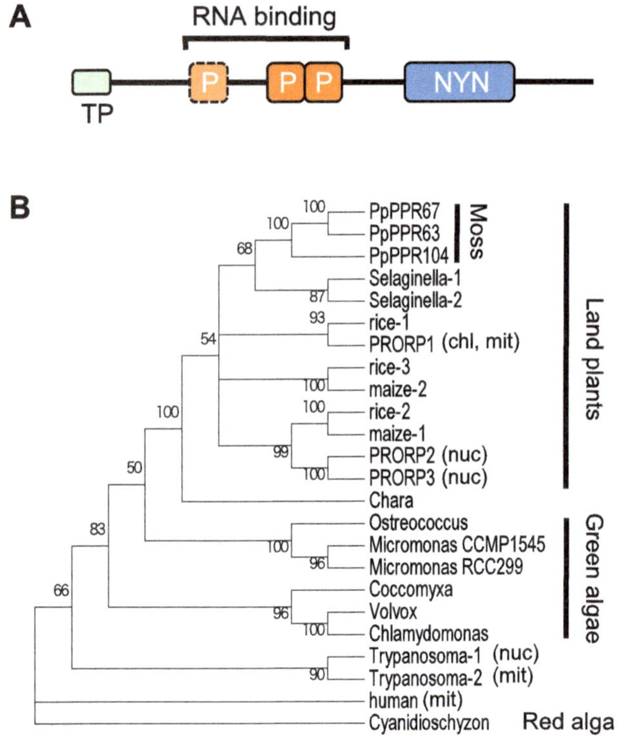

Figure 1. PRORP proteins are widely distributed among eukaryotes. (**A**) Schematic diagram of plant organellar PRORPs with a transit peptide (TP), RNA binding PPR motifs (P), and NYN metallonuclease domain (NYN). (**B**) Neighbor-joining phylogenetic tree of PRORPs and PRORP-like proteins. Representative PRORP protein sequences from evolutionarily distant plants were used for the phylogenic analysis. Experimentally determined localization of PRORPs is presented as chloroplast (chl), mitochondria (mit), or nucleus (nuc). Bootstrap values>50 are indicated along branches. Protein ID or accession numbers of PRORPs and PRORP-like proteins are listed in Table S1.

PpPPR_63, 67, and 104 have RNase P activity

To investigate whether the three *Physcomitrella* PRORP-like PPR proteins are RNase P enzymes, we performed *in vitro* RNase P assays with recombinant proteins composed of the respective PRORP-like protein fused with thioredoxin at the N-terminus and six histidines and a VP5 tag at the C-terminus (Fig. 3A, B). The three recombinant proteins, r67, r104, and r63, removed the 5′ leader sequences from chloroplast pre-tRNA^{Phe}, mitochondrial pre-tRNA^{Cys}, and nuclear pre-tRNA^{Asp} (Fig. 3C). The 5′-end of the tRNA produced was verified by primer extension analysis (Fig. 3D, E). We also confirmed the precise cleavage of pre-tRNAs by circular reverse transcription (cRT)-polymerase chain reaction (PCR) analysis (data not shown). To verify the RNase P activity of these proteins, we also tested two recombinant proteins, rM63 and rC63. rM63 possessed a mutated NYN domain, the two adjacent aspartates of which (D_{460} and D_{461}) were altered to alanines. These two aspartates have been shown to be essential for the activity of *Arabidopsis* PRORP1 [6]. The rC63 was composed of the tag sequences and the NYN domain only. The rM63 and rC63 proteins did not cleave nuclear-encoded pre-tRNA^{Asp} (Fig. 3C).

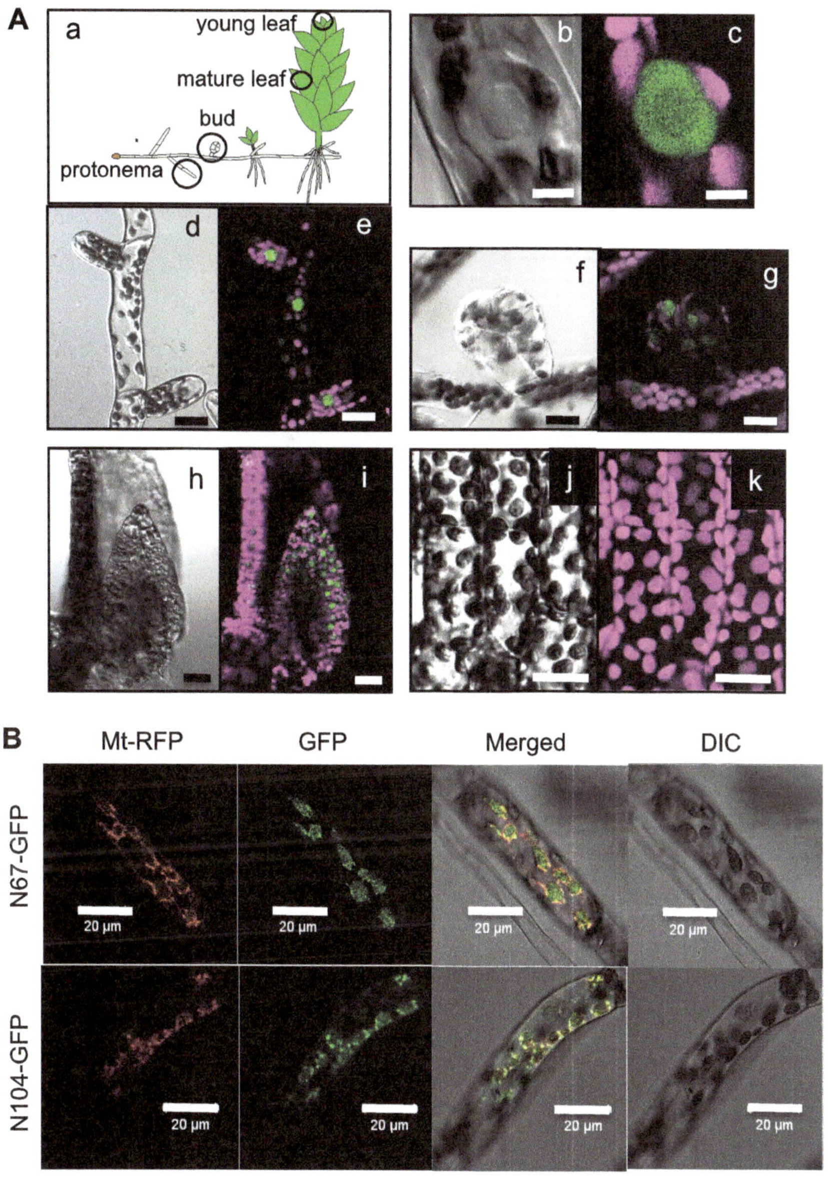

Figure 2. Localization of PpPPR_63, 67, and 104. (**A**) Observations of GFP fluorescence in various cells and tissues of the stable transgenic moss KI-10 line expressing PpPPR_63-GFP fusion protein. Schematic diagram of a moss plant and observed parts are in circles (a). Images are differential interference contrast (DIC) (b, d, f, h, j) and GFP (green) and chlorophyll (magenta) fluorescence (c, e, g, i, k). Single protonemal cell (b, c), protonemal cells (d, e), a bud (f, g), a young leaf (h, i), and mature leaf cells (j, k). Scale bars = 5 µm (b, c) or 20 µm (d–k). (**B**) Subcellular localization of N-terminal PpPPR_67-GFP fusion protein (N67-GFP, upper panels) or N-terminal PpPPR_104-GFP fusion protein (N104-GFP, bottom panels). For mitochondrial localization control, the γ subunit of mitochondrial ATPase and RFP fusion protein (Mt-RFP) was used. Fluorescence of GFP and RFP, the overlay of the two fluorescence images (Merged), and the corresponding Nomarski images (DIC) are shown.

These results indicated that the three moss PPR proteins had RNase P activity similar to that of Arabidopsis PRORPs [6,7,18].

Nuclear PpPPR_63 is dispensable for viability

To investigate the *in vivo* function of PpPPR_63, we generated two independent knockout (KO) mutants of *PpPPR_63*, Δ*63*-1 andΔ*63*-15. Both KO lines were verified to be null mutants. Gene structure of theΔ*63*-1 line is given as Figs. 4A and S2, and its absence of *PpPPR_63* transcripts was verified by reverse transcription (RT)-PCR (Fig. 4B). Therefore, we used theΔ*63*-1 line for further analyses (hereafter named Δ*63*). In addition, to generate the complementation lines of theΔ*63*-1, we transformed the Δ*63* moss with a transgene encoding the full length of PpPPR_63 or the mutated version and generated F63 and M63, respectively (Fig. S3). In M63, two aspartates (D_{460} and D_{461}) in the NYN domain were mutated to alanines (A_{460} and A_{461}). In the complementation lines F63 and M63, the introduced transgene was overexpressed (Fig. 4B). The KO mutantΔ*63* and the complementation line M63 mosses displayed smaller protonemal colonies than the wild type (WT) (Fig. 4C), and they showed abnormal regeneration from leaves (especially defective growth of the sub-apical cells) (Fig. 4D). The complementation line F63 exhibited a phenotype the same as that of the WT. These results

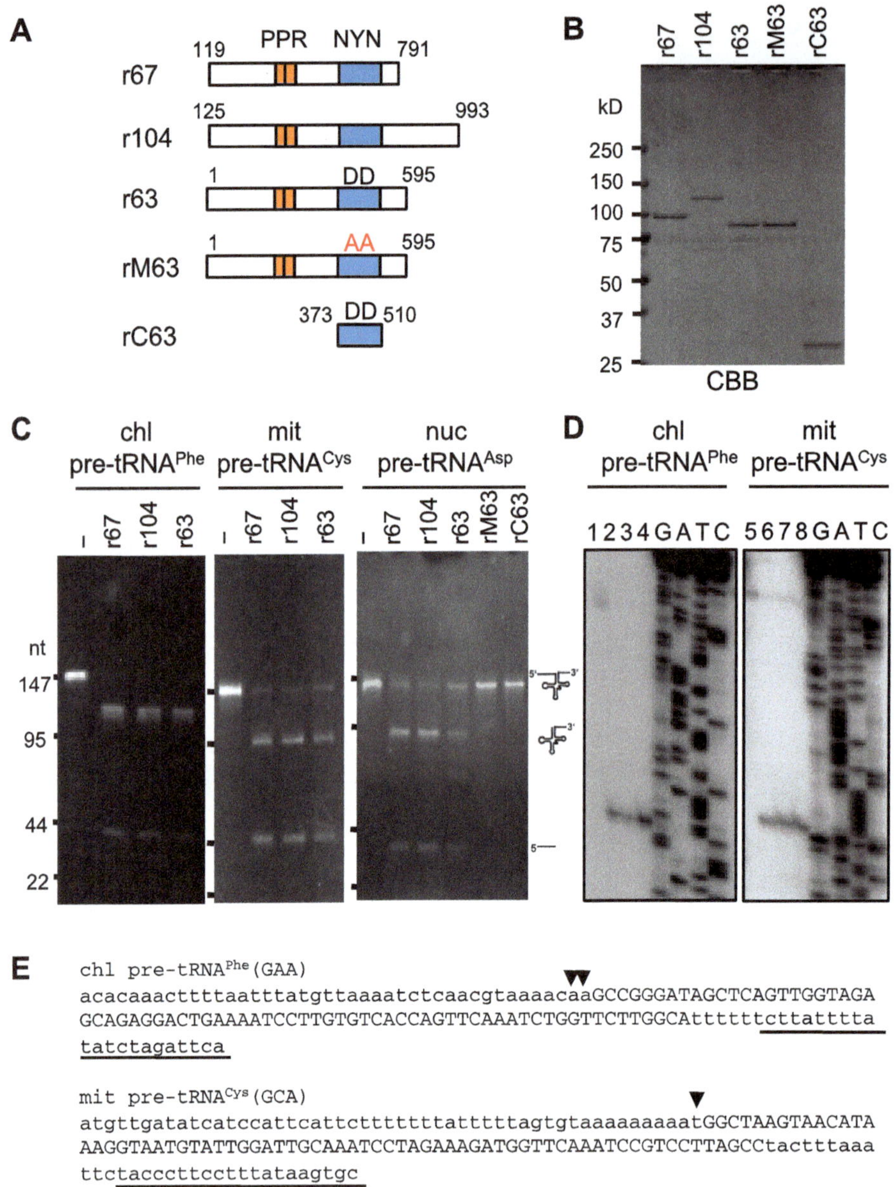

Figure 3. Pre-tRNA cleavage assay of *P. patens* PRORP-like proteins. (**A**) Schematic diagrams of various recombinant proteins. Recombinant proteins r67, r104, r63, and rM63 indicate, respectively, PpPPR_67, PpPPR_104, PpPPR_63, and the mutant PpPPR_63, the two catalytic aspartates (DD) of which were substituted with alanines (AA). rC63 consists of the tag sequences and an NYN domain only of PpPPR_63. (**B**) Coomassie brilliant blue (CBB)-stained recombinant proteins (1 μg each). (**C**) RNase P activity was assayed by using 2 μg of *in vitro*-transcribed chloroplast (chl) pre-tRNA^{Phe}, mitochondrial (mit) pre-tRNA^{Cys}, or nuclear (nuc) pre-tRNA^{Asp} and 100 ng of recombinant proteins. The reaction products were separated by using 8% denaturing PAGE and stained with ethidium bromide. (**D**) The 5′-ends of processed pre-tRNAs in (C) were determined by primer extension analysis. Processed pre-tRNAs without proteins (lanes 1 and 5), with r67 (lanes 2 and 6), with r104 (lanes 3 and 7), or with r63 (lanes 4 and 8) were reverse transcribed from the 5′-end-labeled primers underlined in (E). (**E**) The nucleotide sequences of chloroplast (chl) pre-tRNA^{Phe} and mitochondrial (mit) pre-tRNA^{Cys} are shown with small letters (5′ leader and 3′ trailer sequences) and capital letters (predicted mature tRNA). Arrowheads indicate the 5′-end positions determined in (D). Underlined sequences indicate the position of primers used for primer extension.

indicated that *PpPPR_63* was responsible for the mutant phenotype.

Knockout of *PpPPR_63* slightly affects specific nuclear-encoded tRNAs

If nuclear PpPPR_63 could function as a *bona fide* RNase P, knockout of the *PpPPR_63* gene could result in severely reduced accumulation of nuclear-encoded tRNAs. To investigate this possibility, we analyzed the accumulation levels of several nuclear-encoded tRNAs in the*Δ63* mutant. For this analysis nuclear-encoded tRNA^{Asp} (GUC) and tRNA^{Gln} (CUG) were chosen, because expression of these tRNAs has been demonstrated to significantly decrease in the knockdown mutants of nuclear PRORP2/3 plants [7]. tRNA^{Asp} (GUC) in the*Δ63* mutant and M63 complementation line was slightly reduced to 70% of that in the WT and F63. Whereas, tRNA^{Gln} (CUG) in the *Δ63* mutant, M63 and F63 complementation lines accumulated to the same level as in the WT (Fig. 4E). Similarly, nuclear-encoded tRNA^{Ser}

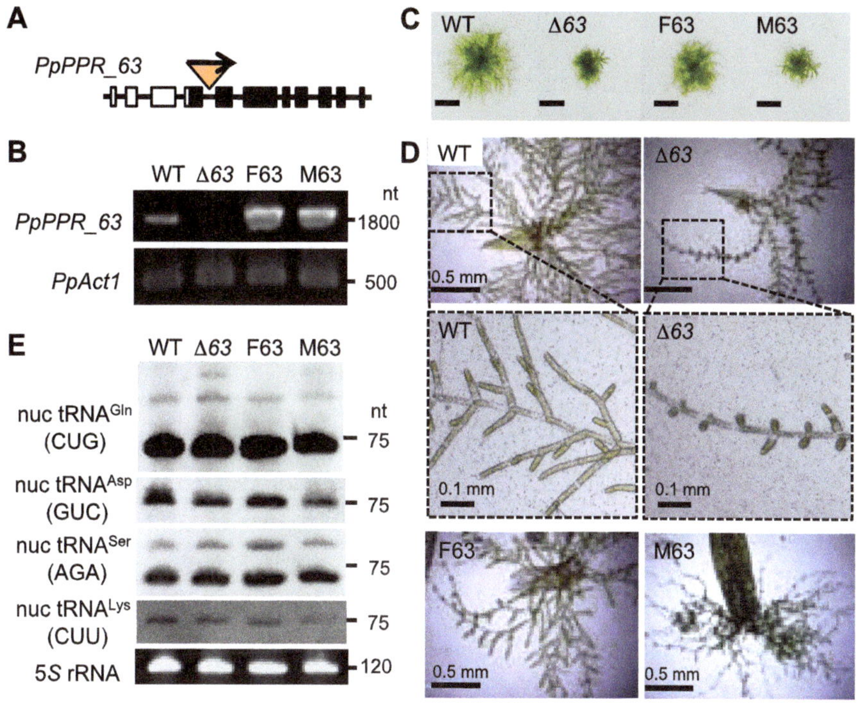

Figure 4. Molecular and morphological phenotypes of the *PpPPR_63* KO and complementation mosses. (**A**) Schematic diagram of *PpPPR_63* and construction of the KO locus. The *nptII* cassette was inserted into the 4th intron of the *PpPPR_63* gene. (**B**) Detection of *PpPPR_63* transcripts by RT-PCR. (**C**) Two-week-old protonemal colonies of the *PpPPR_63* KO line (*Δ63*) and the complementation lines (F63 and M63). Bars =5 mm. (**D**) One-week-old filamentous protonemata regenerated from cut leaves. (**E**) Steady-state levels of nuclear-encoded tRNAs.

(AGA) and $tRNA^{Lys}$ (CUU) levels were not affected in the mutants. This suggests that PpPPR_63 PNase P activity may be limited to specific tRNA(s) including $tRNA^{Asp}$ (GUC).

Knockout of *PpPPR_104* results in a significant reduction in chloroplast $tRNA^{Arg}$ (ACG)

To investigate the *in vivo* function of the *PpPPR_67* and *104* genes, we generated and characterized their KO mutants (Figs. 5A, S4, S5, S6). The single gene KO mutants obtained, *Δ67* and*Δ104*, were null mutants (Fig. 5B), and the protonemal colony of*Δ104* moss was smaller than that of the WT (Fig. 5C). A peunumbra and central part of the protonemal colony mostly consists of caulonemal and chloronemal filaments, respectively. In the*Δ104* moss, caulonemal filaments were poorly induced, resulting in a smaller colony. In contrast, *PpPPR_67* KO moss (*Δ67*) showed little difference from the WT. To generate double KO mutants, we carried out four times transformation experiments, such as introducing the PpPPR_104 KO plasmid into the *PpPPR_67* KO line (*Δ67*-1-17 line) and vice versa. However, no double KO mutants were obtained from the genotyping of 262 moss plants. This indicates that *PpPPR_67* and *104* genes cannot be knocked out simultaneously. Since they may have redundant functions, single KO mutants are viable but double KO is probably lethal.

To investigate whether organellar tRNAs were affected in the *PpPPR_67* and *104* KO mutants, we performed northern blot analyses by using all organelle-encoded tRNA probes. The chloroplast $tRNA^{Arg}$ (ACG) level significantly decreased in*Δ104* but not in*Δ67* or*Δ63* (Fig. 5D). All the other tRNAs, including chloroplast $tRNA^{Phe}$ (GAA) and mitochondrial $tRNA^{Cys}$ (GCA) that were used for the *in vitro* RNase P cleavage assay, were not altered in the WT and the three KO mutants (Figs. 5D, S7, S8). This result suggests that PpPPR_67 and PpPPR_104 have redundant functions as RNase P in moss organelles.

PpPPR_67 and 104 cleave chloroplast pre-$tRNA^{Arg}$ (ACG) with different efficiency

Because chloroplast $tRNA^{Arg}$ (ACG) levels decreased in the *Δ104* but not in theΔ67, we assumed that PpPPR_104 and PpPPR_67 would have distinct RNase P activity against chloroplast pre-$tRNA^{Arg}$ (ACG). To test this possibility, three chloroplast pre-$tRNA^{Arg}$ isoacceptors were used as substrates for *in vitro* RNase P cleavage assay. This assay revealed that PpPPR_67 and 104 efficiently cleaved pre-$tRNA^{Arg}$ (CCG) and pre-$tRNA^{Arg}$ (UCU) but they cleaved pre-$tRNA^{Arg}$ (ACG) with different efficiency (Fig. 6). These results suggest that the function of their RNase P is not completely redundant.

Discussion

The first two questions to be addressed here were: When were plant *PRORP* genes duplicated during plant evolution and where were the respective PRORP proteins localized or targeted to? Our present study provides evidence that *PRORP* genes were duplicated after the emergence of the early land plants, i.e. mosses. The moss *P. patens* has three PRORP-like proteins displaying RNase P activity *in vitro*. Arabidopsis PRORP1 is localized in both the mitochondria and chloroplasts, and PRORP2 and 3 are localized in the nucleus [6]. On the other hand, the moss *P. patens* has two PRORP1-like proteins, PpPPR_67 and 104, both of which are dual-targeted to the mitochondria and chloroplasts, and one nuclear-localized PRORP2/3-like PpPPR_63. Thus, duplicated *PRORP* gene products are likely targeted to either the organelles or the nucleus.

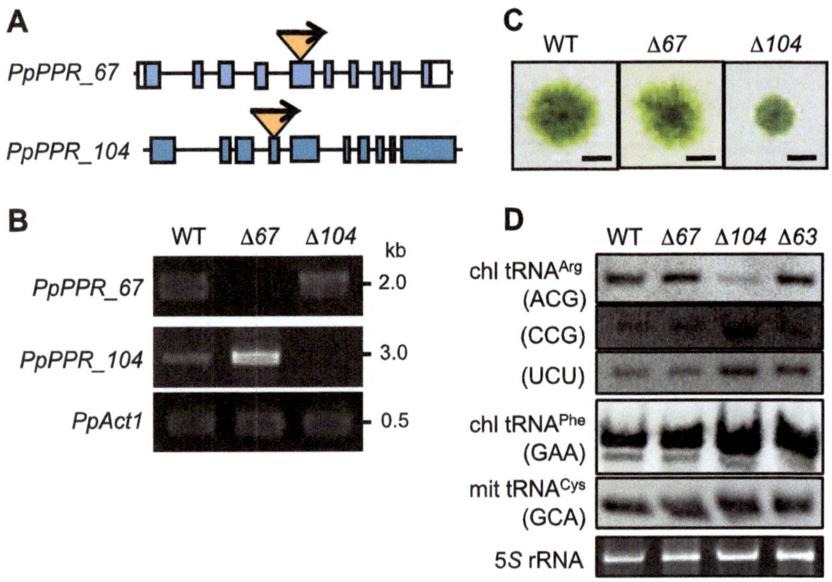

Figure 5. *PpPPR_67* and *PpPPR_104* KO mosses. (**A**) Schematic structure of *PpPPR_67* and *104* KO loci. The *nptII* cassette was inserted into the 5th exon of *PpPPR_67* and the 4th exon of *PpPPR_104*. (**B**) Detection of *PpPPR_67* and *104* transcripts by RT-PCR. (**C**) Two-week-old protonemata colonies of the *PpPPR_67* KO line (Δ*67*) and the *PpPPR_104* KO line (Δ*104*). Scale bars = 5 mm. (**D**) Steady-state levels of chloroplast (chl) $tRNA^{Arg}$ isoacceptors with different anticodons, chl $tRNA^{Phe}$ (GAA) and mitochondrial (mit) $tRNA^{Cys}$ (GCA).

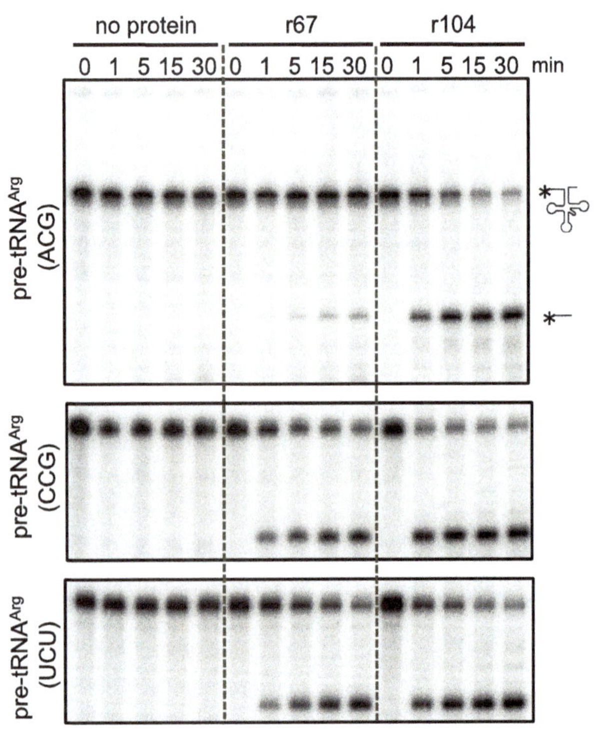

Figure 6. Pre-tRNA cleavage assay of the recombinant proteins r67 and r104. Reactions containing 500 nM of chloroplast 5′-^{32}P-pre-$tRNA^{Arg}$ (ACG) (upper), $tRNA^{Arg}$ (CCG) (middle), or $tRNA^{Arg}$ (UCU) (lower) and 500 nM of recombinant protein were quenched at 1, 5, 15, and 30 min points, resolved on 8% denaturing PAGE, and analyzed with a phosphoimager.

Nuclear PpPPR_63 is involved in normal plant growth and branch formation of protonemata

In Arabidopsis, the double KO mutation of *PRORP2* and *3* resulted in embryonic lethality [6]. In contrast, a null mutant of *PpPPR_63* displayed smaller protonemal colonies than the WT and also showed abnormal branch formation of filamentous protonemata (Fig. 4). This indicates that unlike Arabidopsis nuclear RPORPs, the moss nuclear PpPPR_63 is dispensable for plant viability. The nuclear-encoded $tRNA^{Asp}$ (GUC) level in the *PpPPR_63* KO mutant was slightly decreased (to 70% of that in WT mosses), but most nuclear-encoded tRNA levels were not altered. This result indicates that most of the cytosolic mature tRNAs were produced normally, without proteinaceous RNase P-like PpPPR_63. This does not concur with the results observed from functional analysis of Arabidopsis PRORP2 and 3 [6,7]. In Arabidopsis mutant plants where PRORP3 was absent and the PRORP2 level was decreased, the steady-state levels of five nuclear-encoded tRNAs investigated were significantly reduced, to 30% of the tRNA levels in control plants [7]. These observations indicate that Arabidopsis nuclear PRORP2 and 3 function as master RNase P enzymes, while the moss PpPPR_63 is not the sole RNase P enzyme in the nucleus.

It is unlikely that the slight reduction of the $tRNA^{Asp}$ (GUC) level resulted in growth retardation and abnormal branch formation in the *PpPPR_63* KO mutant. PpPPR_63 may have a versatile function rather than pre-tRNA 5′-end processing. Arabidopsis PRORP proteins are involved in the maturation of not only tRNAs but also mRNA or snoRNA [7]. Similarly, PpPPR_63 is probably involved in the maturation of unknown mRNA or small RNA-containing tRNA-like structures.

Organelle-localized PpPPR_67 and 104 have functions not completely redundant for tRNA accumulation

In this study, we generated single KO mutants of *PpPPR_67* and *PpPPR_104* but failed to obtain double KO mutants. Similarly, KO mutation of *Arabidopsis PRORP1* was shown to result in embryonic lethality [6]. This suggests that PpPPR_67 and 104 have redundant functions. However, it is interesting to observe that a single mutant of *PpPPR_67* and *104* displayed different phenotypes of protonemal growth and chloroplast tRNA^{Arg} (ACG) accumulation. Chloroplast tRNA^{Arg} (ACG) significantly decreased in the *PpPPR_104* KO mutant but not in *PpPPR_67* KO mutants. The *P. patens* chloroplast genome codes for three tRNA^{Arg} isoacceptors [19]. An *in vitro* RNase P cleavage assay revealed that pre-tRNA^{Arg} (ACG) was more efficiently cleaved by PpPPR_104 than by PpPPR_67, whereas pre-tRNA^{Arg} (UCU) and pre-tRNA^{Arg} (CCG) were cleaved with similar efficiency by both proteins. This is the first evidence that PRORP enzymes have substrate specificity of tRNA processing. The anticodon of *P. patens* tRNA^{Arg} (ACG) is modified that including inosine in chloroplasts, and tRNA^{Arg} (ICG) can recognize all four codons. Chloroplast tRNA^{Arg} (CCG) recognizing only the GGC codon is dispensable [20]. Processing of chloroplast pre-tRNA^{Arg} (ACG) by PpPPR_104 is essential for the accumulation of mature tRNA^{Arg} (ACG). Therefore, the *PpPPR_104* KO mutant may have displayed small protonemal colonies because of a great reduction in mature tRNA^{Arg} (ACG) levels.

Unlike Arabidopsis [6,7], pre-tRNAs were hardly detected in the moss organelles (Figs. S7 and S8). It is conceivable that precursors might be turned over faster than mature tRNAs in mosses.

Was RNP-type RNase P enzyme entirely replaced by PRORP protein in mosses?

The last question is whether ribonucleoprotein RNase P enzyme was replaced by PRORP in both the organelles and the nucleus. In this study, we showed that disruption of *PpPPR_63* did not result in a significant reduction in nuclear-encoded tRNAs. This raises the possibility that unidentified RNase P activity may contribute to pre-tRNA maturation in the moss nucleus (Fig. 7). However, it remains unknown whether 5′-end processing of nuclear pre-tRNAs is supported by RNP-type RNase P. In *P. patens*, an RNA component of RNP-type RNase P was not identified, yet but protein components (Pop4, Rpp1, Pop1, and Pop5) are encoded in the genome [21]. Identification of RNase P activity rather than PRORP1-type PpPPR_63 will be a challenge for the future.

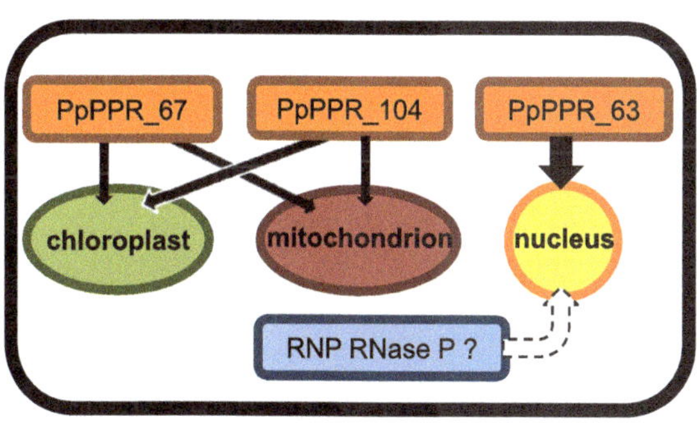

Figure 7. Schematic summarizing the localization of three PRORP homologs, PpPPR_63, 67, and 104 in *P. patens*. This study suggested that unidentified RNase P enzyme(s) including RNP-type RNase P may be present at least in the nucleus.

Materials and Methods

Accession numbers

The sequences reported in this paper have been deposited in the DDBJ/GenBank/EMBL databases with accession numbers: *PpPPR_63* mRNA, AB983707; *PpPPR_67* mRNA, AB983708; *PpPPR_104* mRNA, AB808584.

Plant material and preparation of DNA and RNA

The moss *P. patens* subsp. *patens* was grown at 25°C under continuous light (30 μmol photon $m^{-2}s^{-1}$), and genomic DNA and total cellular RNA were prepared from the moss protonemata as described previously [22].

Preparation of recombinant proteins

DNAs corresponding to the proteins (1–595 residues for PpPPR_63, 119–791 for PpPPR_67, and 125–993 for PpPPR104) were amplified by using the primers listed in Table S2 and cloned into pBAD/Thio-TOPO vector (Invitrogen). Proteins were expressed for 6 h at 18°C in *E. coli* BL21 and purified as described previously [23].

Preparation of pre-tRNA

DNAs representing precursors of chloroplast tRNA^{Phe}, mitochondrial tRNA^{Cys}, and nuclear tRNA^{Asp} were amplified by PCR from *P. patens* genomic DNA with oligonucleotide pairs T7-cpF5 and cpF3, T7-mtC5 and mtC3, and T7-nuDGTC5 and nuDGTC, respectively (Table S2). The 5′ oligonucleotides contained the T7 promoter sequence. The pre-tRNA^{Phe}, pre-tRNA^{Cys}, and nuclear pre-tRNA^{Asp} were respectively 142, 153, and 140 nt long and were designed with respective 5′ leaders of 42, 51, and 40 nt and 3′ trailers of 27, 32, and 25 nt. Amplified DNAs (200 ng) were transcribed with T7 RNA polymerase (TaKaRa) for 1 h at 37°C, then digested with 5 U of RNase-free DNase I (TaKaRa) and 20 U of RNase Inhibitor (TaKaRa) for 15 min at 37°C. Synthesized pre-tRNAs were purified by phenol chloroform extraction.

For preparation of chloroplast pre-tRNA^{Arg} isoacceptors, the corresponding genes were cloned and *in vitro* transcribed. The obtained pre-tRNAs were dephosphorylated with Thermo Sensitive Alkaline Phosphatase (Promega) and were 5′-end radiolabeled with [γ^{32}P] ATP and polynucleotidekinase (TaKaRa). Labeled RNAs were purified from the gel after 8% polyacrylamide gel electrophoresis (PAGE) containing 7M urea.

Pre-tRNA cleavage assay

Cleavage assays were performed as described previously [7]. Recombinant protein (100 ng) and *in vitro*-synthesized pre-tRNA (2 μg) were incubated for 1 h at 25°C in a 20 μl mixture of 20 mM Tris-HCl (pH 8.0), 30 mM KCl, 4.5 mM $MgCl_2$, 20 μg/ml bovine serum albumin, and 2 mM DTT. A half of the reaction mixture was separated on 8% polyacrylamide-7 M urea gels and stained with ethidium bromide.

For cleavage assays of three chloroplast pre-tRNA^{Arg} isoacceptors, the 5′ end-radiolabeled pre-tRNAs (500 nM each) and the recombinant proteins (500 nM each) were incubated and aliquots were withdrawn at the indicated times. Dried PAGE gels were analyzed by Storm 820 (GE Healthcare).

Determination of the 5′-end of tRNA

The ^{32}P 5′-end-labeled oligonucleotides cp-F3 or mt-C3 (Table S2) and 200 ng of *in vitro*-cleaved RNAs were mixed and reverse transcribed with ReverTraAce (ToYoBo) for 1 h at 42°C, then digested with RNase A and precipitated with ethanol. The sequence ladders were obtained by using template DNA, the radiolabled primer cp-F3 or mt-C3, as described previously [22].

Subcellular localization

The DNA region from the 6th intron to the last codon of the *PpPPR_63* gene was inserted in-frame into the GFP coding region of pGFPmutNTPII (http://moss.nibb.ac.jp). The 3′-flanking region of *PpPPR_63* was inserted downstream of the *nptII* cassette of pGFPmutNTPII. The resultant plasmid p63-GFPKI (Fig. S1) was linearized with *Not*I and introduced into *P. patens* by particle bombardment as described previously [23]. For N67-GFP and N104-GFP, cDNAs corresponding to an N-terminal part of the proteins (residues 1–112 and 1–105, respectively) were amplified by using the appropriate primers (Table S2) and cloned into pKSPGFP9 [23]. The resultant plasmid was cotransfected with pMt-RFP into *P. patens* protonemata and GFP and RFP fluorescence were observed as described previously [24].

Generation of KO lines

The 5′ region (2960 bp) of *PpPPR_63*, a 2958-bp region of *PpPPR_67*, and a 3032-bp region of *PpPPR_104* were amplified from moss genomic DNA by PCR using the appropriate primers (Table S2) and cloned into pGEM-T Easy (Promega). The KO plasmid p63KO4-13A (Fig. S2) was generated by insertion of the *nptII* cassette into the 4th intron of the *PpPPR_63* gene by using the GPS-M mutagenesis System (NEB). The KO plasmids p67KO-1 and p67KO-2 (Figs. S4 and S5) were generated by insertion of the *nptII* cassette into *Hin*dIII in the 4th exon, or into *Stu*I in the 5th exon, respectively, of the *PpPPR_67* gene. The KO plasmid p104KO (Fig. S6) was generated by insertion of the *nptII* cassette into the *Bgl*II site in the 4th exon of the *PpPPR_104* gene. These KO plasmids were digested with *Not*I and introduced into *P. patens* by particle bombardment.

Generation of complementation lines

For F63, the *PpPPR_63* cDNA coding region (1785 bp) amplified by PCR using specific primers (Table S2) was cloned into the *Swa*I site of the overexpression vector pOX9WZ1 [25]. The mutated *PpPPR_63* (M63: $D_{460}A/D_{461}A$) was amplified from the F63 construct by using the primers 63DA-F and 63DA-R. Both constructs were linearized with *Not*I and introduced into the KO line *Δ63*.

Northern blot analysis

Total cellular RNA (5 μg) was separated on 10% polyacrylamide-7 M urea gels and transferred to nylon membranes. Hybridization and washing were performed at 42°C. The gene-specific oligonucleotide probes (Table S2) were labeled with DIG-ddUTP and terminal deoxynucleotidyl transferase (Roche). Signals were acquired with EZ-capture (ATTO).

Supporting Information

Figure S1 Generation of *PpPPR_63-GFP* knockin (KI) mosses. (**A**) Schematic of insertion of p63-GFPKI into the 3′ terminal end of the *PpPPR_63* coding region. Translated and untranslated regions are represented by black and white boxes, respectively. Primers P1 to P3 used for PCR, and the expected amplified DNA sizes, are shown. (**B**) PCR analysis using the indicated primer pairs. The predicted 4.5- and 1.8-kb fragments were amplified from the KI lines. (**C**) Protonemal colonies of wild-type (WT) and KI mosses grown for 2 weeks on BCDATG medium plates without antibiotics. KI-10 was used for further analyses (Fig. 2). Scale bars = 5 mm.

Figure S2 Generation of *PpPPR_63* knockout (KO) moss. (**A**) Schematic diagrams of the *PpPPR_63* gene, the KO construct (p63KO4-13), and the generated knockout locus. A Geneticin G418-resistance gene (*nptII*) cassette was inserted into the 4th intron of *PpPPR_63*. Translated and untranslated regions are represented by black and white boxes, respectively. Primers P4 to P7 used for PCR and the expected amplified DNA sizes are shown. The DNA probe used in (C) is represented by grey bars. (**B**) PCR analysis using the indicated primer pairs. The predicted 1.9-, 2.2- and 5.5-kb fragments were amplified from the KO lines; 3.0 kb was amplified from the WT. (**C**) Total cellular DNAs (20 μg) from wild type (WT) and KO mosses were digested with *Sph*I and hybridized with the *PpPPR_63* probe in (A). The probe was labeled with DIG-dUTP. Predicted hybridized fragments were detected in WT and KO mosses.

Figure S3 Generation of *PpPPR_63* complementation lines intoΔ*63*-1. (**A**) Schematic diagrams of *PpPPR_63* loci of wild type andΔ*63*-1. Primers P8 and P9 and the expected fragment sizes for PCR analysis are shown. (**B**) The *PpPPR_63* coding region (1.8 kb) amplified from the wild type cDNA was cloned into the *Swa*I site of the overexpression vector, pOX9WZ1, which harbors a rice actin promoter, a c-myc tag and the Zeocin resistance cassette. The mutated *PpPPR_63* (M63: $D_{460}A/D_{461}A$) gene was modified from the F63 construct. (**C**) The predicted 1.8- and 3.0-kb fragments derived from the transgenic lines and wild type (WT), respectively, were amplified. The predicted 5.0-kb fragments were not amplified fromΔ*63*-1 under the PCR conditions. F63-3 and M63-8 plants were used for further analysis as F63 and M63, respectively, in Fig. 4. Transgenic lines V-7 and V-8 were generated by transferring the vector only into theΔ*63*-1 moss.

Figure S4 Generation of *PpPPR_67* knockout (KO) lines. (**A**) Schematic diagrams of the *PpPPR_67* gene (WT), the KO construct (p67KO-1), and the generated KO locus. A Geneticin G418-resistance gene (*nptII*) cassette was inserted into the *Hin*dIII site of the 4th exon of *PpPPR_67*. Primers used for PCR and the expected amplified DNA sizes are shown. (**B**) The predicted 5.0-, 2.1- and 1.6-kb fragments were amplified from the transgenic lines, and 3.0 kb was amplified from the WT. The KO line *67*-1-17 was used in Figs. S7 and S8.

Figure S5 Generation of *PpPPR_67* knockout (KO) lines. (**A**) Schematic diagrams of the *PpPPR_67* locus, the KO construct (p67KO-2), and the targeted KO locus. A Geneticin G418-resistance gene (*nptII*) cassette was inserted into the *Stu*I site of the 5th exon of *PpPPR_67*. Primers and the amplified DNA sizes for PCR analysis are shown. (**B**) The predicted 5.0-, 2.3- and 1.5-kb fragments were amplified from the transgenic lines, and 3.0 kb was amplified from the WT. The KO line *67*-2-2 was used asΔ*67* in Figs. 5, S7, and S8.

Figure S6 Generation of *PpPPR_104* knockout (KO) line. (**A**) Schematic diagrams of the *PpPPR_104* locus, the KO construct (p104KO), and the targeted KO locus. A Geneticin G418-resistance gene (*nptII*) cassette was inserted into the *Bgl*II site of the 4th exon of *PpPPR_104*. Primers used for PCR and the expected fragment sizes are shown. (**B**) The predicted 5.0-, 1.8-

and 1.3-kb fragments were amplified from the transgenic lineΔ*104*-7, and 3.0 kb was amplified from the wild type (WT). The KO line *104*-7 was used asΔ*104* in Figs. 5, S7, and S8.

Figure S7 Northern blot analysis of chloroplast tRNAs. Total cellular RNAs (5 µg) from wild type (WT), Δ*67*-1-17, Δ*67*-2-2, andΔ*104*-7 protonemata were separated on 8% polyacrylamide containing 7 M urea and transferred to nylon membranes. Chloroplast tRNA gene-specific oligonucleotide probes (Table S2) were labeled with DIG-ddUTP and terminal deoxynucleotidyl transferase (Roche).

Figure S8 Northern blot analysis of mitochondrial tRNAs. Total cellular RNAs (5 µg) were subjected to northern blot analysis as described in Fig. S7. Mitochondrial tRNA gene-specific oligonucleotide probes (Table S2) were labeled as described in Fig. S7.

Table S1 List of PRORP and PRORP-like proteins from various organisms.

Table S2 Oligonucleotides used in this study.

Acknowledgments

We thank Mitsuyasu Hasebe (National Institute of Basic Biology, Okazaki) for the gift of a plasmid pGFPmutNTPII. We are grateful to Tomoaki Nishiyama (Kanazawa University) and Hidetoshi Sakayama (Kobe University) for a TBLASTN search of a PRORP homolog to unpublished genome database of *Chara braunii*. We also thank Haruka Nagashima for technical assistance.

Author Contributions

Conceived and designed the experiments: CS MS. Performed the experiments: CS YK KT KK. Analyzed the data: CS YK KT HS. Wrote the paper: CS MS.

References

1. Frank DN, Pace NR (1998) Ribonuclease P: unity and diversity in a tRNA processing ribozyme. Ann Rev Biochem 67: 153–180.
2. Schön A (1999) Ribonuclease P: the diversity of a ubiquitous RNA processing enzyme. FEMS Microbiol Rev 23: 391–406.
3. Hartmann E, Hartmann RK (2003) The enigma of ribonuclease P evolution. Trends Genet 19: 561–569.
4. Holzmann J, Frank P, Löffler E, Bennett KL, Gerner C, et al. (2008) RNase P without RNA: identification and functional reconstitution of the human mitochondrial tRNA processing enzyme. Cell 135: 462–474.
5. Anantharaman V, Aravind L (2006) The NYN domains: novel predicted RNAses with a PIN domain-like fold. RNA Biol 3: 18–27.
6. Gobert A, Gutmann B, Taschner A, Gössringer M, Holzmann J, et al. (2010) A single Arabidopsis organellar protein has RNase P activity. Nat Struct Mol Biol 17: 740–744.
7. Gutmann B, Gobert A, Giegé P (2012) PRORP proteins support RNase P activity in both organelles and the nucleus in Arabidopsis. Genes Dev 26: 1022–1027.
8. Rosenblad MA, López MD, Piccinelli P, Samuelsson T (2006) Inventory and analysis of the protein subunits of the ribonucleases P and MRP provides further evidence of homology between the yeast and human enzymes. Nucleic Acids Res 18: 5145–5156.
9. Stirewalt VL, Michalowski CB, Löffelhardt W, Bohnert HJ, Bryoant DA (1995) Nucleotide sequence of the cyanelle genome from *Cyanophora paradoxa*. Plant Mol Biol Rep 13: 327–332.
10. Reith ME, Munholland J (1995) Complete nucleotide sequence of the *Porphyra purpurea* chloroplast genome. Plant Mol. Biol Rep 13: 333–335.
11. Turmel M, Lemieux C, Burger G, Lang BF, Otis C, et al. (1999) The complete mitochondrial DNA sequences of *Nephroselmis olivacea* and *Pedinomonas minor*. Two radically different evolutionary patterns within green algae. Plant Cell 11: 1717–1729.
12. Lai LB, Bernal-Bayard P, Mohannath G, Lai SM, Gopalan V, et al. (2011) A functional RNase P protein subunit of bacterial origin in some eukaryotes. Mol Genet Genomics 286: 359–369.
13. Baum M, Cordier A, Schön A (1996) RNase P from a photosynthetic organelle contains an RNA homologous to the cyanobacterial counterpart. J Mol Biol 257: 43–52.
14. Cordier A, Schön A (1999) Cyanelle RNase P: RNA structure analysis and holoenzyme properties of an organellar ribonucleoprotein enzyme. J Mol Biol 289: 9–20.
15. Lai LB, Vioque A, Kirsebom LA, Gopalan V (2010) Unexpected diversity of RNase P, an ancient tRNA processing enzyme: challenges and prospects. FEBS Lett 584: 287–296.
16. O'Toole N, Hattori M, Andres C, Iida K, Lurin C, et al. (2008) On the expansion of the pentatricopeptide repeat gene family in plants. Mol Biol Evol 25: 1120–1128.
17. Sugita M, Ichinose M, Ide M, Sugita C (2013) Architecture of the PPR gene family in the moss *Physcomitrella patens*. RNA Biol 10: 1439–1445.
18. Howard MJ, Lim WH, Fierke CA, Koutmos M (2012) Mitochondrial ribonuclease P structure provides insight into the evolution of catalytic strategies for precursor-tRNA 5′ processing. Proc Natl Acad Sci USA 109: 16149–16154.
19. Sugiura C, Kobayashi Y, Aoki S, Sugita C, Sugita M (2003) Complete chloroplast DNA sequence of the moss *Physcomitrella patens*: evidence for the loss and relocation of *rpoA* from the chloroplast to the nucleus. Nucl Acids Res 31: 5324–5331.
20. Sugiura C, Sugita M (2004) Plastid transformation reveals that moss $tRNA^{Arg}$-CCG is not essential for plastid function. Plant J 40: 314–321.
21. Rensing SA, Lang D, Zimmer AD, Terry A, Salamov A, et al. (2008) The *Physcomitrella* genome reveals evolutionary insights into the conquest of land by plants. Science 319: 64–69.
22. Hattori M, Miyake H, Sugita M (2007) A pentatricopeptide repeat protein is required for RNA processing of *clpP* pre-mRNA in moss chloroplasts. J Biol Chem 282: 10773–10782.
23. Tasaki E, Hattori M, Sugita M (2010) The moss pentatricopeptide repeat protein with a DYW domain is responsible for RNA editing of mitochondrial *ccmFc* transcript. Plant J 62: 560–570.
24. Uchida M, Ohtani S, Ichinose M, Sugita C, Sugita M (2011) The PPR-DYW proteins are required for RNA editing of *rps14*, *cox1* and *nad5* transcripts in *Physcomitrella patens* mitochondria. FEBS Lett 585: 2367–2371.
25. Sugita C, Kato Y, Yoshioka Y, Tsurumi N, Iida Y, et al. (2012) *CRUMPLED LEAF* (*CRL*) homologs of *Physcomitrella patens* are involved in the complete separation of dividing plastids. Plant Cell Physiol 53: 1124–1133.

14

Perspectives on Screening Winter-Flood-Tolerant Woody Species in the Riparian Protection Forests of the Three Gorges Reservoir

Fan Yang*, Yong Wang, Zhulong Chan

Key Laboratory of Plant Germplasm Enhancement and Specialty Agriculture, Wuhan Botanical Garden, Chinese Academy of Sciences, Wuhan, Hubei, PR China

Abstract

The establishment of riparian protection forests in the Three Gorges Reservoir (TGR) is an ideal measure to cope with the eco-environmental problems of the water-level fluctuation zone (WLFZ). Thus, the information for screening winter-flood-tolerant woody plant species is useful for the recovery and re-establishment of the riparian protection forests in the TGR WLFZ. Therefore, we discussed the possibilities of constructing and popularizing riparian protection forests in the TGR WLFZ from several aspects, including the woody plant species distribution in the WLFZ, the survival rate analyses of suitable candidate woody species under controlled flooding conditions, the survival rate investigation of some woody plant species planted in the TGR WLFZ, and the physiological responses of some woody plant species during the recovery stage after winter floods. The results of woody species investigation showed that most woody plant species that existed as annual seedlings in the TGR WLFZ are not suitable candidates for the riparian protection forests. However, arbor species (e.g., *Salix matsudana*, *Populus ×canadensis*, *Morus alba*, *Pterocarya stenoptera*, *Taxodium ascendens*, and *Metasequoia glyptostroboides*) and shrub species (e.g., *Salix variegata*, *Distylium chinensis*, *Lycium chinense*, *Myricaria laxiflora*, and *Rosa multiflora*) might be considered suitable candidates for the riparian protection forests in the TGR WLFZ by survival rate analyses under controlled winter flooding conditions, and survival rate investigations of woody plant species planted in the TGR WLFZ, respectively. Physiological analyses showed that *P. ×canadensis*, *M. alba*, *L. chinense*, and *S. variegata* could develop specific self-repairing mechanisms to stimulate biomass accumulation and carbohydrate synthesis via the increases in chlorophyll pigments and photosynthesis during recovery after winter floods. Our results suggested these woody plant species could endure the winter flooding stress and recover well, and be used as candidate for the construction of riparian protection forests in the TGR WLFZ.

Editor: Yiguo Hong, CAS, China

Funding: The study was sponsored by National Natural Science Foundation of China (No. 31270449), the Executive Office of the State Council Three Gorges Construction Committee (No. SX2008-005), and Hubei Immigration Bureau (No. Y2392639261P0126). The funders had no role in study design, data collection and analysis, decision to publish, or preparation of the manuscript.

Competing Interests: The authors have declared that no competing interests exist.

* Email: yangfan@wbgcas.cn

Introduction

The portion of the Yangtze River Basin between Chongqing City and Yichang City is known as the Three Gorges Reservoir (TGR) Area (TGRA). The Three Gorges Dam (TGD) was designed to control floods, generate electricity, improve navigation, and create tourism opportunities on the Yangtze River [1–3]. The dam was initiated in 1994 and its first impoundment was conducted in 2003 with a water level rising of 60 m above former riverbank of the Yangtze River. The second impoundment was impounded in October 2006 and the water level rose to 156 m. The third impoundment occurred in October 2008 and resulted in a sustained water level at above 170 m for five months. The water was raised to the ultimate planned level of about 175 m above sea level in 2010 [4–6]. To operate the TGR at full capacity, the water level of the TGR fluctuates between 145 and 175 m, i.e., 145 m in summer for flood control and emission sediment and 175 m in winter for energy generation. In October, the water level rises gradually to 175 m. By the following January, the water level starts to fall, finally dropping to 145 m in June [3,5,7–9]. Thus, the peak flows of the TGR occurred during November, December, January, February and March (winter), low flows in June, July, August and September (summer). Therefore, the affecting riparian vegetation factors such as flooding timing, duration, frequency, rate of change, and magnitude [10] in the TGR differed from the natural Yangtze River.

The new hydrological regime, including the reversal of flooding time and the prolonged flooding duration caused by the TGD, dramatically alters the conditions of riparian ecosystems and results in the formation of reservoir water-level fluctuation zone (WLFZ), i.e., the area between the high (175 m) and low (145 m) water levels in the TGR [3,5]. The WLFZ forms a transitional zone between terrestrial and aquatic ecosystems and serves as an important pathway for exchanging of the fluxes of matter, energy, and information between terrestrial and aquatic ecosystems [7,11]. In the WLFZ, plants suffered serial submergence stress with durations as long as 210 days at depths of up to 30 m. Few plant species could tolerate such submergence stress. Therefore, the

Table 1. The coordinate data of 22 sampling sites of the water-level-fluctuation zone in the Three Gorges Reservoir in 2009.

No. sampling site	Sampling site	Longitude (E)	Latitude (N)	Lower elevation (m)	No. of sampling sites	Sampling site	Longitude (E)	Latitude (N)	Lower elevation (m)
1	Longxi River in Changshou	107°05′12.2″	29°48′43.6″	162	**12**	Yangtze River in the old location of Yunyang	108°54′09.0″	30°57′24.2″	158
2	Changming Port in Changshou	107°01′11.7″	29°47′17.3″	160	**13**	Yangxi River in the old location of Yunynag	108°54′07.4″	30°57′45.8″	160
3	Wu River in Fuling	107°23′33.4″	29°40′38.6″	165	**14**	Meixi River in Fengjie	109°31′25.8″	31°03′04.5″	160
4	Yangtze River in Fuling	107°24′47.2″	29°44′39.9″	162	**15**	Yangtze River in Fengjie	109°28′03.9″	31°00′21.7″	158
5	Yangtze River in Fengdu	107°40′51.0″	29°51′48.3″	164	**16**	Yangtze River in Wushan	109°53′29.0″	31°04′19.8″	160
6	Long River in Fengdu	107°44′43.7″	29°52′28.1″	162	**17**	Yangtze River in the Badong	110°23′42.0″	31°02′30.4″	160
7	Yangtze River in Zhong Xian	108°03′11.2″	29°18′08.9″	156	**18**	Padang Port in Badong	110°19′30.6″	31°02′31.3″	160
8	Xintian Town in Wanzhou	108°23′28.4″	30°41′41.7″	155	**19**	Shennong River in Badong	110°49′27.7″	31°03′40.2″	160
9	Wanzhou Port in Wanzhou	108°23′46.7″	30°41′15.1″	160	**20**	Xiangxi River in Zigui	110°40′48.1″	31°03′50.9″	160
10	Pengxi River in Yunyang	108°41′16.1″	30°56′54.2″	160	**21**	Lanling River vegetation recovery test base in Zigui	110°55′11.7″	30°51′55.3″	160
11	Yangtze River in Yunyang	108°42′48.0″	30°55′06.2″	162	**22**	Vegetation recovery test base in Three Gorges Dam in Zigui	110°59′08.9″	30°51′18.3″	160

Table 2. Woody plant species distribution in the TGR WLFZ in 2009.

Species	Elevation (>m)	Plant height (cm)	Life form (Growth age)	Species	Elevation (>m)	Plant height (cm)	Life form (Growth age)
***1 Leguminosae*[a](4[b]: 4[c])**				***2Rosaceae* (2: 2)**			
Bauhinia brachycarpa	173	60–80	Annual shrub	*Rosa multiflora*	172	20–70	Perennial climbing shrub
Indigofera pseudotinctoria	170	100–150	Annual shrub	*Rubus parvifolius*	174	30–50	Annual shrub
Lespedeza cuneata	172	60–780	Annual woody vines	***3 Euphorbiaceae* (1: 1)**			
Robinia pseudoacacia	168	50–100	Annual tree seedlings	*Sapium sebiferum*	165	50–140	Annual tree seedlings
***4 Moraceae* (3: 4)**				***5 Salicaceae* (2: 3)**			
Broussonetia papyrifera	165	30–90	Annual tree seedlings	*Salix matsudana*	168	150–350	Perennial tree
Cudrania tricuspidata	171	100–180	Annual shrub	*S. variegata* Franch.	164	40–130	Perennial shrub
Morus alba	169	90–450	Perennial arbor	*Populus×canadensis*	165	150–500	Perennial tree
M. australis	172	150–250	Perennial arbor	***6 Anacardiaceae* (1: 1)**			
***7 Verbenaceae* (1: 2)**				*Rhus chinensis Mill.*	167	20–80	Annual tree seedlings
Vitex negundo	168	30–140	Annual shrub	***8 Loganiaceae* (1: 1)**			
V. negundo var. *cannabifolia*	168	30–120	Annual shrub	*Buddleja davidii*	170	40–200	Annual shrub
***9 Caprifoliaceae* (1: 1)**				***10 Celastraceae* (1: 1)**			
Sambucus chinensis	172	80–130	Annual subshrub	*Maytenus variabilis*	173	20–30	Annual shrub
***11 Liliaceae* (1: 1)**				***13 Rutaceae* (2: 2)**			
Smilax china	173	25	Annual shrub	*Phellodendron chinense var.glabriusculum*	170	20–70	Annual tree seedlings
***12 Solanaceae* (3: 3)**				*Zanthoxylum armatum*	172	25–45	Annual shrub
Lycium chinense	170	30–90	Perennial Shrub	***15 Coriariaceae* (1: 1)**			
Solanum surattense	172	40–70	Annual subshrub	*Coriaria nepalensis*	170	20–30	Annual shrub
Datura innoxia	173	40–90	Annual shrub	***16 Tamaricaceae* (1: 1)**			
***14 Hamamelidaceae* (1: 1)**				*Myricaria laxiflora*	172	60–110	Perennial shrub
Distylium chinense	166	30–70	Perennial shrub	***18 Sapindaceae* (1: 1)**			
***17 Juglandaceae* (1: 1)**				*Koelreuteria paniculata*	165	20–90	Annual tree seedlings
Pterocarya stenoptera	172	100–450	Perennial tree	***20 Malvaceae* (2: 2)**			
***19 Urticaceae* (1: 1)**				*Hibiscus mutabilis*	174	100–140	Annual tree seedlings
Debregeasia orientalis	173	40–70	Annual shrub	*Urena lobata*	171	30–60	Annual shrub
22 Nyssaceae (1: 1)				**21 Ulmaceae (2: 2)**			
Camptotheca acuminata	172	50–150	Annual tree seedlings	*Trema laevigata*	172	30–50	Annual shrub
24 Araliaceae (1: 1)				*Ulmus pumila*	173	76	Annual tree seedlings
Aralia echinocaulis	173	35	Annual tree seedlings	**23 Linaceae (1: 1)**			
				Reinwardtia indica	171	20	Annual shrub

[a], family name;
[b], the number of genera within each family;
[c], the number of species within each family.

Table 3. Survival rate analyses of 11 woody plant species under controlled flooding conditions.

Plant species	Submergence treatment												
	Submergence depth (m)	**1**			**5**			**15**			**25**		
	Submergence duration (days)	**90**	**150**	**210**	**90**	**150**	**210**	**90**	**150**	**210**	**90**	**150**	**210**
Salix variegata	Number of the test plants	50	50	60	30	30	30	30	30	30	30	30	30
	Number of survival plants	45	37	32	26	18	14	10	6	3	5	3	0
	Survival rate (%)	**90**	**74**	**53**	**87**	**60**	**47**	**33**	**20**	**10**	**17**	**10**	**0**
Distylium chinensis	Number of the test plants	30	30	28	30	30	30	30	30	30	30	30	26
	Number of survival plants	28	24	15	25	17	10	9	6	0	0	0	0
	Survival rate (%)	**93**	**80**	**54**	**83**	**57**	**33**	**30**	**20**	**0**	**0**	**0**	**0**
Rosa multiflora	Number of the test plants	10	10	10	10	10	10	10	10	10	10	10	18
	Number of survival plants	8	6	3	8	5	1	2	0	0	0	0	0
	Survival rate (%)	**80**	**60**	**30**	**80**	**50**	**10**	**20**	**0**	**0**	**0**	**0**	**0**
Salix matsudana	Number of the test plants	90	90	90	30	30	30	30	30	30	30	30	26
	Number of survival plants	75	55	26	25	16	4	5	0	0	0	0	0
	Survival rate (%)	**83**	**61**	**29**	**83**	**53**	**13**	**17**	**0**	**0**	**0**	**0**	**0**
Taxodium ascendens	Number of the test plants	10	10	10	10	10	10	10	10	10	10	10	10
	Number of survival plants	9	7	4	8	6	3	4	2	0	0	0	0
	Survival rate (%)	**90**	**70**	**40**	**80**	**60**	**30**	**40**	**20**	**0**	**0**	**0**	**0**
Metasequoia glyptostroboides	Number of the test plants	10	10	10	10	10	10	10	10	10	10	10	10
	Number of survival plants	9	8	3	8	5	2	3	1	0	0	0	0
	Survival rate (%)	**90**	**80**	**30**	**80**	**50**	**20**	**30**	**10**	**0**	**0**	**0**	**0**
Morus alba	Number of the test plants	20	20	20	25	25	25	25	25	25	25	25	25
	Number of survival plants	16	11	2	18	14	2	3	0	0	0	0	0
	Survival rate (%)	**80**	**55**	**10**	**72**	**56**	**8**	**12**	**0**	**0**	**0**	**0**	**0**
Populus×canadensis	Number of the test plants	30	30	30	30	30	30	30	30	30	30	30	30
	Number of survival plants	26	17	4	22	12	2	3	0	0	0	0	0
	Survival rate (%)	**87**	**57**	**13**	**73**	**40**	**7**	**10**	**0**	**0**	**0**	**0**	**0**
Pterocarya stenoptera	Number of the test plants	10	10	10	10	10	10	15	15	15	15	15	15
	Number of survival plants	8	4	1	4	2	0	1	0	0	0	0	0
	Survival rate (%)	**80**	**40**	**10**	**40**	**20**	**0**	**7**	**0**	**0**	**0**	**0**	**0**
Myricaria laxiflora	Number of the test plants	10	10	10	10	10	10	10	10	10	10	10	10
	Number of survival plants	9	3	1	5	1	0	1	0	0	0	0	0
	Survival rate (%)	**90**	**30**	**10**	**50**	**10**	**0**	**10**	**0**	**0**	**0**	**0**	**0**
Lycium chinense	Number of the test plants	30	30	30	30	30	30	30	30	30	30	30	30
	Number of survival plants	22	12	2	18	5	3	1	0	0	0	0	0
	Survival rate (%)	**73**	**40**	**7**	**60**	**16**	**10**	**3**	**0**	**0**	**0**	**0**	**0**

Table 4. Survival rate analyses of *M. alba* in 2012 planted in the Xiaohe village WLFZ.

		Elevatation(m)				
		175-174	174-173	173-172	172-171	171-170
Total Area (m^2)		600	600	400	400	300
Number of the investigated plants		500	500	500	500	500
Submerged Days		≤83	≤95	≤107	≤118	≤127
Average survival rate (%)		75.7	70.9	42.1	30.3	23.4
Survival rate (%)	Sloping field	82.7	71.9	ND	53.5	ND
	Paddy field	70.8	52.9	42.1	28.3	23.4
Survival rate (%)	With old branches	ND	85.9	47.8	ND	ND
	Without old branches	75.7	61.9	32.1	36.3	23.4

ND, not determined.

TGR WLFZ has become a disturbed ecosystem, leading to eco-environmental problems, such as biodiversity loss [5,6], soil erosion, land-use challenges [3,12], non-point-source pollution [13,14], nutrient accumulation [15,16], and heavy metal pollution [17,18].

Damage to the ecological health and stability of the WLFZ will directly endanger the ecological safety of the whole reservoir area and even the Yangtze River Basin. Riparian protection forests provide numerous ecosystem services including natural scenery improvement, pollutant filtration, wildlife habitat, stream flow mitigation and riverbanks stabilization, and thus maintain the ecological integrity of aquatic and terrestrial ecosystems [19]. However, on one hand, considering the specific hydrological regime in the TGR WLFZ, establishing the riparian protection forest in the whole WLFZ is difficult and even impossible because of prolonged winter flooding durations and navigation disturbances. On the other hand, the riparian protection forest in the upper section of the WFLZ (with an elevation ranging from 170 m to 175 m) can be restored. Therefore, it is important to screen the winter flooding tolerant woody plant species suitable for the construction of riparian protection forests in the TGR WLFZ. For example, mulberry (*Morus alba* L.) had strong adaptation to flooding stress and could survive up to 7 m of flooding in parts of the drawdown zone [20]. The morphological and photosynthetic responses of *P. stenoptera* to a 70 day waterlogging stress showed that this plant species could endure prolonged soil waterlogging with 100% survival rates. Waterlogging stress only slightly affected its photosynthetic apparatus [21]. Compared to unplanted soils, the growth of *P. stenoptera* seedlings significantly increased soil pH value, soil organic matter, total nitrogen, total phosphorus, and total potassium [22]. The seedlings of *Myricaria laxiflora* could survive during the flooding period and recover rapidly after flooding was terminated [23]. The morphological and photosynthetic responses of *Distylium chinense* to simulated autumn and winter floods suggested that this plant species possessed high survival and good recovery growth after a long-term flooding stress [24]. In this paper, we discussed the possibilities of constructing and popularizing riparian protection forests in the TGR WLFZ from several aspects, including the woody plant species distribution in the WLFZ, the survival rate analyses of suitable candidate woody species under controlled flooding conditions (the selection of woody species was determined according to the field survey of pre-dam riparian vegetation in 2001), the survival rate investigation of some woody plant species planted in the TGR WLFZ, and the physiological responses of some woody plant species during the recovery stage after winter floods.

Materials and Methods

Experiment 1: Investigation of woody plant species in the TGR WLFZ

A survey on woody plant species distribution was conducted between July 1, 2009 and August 26, 2009. A total of 22 sampling sites (200 m×30 m, surface area) were plotted in the WFLZ along both sides of the Yangtze River in the Zigui, Badong, Wushan, Fengjie, Yunyang, Wanzhou, Zhongxian, Fengdu, Fuling, and Changshou Districts of Chongqing City (the locations of 22 sampling sites were shown in the figure 2 of reference [1]). Few plant species were distributed below the elevation of 160 m, and majority cliffs existed in the WLFZ. Therefore, these 22 large sampling sites were established possibly above the similar elevation (about 160 m). The coordinate data of 22 sampling sites of the water-level-fluctuation zone in the Three Gorges Reservoir in 2009 was shown in Table 1. Plant specimens were collected in the TGR WLFZ. Basic parameters, such as the lowest distributed elevation among different sample sites of each plant species, plant growth age and height were recorded. No specific permissions were required for all of the field studies. The field study did not involve endangered or protected species.

Experiment 2: Survival rate analysis of the candidate woody plant species after controlled winter flooding

The selection of candidate woody plant species were determined the field survey of the pre-dam riparian vegetation of natural Yangtze River in 2001. The screening experiments of candidate woody plant species for the TGR WLFZ were conducted during 2006 to 2011 at Wuhan Botanical Garden, Chinese Academy of Sciences, and Lanling Creek, which is a Yangtze River tributary. Two year-old seedlings of 11 woody plant species from arbor species (e.g., *Salix matsudana*, *Populus ×canadensis*, *M. alba*, *P. stenoptera*, *Taxodium ascendens*, and *Metasequoia glyptostroboides*) and shrub species (e.g., *Salix variegata*, *D. chinensis*, *Lycium chinense*, *M. laxiflora*, and *Rosa multiflora*) were transplanted into 30 cm×15 cm pot. After a growing season, the branches of all plants were pruned to an identical height, i.e., 40 cm above ground level, and the leaves were removed. According to the the reservoir water level regulation, four submerged depths (1, 5, 15, and 25 m) and three submerged times (90, 150, and 210 days) were arranged as the sub-plot and main plot, respectively. The

Table 5. Survival rate analyses of five woody plants in 2012 planted in the Langling Creek WLFZ.

The test plant species	Elevatation (m)															
	173				171				169				167			
	Number of test plants	Sub. C.	Inv. W.	Suv. R.	Number of test plants	Sub. C.	Inv. W.	Suv. R.	Number of test plants	Sub. C.	Inv. W.	Suv. R.	Number of test plants	Sub. C.	Inv. W.	Suv. R.
P.×canadensis	25	Partial	×	100%	19	Partial	×	100%	20	Partial	×	70%	20	Full	×	30%
S. variegata	10	Full	√	100%	10	Full	√	90%	9	Full	√	78%	5	Full	√	60%
T. ascendens	25	Partial	×	100%	14	Partial	×	100%	ND				15	Full	×	60%
M. glyptostroboides	8	Partial	×	100%	6	Partial	×	100%	ND				6	Full	×	50%
L. chinense	20	Full	√	75%	20	Full	√	55%	ND				20	Full	√	15%

Sub. C., submersion conditions (Full submersion, partial submersion); Inv. W., whether invaded by dominant plant communities or not; Sur. R., survival rate; ND, not determined.

Table 6. Physiological responses of *P.×canadensis* planted in the TGR WLFZ during the recovery stage after winter flooding.

	Ch *a* (mg/g Fw)	Ch *b* (mg/g Fw)	Carotenoids (mg/g Fw)	*A*	g_s	*Ci*	*E*	Fv/Fm	Yield	Y(NPQ)	Y(NO)	qN
Control	1.09±0.035	0.37±0.008	0.23±0.007	12.87±0.381	1.63±0.680	355±12	1.90±0.458	0.71±0.004	0.66±0.017	0.07±0.019	0.27±0.004	0.26±0.059
Flooding-recovery	1.18±0.010	0.38±0.003	0.27±0.003	14.17±0.387	2.64±0.182	348±5	2.78±0.152	0.75±0.004	0.66±0.002	0.04±0.008	0.31±0.006	0.16±0.032
P>Ff	0.083	0.372	0.01	0.53	0.227	0.619	0.142	0.03	0.854	0.233	0.512	0.187

Ch *a*, chlorophyll a; Ch *b*, chlorophyll b; *A*, net photosynthetic rate; g_s, stomatal conductance; C_i, intercellular CO_2 concentration; *E*, transpiration rate. Values are means ± SE (n=6). The values followed by different letters significantly differed at $P<0.05$ according to Duncan's test. *P>Ff* indicate comparisons between treatments; *Ff*, effect of flooding stress.

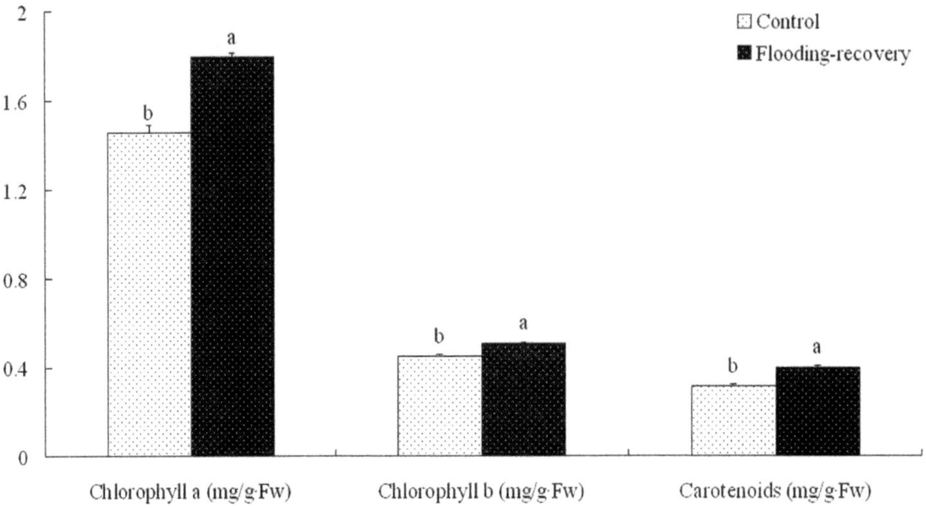

Figure 1. Comparable analyses of pigment contents in *L. chinense* during the recovery stage after winter flooding. Values are means ± SE (n = 6). The values followed by different letters significantly differed at $P<0.05$ according to Duncan's test.

1 m-deep submerged experiment was performed in two pools at Wuhan Botanical Garden, and other three submerged experiments (5, 15, and 25 m) were performed at Lanling Creek. For 5, 15, and 25 m submerged experiments, all the submerged pots were hung at the different pneumatic tires according to their individual submerged depths, and all the tires through the wire rope were fixed at the both sides of the river. The submergence treatment was conducted in batches in coordination with the submergence durations since September each year. All submerged plants were reaped simultaneously for analysis in April next year.

Experiment 3: Investigation of survival rate of some woody plant species planted in the TGR WLFZ

The majority of one-year-old *M. alba* (mulberry) seedlings were planted in March 2010 in the WLFZ of Xiaohe Village, Kai County, Chongqing, Yangtze River tributary, with an elevation ranging from 170 m to 175 m. Their survival rate was more than 95% before the next TGR impoundment. The survival plants were recorded in September 2010, and investigation on the survival rate of *M. Alba* was conducted in May 2011. The majority of seedlings of *S. variegata*, *P.* ×*canadensis*, *T. ascendens*, *M. glyptostroboides*, and *L. chinense* were planted in 2008 in the WLFZ of the Lanling Creek, Yangtze River tributary, with an elevation ranging from 165 m to 175 m. The submergence durations were over 100 days each year since 2010. The survival plants were recorded in September 2011. The investigations on the survival rate of *S. variegata*, *P.* ×*canadensis*, *T. ascendens*, *M. glyptostroboides*, *L. chinense*, and *M. alba* were conducted in May 2012.

Experiment 4: Dertermination the chlorophyll pigments, photosynthesis,chlorophyll a fluorescence of some woody plant species planted in the TGR WLFZ during the recovery stage after winter flooding

Fresh leaves from the six plants of *S. variegata*, *L. chinense*, *P.* ×*canadensis*, and *M. alba* planted at an elevation of 176 m (used as control) and 172 m of the Lanling Creek WLFZ were collected on April 10, 2012. The collected leaves were frozen in liquid nitrogen immediately for pigment content analyses. The contents of chlorophyll a (Ch *a*), chlorophyll b (Ch *b*), carotenoids were determined according to the procedures of [25]. Anthocyanin content was determined according to the procedures of [26] with a slight modification. The anthocyanin was extracted by 1% HCl–methanol solvent (1:99, v:v). Relative content was then determined by the absorbance at 532 nm. To alleviate the effects of chlorophyll, the absorbance of anthocyanin was calculated as follows:

$$A = A_{530} - 0.25 \times A_{652}.$$

The gas exchange and chlorophyll a fluorescence parameter analyses of *P.* ×*canadensis* and *M. alba* were performed according to the procedures of [25]. The same six cuttings and same leaves were selected and used for gas exchange measurements and chlorophyll fluorescence measurements. The net CO_2 assimilation rate (A), stomatal conductance (g_s), intercellular CO2 concentration (Ci) and transpiration (E) were measured from 9:00 to 11:30 am on April 9, 2012 with a LI-COR 6400 portable photosynthesis system (LI-COR Inc. Lincoln, Nebr.). The PAR, provided by a 6400-02 LED light source, was set to 1200 $\mu mol \cdot m^{-2} \cdot s^{-1}$. The flow rate of air through the sample chamber was set at 500 $\mu mol \cdot m^{-2} \cdot s^{-1}$, and the leaf temperature and relative humidity was maintained at 25±0.8°C by thermoelectric coolers and 50%, respectively. Chlorophyll fluorescence kinetics parameters were measured with a portable chlorophyll fluorometer PAM 2500 (Walz, Effeltrich, Germany). The leaf samples were placed in darkness for 30 min by covering with dark leaf clip (DLC-8) followed by measurement of minimum fluorescence (Fo) at 250 $\mu mol\ m^2 s^{-1}$ PPF and Fm at 2400 $\mu mol\ m^2 s^{-1}$ PPF following a saturating pulse of actinic light).

Statistical analyses

Results were expressed as means ± standard errors (n = 6). SPSS 13.0 software was used for statistical analysis. To assess the statistical significance of the treatment differences, a one-way analysis of variance (ANOVA) followed by Duncan's multiple range test was employed. Differences were considered significant at $P<0.05$.

Table 7. Physiological responses of *M. alba* planted in the TGR WLFZ during the recovery stage after winter flooding.

	Ch *a* (mg/g Fw)	Ch *b* (mg/g Fw)	Carotenoids (mg/g Fw)	*A*	g_s	*Ci*	*E*	Fv/Fm	Yield	Y(NPQ)	Y(NO)	qN
Control	2.08±0.058	0.63±0.021	0.49±0.015	11.90±1.021	0.67±0.082	357±15	2.81±0.530	0.70±0.010	0.68±0.005	0.02±0.007	0.30±0.011	0.09±0.032
Flooding-recovery	2.41±0.015	0.81±0.003	0.55±0.002	14.80±2.401	1.50±0.162	355±8	5.38±0.280	0.71±0.014	0.67±0.008	0.02±0.007	0.31±0.014	0.09±0.030
P>*Ff*	0.005	0.001	0.017	0.029	0.010	0.881	0.013	0.488	0.406	0.925	0.673	0.954

Ch *a*, chlorophyll a; Ch *b*, chlorophyll b; *A*, net photosynthetic rate; g_s, stomatal conductance; C_i, intercellular CO_2 concentration; *E*, transpiration rate. Values are means ± SE (n = 6). The values followed by different letters significantly differed at P<0.05 according to Duncan's test. *P*>*Ff* indicate comparisons between treatments; *Ff*, effect of flooding stress.

Results

Woody plant species distribution in the TGR WLFZ

The woody plant species in the TGR WLFZ was obtained from the 22 sampling regions. The results showed that 39 woody plant species including 12 tree species and 27 shrub or sub-shrub species, belonging to 37 genera of 24 families were distributed in the TGR WLFZ (Table 2). Most of these woody species existed as annual seedlings, except for *Sapium sebiferum*, *Ficus tikoua*, *L. chinense*, *M. alba*, *M. australis*, *P.×canadensis*, *S. matsudana*, *R. multiflora*, *S. variegata M. laxiflora*, and *D. chinense*.

Survival rate analyses of 11 woody plant species after controlled winter flooding

Under 1 m-deep submerged condition, all tested plants could survive after 90 days of submergence. Most of the tested plants can endure 90 days of submergence with a survival rate of more than 80%; *L. chinense* and *M. laxiflora* species possessed a survival rate of more than 50% (Table 3). *S. matsudana*, *P.×canadensis*, *M. alba*, *T. ascendens*, *M. glyptostroboides*, *S. variegata*, *R. multiflora*, and *D. chinensis* could endure 150 days of submergence with survival rates of more than 50%. *S. matsudana*, *T. ascendens*, *M. glyptostroboides*, *S. variegata*, *R. multiflora*, and *D. chinensis* could endure 210 days of submergence with a survival rate of more than 20%.

Under 5 m-deep submerged conditions, all tested plants could survive after 90 days of submergence. Most of the tested plants could endure 90 days of submergence with a survival rate of more than 70%; *P. stenoptera*, *M. laxiflora*, *L. chinense* and *M. laxiflora* species possessed a survival rate of more than 30%. *S. matsudana*, *M. alba*, *T. ascendens*, *M. glyptostroboides*, *S. variegata*, *D. chinensis*, and *R. multiflora* could endure 150-days of submergence with a survival rate of more than 50%. *T. ascendens*, *M. glyptostroboides*, *S. variegata*, and *D. chinensis* could endure 210-days of submergence with a survival rate of more than 20% (Table 3).

Under 15 m-deep submerged conditions, *T. ascendens*, *M. glyptostroboides*, *S. variegata*, and *D. chinensis* could endure 90-days of submergence with a survival rate of more than 25%, and they could survive after 150-days submergence but with a low survival rate. Only few plants of *S. variegata* could endure 210-days of submergence and survive (Table 3).

Under 25 m-deep submerged conditions, only few plants of *S. variegata* could endure 150-days of submergence and survive, and all tested woody species could not survive after 210-days of submergence (Table 3).

Investigation of survival rate of some woody plant species planted in the TGR WLFZ

Mulberry seedlings planted in the Xiaohe Village WLFZ could survive after one winter flooding season but possessed different survival rates with variations in elevation from 170 m to 175 m. The plants planted in the sloping fields showed higher survival rates than the planted in the paddy field because of the different drainage property of the landform in the growth season. Plants with old branches also showed higher survival rates than plants without branches. Therefore, the survival rate of mulberry was simultaneously determined by several factors, including submergence duration and depth, landform drainage property, and old branch conditions (Table 4). The mulberry in the WLFZ could endure 5 m-depth submergence for over 120 days and form new crowns.

After four seasonal winter flooding, the survival conditions of *P.×canadensis*, *S. variegata*, *T. ascendens*, *M. glyptostroboides*,

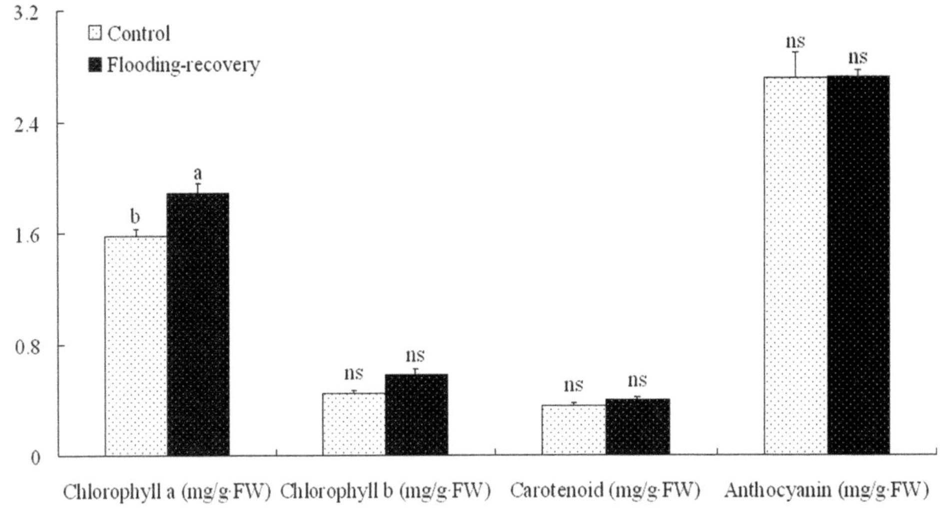

Figure 2. Comparable analyses of pigment contents in *S. variegata* during the recovery stage after winter floods. Values are means ± SE ($n = 6$). The values followed by different letters significantly differed at $P < 0.05$ according to Duncan's test. ns, no significant differences.

and *L. chinense* planted in the Langling Creek WLFZ were detected (Table 5). These five woody plants could endure tremendous winter flooding stress and survive well. *P.* ×*canadensis*, *S. variegata*, *T. ascendens*, and *M. glyptostroboides* possessed equivalent strong survival rates, whereas *L. chinense* had low survival rates. Survival rate was affected by full submersion and invasion. Although some plants died each year for different reasons, including flooding stress, human destruction, ecological competitions, plant diseases, and insect pests, the protection forests of *P.* ×*canadensis*, *T. ascendens*, and *M. glyptostroboides* thrived, specifically the *P.* ×*canadensis* forests. However, the shrub species, including *S. variegata* and *L. chinense*, planted in the WLFZ were seriously threatened by majority weeds such as *Bidens tripartite*, *Bidens bipinnata*, *Xanthium sibiricum*, *Alternanthera philoxeroides*, and *Setaria viridis* because of the dominant plant communities (high height, intensities, and coverage) and rapid growth of such weeds, except for winter flooding stresses.

Physiological responses of some woody plant species planted in the TGR WLFZ during the recovery stage after winter flooding

A slight increase in the contents of Ch *a* and Ch *b*, net photosynthetic rate (*A*), stomatal conductance (*gs*), intercellular CO_2 concentration (C_i,), and transpiration rate (*E*) were found in the submerged *P.* ×*canadensis* during the recovery stage after winter floods compared with the controls. No significant differences on these parameters were found between other submerged plants and the controls. However, the content of carotenoids was significantly increased by flooding stress. The maximum efficiency of PSII (Fv/Fm) in the submerged *P.* ×*canadensis* during the recovery stage after winter floods maintained higher levels in comparison to the controls, whereas other chlorophyll fluorescence parameters were not significantly affected by flooding stress (Table 6).

During the recovery stage, the contents of Ch *a*, Ch *b*, and carotenoids in the submerged *L. chinense* were significantly higher than those in the controls (Fig. 1). The significant increases in the contents of Ch *a*, Ch *b*, carotenoids, *A*, *gs*, and *E* were also detected in the submerged *M. alba* compared with the controls. However, all chlorophyll fluorescence parameters were insignificantly affected by flooding stress (Table 7). The contents of Ch *a*, Ch *b*, carotenoids, and anthocyanin were higher in the submerged *S. variegata* than those in the controls, but significant increases were only found in Ch *a* (Fig. 2).

Discussion

As one of the largest reservoirs in the world, the TGR has a large area and a long riparian line of WLFZ. The total length of the WLFZ riparian line in the TGR is approximately 5711 km. The total area of the upper section (the zone of the elevation ranging from 170 m to 175 m) covers approximately 8000 ha, which is approximately 22% of the whole WLFZ area. However, the WLFZ is encountering several eco-environmental problems as described in the introduction. Therefore, the reasonable management and utilization of the TGR WLFZ has been a highly domestic and international concern. Forested riparian buffers can solve certain eco-environmental problems in the WLFZ, and maintain the ecological integrity of aquatic and terrestrial ecosystems. For example, Zhang *et al.* (2004) indicated that reforestation decreases surface runoff and soil erosion on severely eroded land. The annual total soil loss on the severely eroded bare land varied from 53 tons ha^{-1} to 256 tons ha^{-1}. After reforestation, the annual soil erosion dramatically decreased to 2 tons ha^{-1} to 43 tons ha^{-1} from 1988 to 1990 and was negligible since then [27]. The white poplars planted in contaminated riparian forests could efficiently accumulate the Cd, Zn, and As of the soil [28]. Riparian forests could effectively reduce the nitrate and P concentrations in riparian soils by absorbing mineral N/P for plant growth. Wetlands dominated by plant communities remove N/P more effectively than non-vegetated wetlands because of the plant N/P uptake [29,30]. However, the woody plants planted in the TGR WLFZ must endure winter flooding stress. Therefore, it is very important to screen winter flood-tolerant woody plant species suitable for the construction of riparian protection forests in the TGR WLFZ.

The new hydrological regimes of the TGR can dramatically influence riparian vegetation. About 405 vascular plant species belonging to 240 genera of 83 families were located in the riparian zone of TGRA in 2001 before the TGD impoundment. However, only 231 vascular plant species belonging to 169 genera of 61 families were distributed in the TGR WLFZ in 2009 [5], and only 127 vascular plant species were found in the TGR WLFZ in 2010 [6]. Therefore, hydrological alternation of the TGR caused a significant loss of vascular flora in the TGR WLFZ. Although 39 woody plant species were distributed in the TGR WLFZ, the survival of most of these woody plant species exists in a form of annual seedling and their appearances mainly depend on seed germination from the seed bank [31] without perennial roots. Thus, these species could not survive and recover during or after winter flooding. These woody plant species, in spite of belonging to perennial plant species, were not suitable for re-vegetation in the new riparian zone. Suitable woody plant species for the TGR WLFZ should be selected from flood-tolerant species by controlled experiments.

The survival rate analyses of 11 woody plant species after controlled winter flooding showed that all tested plants could endure long winter floods, thus suggesting that such plant species were suitable candidates for the construction of riparian protection forests in the TGR WLFZ. However, different flooding tolerances were found among the 11 woody plant species. In addition, other tested woody plant species such as *Lespedeza davidii* Franch., *Buxus ichangensis* Hatusima, and *Ficus tikoua* Bur. did not show strong flooding tolerance. The survival rate analyses suggested that the winter flooding tolerance of these plant species was affected by the submergence duration and depth. Thus, different shrub and arbor species should be reasonably distributed according to their flooding tolerance during re-vegetation activities.

The survival rate investigations of the six woody plants planted in the TGR WLFZ showed that these plants could survive and grow well after several winter flooding, thus suggesting that these plants were high-winter flood-tolerant species. Thus, the riparian protection forests in the TGR WLFZ could be popularized. However, the woody plant species planted at different altitudes within the TGR WLFZ must endure different winter flooding durations and depths, thus their reasonable distribution along the elevation of the WLFZ according to their flood tolerance should be concerned. The TGR WLFZ management must be enhanced during restoration, particularly for shrub species, which have slow growth rates. For example, many *L. Chinense*, *D. chinensis*, and *S. variegata* seedlings could survive during the primary recovery stage after winter floods, but parts of the surviving plants died annually because of the invasion of dominant weed communities.

Waterlogging and/or flooding stresses usually decreased chlorophyll pigment contents, photosynthetic efficiency, and PSII maximum efficiency and affect other chlorophyll fluorescence parameters (Yi et al. 2006; Li et al. 2011; Yang et al. 2011). In this study, the increases in the contents of chlorophyll pigments of *L. chinense*, *S. variegata*, *P.×canadensis* and *M. alba* plants, the insignificant variations in the chlorophyll fluorescence parameters of the submerged *P.×canadensis* and *M. alba* plants during the recovery stage after winter flooding stress suggested that such plant species have developed certain self-repairing capabilities during the recovery. The increases in chlorophyll pigments and photosynthesis in the submerged seedlings might efficiently benefit the total biomass accumulation and carbohydrate synthesis during the recovery. All these results suggested that these woody plant species could contribute to the construction of riparian protection forests in different elevations of the TGR WLFZ. In addition, the morphological, photosynthetic, and physiological responses of *P. stenoptera* [21,22], *M. laxiflora* [23], and *D. chinense* [24] to flooding stress were also reported, which suggested these woody plant species could endure the flooding stress and recover well, and be used as candidate for the construction of riparian protection forests in the TGR WLFZ.

Conclusions

Four experiments were used to determine the suitable woody species for the riparian protection forests in the TGR. The woody plant species survey in 2009 suggested that most of woody species existing as annual seedlings in the TGR WLFZ were unsuitable candidates for the riparian protection forests, whereas perennial *S. sebiferum*, *F. tikoua*, *L. chinense*, *M. alba*, *M. australis*, *P.×canadensis*, *S. matsudana*, *R. multiflora*, *S. variegata M. laxiflora*, and *D. chinense* could be found in the TGR WLFZ. In addition, survival rate analyses of the candidate woody plant species after controlled winter flooding showed arbor species such as *S. matsudana*, *P.×canadensis*, *M. alba*, *P. stenoptera*, *T. ascendens*, *M.glyptostroboides*, and shrub species such as *S. variegata*, *D. chinensis*, *L. chinense*, *M. laxiflora*, and *R. multiflora* could endure long winter floods, and should be used for the riparian protection forests in the TGR WLFZ. The survival rate investigations of the six woody plants planted in the TGR WLFZ showed that these plants could survive and grow well after several winter flooding. Thus, the riparian protection forests in the TGR WLFZ should be popularized. In addition, *P.×canadensis*, *M. alba*, *L. chinense*, and *S. variegata* could develop specific self-repairing mechanisms to stimulate biomass accumulation and carbohydrate synthesis via the increases in chlorophyll pigments and photosynthesis during recovery after winter floods. In addition, their reasonable distribution of these plant species along the elevation of the WLFZ and management should be considered during the construction of the riparian protection forests in the TGR WLFZ.

Author Contributions

Conceived and designed the experiments: FY YW. Performed the experiments: FY YW. Analyzed the data: FY YW. Contributed reagents/materials/analysis tools: FY. Wrote the paper: FY. Checked grammar of revised manuscript: ZC.

References

1. Wu J, Huang J, Han X, Gao X, He F, et al. (2004) The Three Gorges Dam: an ecological perspective. Front Ecol Environ 2: 241–248.
2. Stone R (2008) Three Gorges Dam: into the unknown. Science 321: 628–632.
3. Zhang Q, Lou Z (2011) The environmental changes and mitigation actions in the Three Gorges Reservoir region, China. Environ Sci Policy 14: 1132–1138.
4. Gleick PH (2009) Three Gorges Dam Project, Yangtze River, China, The World's Water 2008: 139–150.
5. Yang F, Liu W, Wang J, Liao L, Wang Y (2012) Riparian vegetation's responses to the new hydrological regimes from the Three Gorges Project: Clues to revegetation in reservoir water-level-fluctuation zone. Acta Ecol Sinica 32: 89–98.
6. Wang Q, Yuan X, Willison J, Zhang Y, Liu H (2014) Diversity and above-ground biomass patterns of sascular flora induced by flooding in the drawdown area of China's Three Gorges Reservoir. PLOS One 9(6): e100889.
7. New T, Xie Z (2008) Impacts of large dams on riparian vegetation: applying global experience to the case of China's Three Gorges Dam. Biodivers Conserv 17: 3149–3163.
8. Wang J, Huang J, Wu J, Han X, Lin G (2008) Ecological consequences of the Three Gorges Dam: insularization affects foraging behavior and dynamics of rodent populations. Front Ecol Environ 8: 13–19.
9. Lei S, Zeng B, Yuan Z, Su X (2014) Changes in carbohydrate content and membrane stability of two ecotypes of *Calamagrostis arundinacea* growing at

different elevations in the drawdown zone of the Three Gorges Reservoir. PLOS One 9(3): e91394.

10. Nilsson C, Svedmark M (2002) Basic principles and ecological consequences of changing water regimes: riparian plant communities. Environ Manage 30: 468–480.
11. Naiman RJ, Decamps H (1997) The ecology of interfaces: riparian zones. Annu Review Ecol System 28: 621–658.
12. Shi R (2011) Ecological Environment Problems of the Three Gorges Reservoir Area and countermeasures. Procedia Environ Sci 10: 1431–1434.
13. Ma X, Li Y, Zhang M, Zheng F, Du S (2011) Assessment and analysis of non-point source nitrogen and phosphorus loads in the Three Gorges Reservoir Area of Hubei Province, China. Sci Total environ 412–413: 154–161.
14. Shen Z, Chen L, Hong Q, Xie H, Qiu J, et al. (2013) Vertical variation of nonpoint source pollutants in the Three Gorges Reservoir region. PLOS One 8(8): e71194.
15. Wang Y, Shen Z, Niu J, Liu R (2009) Adsorption of phosphorus on sediments from the Three-Gorges Reservoir (China) and the relation with sediment compositions. J hazard Mater 162: 92–98.
16. Zhang B, Fang F, Guo J, Chen Y, Li Z, et al. (2012a) Phosphorus fractions and phosphate sorption-release characteristics relevant to the soil composition of water-level-fluctuating zone of Three Gorges Reservoir. Ecol Eng 40: 153–159.
17. Song Y, Ji J, Mao C, Yang Z, Yuan X, et al. (2010) Heavy metal contamination in suspended solids of Changjiang River–environmental implications. Geoderma 159: 286–295.
18. Ye C, Li S, Zhang Y, Tong X, Zhang Q (2012) Assessing heavy metal pollution in the water level fluctuation zone of China's Three Gorges Reservoir using geochemical and soil microbial approaches. Environ Monit Assess 185: 231–240.
19. Wasser L, Day R, Chasmer L, Taylor A (2013) Influence of vegetation structure on lidar-derived canopy height and fractional cover in forested riparian buffers during leaf-off and leaf-on conditions. PLOS One 8(1): e54776.
20. Liu Y, Willison JHM (2013) Prospects for cultivating white mulberry (*Morus alba*) in the drawdown zone of the Three Gorges Reservoir, China. Environ Sci Pollut Res 20: 7142–7151.
21. Yi YH, Fan DY, Xie ZQ, Chen FQ (2006) Effects of waterlogging on the gas exchange, chlorophyll fluorescence and water potential of *Quercus variabilis* and *Pterocarya stenoptera*. J Plant Ecol 30: 960–968 (in Chinese).
22. Yang Y, Li C, Li J, Schneider R, Lamberts W (2013) Growth dynamics of Chinese wingnut (*Pterocarya stenoptera*) seedlings and its effects on soil chemical properties under simulated water change in the Three Gorges Reservoir Region of Yangtze River. Environ Sci Pollut Res 20: 7112–7123.
23. Chen FQ, Xie ZQ (2009) Survival and growth responses of *Myricaria laxiflora* seedlings to summer flooding. Aquat Bot 90: 333–338.
24. Li X, Li N, Yang J, Ye F, Chen F, et al. (2011) Morphological and photosynthetic responses of riparian plant *Distylium chinense* seedlings to simulated autumn and winter flooding in Three Gorges Reservoir Region of the Yangtze River, China. Acta Ecol Sinica 31: 31–39.
25. Yang F, Wang Y, Wang J, Deng W, Liao L (2011) Different eco-physiological responses between male and female *Populus deltoides* clones to waterlogging stress. Forest Ecol Manag 262: 1963–1971.
26. Basu P, Chand S (1996) Anthocyanin accumulation in *Hyoscyamus muticus* L. tissue cultures. J Biotech 52: 151–159.
27. Zhang B, Yang Y, Zepp H (2004) Effect of vegetation restoration on soil and water erosion and nutrient losses of a severely eroded clayey Plinthudult in southeastern China. CATENA 57: 77–90.
28. Madejon P, Marañón T, Murillo JM, Robinson B (2004) White poplar (*Populus alba*) as a biomonitor of trace elements in contaminated riparian forests. Environ Pollut 132: 145–155.
29. Fortier J, Gagnon D, Truax B, Lambert F (2010) Nutrient accumulation and carbon sequestration in 6-year-old hybrid poplars in multiclonal agricultural riparian buffer strips. Agr Ecosyst Environ 137: 276–287.
30. Schade JD, Fisher S, Grimm N, Seddon J (2001) The influence of a riparian shrub on nitrogen cycling in a Sonoran Desert stream. Ecology 82: 3363–3376.
31. Lu ZJ, Li LF, Jiang MX, Huang HD, Bao DC (2010) Can the soil seed bank contribute to revegetation of the drawdown zone in the Three Gorges Reservoir Region? Plant Ecol 209: 153–165.

15

Salicylic Acid Alleviates the Adverse Effects of Salt Stress in *Torreya grandis* cv. Merrillii Seedlings by Activating Photosynthesis and Enhancing Antioxidant Systems

Tingting Li[1,2☯], Yuanyuan Hu[1,2☯], Xuhua Du[3], Hui Tang[1,2], Chaohua Shen[1,2], Jiasheng Wu[1,2*]

1 Nurturing Station for the State Key Laboratory of Subtropical Silviculture, Zhejiang A & F University, Lin'an, Hangzhou, P. R. China, **2** School of Forestry & Biotechnology, Zhejiang A & F University, Lin'an, Hangzhou, P. R. China, **3** China National Bamboo Research Center, Hangzhou, P. R. China

Abstract

Background: Salt stress is a major factor limiting plant growth and productivity. Salicylic acid (SA) has been shown to ameliorate the adverse effects of environmental stress on plants. To investigate the protective role of SA in ameliorating salt stress on *Torreya grandis* (*T. grandis*) trees, a pot experiment was conducted to analyze the biomass, relative water content (RWC), chlorophyll content, net photosynthesis (Pn), gas exchange parameters, relative leakage conductivity (REC), malondialdehyde (MDA) content, and activities of superoxide dismutase (SOD) and peroxidase (POD) of *T. grandis* under 0.2% and 0.4% NaCl conditions with and without SA.

Methodology/Principal Findings: The exposure of *T. grandis* seedlings to salt conditions resulted in reduced growth rates, which were associated with decreases in RWC and Pn and increases in REC and MDA content. The foliar application of SA effectively increased the chlorophyll (chl (a+b)) content, RWC, net CO_2 assimilation rates (Pn), and proline content, enhanced the activities of SOD, CAT and POD, and minimized the increases in the REC and MDA content. These changes increased the capacity of *T. grandis* in acclimating to salt stress and thus increased the shoot and root dry matter. However, when the plants were under 0% and 0.2% NaCl stress, the dry mass of the shoots and roots did not differ significantly between SA-treated plants and control plants.

Conclusions: SA induced the salt tolerance and increased the biomass of *T. grandis* cv. by enhancing the chlorophyll content and activity of antioxidative enzymes, activating the photosynthetic process, and alleviating membrane injury. A better understanding about the effect of salt stress in *T. grandis* is vital, in order gain knowledge over expanding the plantations to various regions and also for the recovery of *T. grandis* species in the future.

Editor: Jin-Song Zhang, Institute of Genetics and Developmental Biology, Chinese Academy of Sciences, China

Funding: This project was supported in part by the Fruit Innovation Team Project of Zhejiang Province (2009R50033-7) and the Major Project of National Spark Plan of China (2012GA700001). The funders had no role in study design, data collection and analysis, decision to publish, or preparation of the manuscript.

Competing Interests: The authors have declared that no competing interests exist.

* Email: wujs@zafu.edu.cn

☯ These authors contributed equally to this work.

Introduction

Salt stress is one of the major abiotic stress factors that limit plant growth and productivity [1]. Salt stress alters various biochemical and physiological responses in plants, thereby affecting almost all plant processes [2]. The major effects of salinity are significant reductions in growth, such as decreased leaf area, leaf length and root and shoot dry weight [3]. High salt stress can also restrict photosynthesis by decreasing the abundance of green pigments and causing stomatal closure and oxidative stress, thereby resulting in the formation of reactive oxygen species (ROS) [4,5,6]. Excessive amounts of ROS can enhance membrane lipid peroxidation, cause electrolyte leakage, and damage the chloroplast, thus inhibiting photochemical reactions and decreasing photosynthesis [7,8]. To alleviate the harmful effects of these ROS, plants have evolved an effective scavenging system composed of non-enzymatic antioxidants (carotenoids, ascorbate, and tocopherol) and enzymatic antioxidants, such as catalase (CAT) and peroxidase (POD). Gossett et al. [9] reported that treatment with 150 mM NaCl resulted in a 72% increase in the POD activity in salt-tolerant cotton. Therefore, it is necessary to improve the salinity tolerance of plants and exploit various compounds to alleviate plant stress.

As an important signaling molecule, salicylic acid (SA) influences various physiological and biochemical functions in plants and has diverse effects on the tolerance to biotic and abiotic stress [10,11,12]. The role of SA in plant tolerance to abiotic stresses such as ozone, UV-B, heat, heavy metal, and osmotic stress has been reported [12,13]. Khan et al. [14] reported that SA application alleviated the adverse effects of salt stress in mungbean through the improvement of plant photosynthesis and growth and also enhancing antioxidant system. However, the improvement of salt tolerance by exogenous SA depends on genotype and the concentration of SA used. For example, Arfan et al. [10] reported that exogenous SA counteracted the salt stress-induced growth inhibition in a salt-tolerant wheat genotype, although no improvement occurred in a salt-sensitive cultivar. Therefore, the questions of whether the effectiveness of SA in the alleviation of salt stress is dependent on the plant species must be further elucidated.

The genus Torreya is an endangered and primitive member of the yew family (Taxaceae) and consists only of six species with a restricted worldwide distribution [15]. The conservation and recovery of these species have received increasing attention in recent years. *Torreya grandis* (*T. grandis*) cv. Merrillii, belonging to the genus Torreya family Taxaceae, is an important economic tree species and has drupe-like fruits with nut seeds that were used as food for thousands years in China. *T. grandis* has also important medicinal value due to its anthelmintic, antitussive, carminative, laxative, antifungal, antibacterial, and antitumor activity [16]. Currently, *T. grandis* is only planted on hills and mountain slopes in south China. The expansion of *T. grandis* plantations on new sites is important for the recovery of *T. grandis* species. However, throughout the entire country, 99 million hm^2 arable land is adversely affected by high salt concentrations, and the soil salinization is forecasted to increase drastically in the coming decades due to the worldwide ongoing climate change [17]. Therefore, it is necessary to improve the salinity tolerance of *T. grandis* species to expand its plantation under stress conditions. For the past years, numerous studies have been focused over *T. grandis* app:addword:concentratespecies for their ecological characterization; chemical constituents, fatty acid composition and its distribution in the seeds and the therapeutic use of the seed extracts [18,19,20,21]. Unfortunately, few published studies are available on the effect of salt stress on the growth and physiology of *T. grandis* trees. Additionally, it remains unclear whether SA plays a significant protective role in ameliorating the influence of salt stress on *T. grandis* trees to increase its salt tolerance.

The objective of this study was to investigate the effects of exogenous SA application on *T. grandis* growth and physiology under 0.2% and 0.4% NaCl conditions by analyzing the biomass, photosynthesis, membrane injury, and antioxidative enzyme activity. The information obtained is essential for propagation and cultivation of T. grandis and the other Torreya species under salt stress condition.

Materials and Methods

Plant materials and growth conditions

The pot experiments under different salt stress treatments were conducted in a controlled environment room of the College of Forest and Biotechnology, Zhejiang Agriculture and Forestry University, Zhejiang Province, China (N30°23′, E119°72′) in 2012. In March, two-year-old healthy and homogenous *T. grandis* seedlings were transplanted from Chunan, Zhejiang Province, China, and transferred to plastic pots (13.5 cm inner-diameter and 16 cm height with holes in the bottom, one seedling per pot) that were filled with 4 kg mixture of silt and perlite soil (3:1, v/v, pH 6.40). The soil was a loam with an organic matter content of 2.90%, total N content of 0.255%, total P content of 0.125%, and total K content of 1.574%. Four weeks after transplanting, a completely randomized design with 3 replications per treatment and 5 plants per replication was adopted. Six treatments with different salt concentrations (0, 0.2 and 0.4% NaCl) and different SA concentrations (0 and 0.5 mmol) were implemented as follows: T1 (distilled water without SA); T2 (distilled water with 0.5 mmol SA); T3 (0.2% NaCl without 0.5 mmol SA); T4 (0.2% NaCl with 0.5 mmol SA); T5 (0.4% NaCl without 0.5 mmol SA); and T6 (0.4% NaCl with 0.5 mmol SA). To avoid osmotic shock, the NaCl solution was gradually added to the soil in eight steps to achieve the final concentrations of 0.2% and 0.4% in 4 days. SA was dissolved in ethanol and then added to "Tween-20" (0.1% dilute solution) to facilitate spreading of the solution on the plant-leaf surface. The desired SA concentration was attained using double-distilled water (DDW). SA was sprayed over the leaves 3 days before NaCl treatment and also continued during the NaCl application. SA was sprayed over the adaxial and abaxial surfaces of a leaf until dripping a common insecticidal atomizer and it was done for twice a day at the time interval of 07:00 and18:00. When SA solution was sprayed, the surface of the potting soil was covered with plastic film in case SA solution dripped to the root. The plants were then collected for biomass analysis after 60 days, and measurements of other indexes were collected after 30 days of treatments. Twice in week irrigation were done for different treatmental plants with their corresponding solution in order to maintain the field capacity at 70–75%. The environmental conditions were maintained as same throughout the experiment.

Analysis of growth and biomass

Sixty days after the treatments, one intact plant from each replicate of treatments was uprooted for biomass analysis. The green plant tissues and roots were oven-dried at 65°C to a constant weight and weighed using an electronic scale to determine the biomass.

Determination of the leaf relative water content (RWC) and proline content

A total of 0.5 g of fresh leaf tissue was immediately weighed (fresh weight, WF) after rehydrating for 24 h in the dark (saturated weight, WS) and after oven-drying at 85°C for 24 h to a constant weight (WD). The RWC was calculated using the following formula:

$$RWC(\%) = [(WF - WD)/(WS - WD)] \times 100.$$

Proline accumulation was determined as described by Bates et al. [22] with slight modifications. Approximately 0.2 g of fresh leaf material was homogenized in 5 mL 3% aqueous sulfosalicylic acid. The samples were mixed, heated in boiling water for 30 min, and filtered through Whatman's No. 2 filter paper. A total of 2 mL of filtrate was mixed with 2 mL of acid-ninhydrin and 2 mL of glacial acetic acid in a test tube. The mixture was placed in a water bath for 1 h at 100°C and cooled to room temperature. The reaction mixture was extracted with 5 mL of toluene, and the chromophore containing toluene was aspirated. The absorbance was measured at 520 nm using toluene as a blank. The proline concentration was calculated using proline standards (0–50 mg/mL).

Analysis of gas exchange

The youngest healthy, fully developed leaves from a randomly selected branch were used for the gas exchange measurements on recently expanded leaves using a portable photosynthesis measuring instrument (LI-6400, LiCor, Inc. Lincoln, NE, USA) equipped with an 6400-05 conifer chamber at concentration of 400 μmol mol^{-1} CO_2, 21% O_2, and 50% relative humidity with a natural light source. The temperature of the chamber was maintained at

29–32°C. Photosynthetic gas exchange measurements were made during 9:00–10:30 when the light intensity was between 1200–1600 μmol $m^{-2}s^{-1}$. This irradiance was higher than the saturation irradiance for current year leaf of *T. grandis* (452–514 μmol m^{-2} s^{-1}, data not shown). Therefore, light was not the limitation for photosynthesis in these gas exchange measurements.

Determinations of the chlorophyll concentration and soluble protein content

A total of 0.1 g of leaf discs was weighed, placed into a test tube, and then repeatedly extracted with 8 mL of 95% (V/V) ethanol. The photosynthetic pigments were extracted at 4°C for 48 h in the darkness. The absorbance of the supernatant was measured at 646 and 663 nm. The chlorophyll a (Chla) and Chlb concentration was determined according to the methods of Lichtenthaler [23].

The soluble protein content was measured according to the procedure of Bradford [24] with a slight change. A 0.5-g frozen sample was extracted in sodium phosphate buffer (50 mmol, pH 7.8). The extracts were centrifuged at 4000 ×*g* for 10 min at 4°C. The soluble proteins were quantified using the supernatants of samples with bovine serum albumin as a standard.

Determinations of the relative electrolyte leakage rate and lipid peroxidation

To measure the electrolyte leakage, 10 leaf discs (10 mm diameter) from young, fully expanded leaves were placed in 50-mL glass vials and rinsed with distilled water to remove electrolytes that were released during the leaf disc excision. The vials were then filled with 30 mL of distilled water and allowed to stand in the dark for 24 h at room temperature. The electrical conductivity (EC_1) of the bathing solution was determined at the end of incubation period. The vials were heated in a temperature-controlled water bath at 95°C for 20 min and then cooled to room temperature, after which the electrical conductivity (EC_2) was measured. The relative electrolyte leakage (REC) was calculated using the following formula [25]:

$$REC(\%)=\left(\frac{EC_1}{EC_2}\right)\times 100.$$

The degree of lipid peroxidation was assessed based on the malondialdehyde (MDA) contents according to Zheng et al. [26]. A volume of 2 mL of the supernatant sample was combined with 2 mL of thiobarbituric acid (TBA) incubated in boiling water for 30 min and then quickly cooled in an ice bath. The mixture was centrifuged at 5000 × g for 10 min, and the absorbance of supernatant was monitored at 532, 600 and 450 nm. The MDA concentration was expressed as μmol g^{-1} using the following formula:

$$MDA(\mu mol g^{-1})=6.45\times(A_{532}-A_{600})-0.56\times A_{450}$$

.

Determination of antioxidative enzyme activities

The enzymes were extracted at 4°C from approximately 0.3 g of tissue from the second leaves using a mortar and pestle with 5 mL of extraction buffer, containing 0.1 mol phosphate buffer (pH 7.8), 0.1 mol ethylenediaminetetraacetic acid (EDTA), and 1% polyvinylpyrrolidone (PVP). The extracts were centrifuged at 6000 × g for 30 min, and the supernatants were used for the enzyme assays.

The superoxide dismutase (SOD) activity was assayed by the inhibition of the photochemical reduction of β-nitroblue tetrazolium chloride (NBT). The reaction mixture (3 mL) consisted of 50 mol potassium phosphate buffer (pH 7.8) containing 0.3 μmol EDTA, 39.15 μmol methionine, 0.225 μmol nitroblue tetrazolium, 0.006 μmol riboflavin, and 0.05 mL of enzyme extract. The reaction was conducted for 15 min at 25°C under 4000 l×. One unit of SOD activity was defined as the amount of enzyme that inhibited the NBT reduction by 50% at 560 nm [27].

The CAT activity was assayed by monitoring the disappearance of hydrogen peroxide (H_2O_2) [28]. This disappearance can be detected by measuring the decrease in the absorbance at 240 nm of a reaction mixture consisting of 1.5 mL of 50 mmol sodium phosphate buffer (pH 7.8), 0.3 mL of 100 mmol H_2O_2, and 0.2 mL of enzyme extract. One CAT unit was defined as the amount of enzyme that was necessary to decompose 1 mmol H_2O_2 per minute under the aforementioned assay conditions, and the specific activity was expressed as U g^{-1} FW min^{-1}.

The POD activity was determined based on guaiacol oxidation measured at 470 nm [29]. The reaction mixture (3 mL) contained 0.05 mL of enzyme extract, 2.75 mL of 50 mmol phosphate buffer (pH 7.0), 0.1 mL of 1% H_2O_2, and 0.1 mL of a 4% guaiacol solution. The increase in absorbance at 470 nm due to guaiacol oxidation was recorded for 2 min. One unit of enzyme activity was defined as the amount of the enzyme that caused a change of 0.01 absorbance unit per minute. The specific POD activity was expressed as U g^{-1} FW min^{-1}.

Statistical analyses

The experiment was conducted according to a simple randomized block design. Each treatment was replicated three times. The data were statistically analyzed by an analysis of variance (ANOVA) using SAS 9.2 Institute Inc. (2008) software. The least significant difference (LSD) was calculated to separate the means. A simple correlation analysis was performed using the SAS program to determine the relationship between each physiological variable.

Results

Dry mass of the shoots and roots of *T. grandis* under different salt stress and SA treatments

Salt stress significantly inhibited the plant growth, leading to a pronounced reduction in the dry mass of the shoots and roots. Compared with the non-salt-treated *T. grandis* seedlings (T1), the 0.2% and 0.4% NaCl treatments significantly reduced the dry mass by 12.1% and 29% in the shoots and by 9.9% and 25.3% in the roots after 60 days, respectively (Table 1). The SA treatment reduced the decrease in the growth of the salt-stressed plants. Compared with the salt-stressed plants, SA treatment increased the dry mass of the shoots, roots, and shoot+roots by 6.2% ($P<0.05$), 5.6% ($P<0.05$) and 5.9% ($P<0.05$) under 0.2% NaCl conditons, and 16.8% ($P<0.05$), 18.2% ($P<0.05$), and 17.7% ($P<0.05$) under 0.4% NaCl conditions, respectively (Table 1). However, there were no significant differences in the dry mass of shoots and roots between different treatments with or without SA treatment under 0% NaCl stress condition. In addition, the salt-stressed plants grown under SA were larger than those without SA (Fig. 1).

Table 1. Effects of SA on the dry matter of the shoots, roots, and roots+shoots in *T. grandis* grown under salt stress (means±SD).

Treatments	Shoots (g)	Roots (g)	Shoots+Roots (g)
T1	18.4 ± 1.15^{a}	11.9 ± 0.68^{a}	30.3 ± 0.49^{a}
T2	18.7 ± 0.93^{a}	12.0 ± 0.39^{a}	30.7 ± 0.98^{a}
T3	16.2 ± 1.17^{b}	10.7 ± 0.37^{b}	26.9 ± 1.54^{bc}
T4	17.2 ± 1.31^{ab}	11.3 ± 0.74^{ab}	28.5 ± 1.62^{ab}
T5	13.1 ± 0.92^{c}	8.88 ± 0.56^{c}	22.0 ± 1.47^{d}
T6	15.3 ± 1.34^{b}	10.5 ± 0.45^{b}	25.9 ± 1.10^{c}

Treatments: T1, distilled water without SA; T2, distilled water with 0.5 mmol SA; T3, 0.2% NaCl without SA; T4, 0.2% NaCl with 0.5 mmol SA; T5, 0.4% NaCl without SA; and T6, 0.4% NaCl with 0.5 mmol SA. Numbers followed by different letters indicate significant differences ($P<0.05$) according to an LSD test; the same letter indicates no significant differences between the treatments, n = 5.

Relative water content and proline content of *T. grandis* under different salt stress and SA treatments

The RWC is a useful parameter when evaluating the physiological water status of plants [24]. Compared to the distilled water-treated seedlings (T1), the RWC in the *T. grandis* seedling leaves decreased by 4.50% ($P<0.05$) under 0.2% salt stress conditions and 16.70% ($P<0.05$) under 0.4% salt stress conditions. The 30-day treatment with 0.5 mmol SA reduced the effects of salt stress on the relative water content decrease in the presence of 0% and 0.2% NaCl; however, these changes were not significant ($P>0.05$). Under 0.4% NaCl stress conditions, the plants that were treated with SA had a higher RWC than those without SA treatment (Fig. 2A, $P<0.05$). The salt stress dramatically induced the accumulation of proline in leaves. SA treatment further increased the proline content. In the SA treatments, the proline content in the 0.2% and 0.4% salt-stressed plants increased by 27.6% ($P<0.05$) and 21.8% ($P<0.05$), respectively (Fig. 2B).

Gas exchange parameters in *T. grandis* under different salt and SA treatments

As shown in Table 2, salt stress reduced the gas exchange parameters, with decreases in the net assimilation rate (Pn), stomatal conductance (Gs), and transpiration rate (Tr) of the leaves of 16.2% ($P<0.05$), 50% ($P<0.05$), and 30.2% ($P<0.05$), respectively, under 0.2% NaCl conditions and 65.0% ($P<0.05$), 50% ($P<0.05$), and 59.3% ($P<0.05$) under 0.4% NaCl condition, respectively. Interestingly, the intercellular CO_2 (Ci) did not change significantly with 0.2% NaCl ($P>0.05$) but was markedly increased with 0.4% NaCl ($P<0.05$). The SA treatment increased the Pn, Tr, and Gs in leaves under salt stress. The Pn, Tr, and Gs in SA-treated plants increased by approximately 65%, 25%, and 100%, respectively, under 0.2% NaCl conditions and 146%, 95%, and 100%, respectively, under 0.4% NaCl conditions compared with the non SA-treated plants. The Ci in the SA-treated plants under 0.2% NaCl conditions was similar to that in the non SA-treated plants ($P>0.05$). However, the Ci in the SA-treated plants under 0.4% NaCl stress was markedly lower (22%) than that of the plants without SA treatment ($P<0.05$). Additionally, there were no significant differences in the Pn, Tr, Gs, and Ci values between the SA- and distilled water-treated plants.

Contents of chlorophyll and soluble protein of *T. grandis* under different salt stress and SA treatments

Chlorophyll is a biologically important pigment that is used for the photosynthetic conversion of inorganic molecules or ions into organic bio-molecules. Salt stress caused a decrease in the chl(a+b) content compared with that of the non-salt-treated seedlings. The contents of chl(a+b) decreased by 9.9% under 0.2% saline conditions and 24.1% under 0.4% saline conditions app:addword:respectively (Fig. 3A). The foliar spray of SA under salt stress conditions significantly reduced the decreases in the contents of chl(a+b) by approximately 10% ($P<0.05$) under 0.2% saline conditions and 17% ($P<0.05$) under 0.4% saline conditions compared that of the non SA-treated plants (Fig. 3A). In contrast to the change in the chlorophyll contents under saline condition, the salt stress increased the total soluble protein content. The 0.2% NaCl treatment significantly increased the soluble protein content ($P<0.05$); however, no obvious increase was found between the 0.4% NaCl- and non NaCl-stressed plants ($P>0.05$) (Fig. 3B). The addition of the SA solution significantly increased the soluble protein contents under NaCl conditions, and the soluble protein content in the SA-treated plants increased by 13.3% ($P<0.05$) under 0.2% saline conditions and 15.9% ($P<0.05$) under 0.4% saline conditions (Fig. 3B). However, the SA treatment did not affect the contents of chl(a+b) and soluble protein compared with those of the SA- and distilled water-treated seedlings ($P>0.05$).

Activities of antioxidative enzymes and cellular membrane damage of *T. grandis* under different salt stress and SA treatments

The activities of SOD, CAT, and POD in *T. grandis* seedlings were significantly affected by the salt stress and SA treatments. With the increasing extent of salt stress, the activities of SOD, CAT, and POD increased markedly and then significantly

T1 T2 T3 T4 T5 T6

Figure 1. The appearance of whole plants in *T. grandis* seedlings. Treatments: T1, distilled water without SA; T2, distilled water with 0.5 mmol SA; T3, 0.2% NaCl without SA; T4, 0.2% NaCl with 0.5 mmol SA; T5, 0.4% NaCl without SA; and T6, 0.4% NaCl with 0.5 mmol SA.

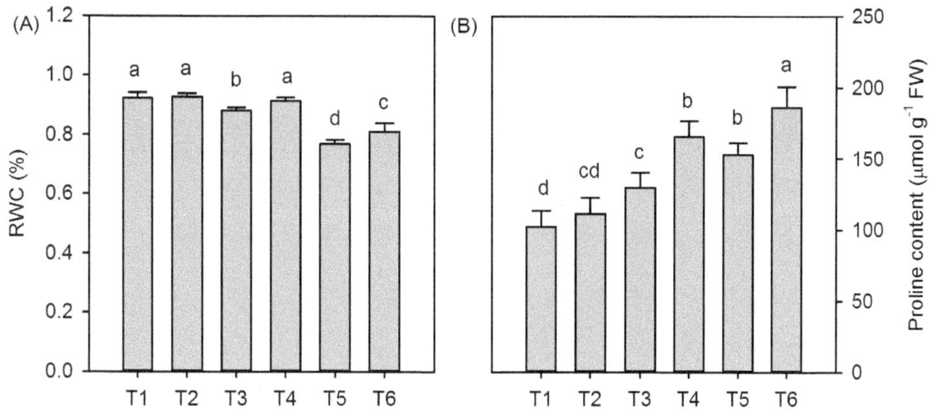

Figure 2. Effects of SA on the RWC (A) and proline content (B) in *T. grandis* grown under salt stress conditions (means ± SD). Treatments: T1, distilled water without SA; T2, distilled water with 0.5 mmol SA; T3, 0.2% NaCl without SA; T4, 0.2% NaCl with 0.5 mmol SA; T5, 0.4% NaCl without SA; and T6, 0.4% NaCl with 0.5 mmol SA. Different letters indicate significant differences ($P<0.05$) according to an LSD test; the same letter indicates no significant differences between the treatments, n = 5.

decreased (Fig. 4). The SA treatment significantly enhanced the activities of SOD, CAT, and POD under salt stress, and the increases in the activities of SOD, CAT, and POD were approximately 9.8%, 20.6%, and 14.4% under 0.2% NaCl conditions and 13.4%, 38.2%, and 18.9% at 0.4% under NaCl conditions, respectively (Fig. 4).

Salt stress caused membrane injury in the leaves of *T. grandis* seedlings (Fig. 5). The relative electrolyte leakage rate (REC) increased by 6.9% and 17.4% under 0.2% and 0.4% NaCl conditions, respectively, and the MDA content increased by 13.5% and 43.3% under 0.2% and 0.4% NaCl conditions, respectively. The SA treatments prevented lipid peroxidation and then alleviated the membrane injury of *T. grandis* seedlings. Compared with the non-SA-treated plants, the reductions in the REC and MDA content in the SA-treated seedlings were approximately 9.0% ($P<0.05$) and 12.6% ($P<0.05$) under 0.2% NaCl conditions and 12.0% ($P<0.05$) and 26.4% ($P<0.05$) under 0.4% NaCl conditions, respectively. Interestingly, no significant difference was found between the REC and MDA content between the SA- and distilled water-treated seedlings ($P>0.05$).

Relationships between antioxidant enzymatic activities and cellular membrane damage

MDA content shows linear and negative correlation with SOD and POD activities (Fig. 6A, C, $P<0.05$) likewise, REC content also shows negative correlation with SOD and POD activities (Fig. 6D, F, $P<0.05$). However, there were no significant negative relationships between MDA content and CAT activity, and also between membrane permeability and CAT activity, respectively (Fig. 6B, E, $P>0.05$).

Discussion

Salt stress constrains plant growth by adversely affecting various physiological and biochemical processes, such as photosynthesis, antioxidant phenomena, proline metabolism, and osmolyte accumulation [30,31,32,33]. Given that SA plays key roles in the regulation of plant growth, development, the interaction with other organisms, and the responses to environmental stresses [34,35]. However, there is some controversy in the literature regarding whether SA can counteract the adverse effects of salt stress [10,30]. In this study, a marked reduction in the growth of both the shoots and roots of *T. grandis* seedlings was observed under increasing salt stress (Fig. 1). These studies show that SA treatment recovered the growth rate of *T. grandis* seedlings from high salt conditions (0.4% NaCl), when compared with their corresponding non-SA applied plants (Table 1). This is in consistent with the reports of Idrees et al. [31] where it was shown that SA treatment ameliorated the adverse effects of salt stress in terms of growth parameters in *Catharanthus roseus*. Coronado et al. [36] also showed that applying aqueous solutions

Table 2. Effects of SA treatments on the net photosynthetic rate (Pn), internal carbon dioxide concentration (Ci), transpiration rate (Tr), and stomatal conductance (Gs) in *T. grandis* grown under salt stress conditions.

Treatments	Pn (μmol m^{-2}s^{-1})	Ci (μmol mol^{-1})	Tr (μmol m^{-2}s^{-1})	Gs (μmol m^{-2}s^{-1})
T1	1.44 ± 0.14^{bc}	251.8 ± 15.8^{ab}	1.37 ± 0.098^{a}	0.02 ± 0.002^{a}
T2	1.50 ± 0.11^{b}	257.5 ± 10.7^{ab}	1.43 ± 0.086^{a}	0.02 ± 0.016^{a}
T3	1.20 ± 0.12^{d}	230.5 ± 25.4^{bc}	0.96 ± 0.068^{c}	0.01 ± 0.001^{c}
T4	1.98 ± 0.12^{a}	243.0 ± 16.0^{bc}	1.20 ± 0.129^{b}	0.02 ± 0.001^{a}
T5	0.50 ± 0.05^{e}	276.2 ± 22.5^{a}	0.56 ± 0.035^{d}	0.01 ± 0.001^{c}
T6	1.23 ± 0.15^{cd}	214.7 ± 14.7^{c}	1.09 ± 0.108^{bc}	0.02 ± 0.002^{b}

Treatments: T1, distilled water without SA; T2, distilled water with 0.5 mmol SA; T3, 0.2% NaCl without SA; T4, 0.2% NaCl with 0.5 mmol SA; T5, 0.4% NaCl without SA; and T6, 0.4% NaCl with 0.5 mmol SA. Numbers followed by different letters indicate significant differences ($P<0.05$) according to an LSD test, n = 5.

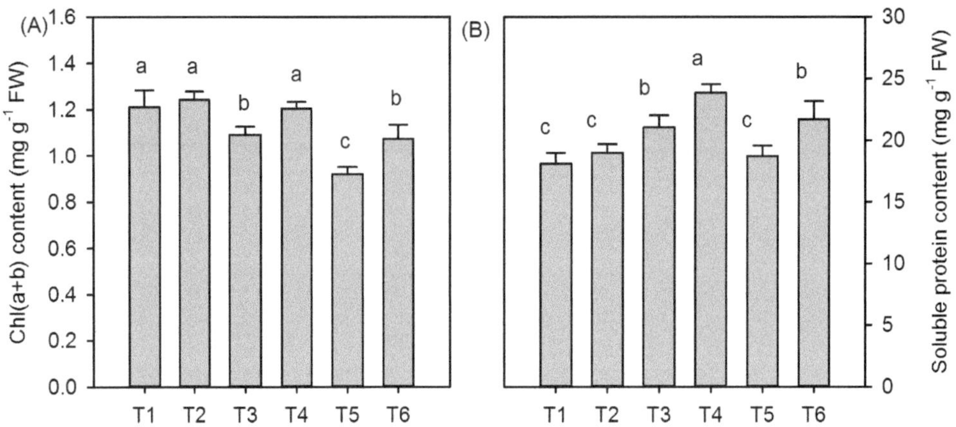

Figure 3. Effects of SA on the chl (a+b) (A) and soluble protein contents (B) in *T. grandis* grown under salt stress conditions (means ± SD). Treatments: T1, distilled water without SA; T2, distilled water with 0.5 mmol SA; T3, 0.2% NaCl without SA; T4, 0.2% NaCl with 0.5 mmol SA; T5, 0.4% NaCl without SA; and T6, 0.4% NaCl with 0.5 mmol SA. Different letters indicate significant differences ($P<0.05$) according to an LSD test; the same letter indicates no significant differences between the treatments, n = 5.

of SA as a spray solution to the shoots of soybean significantly increase the growth of the shoots and roots under greenhouse or field conditions. Interestingly, no significant differences were found in the dry mass of the shoots and roots between the plants with and without SA treatment under 0% NaCl stress conditions (Table 1), suggesting that exogenous SA did not positively affect the growth of *T. grandis* seedlings that were grown under non-saline and low-saline conditions. This result was not consistent with the report of Kaydan et al. [37], who found that a pre-sowing soaking treatment of the seeds that were treated with SA positively affected the shoot and root dry mass in wheat seedlings under both saline and non-saline conditions.

Growing leaf cells can be maintained through the accumulation of solutes. Proline accumulation is one of the most frequently reported adjustments that is induced by salt stress in plants and is often considered to be involved in maintaining the water content. Increases in the leaf RWC and accumulation of free proline in *T. grandis* seedlings after 30 days of salt stress under SA treatment might be an adaptive feature in improving its succulence and maintaining the water balance in response to salinity-induced osmotic stress (Fig. 2). A similar result was reported by Misra and Saxena [32], who found that SA alleviated the salt stress of lentil by improving the proline metabolizing system. Moreover, proline enhances the salt-tolerance of *Pancratium maritimum* L. by up-regulating stress-protective proteins [38]. The application of SA induced an increase in the protein content under both saline stress conditions (Fig. 3B). Therefore, it has been concluded that SA ameliorates the stress generated by NaCl through proline accumulation, which was involved in the synthesis of protective proteins that are necessary for the stress response. However, to test this hypothesis, further studies are necessary to determine which stress-protective protein would increase under a stress environment.

Stomatal closure is one of the earliest responses to salt stress. Compared to the non-salt-treated *T. grandis* seedlings, both the Pn and Gs in *T. grandis* seedlings under 0.2% NaCl conditions decreased by approximately 16% and 39%, respectively, suggesting that the decrease in the photosynthetic rate was associated with the closure of stomata under 0.2% NaCl stress. SA alleviated the negative effects of 0.2% NaCl stress on leaf photosynthesis by increasing Pn and Gs. Fariduddin et al. [33] also found that the exogenous application of SA could enhance the net photosynthetic rate, stomatal conductance, and transpiration rate in *B. juncea*. However, non-stomatal factors, such as enzyme-related activities at the chloroplast level, the efficiency of PSII, the concentration and activity of the Rubisco enzyme, and the supply of ATP and NADPH to the photosynthetic carbon reduction cycle might also affect the Pn in the leaves under stressful conditions [39]. Furthermore, Brugnoli and Bjökman [40] noted that both stomatal and non-stomatal limitations accounted for a reduction in photosynthesis. In this study, compared with the 0.2% NaCl treatment, the Pn further decreased, although the Gs did not obviously change in *T. grandis* seedlings under 0.4% NaCl conditions, suggesting that other non-stomatal limiting factors might affect photosynthesis under 0.4% NaCl conditions. The enhancement of the effects of SA on the Pn was attributable to its stimulatory effects on the pigment contents and Rubisco enzyme activity [41]. In this study, the SA-treated seedling leaves exhibited higher Pn and lower Ci contents, consistent with higher levels of Gs, chlorophyll, and soluble proteins than those of the 0.4% NaCl-treated plants that were treated with SA (Table 2; Fig. 3A,B). These results indicate that the application of SA induced the severely salt-stressed *T. grandis* seedlings to effectively use the available CO_2 inside the leaves and increased the Pn by up-regulating the related photosynthetic enzyme activities at the chloroplast level.

The photosynthetic electron transport system is a major source of ROS in plant tissues and has the potential to generate singlet oxygen $^1O_2$ and superoxide ($O_2^{\cdot-}$). Consequently, the accumulation of reduced intermediates in the photosynthetic electron transport chain results in the formation of ROS, thereby damaging the membrane system and photosynthetic complexes [42]. Plants can activate antioxidative defense systems to protect themselves from the harmful effects of salinity-induced oxidative stress. In this study, salinity stress induced drastic changes in the activities of antioxidative enzymes, including SOD, CAT, and POD. There were dramatic increases in the activities of SOD, CAT, and POD under 0.2% salt stress conditions, although there was only a slight increase under 0.4% salt stress conditions compared to the distilled water-treated seedlings (Fig. 3), suggesting that with increasing salt stress, the ability of *T. grandis* seedlings to resist salt stress decreased. SOD and CAT, which act synergistically to resist oxidative damage induced by salt stress in *T. grandis* seedlings, were also observed in other plant tissues under NaCl stress [43].

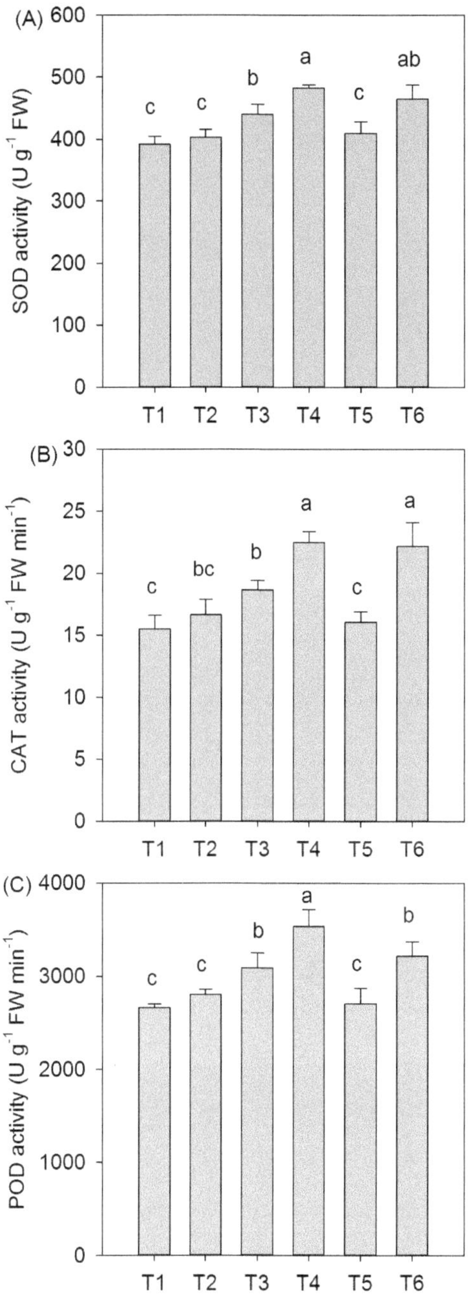

Figure 4. Influence of the SA treatments on the SOD (A), CAT (B), and POD (C) activities in *T. grandis* grown under salt stress conditions. Treatments: T1, distilled water without SA; T2, distilled water with 0.5 mmol SA; T3, 0.2% NaCl without SA; T4, 0.2% NaCl with 0.5 mmol SA; T5, 0.4% NaCl without SA; and T6, 0.4% NaCl with 0.5 mmol SA. Different letters indicate significant differences (*P<0.05*) according to an LSD test, n = 5.

After the application of exogenous SA, the plants showed higher activities of SOD, CAT, and POD than those of the non-SA-treated plants grown under 0.2% and 0.4% salt stress conditions (Fig. 4). Similarly, SA enhanced the activities of CAT, POD, and SOD when sprayed exogenously on salinity-stressed *B. juncea* plants [44].

One of the major injuries due to salinity stress is the damage of plant cell membranes, which can impair their ion selectivity. This deficiency results in the exosmosis of intracellular ions and organic matter and the infiltration of extracellular toxic ions, ultimately disrupting plant physiological and biochemical processes. The maintenance of cellular membrane integrity under salt stress is considered to be an integral part of the salinity tolerance mechanism [45]. The membrane damage caused by different abiotic stresses, including salinity, is largely mediated through the end product of lipid peroxidation (MDA) [46]. In this study, a significant positive correlation (r = 0.98, p = 0.0029) was observed between the MDA concentration and membrane permeability. An increase in the electrical conductivity with increasing salt stress indicated an elevated leakiness of ions due to a loss of membrane integrity (Fig. 4), which is in agreement with results described by Parida and Das [47]. The application of exogenous SA could ameliorate the membrane deterioration in plants under salt stress and facilitate the maintenance of membrane functions [7,45]. We also found that the SA treatment significantly reduced the increase in the electrolyte leakage rate and MDA content in *T. grandis* seedlings under salt stress (Fig. 5).

The photosynthetic pigment chlorophyll plays an important role in light absorption and energy transduction and is essential for photosynthesis. The reduction of leaf chlorophyll under high salinity has been attributed to the destruction of pigments and the instability of the pigment-protein complex [48]. In this study, NaCl treatment reduced the chlorophyll content of *T. grandis* seedlings, and there was a significantly positive and linear relationship between Pn and the chlorophyll content (r = 0.96, p = 0.0099) (Fig. 3A). The addition of SA to NaCl-stressed plants markedly increased the chlorophyll content, suggesting that the increase in chloroplast pigments under the SA treatment might be due to the ability of SA to increase the activity of certain enzymes, thereby stimulating chlorophyll biosynthesis or reducing chlorophyll degradation, leading to increased net photosynthesis under salt stress tolerance. The reduction in chlorophyll due to osmotic stress has been ascribed to the strong damage to chloroplast membranes, which increases the membrane permeability or loss of membrane integrity [49]. Indeed, a strong negative correlation between the chlorophyll and MDA contents (r = 0.99, *p* = 0.001) was observed in this study. Membrane oxidative damage can be prevented by antioxidative enzymes. Correlation studies were demonstrated for MDA content and REC content against the enzymatic activities of SOD, CAT and POD in *T. grandis* seedlings grown under 0.2% and 0.4% salt stress conditions. We found that both MDA content and membrane permeability shows significant negative correlation against the activities of SOD and POD (Fig. 6A,C,D,F, $P<0.05$), while there is no significant negative correlation between MDA content and CAT activity, and also between membrane permeability and CAT activity (Fig. 6B,E, $P>0.05$). Therefore, the observed increase in the growth and CO_2 assimilation under salt stress and SA treatment might be due to the enhanced activities of SOD and POD, which protect membrane integrity, leading to an increase in the amount of photosynthesis pigments and net photosynthesis rate.

In conclusion, our study showed that salt stress decrease the vegetative growth and CO_2 assimilation in *T.grandis* together with increased membrane damage and lipid peroxidation. SA efficiently ameliorated the negative effects of salt stress over the growth and photosynthesis of *T. grandis* by increasing the chlorophyll content and also enhancing the plants antioxidant enzyme mechanisms, thus alleviate the membrane oxidative damage from salinity. To our knowledge, this is the first report of salt tolerance mechanisms in *T. grandis* seedlings. The results of this study can also provide a useful guide for the conservation and large-scale plantation of *T. grandis*.

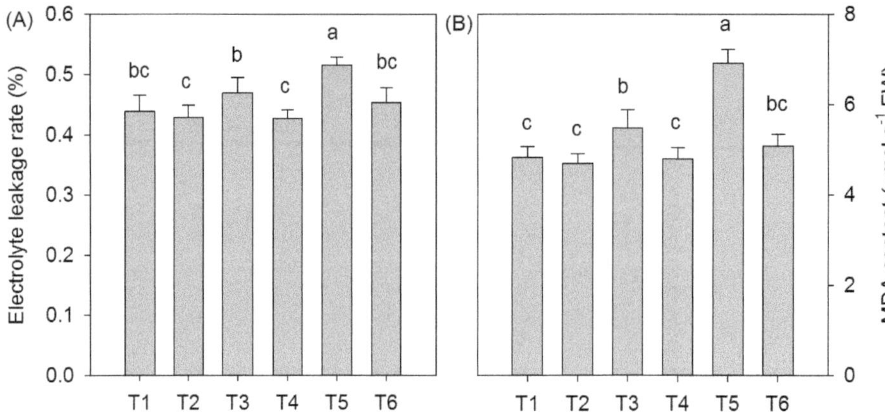

Figure 5. Effects of the SA treatments on the electrolyte leakage rate (A) and MDA contents (B) in *T. grandis* seedlings grown under salt stress conditions. Treatments: T1, distilled water without SA; T2, distilled water with 0.5 mmol SA; T3, 0.2% NaCl without SA; T4, 0.2% NaCl with 0.5 mmol SA; T5, 0.4% NaCl without SA; and T6, 0.4% NaCl with 0.5 mmol SA. Different letters indicate significant differences ($P<0.05$) according to an LSD test, n=5.

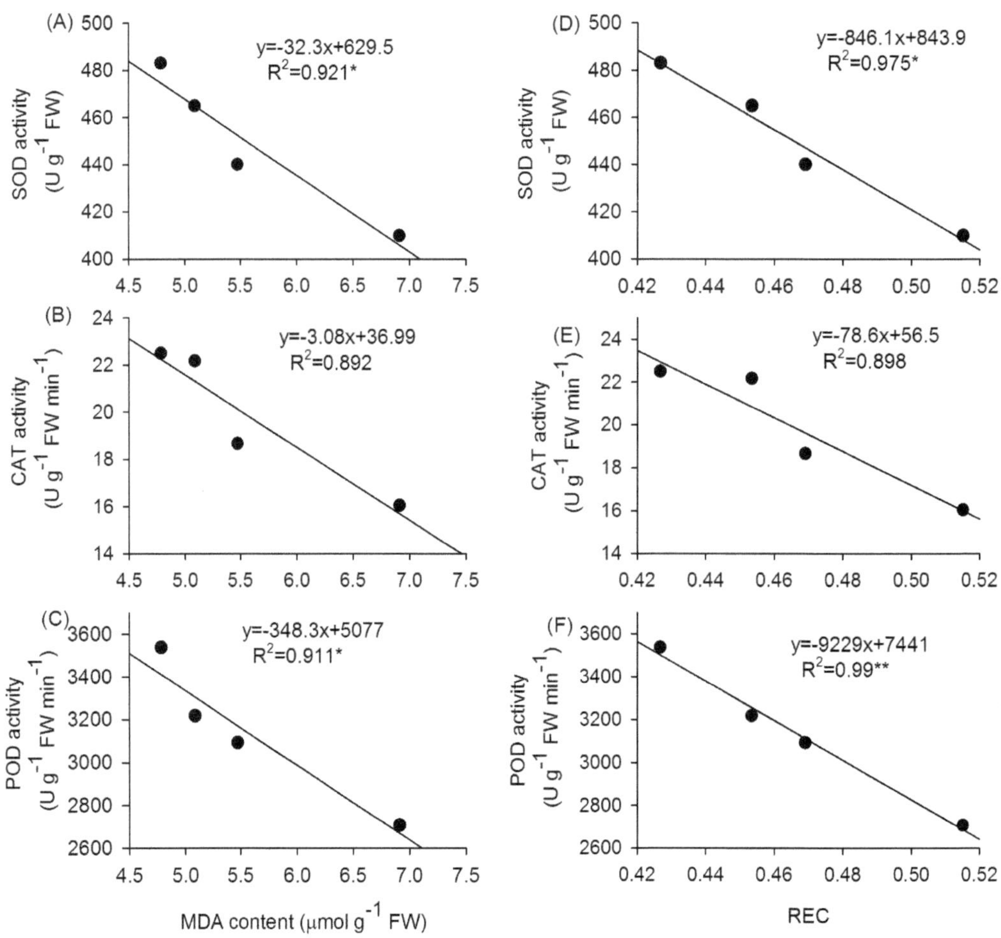

Figure 6. Relationship between MDA content and activities of SOD (A), CAT (B) and POD (C), and between REC and activities of SOD (D), CAT (E) and POD (F), respectively in *T. grandis* seedlings grown under 0.2% and 0.4% salt stress conditions. Each point represents the mean of 5 seedlings. *and **Significant at $P<0.05$ and $P<0.01$, respectively.

Acknowledgments

This work was funded by the Fruit Innovation Team Project of Zhejiang Province (2009R50033-7) and the Major Project of National Spark Plan of China (2012GA700001).

Author Contributions

Conceived and designed the experiments: JSW. Performed the experiments: TTL YYH XHD HT CHS. Analyzed the data: TTL YYH. Contributed reagents/materials/analysis tools: TTL YYH JSW. Wrote the paper: TTL YYH JSW.

References

1. Allakhverdiev SI, Sakamoto A, Nishiyama Y, Inaba M, Murata N (2000) Ionic and osmotic effects of NaCl-induced inactivation of photosystems I and II in *Synechococcus* sp. Plant Physiol 123: 1047–1056.
2. Iqbal M, Ashraf M, Jamil A, Rehman S (2006) Does seed priming induce changes in the levels of some endogenous plant hormones in hexaploid wheat plants under salt stress? Journal of Integrative Plant Biol 48: 181–189.
3. Saleh B (2012) Salt stress alters physiological indicators in cotton (Gossypium hirsutum L.). Soil Environ 31(2): 113–118.
4. He Y, Yu CL, Zhou L, Chen Y, Liu A, et al. (2014) Rubisco decrease is involved in chloroplast protrusion and Rubisco-containing body formation in soybean (*Glycine max.*) under salt stress. Plant Physiol Biochem 74: 118–124.
5. Nazar R, Iqbal N, Syeed S, Khan NA (2011) Salicylic acid alleviates decreases in photosynthesis under salt stress by enhancing nitrogen and sulfur assimilation and antioxidant metabolism differentially in two mungbean cultivars. J plant physiol 168(8): 807–815.
6. Sudhir P, Murthy SDS (2004) Effects of salt stress on basic processes of photosynthesis. Photosynthetica 42 (2): 481–486.
7. Gunes A, Inal A, Alpaslan M, Eraslan F, Bagci EG, Cicek N (2007) Salicylic acid induced changes on some physiological parameters symptomatic for oxidative stress and mineral nutrition in maize (*Zea mays* L.) grown under salinity. J Plant Physiol 164: 728–736.
8. Steduto P, Albrizio R, Giorio P, Sorrentino G (2000) Gas exchange response and stomatal and non-stomatal limitations to carbon assimilation of sunflower under salinity. Environ Exp Bot 44: 243–55.
9. Gossett DR, Millhollon EP, Lucas MC (1994) Antioxidant Response to NaCl Stress in Salt-Tolerant and Salt-Sensitive Cultivars of Cotton. Crop Sci 34: 706–714.
10. Arfan M, Athar HR, Ashraf M (2007) Does exogenous application of salicylic acid through the rooting medium modulate growth and photosynthetic capacity in two differently adapted spring wheat cultivars under salt stress? J Plant Physiol 6: 685–694.
11. Syeed S, Anjum NA, Nazar R, Iqbal N, Masood A et al. (2011) Salicylic acid-mediated changes in photosynthesis, nutrients content and antioxidant metabolism in two mustard (*Brassica juncea* L.) cultivars differing in salt tolerance. Acta Physiol Plant 33 (3): 877–886.
12. Wang LJ, Fan L, Loescher W, Duan W, Liu GJ et al. (2010) Salicylic acid alleviates decreases in photosynthesis under heat stress and accelerates recovery in grapevine leaves. BMC Plant Biol 10: 34.
13. El-Tayeb MA (2005) Response of barley grains to the interactive effect of salinity and salicylic acid. Plant Growth Regul 45: 215–24.
14. Khan MIR, Asgher M, Khan NA (2014) Alleviation of salt-induced photosynthesis and growth inhibition by salicylic acid involves glycinebetaine and ethylene in mungbean (*Vigna radiata* L.). Plant Physiol Biochem 80: 67–74.
15. Zhou XR, Zhang W, He XJ, Zhou SD, Yu Y (2012) Resources and Conservation of Torreya fargesii. J Northeast For Univ 40 (2): 42–46.
16. Huang Y, Wang J, Li G, Zheng Z, Su W (2001) Antitumor and antifungal activities in endophytic fungi isolated from pharmaceutical plants Taxus mairei, Cephalataxus fortunei and Torreya grandis. FEMS Immunol Med Mic 31: 163–167.
17. Liu AR, Zhang YB, Wang JF, Jiao M, Liu XM (2013) Appropriate nitrogen fertilizer strengthens growth and antioxidative ability of Festuca arundinacea under salt stress. Trans Chin Soc Agric Eng 29 (15), 126–135. (in Chinese).
18. Chen BQ, Cui XY, Zhao X, Zhang YH, Piao HS et al. (2006) Antioxidative and acute antiinflammatory effects of *Torreya grandis*. Fitoterapia 77:262–267.
19. Cheng XJ, Li ZJ, Yu WW, Dai WS, Fu QG (2007) Distribution and ecological characteristics of *Torreya grandis* in China. Journal of Zhejiang Forestry College 24: 383–388. (In Chinese).
20. Saeed MK, Deng Y, Parveen Z, Dai R, Ahmad W et al. (2007) Studies on the chemical constituents of *Torreya grandis* Fort. Ex Lindl. J App Sci 7: 269–273.
21. Wolff RL, Pédrono F, Marpeau AM, Christie WW, Gunstone FD (1998) The seed fatty acid composition and the distribution of Δ5-olefinic acids in the triacylglycerols of some taxaceae (*Taxus and Torreya*). J Am Oil Chem Soc 75: 1637–1641.
22. Bates LS, Waldren RP, Teare ID (1973) Rapid determination of proline for water stress studies. Plant Soil 39: 205–207.
23. Lichtenthaler HK (1987) Chlorophylls and carotenoids: Pigments of photosynthetic biomembranes. Method enzymol 148: 350–382.
24. Bradford MM (1976) A rapid and sensitive method for the quantitation of microgram quantities of protein utilizing the principle of protein-dye binding. Anal biochem 72 (1): 248–254.
25. Shi QH, Bao ZY, Zhu ZJ, Ying QS, Qian QQ (2006) Effects of different treatments of salicylic acid on heat tolerance, chlorophyll fluorescence, and antioxidant enzyme activity in seedlings of Cucumis sativa L. Plant Growth Regul 48: 127–135.
26. Zheng B, Jing S, Yan Y (2006) Research methodology of modern plant physiology and biochemistry. China Meteorological Press, Beijing, pp 189–198.
27. Costa H, Gallego SM, Tomaro ML (2002) Effect of UV-B radiation on antioxidant defense system in sunflower cotyledons. Plant Sci 162(6): 939–945.
28. Fu J, Huang B (2001) Involvement of antioxidants and lipid peroxidation in the adaptation of two cool-season grasses to localized drought stress. Environ Exp Bot. 45: 105–114.
29. González L, González-Vilar M (2003) Determination of relative water content. In: Handbook of plant ecophysiology techniques. Springer Netherlands, pp 207–212.
30. Borsani O, Valpuesta V, Botella MA (2001) Evidence for a role of salicylic acid in the oxidative damage generated by NaCl and osmotic stress in *Arabidopsis* seedlings. Plant Physiol 126: 1024–1030.
31. Idrees M, Naeem M, Aftab T, Khan MMA, Moinuddin (2011) Salicylic acid mitigates salinity stress by improving antioxidant defense system and enhances vincristine and vinblastine alkaloids production in periwinkle (*Catharanthus roseus* L. G. Don). Acta Physiol Plant 33:987–999.
32. Misra N, Saxena P (2009) Effect of salicylic acid on proline metabolism in lentil grown under salinity stress. Plant Sci 177: 181–189.
33. Fariduddin Q, Hayat S, Ahmad A (2003) Salicylic acid influences net photosynthetic rate, carboxylation efficiency, nitrate reductase activity and seed yield in Brassica juncea. Photosynthetica 41: 281–284.
34. Hayat Q, Hayat S, Irfan M, Ahmad A (2010) Effect of exogenous salicylic acid under changing environment: A review. Environmental and Experimental Botany 68: 14–25.
35. Senaratna T, Touchell D, Bunn E, Dixon K (2000) Acetyl salicylic acid (aspirin) and salicylic acid induce multiple stress tolerance in bean and tomato plants. Plant Growth Regul 30: 157–161.
36. Coronado MAG, Lopez CT, Saavedra AL (1998) Effects of salicylic acid on the growth of roots and shoots in soybean, Plant Physiol Biochem 8: 563–565.
37. Kaydan D, Yagmur M, Okut N (2007) Effects of Salicylic acid on the growth and some physiological characters in salt stressed wheat (*Triticum aestivum* L.). Tarim Bilim Derg 13 (2): 114–119.
38. Khedr AHA, Abbas MA, Wahid AAA, Quick WP, Abogadallah GM (2003) Proline induces the expression of salt-stress-responsive proteins and may improve the adaptation of *Pancratium maritimum* L. to salt-stress. J Exp Bot 54 (392): 2553–2562.
39. Lawlor DW, Cornic G (2002) Photosynthetic carbon assimilation and associated metabolism in relation to water deficits in higher plants. Plant Cell Environ 25: 275–94.
40. Brugnoli E, Björkman O (1992) Growth of cotton under continuous salinity: influence on allocation pattern, stomatal and light energy. Planta 187: 335–347.
41. Khodary SEA (2004) Effect of salicylic acid on the growth, photosynthesis and carbohydrate metabolism in salt-stressed maize plants. Inter J Agric Biol 6(1): 5–8.
42. Asada K (2006) Production and scavenging of reactive oxygen species in chloroplasts and their functions. Plant Physiol 141: 391–396.
43. Tang D, Shi S, Li D, Hu C, Liu Y (2007). Physiological and biochemical responses of Scytonema javanicum (*cyanobacterium*) to salt stress. J Arid Environ 71(3): 312–320.
44. Yusuf M, Hasan SA, Ali B, Hayat S, Fariduddin Q et al. (2008) Effect of salicylic acid on salinity-induced changes in Brassica juncea. J Integr Plant Biol 50: 1096–1102.
45. Stevens J, Senaratna T, Sivasithamparam K (2006) Salicylic acid induces salinity tolerance in tomato (*Lycopersicon esculentum cv. 'Roma'*): associated changes in gas exchange, water relations and membrane stabilisation. Plant Growth Regul 49: 77–83.
46. Mishra A, Choudhuri MA (1999) Effects of salicylic acid on heavy metal-induced membrane deterioration mediated by lipoxygenase in rice. Biol Plantarum 42: 409–15.
47. Parida AK, Das AB (2005) Salt tolerance and salinity effects on plants: a review. Ecotox Environ Safe 60: 324–349.
48. Jaleel CA, Sankar B, Sridharan R, Panneerselvam R (2009) Soil salinity alters growth, chlorophyll content, and secondary metabolite accumulation in *Catharanthus roseus*. Turk J Biol 32(2): 79–83.
49. Kaiser WM, Heber U (1981) Photosynthesis under osmotic stress: effect of high solute concentrations on the permeability properties of the chloroplast envelope and on activity of stroma enzymes. Planta 153: 423–429.

The Diurnal Logic of the Expression of the Chloroplast Genome in *Chlamydomonas reinhardtii*

Adam D. Idoine, Alix Boulouis, Jens Rupprecht, Ralph Bock*

Max-Planck-Institut für Molekulare Pflanzenphysiologie, Potsdam, Golm, Germany

Abstract

Chloroplasts are derived from cyanobacteria and have retained a bacterial-type genome and gene expression machinery. The chloroplast genome encodes many of the core components of the photosynthetic apparatus in the thylakoid membranes. To avoid photooxidative damage and production of harmful reactive oxygen species (ROS) by incompletely assembled thylakoid protein complexes, chloroplast gene expression must be tightly regulated and co-ordinated with gene expression in the nucleus. Little is known about the control of chloroplast gene expression at the genome-wide level in response to internal rhythms and external cues. To obtain a comprehensive picture of organelle transcript levels in the unicellular model alga *Chlamydomonas reinhardtii* in diurnal conditions, a qRT-PCR platform was developed and used to quantify 68 chloroplast, 21 mitochondrial as well as 71 nuclear transcripts in cells grown in highly controlled 12 h light/12 h dark cycles. Interestingly, in anticipation of dusk, chloroplast transcripts from genes involved in transcription reached peak levels first, followed by transcripts from genes involved in translation, and finally photosynthesis gene transcripts. This pattern matches perfectly the theoretical demands of a cell "waking up" from the night. A similar trend was observed in the nuclear transcripts. These results suggest a striking internal logic in the expression of the chloroplast genome and a previously unappreciated complexity in the regulation of chloroplast genes.

Editor: Frederik Börnke, Leibniz-Institute for Vegetable and Ornamental Plants, Germany

Funding: This work was supported by grants from the BMBF (Systems Biology Initiative FORSYS: GoFORSYS) and the Deutsche Forschungsgemeinschaft (FOR2092, BO 1482/17-1) to RB. The funders had no role in study design, data collection and analysis, decision to publish, or preparation of the manuscript.

Competing Interests: The authors have declared that no competing interests exist.

* Email: rbock@mpimp-golm.mpg.de

Introduction

The unicellular green alga *Chlamydomonas reinhardtii* has become a model organism for many research areas, including chloroplast biology and photosynthesis research [1], [2], [3], [4], [5]. Moreover, its capacity to produce molecular hydrogen and to synthesize and accumulate considerable amounts of oil also have made *C. reinhardtii* an attractive target organism of many efforts to develop algae as an efficient and sustainable production platform for biofuels [6], [7], [8], [9]. Since all algal biofuels and renewable green chemical feedstocks are directly or indirectly derived from photosynthesis and many of the core components of the photosynthetic apparatus are encoded in the chloroplast (plastid) DNA, a comprehensive understanding of the mechanisms that govern expression of the chloroplast genome is of great importance to all future attempts to rationally engineer primary metabolism. However, our knowledge about the regulatory mechanisms operating in the control of chloroplast gene expression in response to endogenous cues and environmental stimuli and the signaling pathways co-ordinating expression of the chloroplast genome with that of the nuclear genome is still far from complete.

The chloroplast genome of *C. reinhardtii* is 204 kb in size and contains 99 genes [10]. The vast majority of these genes belong to the two main gene classes found in plastid genomes of algae and embryophyte plants: photosynthesis genes and genetic system genes [11]. The photosynthesis-related genes comprise genes for subunits of the major multiprotein complexes in the thylakoid membrane that participate in photosynthetic electron transfer and ATP generation (photosystem II, cytochrome b_6f complex, photosystem I, ATP synthase). The genetic system genes encode components of the transcriptional apparatus (subunits of the plastid-encoded RNA polymerase, PEP) and the translational machinery (rRNAs, tRNAs, ribosomal proteins).

Genes in the chloroplast genome of *C. reinhardtii* are transcribed by a multisubunit RNA polymerase that resembles bacterial RNA polymerases of the $\alpha_2\beta\beta'$ type. All of its core subunits are encoded in the chloroplast DNA and, for this reason, the enzyme has been dubbed PEP (for plastid-encoded RNA polymerase). In *Chlamydomonas*, there appears to be no nucleus-encoded plastid RNA polymerase that has been shown to be additionally present in plastids of embryophyte plants [12], [13]. The PEP core complex interacts with the single sigma factor found in the nuclear genome of *C. reinhardtii* (RPOD) to confer sequence specificity in promoter recognition [14], [15]. TATA-like -10 box sequences (and, in some cases, the extended -10 box motif TATAATAT) are found upstream of many reading frames in the *C. reinhardtii* plastid DNA and are thought to act as key promoter elements. However, several other types of promoters have been identified, and a consensus is yet to emerge [16], [17], [18].

Natural conditions for *C. reinhardtii* involve several environmental variables which vary reliably over a day. The strongest signal over a day is light, which is often used in laboratory conditions to simulate a diurnal cycle. Many cellular processes respond to light-dark cycles, such as regulation of the cell cycle, as well as various steps of chloroplast gene expression. As examples, the structure of the entire plastid genome is most accessible early in the light period, and the half-life of several mRNAs is much shorter in the light period than in the dark [19], [20].

So far, nine chloroplast transcripts have been quantified over a diurnal cycle in at least one study (Figure S1 in File S1). The majority of these attain their highest levels in the light period. However, the majority of the mRNAs which have been studied so far are involved in photosynthesis. Neither RNA polymerase subunits nor any subunits of the ribosome have been analyzed. Interestingly, when the diurnal regulation of chloroplast transcription in *C. reinhardtii* was studied, it was found to be accompanied by little or no fluctuation in RNA polymerase activity or in the level of RPOD, the only sigma factor in chloroplasts of *Chlamydomonas* [21].

Here, we have undertaken a systematic study of the expression of the chloroplast genome in *Chlamydomonas* under diurnal conditions. Our work has uncovered a surprising internal logic in the expression of the chloroplast genome in that transcripts of the transcriptional apparatus peak first, followed by transcripts of the translational machinery, and finally by photosynthesis gene transcripts.

Results

Development of a qRT-PCR platform for transcript quantification in *Chlamydomonas*

We have previously designed a microarray for profiling of chloroplast transcripts in the model alga *C. reinhardtii* and used it to determine changes in RNA accumulation in response to changing environmental parameters [22]. While the microarray faithfully detected expression changes in genes responding strongly to environmental cues, its sensitivity and reproducibility was not high enough to accurately determine more subtle changes in transcript levels as they occur, for example, over a diurnal cycle. Since PCR-based transcript profiling assays are potentially more sensitive and less prone to technical variation between experiments [23], we sought to establish a qRT-PCR platform for the quantitative analysis of chloroplast transcripts. To this end, primers were designed for all plastid genes and open reading frames (see Material and Methods; Dataset S1). In addition, the mitochondrial genes and a selected set of nuclear genes were included in the platform (Table S1 in File S1; Dataset S1). Test PCRs confirmed that amplicons of the expected sizes could be generated for most genes and cDNAs derived from their transcripts. Where necessary, several oligonucleotides were tested to optimize the PCRs and prevent amplification of non-specific products. Genes which repeatedly failed quality criteria were excluded from the platform. These included the tRNA genes and several genes encoding small polypeptides, owing to difficulty in designing suitable primers for small sequences. The final design of the fully optimized platform comprises 68 chloroplast, 21 mitochondrial and 51 nuclear transcripts (Table S1 in File S1; Dataset S1).

The qRT-PCR platform was then validated against our previously designed microarray for profiling of chloroplast

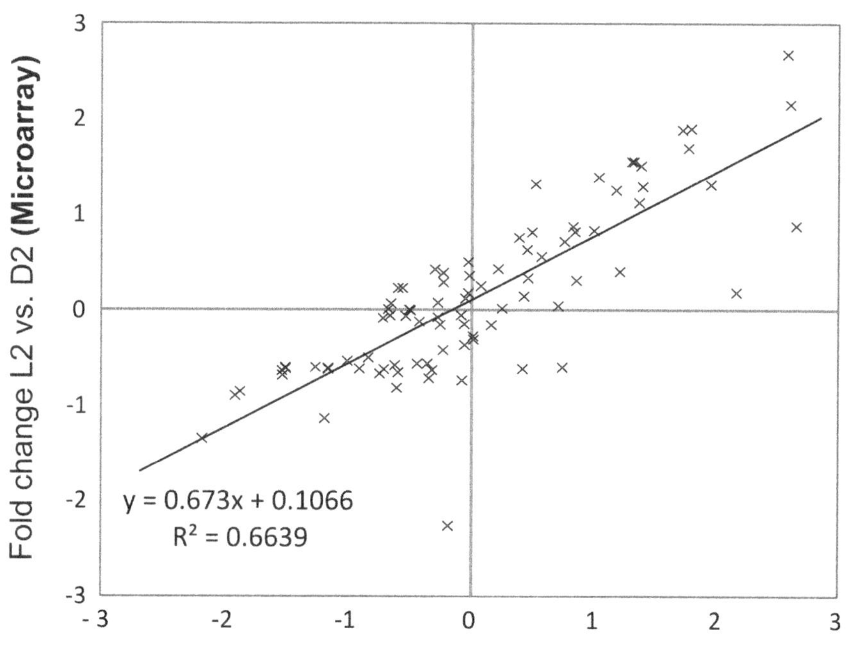

Figure 1. Comparison of microarray and qRT-PCR-based quantification of transcripts in a diurnal rhythm. Changes in organelle (chloroplast and mitochondrial) transcript levels between time points L2 and D2 in a wild-type strain of *C. reinhardtii* (CC-124) were determined by microarray and by qRT-PCR. The resulting fold-changes correlate with an R^2 value of 0.66. It is lower than 1.0, mainly because the microarray produces less accurate results for lowly expressed genes. All data points represent the average of three independent experiments.

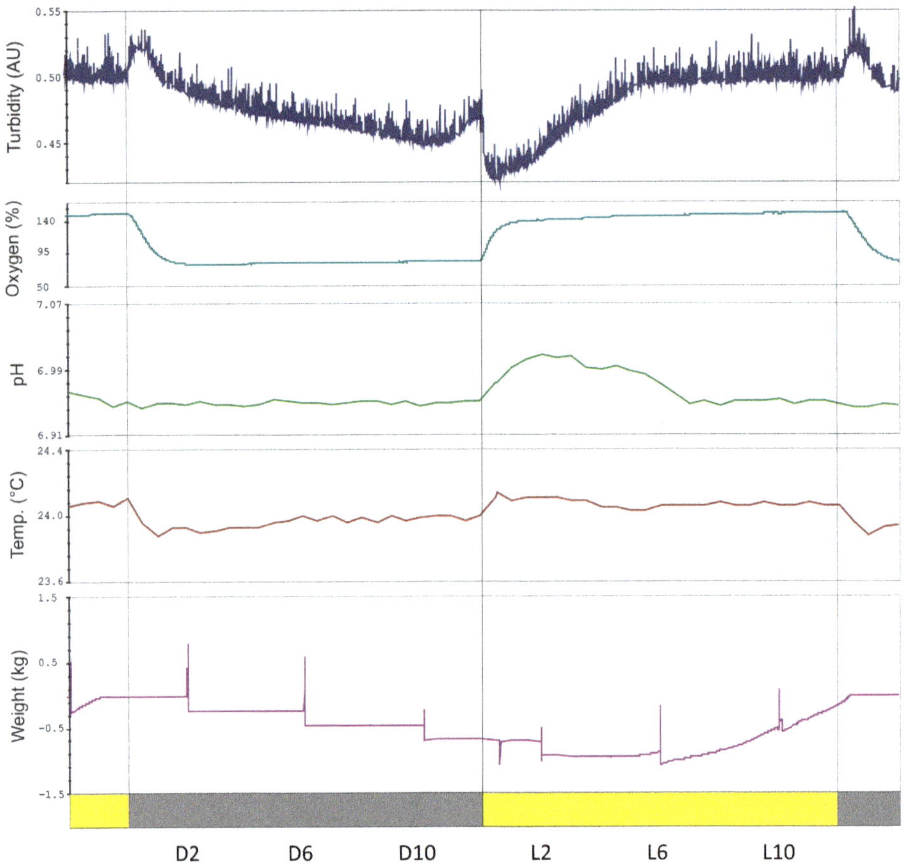

Figure 2. Key bioreactor parameters from a representative bioreactor run (R2) in diurnal conditions. Turbidity (navy blue) slightly decreased during the dark period and increased during the light. The rapid peak after dusk and rapid drop after dawn are technical artefacts of the turbidimeter. The peaks in the weight graph correspond to the sampling times, when the fermenter was disturbed. Dissolved oxygen concentration increases from ca. 80% to ca. 150% very rapidly in the light. The pH of the culture is controlled by titration of acid or base, temperature is controlled by dynamic cooling to maintain reasonably stable values for these parameters. Samples were taken every 4 h, these points are indicated and also visible as peaks in the weight data (purple). Yellow bars represents the 12 h light period (L), flanked by dark periods (D; grey bars).

transcripts in *Chlamydomonas* [22]. For this purpose, total RNA was isolated from algal cells grown in diurnal conditions in batch cultures, converted into cDNA by random priming and analyzed by both methods. The fold changes observed between the time points 2 hours into the dark period (D2) and 2 hours into the light period (L2) for each gene agree excellently between the two methods ($R^2 = 0.66$; Figure 1). However, overall we found the qRT-PCR platform to produce better results for lowly expressed genes (Figure 1).

To validate the qRT-PCR platform to more diverse studies, we compared the absolute expression levels determined in the light with the data generated by three previous *C. reinhardtii* chloroplast transcriptomic studies. The data obtained in this study were compared to data from two independent microarrays [22], [24] and to data obtained from NanoString technology [25]. Excellent correlations were obtained between this work, one of the microarray studies [22] and the NanoString study [25] with correlation coefficients of 0.82 and 0.60, respectively (Figure S1 in File S1). The data produced in the second microarray study [24] showed a lower correlation coefficient ($R^2 = 0.47$), but a similar trend was still visible.

Taken together, these results prove that the qRT-PCR platform developed for this work is of high quality. It, therefore, was used to analyze the accumulation of transcripts in cells grown in fully controlled diurnal conditions.

Diurnal growth

Cells were grown in highly controlled 5 L bioreactors [22] (see Materials and Methods) which allow very precise control and measurement of culture parameters, such as oxygen concentration, temperature, pH and density of the culture (Figure 2). As expected, the dissolved oxygen concentration in the culture medium increased rapidly at the beginning of the light period to approximately 140 % air saturation due to the rapid onset of photosynthetic oxygen evolution. During the dark period, the dissolved oxygen decreased to around 80 % as the cells consumed some of the dissolved oxygen through respiration, and ceased to generate oxygen from photosynthetic water splitting (Figure 2). Constant bubbling with air ensured that the cells did not experience hypoxia. A slight increase in the medium temperature in the light due to heat produced by the light-emitting diode light shells was also observed. Although the temperature of the bioreactor was dynamically controlled with circulating water, a small increase of 0.15–0.20°C was observed in the light. The pH also fluctuated slightly over the diurnal cycle. It increased at the start of the light period from 6.95 to around 7.0 (Figure 2). This is potentially due to the increase in dissolved oxygen and/or the higher amount of CO_2 consumed by photosynthesis. The pH then slowly decreased over the rest of the day until a time point between midday (L6) and dusk (L12), when the pH dropped below 6.95. At

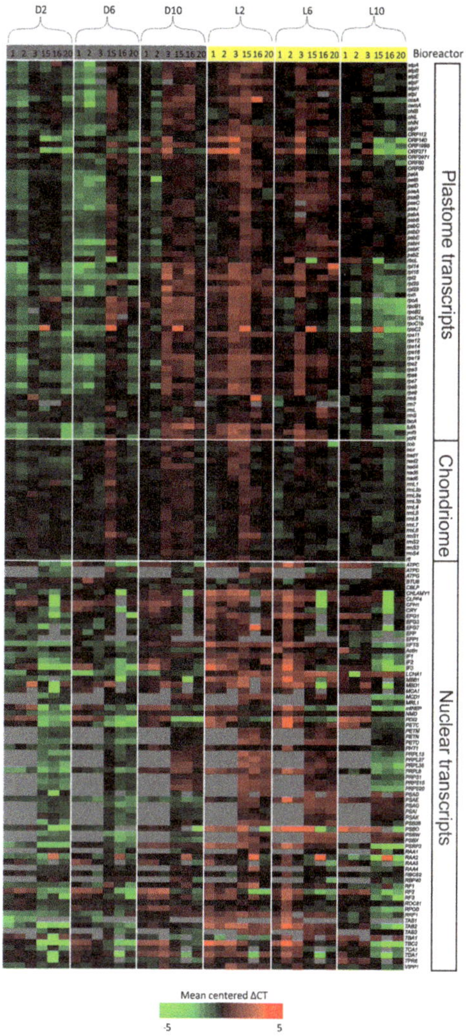

Figure 3. Diurnal transcriptomics on bioreactor cultures of *Chlamydomonas*. The effect of 12 h light/12 h dark cycles on organelle (and selected nuclear) transcripts was analyzed by qRT-PCR. *C. reinhardtii* cells were grown in six independent bioreactor runs (run numbers 1, 2, 3, 15, 16, 20) and harvested at the time points indicated. Data obtained for plastome transcripts is shown in the upper portion of the heatmap, chondriome transcripts in the central region, and nuclear transcripts in the lower region. Within each genome, transcripts are listed alphabetically. Data are normalized to housekeeping transcripts and then to the average across all samples for that gene. Red boxes indicate up-regulation, green boxes: down-regulation, grey boxes: no data. Visualization by Multiexperiment Viewer [38].

this point KOH was titrated into the bioreactor to prevent the pH dropping further. KOH titration continued until shortly before midnight (D6), when the pH stabilized and remained at 6.95 until dawn (D12).

The growth of the cells was monitored by the optical density of the culture. The optical density was set to a target value (0.4 or 0.5), but it still showed a complex diurnal pattern. At dawn, a rapid decrease in the optical density was caused directly by light acting on the turbidimeter. The turbidimeter was positioned in the bioreactor to minimize the magnitude of this effect and no changes greater than 0.06 units were observed. The optical density of the culture increased between approximately L1 and L6 as the cells grew, until the target optical density was attained. The culture was then diluted dynamically with fresh medium to maintain a steady state. After the light was turned off, a small peak was observed (again due to a technical effect on the turbidimeter), then the optical density decreased for most of the dark period (presumably due to degradation of intracellular storage compounds). Shortly before the end of the dark period, the optical density began to increase modestly (Figure 2), most likely due to the release of daughter cells.

Transcript levels over the diurnal cycle determined by qRT-PCR

RNA was extracted from cells which were sampled from the bioreactors every 4 hours in diurnal conditions. There is a general trend for samples harvested in the light to yield greater amounts of RNA than samples harvested during the dark period (data not shown). To determine changes in transcript levels, the RNA was analyzed by qRT-PCR (Figure 3). Most chloroplast and nuclear transcripts show high levels between at the end of the night and first half of the day (D10 to L6), whereas transcripts derived from the mitochondrial genome (chondriome) do not vary greatly over a diurnal cycle (Figure 3). Although there was some variation between the replicates, the altogether six biological replicates performed allow strong conclusions to be drawn.

k-means clustering was used to group transcripts according to the patterns of their responses to the diurnal conditions. Within the chloroplast transcripts, k-means clustering separates the transcripts into three classes: one containing all PEP transcripts (i.e., mRNAs encoding subunits of the plastid-encoded RNA polymerase), a second one strongly enriched in ribosome-related transcripts, and a third one enriched in photosynthesis-related transcripts (Figure 4). These are the three major classes of protein-coding genes encoded in the *C. reinhardtii* plastid genome (plastome), and the degree of separation is remarkable. Interestingly, a similar separation of photosynthesis- and translation-related transcripts was observed in the subset of nuclear-encoded transcripts which were investigated (Figure 4; Dataset S1), but not in the mitochondrial genes.

Figure 4 also shows that most transcripts resemble a sine curve in their diurnal fluctuations. This allowed the quantitation of their phases by comparing each transcript profile to a series of sine curves. The sine curve to which the data for each transcript fitted best, was used to classify the time of the day at which that transcript peaks. This analysis revealed that the defining characteristic of each of the three k-means clusters (Figure 4) is the phase at which the transcripts peak. The PEP cluster peaks first, followed by the ribosome cluster, finally followed by the photosynthesis cluster. Remarkably, in all six experiments, using two different *C. reinhardtii* strains, this order in which the three groups of genes responded remained consistent (Figure 5).

Northern blot analyses confirm the diurnal patterns in chloroplast transcript levels

In order to confirm the observed striking sequence in diurnal expression of the different classes of chloroplast genes, two representative transcripts from each of the three classes (PEP genes: *rpoA* and *rpoC2*; ribosome-related genes: *rps2* and *rps3*; photosynthesis-related genes: *psaB* and *psbB*) were tested by northern blot analysis (Figure 6). In addition, *tufA*, the gene encoding translation elongation factor EF-Tu [26], [27] was investigated. While being a component of the translational apparatus (and thus belonging to the ribosome-related genes), *tufA* is special in that its expression is known to be under stringent circadian control [28], [29].

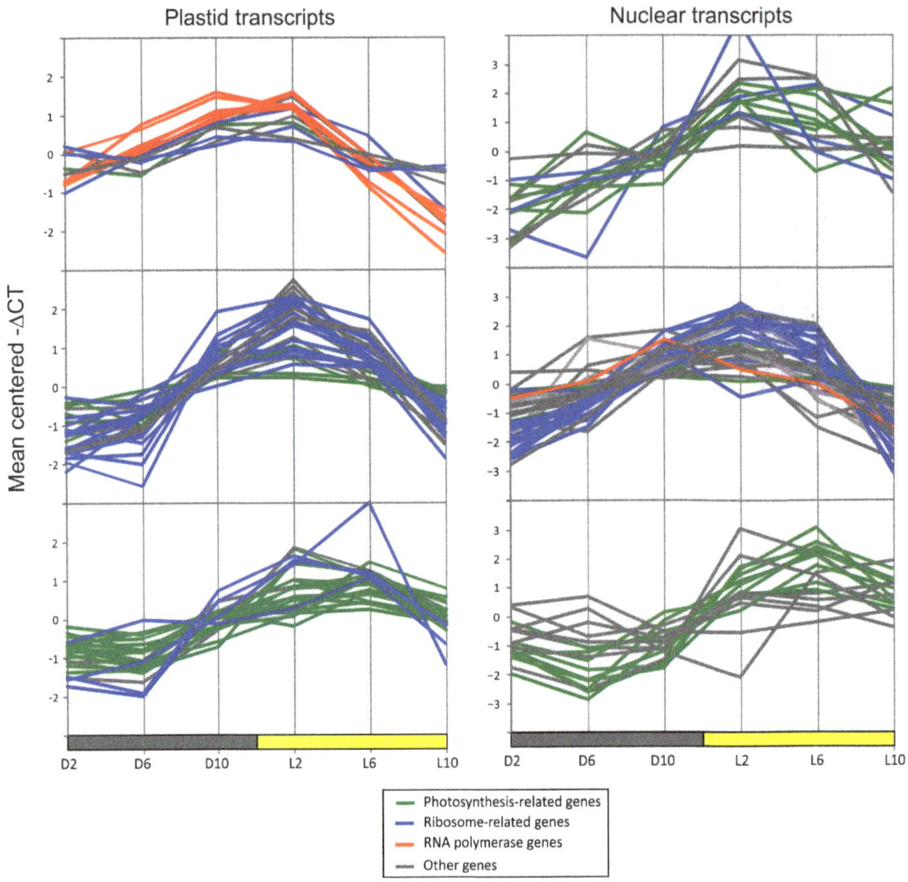

Figure 4. k-means clustering of diurnal qRT-PCR data for chloroplast transcripts (left) and a subset of nuclear transcripts (right). The averaged data for six biological replicates were grouped into three clusters with the Pearson correlation as the distance metric. Gene profiles are colored according to the gene product's function. Green: photosynthesis-related genes, blue: ribosome-related genes, red: plastid-encoded RNA polymerase genes, grey: miscellaneous genes. Clustering and visualization by Multiexperiment Viewer [38].

The expression profiles determined by northern blot analyses for all seven transcripts agreed excellently with the data obtained from qRT-PCR analysis. *rpoC2* and *rpoA* both reached their highest levels at the end of the night, *rps2* and *rps3* reached their highest levels in the first half of the day, and *psaB* and *psbB* reached high levels in the second half of the day. These results confirm that, during the diurnal cycle, chloroplast transcripts are differentially regulated depending on their function. They also confirm the striking internal logic of chloroplast genome expression (Figures 4–6): the genes encoding the components of the transcriptional machinery are induced and peak shortly before dawn, followed by the genes encoding components of the translational machinery, and finally followed by the genes encoding the components of the photosynthetic apparatus.

Discussion

In this work, we have used the model alga *C. reinhardtii* to study the expression of the chloroplast and mitochondrial genomes in comparison to a subset of nuclear genes over the diurnal cycle. The qRT-PCR platform developed for this purpose provides a useful tool for systematic analyses of organellar gene expression in *Chlamydomonas* in response to environmental conditions and developmental cues. With respect to sensitivity and reproducibility, the platform compares favourably to other methods for plastid transcript profiling.

This platform was used to quantify the changes in an unbiased group containing most chloroplast transcripts over the diurnal cycle in *C. reinhardtii*. Strikingly, this analysis revealed that transcripts behave differently depending on the function of their gene product. Unbiased clustering into three groups reliably separated photosynthesis-related transcripts, ribosome-related transcripts and RNA polymerase (PEP) transcripts.

Even more strikingly, these groups of transcripts fluctuate in a temporally ordered manner. PEP transcripts peak first at the end of the night, ribosome-related transcripts peak early in the day, and photosynthesis transcripts attain high levels at the end of the day. This agrees excellently with the hypothetical demands of a cell "waking up" from the night. Assuming that protein complexes decline during the night (which is likely due to the fast growth rate of *C. reinhardtii*, causing "dilution by division"), the cells will have to build up all complexes again for the day. Hence, remaining RNA polymerase can first transcribe its own genes, which can then be translated by remaining ribosomes in the chloroplast. In this way, a feed-forward loop creates more RNA polymerase. After a temporal delay, ribosome transcripts begin to increase, which are in turn translated by remaining ribosomes, creating another feed-forward loop on ribosome content. Finally, after a longer temporal delay, photosynthesis transcripts start to increase, which can then be translated by the large pool of ribosomes already built up.

Although substantial regulation of chloroplast gene expression also occurs at the translational level [30], the observed patterns in

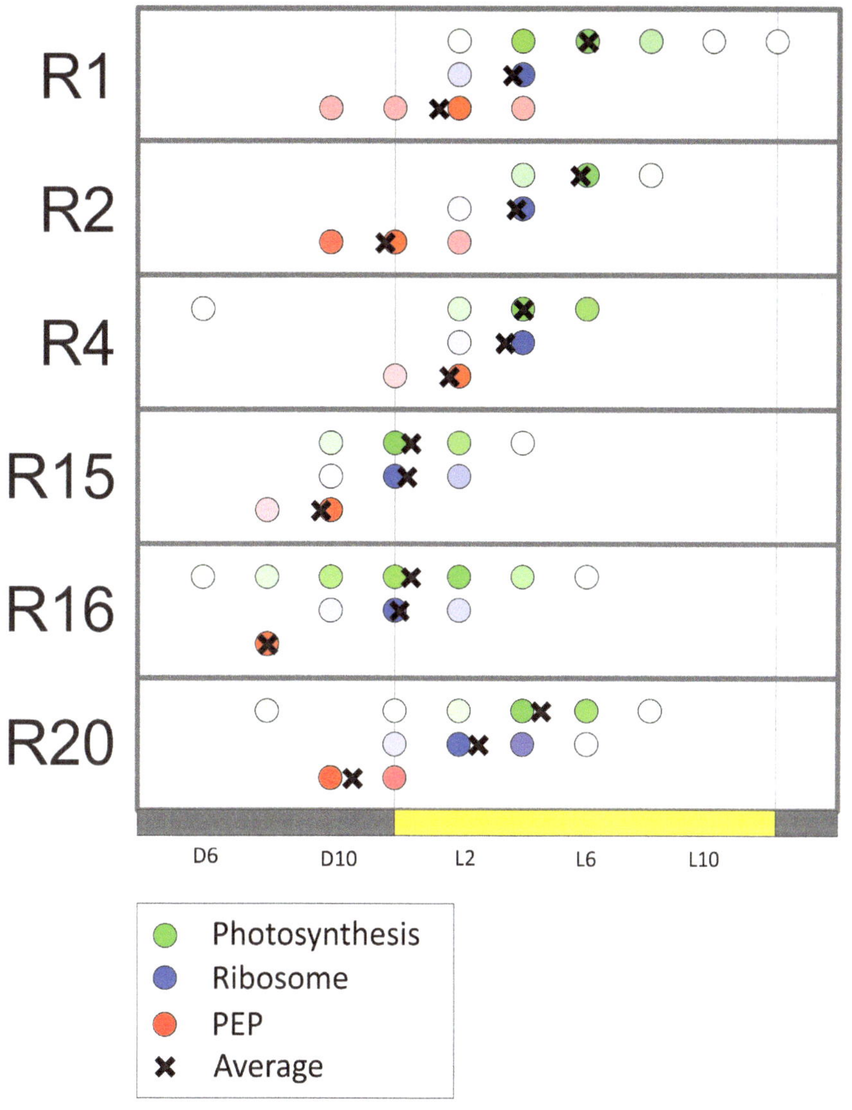

Figure 5. Behavior of functionally related chloroplast transcripts in diurnal conditions. The average peak time of all transcripts belonging to the three functional classes (red: plastid-encoded RNA polymerase genes, PEP; blue: ribosome-related genes; green: photosynthesis-related genes) is shown for each bioreactor experiment. The intensity of the circle represents the number of transcripts peaking at the time, and the black x represents the average of all transcripts. The yellow bar represents the 12 h light period, flanked by dark periods (grey bars).

transcript levels are unlikely to be insignificant. It is also interesting in this respect that the window of peak transcript levels corresponds to the window of increased supercoiling of the chloroplast DNA [19]. Furthermore, our finding that a similar temporal order of gene expression over the diurnal cycle exists in the nucleus illustrates the tight co-regulation of plastid and nuclear gene expression through anterograde and retrograde signalling pathways [31], [32], [33]. In fact, our set of nucleus-encoded genes for chloroplast proteins included several translation factors (Dataset S1), whose expression also cycled diurnally (Figures 3 and 4), raising the possibility that translational regulation reinforces the changes occurring at the transcriptional level. Finally, it is also important to realize that the changes in transcript accumulation measured by microarray hybridization or qRT-PCR represent the result of transcriptional regulation and regulation at the level of RNA turnover. It is noteworthy in this respect that at least some chloroplast genes are also diurnally regulated at the level of mRNA degradation [34], [35], [36]. Consistent with this, several known (nucleus-encoded) proteins that determine the stability of specific chloroplast mRNAs were included in our analyses (Figure 3; Dataset S1) and showed diurnal rhythmicity.

In summary, our data reported here uncover a striking internal logic in the diurnal gene expression program of the model alga *C. reinhardtii*. The peak in the expression of genes related to transcription is followed by the peak in the expression of genes related to translation which in turn is followed by the peak in the expression of genes involved in photosynthesis. It will be interesting to see whether this pattern is conserved also in seed plants. Also, the identification of the *trans*-factors that confer this intriguing pattern of diurnal regulation represents an attractive challenge. In combination with the *cis*-elements they bind, these factors could provide useful tools for controlling gene and transgene expression in both the nuclear and the chloroplast genome.

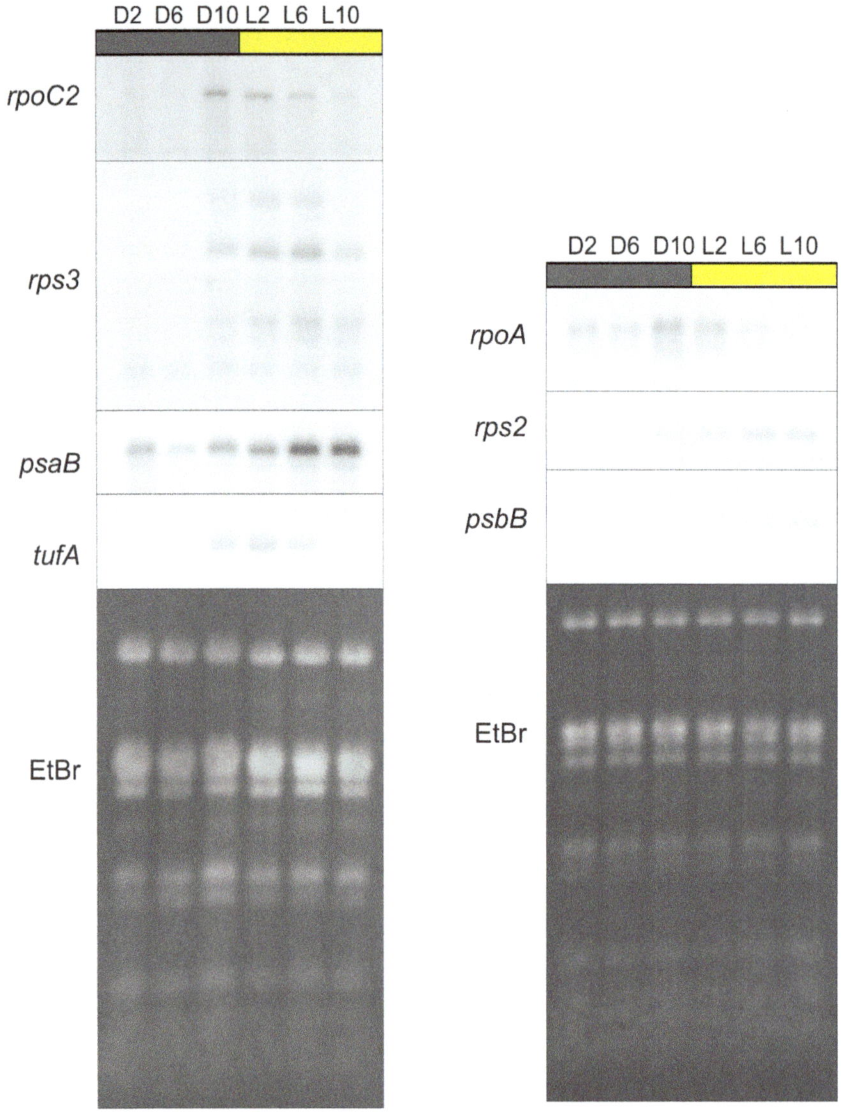

Figure 6. Confirmation of the diurnal expression patterns observed in the pRT-PCR experiments by northern blot analyses. To avoid variation from RNA loading, each membrane was hybridized and stripped multiple times. The order of probing corresponds to the vertical order of the images (and the increasing signal intensity known to be obtained with the different probes). The ethidium bromide fluorescence (EtBr) remaining on the membrane after blotting is shown as a loading control.

Materials and Methods

Algal strains and growth conditions

Chlamydomonas reinhardtii strains CC-124 and SAG73.72 (equivalent to CC-3348 in the Chlamydomonas Resource Center; www.chlamycollection.org) were grown in bioreactors which allow the growth of 5 L of continuous culture in highly controlled conditions [22]. The reactors were bubbled with 200 $cm^3 min^{-1}$ of filtered 5% CO_2 in air and were stirred at 50 rpm with three impellers. Light was supplied from two custom-made light-emitting diode half-shells, and the light was focused with individual lenses to the center of the bioreactor, to minimize light gradients within the bioreactor. Cell density was monitored with a turbidimeter, and when the threshold of 0.4 or 0.5 was exceeded, sterile medium was pumped into the bioreactor to dilute the culture. The addition of medium was sensed as an increase in weight of the bioreactor, and a corresponding volume of culture was removed to maintain a steady volume. When the pH of the culture dropped below 6.95, or above 7.05, sterile 1 M KOH or 1 M HCl was slowly pumped into the bioreactor until pH 7.00 (+/−0.05) was attained. Temperature was maintained at 24.0°C by cooled water circulating within an outer mantle. The dissolved oxygen concentration was also measured. All of these parameters were logged in the BioPAT MFCS/win software (Sartorius; http://www.sartorius.com/en/products/process-control-tools-software/biopat-mfcs/), which also controlled all parameters of the bioreactor, including light in conjunction with a second software program, Quattro Color (ProBioData GmbH, Berlin, Germany).

Cells were grown in H_5P media (Tris-minimal medium with 5 mM HEPES in place of 20 mM Tris; [37]). Cells were precultured in H_5AP (Tris-acetate medium with 5 mM HEPES in place of 20 mM Tris) in continuous light, before bioreactor inoculation. Cultures were grown in constant light until they reached their target optical density, then they were allowed to grow for at least an additional two days in constant light. The cells were then exposed to three days of 12 h light/12 h dark cycles before samples were taken on the fourth day of diurnal conditions. Samples were centrifuged at 4000×*g* for 4 min at 4°C, and the cell pellet was frozen in liquid nitrogen until use.

RNA extraction

RNA was extracted from 3–5×10^7 cells with the Trifast reagent according to the manufacturer's instructions (Peqlab, Erlangen, Germany). Contaminating DNA was removed by treatment with the TURBO DNA-free kit (Invitrogen, Paisley, UK). RNA quantity was estimated by measuring the absorbance of the RNA solution at 260 nm. RNA quality was determined by denaturing electrophoresis in formaldehyde-containing agarose gels using standard protocols.

Primer design for qRT-PCR

Primers for qRT-PCR were designed with the online software QuantPrime (www.quantprime.de) after the chloroplast and mitochondrial genomes were added to a custom database. Additional primers were designed with Primer3 (http://primer3.ut.ee/). All primers were tested for acceptable melting curves and product sizes. Primer pairs which failed these tests were discarded and redesigned. After testing and redesign, primers for 68 chloroplast, 21 mitochondrial, and 51 nuclear transcripts were retained. These primers are listed in Table S1 in File S1.

qRT-PCR experiments

cDNA was synthesized from 2 μg of total cellular RNA with the SuperScript III First-Strand Synthesis System for RT-PCR (Invitrogen) according to the manufacturer's instructions. Each 5 μL qPCR reaction contained 2.5 μL Power Sybr Green PCR Master Mix (Applied Biosystems), 0.5 μL of diluted cDNA (corresponding to cDNA obtained from approximately 5 ng total RNA), and 2 μL of 5 pmol μL^{-1} forward and reverse primers). Reactions were conducted in opaque 384 well plates in a 7900HT Fast Real-Time PCR System Instrument (Applied Biosystems).

Northern blotting and mRNA detection

Denaturing electrophoresis in formaldehyde-containing agarose gels was performed according to standard protocols. Electrophoretically separated RNAs were blotted onto nylon membranes (GE Healthcare) by capillary transfer. RNA was immobilized on the nylon membrane by UV crosslinking. Specific signals were detected with probes generated by PCR (Table S2 in File S1), purified from excised gel slices and labelled with ^{32}P dCTP using the Megaprime DNA Labeling System (GE Healthcare). Membranes were stripped by washing several times with boiling 1% SDS.

qRT-PCR data analysis

CT values were normalized to the geometric average of three housekeeping genes (*CBLP*, *BTUB* and *RBCS2*) to obtain ΔCT values. ΔCT values were mean centered per gene to the average across all time points.

k-means clustering was performed with the Pearson correlation metric using the Multiexperiment Viewer software package [38], after Figure of Merit analysis was performed to determine that 3 was a suitable value of k.

To classify transcript profiles to a sine curve, each individual transcript profile was compared to twelve sine curves (period 24 h, amplitude 0.5, phase offset by 2 h) using the Pearson correlation metric. The best match was used to classify the time at which that particular transcript peaks.

Supporting Information

File S1 Contains the following files: **Table S1.** List of primers for qPCR used in this study. **Table S2.** List of PCR primers used to generate hybridization probes. **Figure S1.** Average expression levels of plastid transcripts across multiple experiments. The average expression level for each transcript was calculated for each experiment. The expression levels determine by qRT-PCR are shown on the left, and were used to sort the heatmap. This dataset is compared to two microarray experiments [22], [24] and one NanoString experiment [25]. Expression levels are all in log2 and are mean centered. High expression is indicated by red, low expression by green, and grey boxes indicate missing data. All experiments were performed with light-grown algal cultures, but the growth conditions were likely not identical in all details (e.g., light intensity, spectral quality, cell density, media composition), thus potentially explaining part of the variation.

Dataset S1 List of genes, PCR primers, gene functions, gene classifications, and associations with clusters.

Acknowledgments

We are grateful to Dr. Maria Mittag (University of Jena, Germany) and the Chlamydomonas Resource Center for providing algal strains. We thank Mrs. Liliya Yaneva-Roder for excellent technical assistance, and Dr. Michael Schroda (University of Kaiserslautern) and the other members of the *Go*FORSYS consortium for helpful discussion. This work was supported by grants from the BMBF (Systems Biology Initiative FORSYS: GoFORSYS) and the Deutsche Forschungsgemeinschaft (FOR2092, BO 1482/17-1) to RB.

Author Contributions

Conceived and designed the experiments: RB ADI AB JR. Performed the experiments: ADI JR. Analyzed the data: ADI AB JR RB. Contributed to the writing of the manuscript: ADI RB.

References

1. Harris EH (2001) Chlamydomonas as a model organism. Annu Rev Plant Physiol Plant Mol Biol 52: 363–406.
2. Rochaix J-D (2001) Assembly, functions, and dynamics of the photosynthetic machinery in Chlamydomonas reinhardtii. Plant Physiol 127: 1394–1398.
3. Merchant SS, Prochnik SE, Vallon O, Harris EH, Karpowicz SJ, et al. (2007) The Chlamydomonas genome reveals the evolution of key animal and plant functions. Science 318: 245–251.
4. Day A, Goldschmidt-Clermont M (2011) The chloroplast transformation toolbox: selectable markers and marker removal. Plant Biotechnol J 9: 540–553.
5. Terashima M, Specht M, Hippler M (2011) The chloroplast proteome: a survey from the Chlamydomonas reinhardtii perspective with a focus on distinctive features. Curr Genet 57: 151–168.
6. Rupprecht J, Hankamer B, Mussgnug JH, Ananyev G, Dismukes C, et al. (2006) Perspectives and advances of biological H_2production in microorganisms. Appl Microbiol Biotechnol 72: 442–449.
7. Hu Q, Sommerfeld M, Jarvis E, Ghirardi M, Posewitz M, et al. (2008) Microalgal triacylglycerols as feedstocks for biofuel production: perspectives and advances. Plant J 54: 621–639.
8. Beer LL, Boyd ES, Peters JW, Posewitz MC (2009) Engineering algae for biohydrogen and biofuel production. Curr Op Biotechnol 20: 264–271.
9. Esquível MG, Amaro HM, Pinto TS, Fevereiro PS, Malcata FX (2011) Efficient H_2 production via Chlamydomonas reinhardtii. Trends Biotechnol 29: 595–600.
10. Maul JE, Lilly JW, Cui L, dePamphilis CW, Miller W, et al. (2002) The Chlamydomonas reinhardtii plastid chromosome: islands of genes in a sea of repeats. Plant Cell 14: 2659–2679.

11. Shimada H, Sugiura M (1991) Fine structural features of the chloroplast genome: comparison of the sequenced chloroplast genomes. Nucleic Acids Res 19: 983–995.
12. Shiina T, Tsunoyama Y, Nakahira Y, Khan MS (2005) Plastid RNA polymerases, promoters, and transcription regulators in higher plants. Int Rev Cytol 244: 1–68.
13. Liere K, Börner T (2007) Transcription and transcriptional regulation in plastids. Top Curr Genet 19: 121–174.
14. Carter ML, Smith AC, Kobayashi H, Purton S, Herrin DL (2004) Structure, circadian regulation and bioinformatic analysis of the unique sigma factor gene in Chlamydomonas reinhardtii. Photosynth Res 82: 339–349.
15. Bohne A-V, Irihimovitch V, Weihe A, Stern DB (2006) Chlamydomonas reinhardtii encodes a single sigma^{70}-like factor which likely functions in chloroplast transcription. Curr Genet 49: 333–340.
16. Klein U, De Camp JD, Bogorad L (1992) Two types of chloroplast gene promoters in Chlamydomonas reinhardtii. Proc Natl Acad Sci USA 89: 3453–3457.
17. Klein U, Salvador ML, Bogorad L (1994) Activity of the Chlamydomonas chloroplast rbcL gene promoter is enhanced by a remote sequence element. Proc Natl Acad Sci USA 91: 10819–10823.
18. Blowers AD, Ellmore GA, Klein U, Bogorad L (1990) Transcriptional analysis of endogenous and foreign genes in chloroplast transformants of Chlamydomonas. Plant Cell 2: 1059–1070.
19. Salvador ML, Klein U, Bogorad L (1998) Endogenous fluctuations of DNA topology in the chloroplast of Chlamydomonas reinhardtii. Mol Cell Biol 18: 7235–7242.
20. Salvador ML, Klein U (1999) The redox state regulates RNA degradation in the chloroplast of Chlamydomonas reinhardtii. Plant Physiol 121: 1367–1374.
21. Kawazoe R, Mahan KM, Venghaus BE, Carter ML, Herrin DL (2012) Circadian regulation of chloroplast transcription in Chlamydomonas is accompanied by little or no fluctuation in RPOD levels or core RNAP activity. Mol Biol Rep 39: 10565–10571.
22. Mettler T, Mühlhaus T, Hemme D, Schöttler MA, Rupprecht J, et al. (2014) Systems analysis of the response of photosynthesis, metabolism and growth to an increase in irradiance in the photosynthesis model organism Chlamydomonas reinhardtii. Plant Cell 26: 2310–2350.
23. Czechowski T, Bari RP, Stitt M, Scheible W-R, Udvardi MK (2004) Real-time RT-PCR profiling of over 1400 Arabidopsis transcription factors: unprecedented sensitivity reveals novel root- and shoot-specific genes. Plant J 38: 366–379.
24. Erickson B, Stern DB, Higgs DC (2005) Microarray analysis confirms the specificity of a Chlamydomonas reinhardtii chloroplast RNA stability mutant. Plant Physiol 137: 534–544.
25. Ramundo S, Rahire M, Schaad O, Rochaix J-D (2013) Repression of essential chloroplast genes reveals new signaling pathways and regulatory feedback loops in Chlamydomonas. Plant Cell 25: 167–186.
26. Baldauf SL, Manhart JR, Palmer JD (1990) Different fates of the chloroplast tufA gene following its transfer to the nucleus in green algae. Proc Natl Acad Sci USA 87: 5317–5321.
27. Hwang S, Kawazoe R, Herrin DL (1996) Transcription of tufA and other chloroplast-encoded genes is controlled by a circadian clock in Chlamydomonas. Proc Natl Acad Sci USA 93: 996–1000.
28. Matsuo T, Onai K, Okamoto K, Minagawa J, Ishiura M (2006) Real-time monitoring of chloroplast gene expression by a luciferase reporter: Evidence for nuclear regulation of chloroplast circadian period. Mol Cell Biol 26: 863–870.
29. Matsuo T, Okamoto K, Onai K, Niwa Y, Shimogawara K, et al. (2008) A systematic forward genetic analysis identified components of the Chlamydomonas circadian system. Genes & Dev 22: 918–930.
30. Eberhard S, Drapier D, Wollman F-A (2002) Searching limiting steps in the expression of chloroplast-encoded proteins: relations between gene copy number, transcription, transcript abundance and translation rate in the chloroplast of Chlamydomonas reinhardtii. Plant J 31: 149–160.
31. Bräutigam K, Dietzel L, Pfannschmidt T (2007) Plastid-nucleus communication: anterograde and retrograde signalling in the development and function of plastids. Top Curr Genet 19: 409–455.
32. Berry JO, Yerramsetty P, Zielinski AM, Mure CM (2013) Photosynthetic gene expression in higher plants. Photosynth Res 117: 91–120.
33. Jarvis P, López-Juéz E (2013) Biogenesis and homeostasis of chloroplast and other plastids. Nat Rev Mol Cell Biol 14: 787–802.
34. Herrin DL, Michaels AS, Paul AL (1986) Regulation of genes encoding the large subunit of ribulose-1,5-bisphosphate carboxylase and the photosystem II polypeptides D-1 and D-2 during the cell cycle of Chlamydomonas reinhardtii. J Cell Biol 103: 1837–1845.
35. Leu S, White D, Michaels A (1990) Cell cycle-dependent transcriptional and post-transcriptional regulation of chloroplast gene expression in Chlamydomonas reinhardtii. Biochim Biophys Acta 1049: 311–317.
36. Salvador ML, Klein U, Bogorad L (1993) Light-regulated and endogenous fluctuations of chloroplast transcript levels in Chlamydomonas. Regulation by transcription and RNA degradation. Plant J 3: 213–219.
37. Harris EH (1989) The Chlamydomonas Sourcebook. Academic Press, San Diego, CA.
38. Saeed AI, Sharov V, White J, Li J, Liang W, et al. (2003) TM4: a free, open-source system for microarray data management and analysis. BioTechniques 34: 374–378.

17

Evolutionary Diversification of New Caledonian *Araucaria*

Mai Lan Kranitz[1¶], Edward Biffin[4*¶], Alexandra Clark[1], Michelle L. Hollingsworth[1], Markus Ruhsam[1], Martin F. Gardner[1], Philip Thomas[1], Robert R. Mill[1], Richard A. Ennos[2], Myriam Gaudeul[3], Andrew J. Lowe[4], Peter M. Hollingsworth[1]

1 Royal Botanic Garden Edinburgh, Edinburgh, United Kingdom, **2** University of Edinburgh, Institute of Evolutionary Biology, Edinburgh, United Kingdom, **3** UMR CNRS MNHN UPMC EPHE 7205 Institut de Systématique, Evolution, Biodiversité, Muséum National d'Histoire Naturelle, Paris, France, **4** Australian Centre for Evolutionary Biology and Biodiversity, Environment Institute, School of Earth and Environmental Science, University of Adelaide, Adelaide, Australia

Abstract

New Caledonia is a global biodiversity hotspot. Hypotheses for its biotic richness suggest either that the island is a 'museum' for an old Gondwana biota or alternatively it has developed following relatively recent long distance dispersal and *in situ* radiation. The conifer genus *Araucaria* (Araucariaceae) comprises 19 species globally with 13 endemic to this island. With a typically Gondwanan distribution, *Araucaria* is particularly well suited to testing alternative biogeographic hypotheses concerning the origins of New Caledonian biota. We derived phylogenetic estimates using 11 plastid and rDNA ITS2 sequence data for a complete sampling of *Araucaria* (including multiple accessions of each of the 13 New Caledonian *Araucaria* species). In addition, we developed a dataset comprising 4 plastid regions for a wider taxon sample to facilitate fossil based molecular dating. Following statistical analyses to identify a credible and internally consistent set of fossil constraints, divergence times estimated using a Bayesian relaxed clock approach were contrasted with geological scenarios to explore the biogeographic history of *Araucaria*. The phylogenetic data resolve relationships within Araucariaceae and among the main lineages in *Araucaria*, but provide limited resolution within the monophyletic New Caledonian species group. Divergence time estimates suggest a Late Cretaceous-Cenozoic radiation of extant *Araucaria* and a Neogene radiation of the New Caledonian lineage. A molecular timescale for the evolution of Araucariaceae supports a relatively recent radiation, and suggests that earlier (pre-Cenozoic) fossil types assigned to *Araucaria* may have affinities elsewhere in Araucariaceae. While additional data will be required to adequately resolve relationships among the New Caledonian species, their recent origin is consistent with overwater dispersal following Eocene emersion of New Caledonia but is too old to support a single dispersal from Australia to Norfolk Island for the radiation of the Pacific *Araucaria* sect. *Eutacta* clade.

Editor: William J. Etges, University of Arkansas, United States of America

Funding: This work was supported by funds from the Darwin Trust of Edinburgh and the ANR project BIONEOCAL, the PPF MNHN 'Structure et évolution des écosystèmes' led by P. Grandcolas. The work of AJL and EB was supported by the Australian Research Council through a Super Science Fellowship (FS1102 00051) and Discovery Grant (DP0665859) awarded to AJL. The Royal Botanic Garden Edinburgh is supported by the Scottish Government's Rural and Environment Science and Analytical Services Division. The funders had no role in study design, data collection and analysis, decision to publish, or preparation of the manuscript.

Competing Interests: The authors have declared that no competing interests exist.

* Email: ed.biffin@adelaide.edu.au

¶ These authors are joint first authors on this work.

Introduction

New Caledonia, in the tropical south-western Pacific, is noted for a remarkably rich biota with outstanding levels of endemism [1]. The island has been considered both a 'museum' for phylogenetic relicts and a natural laboratory for the study of island radiations [2]. A long held view is that New Caledonia (as part of Zealandia) was rifted from the eastern Australian margin some 80 million years before present (Ma), carrying a complement of Gondwanan lineages that have since evolved there in relative isolation [3]. Alternatively, geological reconstructions for the region indicate lengthy deep-water marine transgressions (Palaeocene-Eocene) with limited evidence of a continual landmass prior to the emersion of New Caledonia in the Late Eocene-Oligocene (c. 37 Ma) [4]. This implies that dispersal events must be central in the assembly of the modern biota [2]. Phylogenetic and molecular dating methods, within the context of a well-constrained geological scenario, can be used to explore these alternative biogeographic scenarios [5]. With respect to New Caledonia such studies provide strong support for the latter view [2,6].

The conifer genus *Araucaria* Juss. (Araucariaceae) provides a good model to explore alternative hypotheses regarding the age, origins and diversification of the New Caledonian biota [7,8]. *Araucaria* has been considered an old lineage [9] and for instance, fossil data from both the Northern and Southern Hemisphere of the early Mesozoic (Jurassic, c. 190 Ma, or earlier) have been taken to indicate that at least some of the extant sections of *Araucaria* had begun to diversify and were widespread by this time. Thirteen of the 19 recognised *Araucaria* species are now endemic to New Caledonia, with the other species occurring in eastern Australia and New Guinea (3 spp., 1 shared), Norfolk Island (1 sp.) and southern South America (2 spp.). In light of these data extant *Araucaria* have been considered a relict of a formerly widespread Gondwanan distribution [10].

A key question is the biogeographic origin of the New Caledonian *Araucaria* – an old lineage isolated by vicariance that has diversified *in situ* or a relatively recent radiation following long distance dispersal? There have been some indirect tests of these hypotheses. Setoguchi et al. [8] used plastid *rbcL* sequence data to infer the phylogeny of Araucariaceae and found support for the monophyly of the New Caledonian *Araucaria*. In this study, however, the morphological species were characterised by very low levels of DNA sequence divergence suggestive of a recent radiation. Gaudeul et al. [7] also argued for a recent diversification of the New Caledonian *Araucaria* based on a low ratio of among versus within species genetic divergence using AFLP markers. In the context of these findings it has been proposed that the origin of this lineage may be maximally constrained by the age of Norfolk Island (c. 1000 km to the south of New Caledonia) [2], a volcanic island that in its present form emerged c. 3.7 Ma. This island is home to the sister species (*A. heterophylla* (Salisb.) Franco; Norfolk Island Pine) of the New Caledonian *Araucaria* lineage [8,11]. However, the use of island age to date island-endemic lineages can be problematic as these taxa could represent historically widespread lineages that have been isolated by extinction [12]. Limitations of the molecular clock have also been well-characterised [13] and there may be a poor correlation between levels of DNA sequence divergence and divergence times if rates have been slow relative to related taxa. Furthermore, finding low levels of divergence among molecular markers does not rule out the possibility of an ancient lineage and a recent diversification in the absence of a timescale for the origin of the New Caledonian lineage [14]. Recent analyses of araucarian and conifer divergence times [15–19] for the most part suggest a relatively recent radiation of extant Araucariaceae and *Araucaria* but as yet there has been no formal attempt to date the diversification of *Araucaria* on New Caledonia.

In the present study, we develop a phylogenetic estimate for *Araucaria* using rDNA internal transcribed spacer 2 (ITS2) sequences and a mix of 11 coding and non-coding plastid data. We include a complete sampling of species and several accessions of each of the 13 New Caledonian representatives. Our overall aims are: (1), to clarify relationships among *Araucaria*, particularly the New Caledonian species; (2), using well-constrained fossil dates and molecular clock methods, to estimate a timescale for *Araucaria* evolution focussing on the origins of the New Caledonian radiation.

Methods

Plant material and DNA extractions

Collection, voucher and permit details of all samples used for the phylogenetic analyses are listed in Table S1. Permits allowed for the collection of samples from threatened species and for collection of samples from within protected areas. CITES permits were obtained for *Araucaria araucana*. DNA was extracted using the Qiagen DNeasy Mini Plant Kit following the manufacturer's protocol.

PCR amplification and sequencing

DNA sequences were obtained from 11 chloroplast regions (*matK*, *rcbL*, *rpoB*, *rpoC1*, *atpF-atpH*, *atpH-atpI*, *rps12-rpl20*, *trnC-ycf6*, *trnH-psbA*, *trnS-trnfm*, *trnS-trnG*) and one nuclear region (ITS2). For a subset of 21 taxa (nine *Agathis* species, and outgroup taxa including a single representative each of the genera *Pinus* L. (Pinaceae), *Taxus* L. (Taxaceae), *Cryptomeria* D.Don (Cupressaceae), *Taxodium* Rich. (Taxodiaceae), *Lepidothamnus* Phil., *Halocarpus* Quinn, *Falcatifolium* de Laub., *Dacrycarpus* de Laub., *Phyllocladus* Rich. ex Mirb., *Podocarpus* L'Hér. ex Pers., *Retrophyllum* C.N.Page and *Sundacarpus* (J.Buchholz and N.E. Gray) C.N. Page (all Podocarpaceae)) we assembled a data set comprising *rbcL*, *matK* and non-coding plastid regions *atpF-atpH* and *rps12-rps20* to supplement the *Araucaria* data for molecular dating analyses (see below). These regions were chosen because they could be unambiguously aligned in all taxa.

PCR primers and conditions for *matK*, *rcbL*, *rpoB*, *rpoC1*, *atpF-atpH* and *trnH-psbA* are detailed in Appendix 4 of the supplementary material of [20]. For the remaining regions primer sequences were taken from the following publications: *atpH-atpI* [21], *rps12-rpl20* and *trnC-ycf6* [22], *trnS-trnfm* [23], *trnS-trnG* [24] and ITS2 [25].

PCR reactions for *atpH-atpI*, *rps12-rpl20*, *trnC-ycf6*, *trnS-trnG* and ITS2 were performed in volumes of 25 μl using the following protocol: 1xbuffer (Bioline, London, UK), 0.2 mM dNTPs, 2.5 mM $MgCl_2$, 0.3 μM of each forward and reverse primer, (4% Dimethyl Sulfoxide, DMSO, ITS2 only), 0.04 U BioTaq (Bioline, London, UK) and 1 μl of unquantified DNA. The mixture was then cycled through the profile: 2 min at 95°C, 40 cycles of 30 sec at 95°C, 30 sec at 50°C and 1 min at 72°C, ending with 10 min at 72°C to complete extension and subsequent storage at 4°C.

PCR reactions for *trnS-trnfm* followed this protocol (25 μL): 1x buffer (Bioline, London, UK), 0.2 mM dNTPs, 2.5 mM $MgCl_2$, 0.3 μM of each forward and reverse primer, 2 M Betaine, 0.03 U BioTaq (Bioline, London, UK) and 2 μl of unquantified DNA. The mixture was then cycled through the following profile: 4 min at 94°C, 30 cycles of 45 sec at 94°C, 45 sec at 62°C and 2 min at 72°C, ending with 10 min at 72°C and subsequent storage at 4°C.

PCR products were cleaned using 2 μl exonuclease I and shrimp alkaline phosphatase (ExoSAP) for 5 μl of product and sequenced in two reactions using each of the two PCR primers, following Big Dye v. 3.1 chemistry (Applied Biosystems, Warrington, UK). Sequences were then assembled and aligned using Sequencher v. 3.7 (GeneCodes Corp., Ann Arbor, Michigan, USA).

Phylogeny

Araucaria phylogeny was inferred separately for the plastid (53 taxa), ITS2 (48 taxa) and a combined plastid and ITS2 DNA sequence data set (48 taxa) using both maximum likelihood (ML) and Bayesian inference (BI) optimisation criteria (Table S1; vouchers).

We used GARLI 0.951 [26] for the ML analyses with a general time reversible model of sequence evolution, gamma distributed rate variation and a proportion of invariant sites (GTR+I+Γ). Clade support was assessed using 250 non parametric bootstrap (BS) pseudoreplicates.

Bayesian inference was performed using MrBayes 3.1.2 [27,28]. For the plastid and combined data sets we used a mixed model approach with GTR+I+Γ model parameters estimated separately for each locus in the concatenated matrices. For the analysis of the ITS2 sequences a GTR+I+Γ model was used. For each analysis we performed four independent runs of 2×10^6 generations, sampling topology and parameter values every hundredth generation, with default (flat) priors, a random starting tree and four starting chains (one cold, three heated). Convergence between runs was assessed relative to the variance in parameter estimates (average standard deviation of split frequencies <0.01; Ronquist 2005) and by inspection of parameter estimates in Tracer v1.5 [29]. Tracer was used to estimate the burn in proportion, and 50% majority rule consensus topologies were generated from the pooled post-burn in topologies from the four independent runs.

Molecular Dating

Molecular data. To estimate topology and divergence times for Araucariaceae, we assembled a data set comprising partial sequences of the plastid genes *rbcL, matK* and non-coding plastid regions *atpH-atpI* and *rps12-rps20* for 45 taxa). Taxon sampling includes all of the 19 currently recognised species of *Araucaria*, 11 species of *Agathis* Salisb., *Wollemia* (monotypic), and outgroup taxa including Cupressaceae, Taxaceae, Podocarpaceae and Pinaceae (Table S1).

Fossil constraints. Five fossil dates were used to calibrate molecular evolutionary rates amongst Araucariaceae. These comprise two fossil constraints that have been assigned to the Araucariaceae crown group, two within the Podocarpaceae as well as a constraint upon the age of the Cupressaceae *sensu lato* (Table 1). The araucarian fossils include the oldest unequivocal macrofossil remains of *Agathis* from south-eastern Australia of the Late Palaeocene (*A. vittata* R.S.Hill et al., c. 55 Ma) [30] and a recently described bract-scale complex from the Lower Jurassic (c. 190 Ma) of eastern North America, which has been proposed as the earliest known occurrence of *Araucaria* section *Eutacta* (Link) Endl. [31]. It should be noted that this fossil age is not inconsistent with several other well documented fossil types (e.g. *Araucaria mirabilis* (Spegazzini) Windhausen, *A. sphaerocarpa* Carruthers) that have suggested affinities within extant *Araucaria* [32]. Several extant podocarp lineages have reliable macrofossil records including *Prumnopitys* Phil. [33,34] and *Acmopyle* Pilg. [33,35]. We used macrofossil remains from K-T boundary exposures (c. 65 Ma) of New Zealand [34] to constrain the crown group age for *Prumnopitys sensu lato* (most recent common ancestor, hereafter MRCA, of *Prumnopitys* + *Sundacarpus*; see [36]) and defined a minimum age of 55 Ma to constrain the *Acmopyle* stem (MRCA *Podocarpus* + *Retrophyllum* + *Dacrycarpus* + *Falcatifolium* + *Acmopyle*, based on *A. florinii* R.S.Hill and Carpenter). The fossil taxon *Austrohamia minuta* Escapa, Cúneo and Axsmith, from the Middle-Late Jurassic has been placed within Cupressaceae *sensu lato* following a cladistic analysis of morphological characters from fossil and living taxa [37]. On this basis we assumed a minimum age of 160 Ma to constrain the Cupressaceae stem (MRCA Cupressaceae s.l. + Taxaceae) (Table 1; Figure 1).

Assessment of fossil constraints. We adopted the approach of Dornburg et al. [38] to provide an assessment of the palaeontological data described above. These authors extend the methods of Marshall [39], which use minimum node age estimates derived from the fossil record and an uncalibrated ultrametric tree to bracket absolute divergence times from molecular phylogenies. Given that for a particular lineage the known fossil record is generally incomplete, Marshall [39] provides an approach to estimate the coverage of a fossil assigned to that clade (empirical scaling factor, S_i, i.e. the proportion of the true duration of the lineage represented by its oldest known fossil) which can be used to estimate the true scaling factor, S, i.e. the actual time of divergence of the root. To incorporate uncertainty in S_i estimates, Dornburg et al. [38] sample a distribution of ultrametric tree topologies and relative branch lengths. Rather than applying an age bracket to the single fossil with the highest coverage [39], they identify a set of internally consistent fossil constraints based upon the S_i estimates and this set of consistent calibrations is incorporated in molecular dating analyses to derive absolute estimates of node heights [38].

We estimated the posterior distribution of uncalibrated ultrametric trees (i.e. branch lengths in units of substitutions/site) from the molecular data using the Bayesian relaxed clock (BRC) application BEAST (v.1.6.1) [40] with a GTR+I+Γ model of sequence evolution, a Yule prior on branch rates and an uncorrelated log normal relaxed clock model [41]. For these analyses two independent Markov-chain Monte Carlo runs were performed, each of 4×10^7 steps (sampling topology and parameter values every 5000 steps). Tracer v1.5 [29] was used to assess convergence between runs and estimate an appropriate burn-in proportion, the mean and 95% highest posterior density (HPD) of parameters sampled from the posterior distribution of the combined runs, and to ensure that the effective sample size was sufficient to provide reasonable estimates of model parameter variance (i.e.>200).

For each fossil constraint (Table 1) S_i was estimated using Equation 1 of Marshall [39] and the distribution of S_i values was estimated from 1000 topologies sampled from posterior distribution of uncalibrated ultrametric trees. In order to assess the internal consistency of the fossil constraints, we simultaneously compared the interval encompassing 95% of the values for the fossil with the highest S_i estimates (i.e. the highest coverage) against all other S_i intervals for each fossil. The consistent fossil set is defined as those with 95% S_i intervals that overlap the interval with the highest S_i values [38].

Where possible, we assessed the relative placement of the fossil age within a lineage (i.e. crown or stem node). For three of the constraints, taxon sampling (*Acmopyle* and Cupressaceae) and/or phylogenetic uncertainty (*Prumnopitys* s.l.) excluded this possibility, but for the *Agathis* fossil age, we compared crown and stem node placement (MRCA *Agathis* + *Wollemia*) (nodes 4 and 5, respectively; Table 1, Figure 1). For the *Araucaria* section *Eutacta* fossil age we considered alternative placements on the section *Eutacta* crown (node 6), *Araucaria* crown (node 7), Araucariaceae crown (node 8), and on the Araucariaceae stem (MRCA of Araucariaceae + Podocarpaceae; node 9). This is in the light of questions, on a number of grounds [8,11,15,42] of generic and/or sectional affinities of Early Mesozoic araucarians.

We placed a caveat on the selection of the fossil with the highest coverage. Fossil prepollen of *Potonieisporites* D.C.Bhardwaj, a monosaccate type that was produced by walchian conifers, is known from at least as early as the Langsettian (Upper Carboniferous >313–314 Ma) from Scotland [43] while macrofossils of the walchian conifer *Hermitia* Kerp and Clement-Westerhof are known from at least as early as the Asturian (= Westphalian D, 308–306 Ma) of the Netherlands [44]. Extant conifer families have fossil records extending from the Triassic [45,46] providing an expectation on the 'upper' distribution of reasonable S_i values. Estimates that consistently exceeded the age of the earliest known conifers were treated as outliers.

Divergence time estimates. We used the consistent set of fossils identified following the procedure outlined above to estimate araucarian divergence times with BEAST. For each fossil datum, we defined a log normal prior probability distribution on the age of the fossil constrained node, with a hard minimum age equivalent to the age of the associated fossil. The other parameters of the log normal distribution are the mean and standard deviation. For all analyses, the standard deviation was set to 0.5, and to model taphonomic bias of the fossil record, the log normal prior mean was set at c. 10% and 50% older than each minimum fossil age, in separate sets of analyses (Table 1). At the highest value of the mean, the upper bound of the 95% confidence

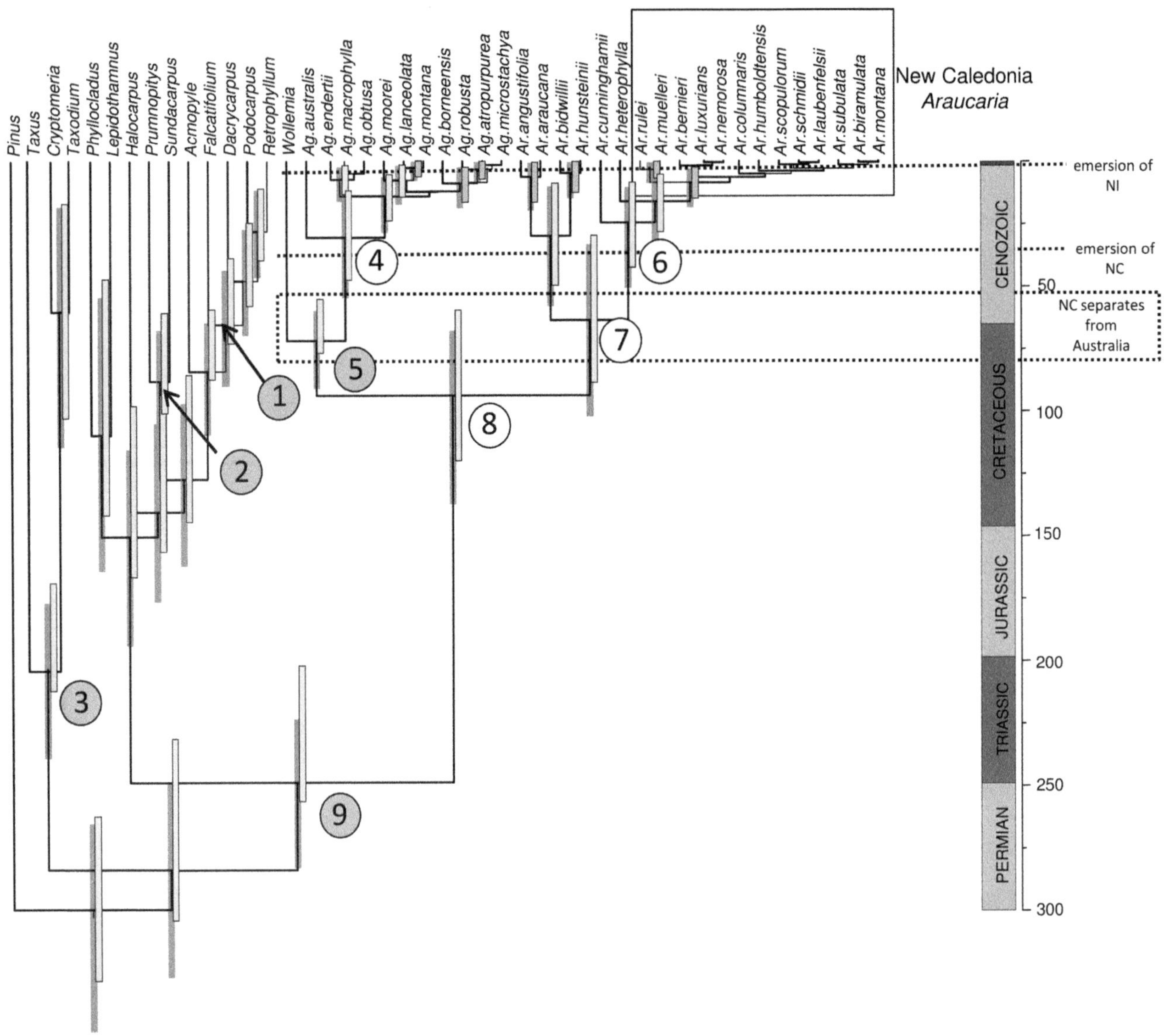

Figure 1. Divergence time estimates for Araucariaceae. The topology presented is the maximum credibility tree (median node heights) obtained by pooling four separate runs under two constraint scenarios (see Table 1, and text for details). Circles indicate nodes that were included in the assessment of candidate nodal constraints, and the dark shaded circles indicate nodes that were constrained for the BRC analyses (see Table 1 for details). Node bars (nodes with a posterior probability ≥0.75) indicate the 95% HPD of divergence times for the two prior scenarios: left, log normal prior mean = minimum fossil age +50%; right, log normal prior mean = minimum fossil age +10% (NC, New Caledonia; NI, Norfolk Island).

interval (CI) of the log normal constraint priors included values estimated using Marshall's equation 11 [39] approach for estimating a 95% CI for the true time of origin of a lineage, given its oldest known fossil (with $n = 5$, i.e. the number of fossils included in the consistent set of fossil calibrations) [38]. For each value of the log normal prior mean, two separate BEAST analyses were performed, with model parameters and settings as outlined above. For each mean value, the two separate runs were combined after excluding the burn in fraction, and topologies and parameter values were summarised on the 'maximum credibility' tree using Tree Annotator v. 1.6. [40]. For these analyses we additionally constrained the root age with a log normal prior with a mean of c. 300 Ma and a 95% CI ranging from 260–350 Ma approximating the age of stem group conifers [46].

Results

Sequence data

The alignment of the 11 used plastid regions and the nuclear ribosomal ITS2 comprised 9970 bp and 500 bp, respectively, totalling 10470 bp in the combined dataset. For all *Araucaria* samples (except *A.schmidii*), *Agathis moorei* (Lindl.) Mast., *Agathis lanceolata* Warb., *Wollemia nobilis* W.G.Jones et al., *Prumnopitys ferruginoides* (Compton) de Laub. and *Acmopyle pancheri* (Brongn. and Gris) Pilg., sequencing was successful for all individuals and regions (Table S2). Certain regions could not be obtained for *Prumnopitys ferruginoides* (*rpoB*, *trnH-psbA*, *trnS-trnfm*), *Acmopyle pancheri* (*trnH-psbA*), *A. schmidii* de Laub. (ITS2), *Taxodium distichum* (*atpF-atpH*), *Podocarpus gnidioides* (*rps12-rpl20*) as well as *Lepidothamnus fonkii* and *Retrophyllum rospigliosii* (both *atpF-atpH* and *rps12-rpl20*) (Table S2).

Table 1. Details of fossil constraints used to estimate araucarian divergence times (Prior 1, log normal prior mean = minimum fossil age +10%; Prior 2, log normal prior mean = minimum fossil age +50%) and empirical scaling factor (S_i) estimates inferred from the unconstrained BRC analyses of the DNA sequences.

Node	Fossil	Minimum age constraint (Ma)	S_i (mean[95% PD])	Prior 1	Prior 2
1	*Acmopyle florinii*	55	**186 (250–133)**	**0.5, 2**	**0.5, 3.3**
2	*Prumnopitys* 'Mt. Somers'	60	**175 (247–170)**	**0.5, 2**	**0.5, 3.3**
3	*Austrohamia minuta*	160	**284 (413–210)**	**0.5, 2.8**	**0.5, 4.4**
4			784 (1186–415)	-	-
5	*Agathis vittata*	55	**368 (565–198)**	**0.5, 2**	**0.5, 3.3**
6			3083 (5426–1305)	-	-
7	'araucarian bract scale complex' (cf. *Araucaria* section *Eutacta*)		1260 (1928–698)	-	-
8		190	842 (1191–488)	-	-
9			**221 (264–196)**	**0.5, 3.2**	**0.5, 4.5**

Node numbers are as per Figure 1. Values highlighted in bold were selected on the basis of their S_i values for the molecular dating analyses. PD, marginal posterior distribution of S_i; SD, standard deviation of the log normal prior. Values in the Prior1 and Prior 2 column denote SD and log normal mean, respectively.

Phylogeny of *Araucaria*

Phylogenies estimated from the plastid (Figure 2) and ITS2 (Figure 3) data resolved broadly similar groupings including *Wollemia* + *Agathis* and *Araucaria*. Within *Araucaria*, section *Eutacta* is resolved as sister to a clade including the sections *Araucaria* and *Bunya* Wilde and Eames + *Intermedia* C.T.White. ITS2 data (Figure 3) resolved *A. heterophylla* (Salisb.) Franco as sister to the remaining *A.* section *Eutacta* clade, but this placement is relatively weakly supported (BS<50%; PP = 0.93). In the combined plastid and ITS2 data, there is robust support for *A. cunninghamii* Mudie as sister to the Pacific *A.* section *Eutacta* (*A. heterophylla*+New Caledonian species) (Figure 4). There is strong support for the monophyly of New Caledonian *A.* section *Eutacta*, but low resolution within this clade. In the combined data, well-supported groupings (here, defined as having a BS≥75% and PP≥ 0.95) include *A. nemorosa* de Laub. + *A. luxurians* (Brongn. and Gris.) de Laub. + *A. columnaris* (G.Forst.) Hook., and multiple accessions of some species including *A. nemorosa*, *A. humboldtensis* J.Buchholz and *A. subulata* Vieill. In the plastid data set, the three accessions of *A. schmidii* are also resolved with strong support, but were not included in the combined analyses because ITS2 sequences could not be obtained (Figures 2–4).

Molecular dating of Araucariaceae

Assessment of fossil constraints. We used an expanded sampling of taxa and four plastid loci to infer araucarian phylogeny using a BRC model (Figure 1). With respect to relationships, these data are largely consistent with the results reported above.

Scaling factor distributions estimated under a BRC approach with unconstrained branch lengths are shown in Figure 5. Of the potential calibrations tested, the highest S_i estimates were associated with the '*Eutacta*' fossil constraint, although assigning an Early Jurassic age to the *Araucaria* section *Eutacta* crown, *Araucaria* crown or Araucariaceae crown (nodes 6, 7 and 8, respectively; Table 1) results in a distribution of S_i values that consistently exceeds the age of the earliest known conifer fossils (i.e.> c. 310 Ma), or indeed, the oldest known land plants. Similarly, constraining the *Agathis* crown node to a Late Paleocene age returned unrealistic S_i estimates. After excluding these outliers, we identified the constraint on node 5 as returning the highest empirical scaling factor, and 4 potential clade age calibrations (fossils associated with nodes 1, 2 3 and 9) with S_i intervals that substantially overlap with node 5.

Araucarian divergence times. Five fossil constraints with overlapping S_i intervals (Figure 5; Table 1) were used to estimate araucarian divergence times, varying the mean of the log normal calibrations priors to model taphonomic bias. In general, node age estimates substantially overlapped, irrespective of the log normal prior mean but differed substantially in terms of the upper versus lower values encompassed within the estimated 95% HPD. As there is no strong basis to favour a particular distribution of node age priors, we consider the full range of posterior value estimates as equally plausible (Table 2). Our estimates for the Araucariaceae stem (node 9) range from c. 284–202 Ma; the Araucariaceae crown (node 8), c. 138–60 Ma; the MRCA of *Agathis* + *Wollemia* (node 5) is estimated to have diversified at c. 91–55 Ma, while the *Agathis* crown (node 4) ranges from c. 55–12 Ma. The *Araucaria* crown (node 7) is estimated to have diverged c.103–29 Ma and our analyses support a preponderance of Cenozoic radiations among extant *Araucaria*. For instance, the MRCA of sections *Araucaria*, *Bunya* and *Intermedia* radiated between c. 58–9 Ma, while the diversification of crown *Eutacta* (node 6) is estimated at between c. 51–9 Ma. The radiation of the Pacific *Eutacta* (MRCA *A. heterophylla* + New Caledonian *Eutacta*) is Oligocene-Miocene age (c. 33–5 Ma), and the New Caledonian species appear to have diversified from the Miocene-Pliocene (c. 19–3 Ma) (Table 2).

Discussion

Phylogenetic relationships

Several studies have used molecular data to infer Araucariaceae phylogeny [8,11,15,17–19,47]. These are generally consistent regarding relationships among genera and among the major lineages within *Araucaria*. While the New Caledonian species of *Araucaria* section *Eutacta* consistently form a clade, establishing a robust understanding of interspecific relationships has proven difficult [7,8,37]. AFLP markers resolved three main genetic groups including a small leaved, large leaved and a coastal species group [7]. Escapa and Catalano [11] presented an analysis of a combined morphological and molecular data set, and found reasonable agreement with these interspecific groupings but their hypothesis differed from Gaudeul et al. [7] in the resolution of group inter-relationships. In the present study, employing a

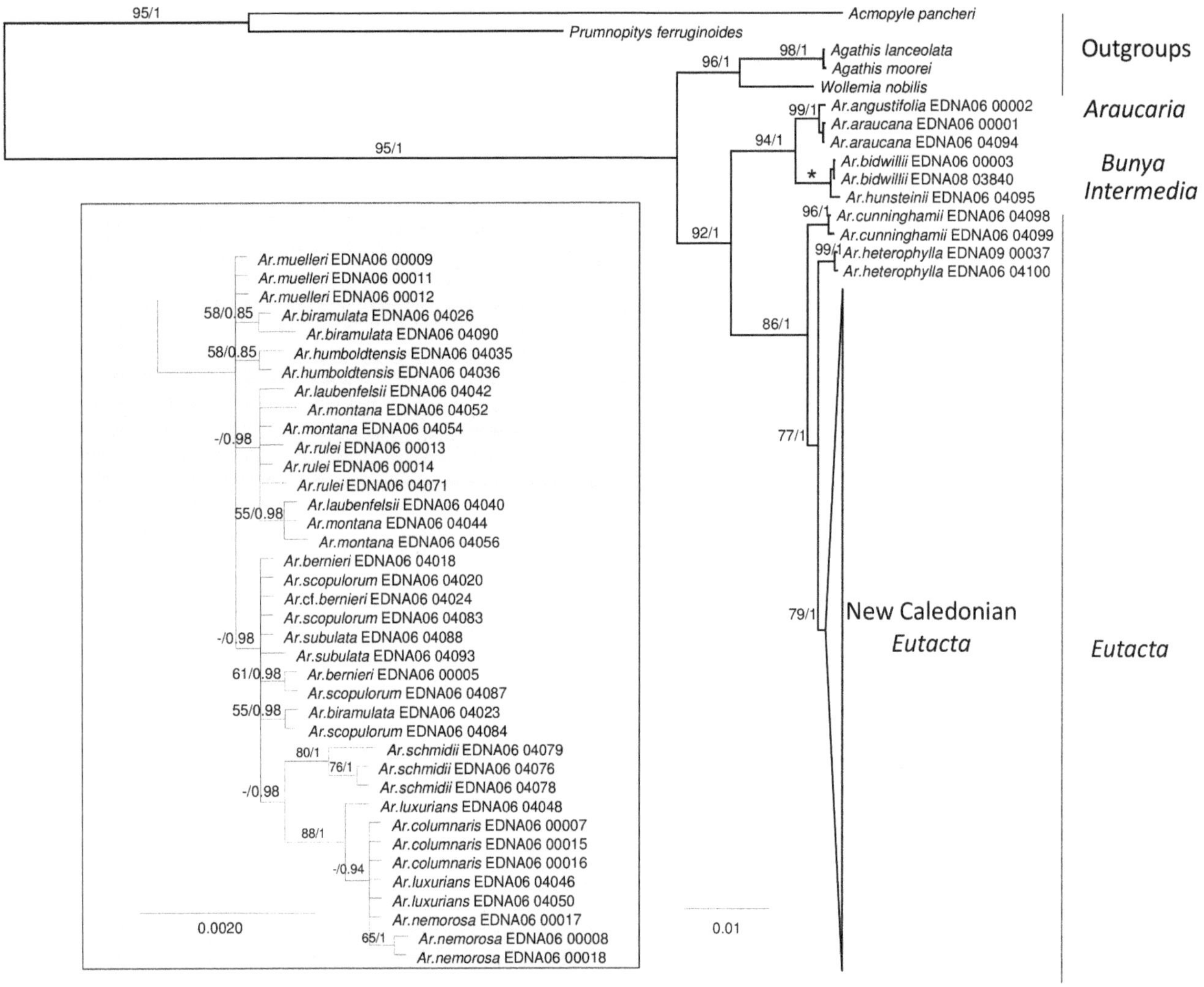

Figure 2. Araucariaceae phylogeny inferred from plastid DNA sequences using Bayesian optimisation criteria (50% majority rule topology). Support values, below the branch, are: ML bootstrap/Bayesian posterior probability. Branches marked with an asterisk have a BS of 100% and a PP of 1.0. Detail of the relationships among New Caledonian *Araucaria* is shown at the bottom left. Branch lengths are proportional to the inferred number of substitutions along that branch.

relatively large amount of sequence data (Table S1) did return some well supported resolutions among the New Caledonian *Araucaria* including the coastal group (*A. nemorosa, A. luxurians, A. columnaris*) as noted above. However, additional lines of evidence are clearly required to more fully resolve relationships and to test the monophyletic status of some of the species (Figures 3 and 4). Low levels of genetic divergence in previous studies and ours argue for a relatively recent and rapid radiation of the New Caledonian species [2,7,8].

Assessment of fossil constraints

Molecular dating has become a routine approach in molecular phylogenetic studies, and is central to evolutionary interpretation [16]. While there have been several advances in molecular clock methods, a key issue is the placement of fossil constraints, which, regardless of the clock methods employed, can strongly influence branch lengths. Several recent studies have used cross-validation approaches to identify sets of internally consistent constraints [38,48,49].

The fossil record of Araucariaceae extends to the Triassic, although in many cases the affinities of pre-Jurassic remains are doubtful [32]. Nevertheless, several fossil types from the Early Mesozoic have been assigned to extant araucarian lineages [9,10,33]. In the light of our analyses using unconstrained branch lengths, we find little support for this approach: assigning an Early Jurassic age to the *Araucaria* section *Eutacta* crown, *Araucaria* crown or Araucariaceae crown (nodes 6, 7 and 8, respectively; Table 1) results in a distribution of S_i values that are unrealistic in the context of the evolutionary time scale of land plants [50]. We interpret these values as reflecting incorrect phylogenetic placement of the '*Eutacta*' fossil constraint which would bias age estimates to be too old if retained in molecular clock analyses. Indeed, Biffin et al. [15] found that minimally constraining *Araucaria* to a Middle Cretaceous age resulted in implausibly high levels of molecular rate variation between stem and crown Araucariaceae or alternatively, unrealistically old estimates for the divergence of extant conifers. In the present study, assuming an Early Jurassic age for stem Araucariaceae (node 9) returned a distribution of S_i estimates that substantially overlapped with the

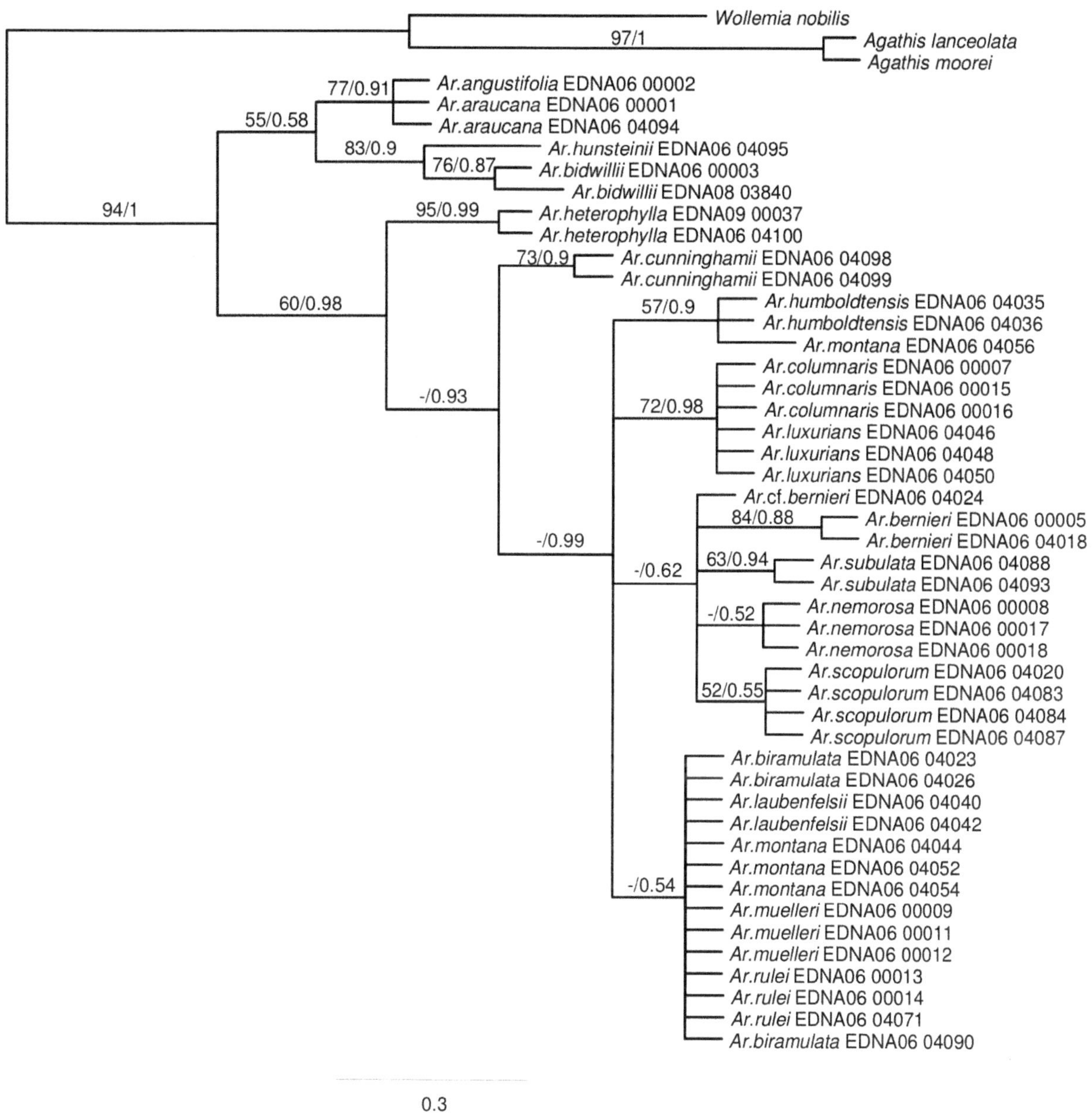

Figure 3. Araucariaceae phylogeny inferred from ITS2 DNA sequences using Bayesian optimisation criteria (50% majority rule topology). Support values are as detailed for Figure 2.

intervals for four other fossils, and in light of the conifer fossil record, provides a credible estimate of the age of the root [18] (Figures 1 and 5; Table 1).

Palaeobotanical implications

If the *Araucaria* crown group radiated not more than c. 100 Ma (Figure 1; Table 2) this casts doubt on the status of several relatively ancient fossil types with putative affinities to extant taxa (e.g. Axsmith et al., 2008). However, aspects of this record have previously been questioned on a number of grounds. Firstly, several authors have noted that the extant diversity of Araucariaceae almost certainly underestimates past diversity and morphological variation within the family [8,10,42,51]. In this context, araucarian fossils have been described that possess unique combinations of character states but these overlap with extant variation, e.g. *Yezonia* Stopes and Fujii [52], making the placement of fragmentary fossil material problematic. Secondly, in the light of the discovery of *Wollemia* and its phylogenetic placement in the family, it has been suggested that important characters used to diagnose *Araucaria* and its extant sections may be plesiomorphic for the family [11,42,51]. Thirdly, the low levels of molecular sequence divergence among extant *Araucaria* are difficult to reconcile with a Jurassic fossil record for these lineages [15]. Taken together, it has been suggested that the pre-Cretaceous fossil record of putative *Araucaria* may entirely represent stem Araucariaceae, or (at best) stem lineages of the extant genera [11,42]. However, the molecular dates presented here are not inconsistent with a hypothesis that crown *Araucaria* had evolved by the late Early Cretaceous as indicated by fossil taxa assigned to living sections, e.g. *A. grandiflora* Feruglio and *A.*

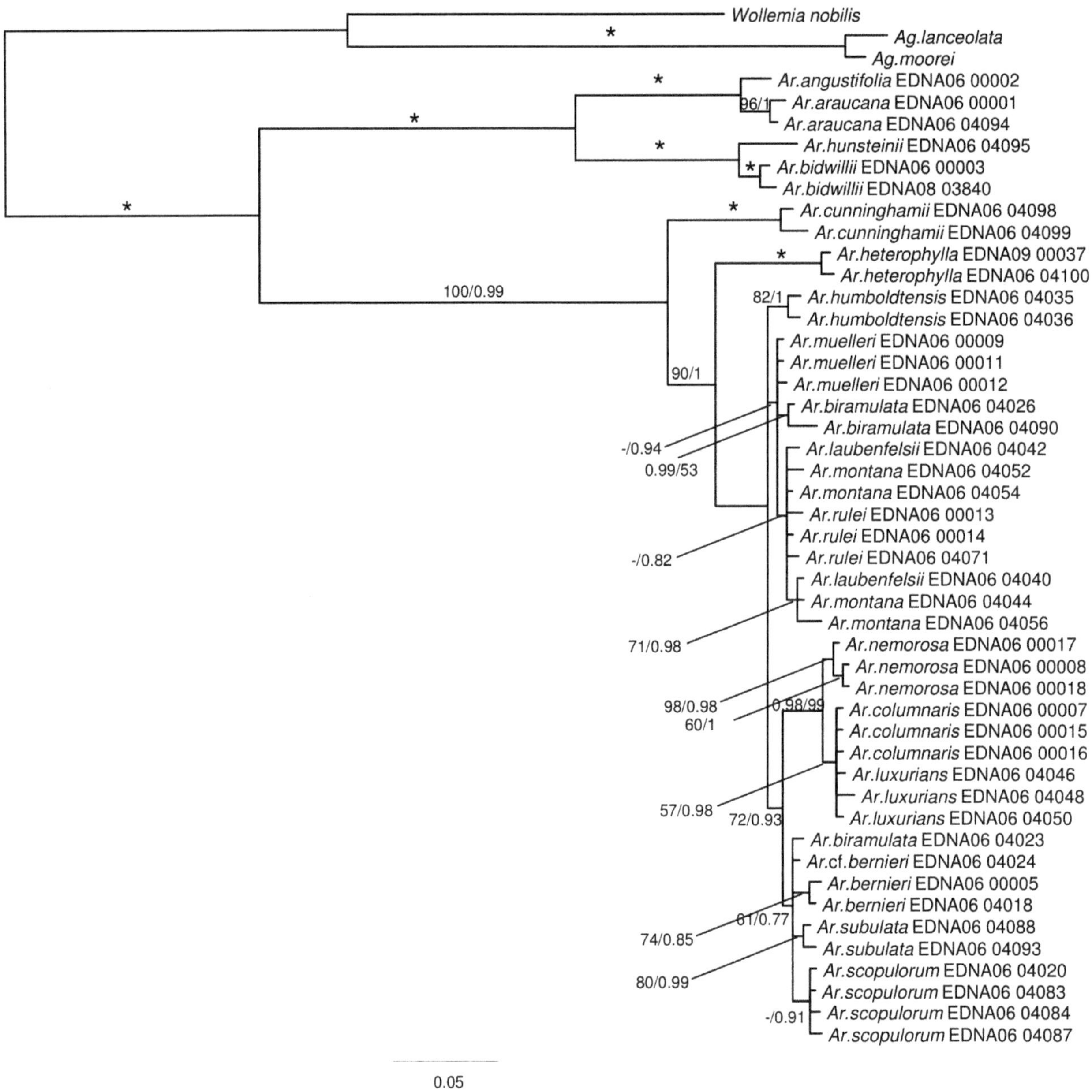

Figure 4. Araucariaceae phylogeny inferred from a combined plastid and ITS2 DNA data set using Bayesian optimisation criteria (50% majority rule). Support values are as detailed for Figure 2.

otwayensis Cantrill, estimated at c. 105–95 Ma from Patagonia and southern Australia and referred to sections *Araucaria* and *Eutacta*, respectively [33,53,54] (Table 2, Figure 1).

The fossil record for the crown *Agathis* clade (i.e. *Agathis*+ *Wollemia*) extends from the Late Cretaceous in reasonable agreement with our molecular age estimates for the corresponding nodes (Figure 1, Table 2). While there are no formal descriptions from the macrofossil record, pollen that is almost indistinguishable from that of extant *Wollemia nobilis* (*Dilwynites* W.K.Harris sp.) is first recorded in Turonian (c. 93–89 Ma) strata of Australia and the Maastrichtian (c. 70–65 Ma) of New Zealand, representing the oldest probable record of that genus [51] (but see [55]). In a recent review of the macrofossil record of *Agathis* it was concluded that the oldest unequivocal remains for that genus are from the Late Palaeocene (c. 55–50 Ma) of Southern Australia [30]. Putative *Agathis* macrofossils have been recorded from the Cretaceous but lack organic preservation, e.g. *A. victoriensis* Cantrill of the Lower Cretaceous, c. 110 Ma, of Southern Australia [53] or include characters that are atypical of modern *Agathis* [56,57] and therefore remain equivocal [30]. Using a cladistic analysis of morphological characters, the fossil taxa *Wairarapaia mildenhallii* Cantrill and Raine, and *Emwadea macrocarpa* Dettmann, Clifford and Peters, were resolved as stem group lineages of the *Agathis*-*Wollemia* clade [11] and date from the Early Cretaceous of New Zealand and Australia, respectively [42,58].

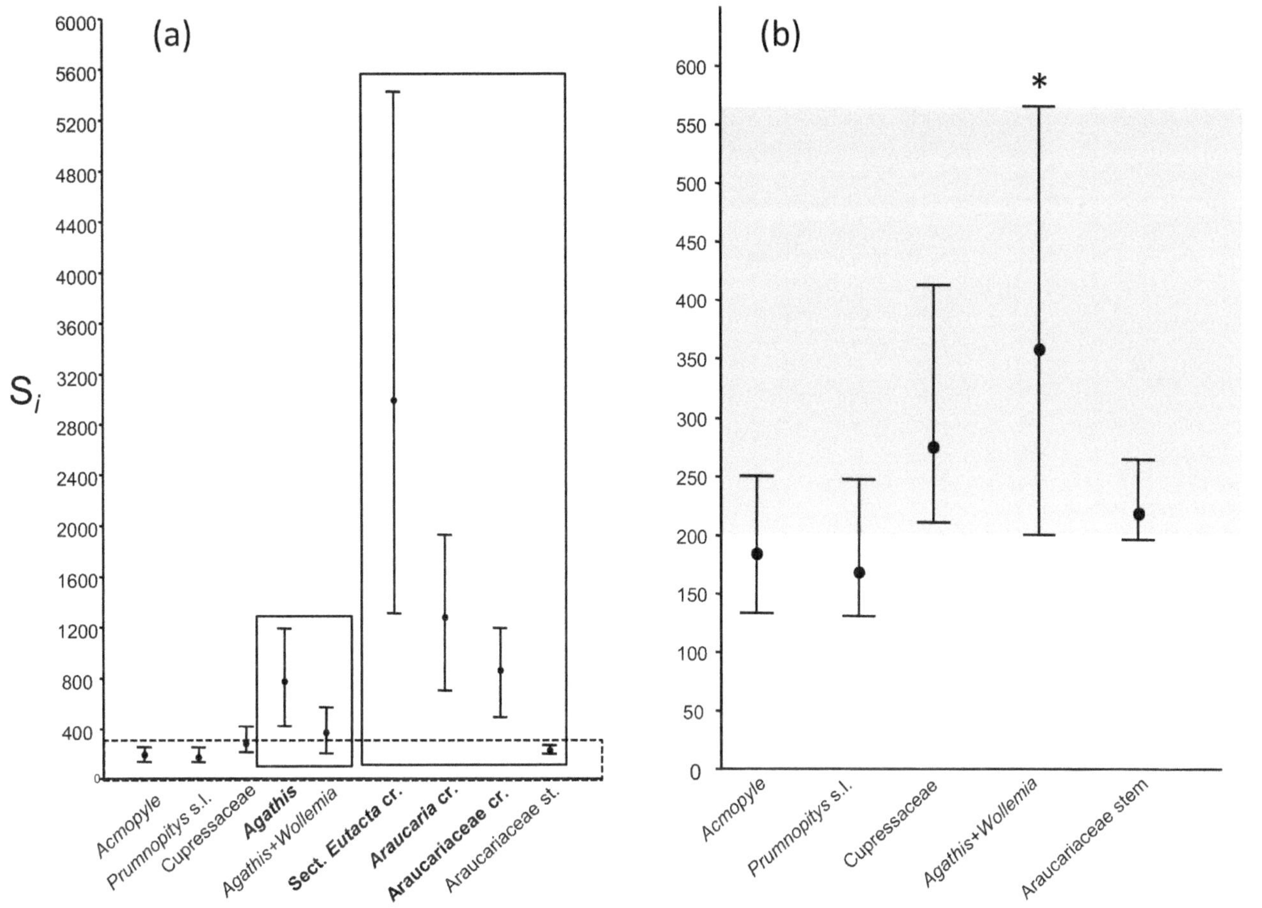

Figure 5. Assessment of calibration constraints using empirical scaling factor (S_i) estimates. The mean (circle) and 95% highest posterior density interval (lines) of S_i is shown for each proposed calibration, including: (a) alternative nodal placements of fossil *Agathis* and *Araucaria* (vertical boxes). In (a), the S_i intervals are contrasted with the estimated age of conifers (c. 320 Ma; horizontal box). In (b), candidate fossils with S_i estimates that consistently exceeded the maximum age of conifers have been removed. The grey shading indicates the region of overlap in S_i intervals for the five retained calibrations.

Biogeographic implications

The present distribution of *Araucaria*, which is Gondwanan in character, is often interpreted as relictual, reflecting regional (Northern Hemisphere) extinction, localised (Southern Hemisphere) contraction and Gondwanan tectonic events [9,10]. Our molecular dating scenario is consistent with aspects of this view. For instance, there is evidence for the presence of *Araucaria* in Europe until c. 66 Ma, which has been proposed as the youngest

Table 2. Node age estimates (millions of years) for Araucariaceae estimated from the BRC analyses of DNA sequences (Prior 1, log normal prior mean = minimum fossil age +10%; Prior 2, log normal prior mean = minimum fossil age +50%).

node	age estimate prior 1 (median[95%HPD])	age estimate prior 2 (median[95%HPD])	range
node 4 (*Agathis* crown)	26 (47–12)	31 (55–13)	**55–12**
node 5 (*Agathis*+*Wollemia*)	61 (77–55)	72 (91–60)	**91–55**
node 6 (*Eutacta* crown)	21 (43–9)	25 (51–11)	**51–9**
node 7 (*Araucaria* crown)	55 (87–29)	64 (103–34)	**103–29**
node 8 (Araucariaceae crown)	81 (122–60)	94 (138–68)	**138–60**
node 9 (Araucariaceae stem)	225 (233–202)	250 (284–224)	**284–202**
Pacific *Eutacta* (*A. heterophylla*+New Caledonia)	14 (28–5)	16 (33–7)	**33–5**
New Caledonian *Eutacta*	7 (15–3)	9 (19–3)	**19–3**

The preferred node age estimates (values in bold) encompass the full range of values included in the 95% HPD of both prior scenarios. All tabulated nodes are supported by a PP of 1.0.

reliable record of the genus for the Northern Hemisphere [59]. In light of our findings, this date does not reject placement of these fossils within crown *Araucaria* (Figure 1) although in contrast to some interpretations of the Northern Hemisphere fossil record, our dated phylogeny suggests that the crown groups of the extant sections are derived entirely from Cenozoic Southern Hemisphere radiations. The regional (Northern Hemisphere) extinction of Araucariaceae was probably multicausal [10]. It has been linked to palaeoclimatic and vegetation change [59] as well as to short term environmental perturbations at the Cretaceous-Palaeogene (K-Pg) boundary (c. 65 Ma) following the asteroid impact at Yucatan (see also [34] where the virtual disappearance of *Araucaria* at the K-Pg boundary of New Zealand is noted). A recent study documents the survival of a seed fern lineage (*Komlopteris* Barbacka) into the Eocene of Tasmania [60]. These authors hypothesise that a delayed radiation of the angiosperms at high latitudes, and the remoteness of southern Gondwana from the Yucatan impact site may have afforded refuge to *Komlopteris* and other 'archaic' gymnosperm lineages into the Cenozoic [60]. Given this hypothesis, *Araucaria*, too might have persisted in isolated Gondwanan refuges following the K-Pg boundary event and radiated from these during the Palaeocene [34]. In this case, one implication is that the pre-Cretaceous Northern Hemisphere fossil record of Araucariaceae may have little direct bearing on the evolution of extant *Araucaria*.

The importance of Gondwanan tectonic vicariance versus long distance dispersal in the assembly of the Southern Hemisphere flora has been widely debated and generally, the latter appears to be prominent [61]. Dispersal is implicated by clade origination times (i.e. stem group age estimates) that are younger than the timing of geological events [5]. In the context of our data, vicariance is not rejected for the divergence of the clade that includes *Araucaria* section *Araucaria* (southern South America) and sections *Bunya* + *Intermedia* (Australia and New Guinea, respectively). Southern South America and Australia remained connected via Antarctica until the Neogene [62,63] and potentially harboured a widespread ancestor of these lineages [64].

The biogeography of New Caledonia has been considered as an example where geological and biological evidence are at odds [2,5,6,65]. New Caledonia supports an apparently ancient biota [3] despite the lack of firm geological evidence for a continuously emergent land surface following the Late Cretaceous-Palaeogene rifting of New Caledonia (as part of the continental fragment, Zealandia) from East Gondwana [4,65]. The re-emergence of New Caledonia c. 37 Ma places an upper limit on the age of the island's biota. However, the radiation of *Araucaria* section *Eutacta* on New Caledonia may be younger still if the emersion of Norfolk Island (c. 3.7 Ma) is assumed to constrain the possible age of Pacific *Araucaria* section *Eutacta* [2] (Figures 1 and 4). In the present study, we find that the relevant divergence (c. 33–5 Ma) is too old to support this hypothesis and implicitly rejects a single long distance dispersal event from Australia to Norfolk Island for the MRCA of Pacific *Araucaria* section *Eutacta*. Alternative plausible scenarios for the origin of *A. heterophylla* involve a dispersal event from Australia to New Caledonia and a subsequent dispersal from either New Caledonia or Australia. In either case, the second dispersal event requires an extinction event post-dating the emersion of Norfolk Island involving the *A. heterophylla* lineage in New Caledonia, or the Pacific *Araucaria* section *Eutacta* lineage in Australia. Given the uncertainty in divergence time estimates for the *Eutacta* crown (51-9 Ma), the origin of New Caledonian *Araucaria* could be explained by a single dispersal or alternatively by short range dispersals among ephemeral landmasses such as those identified or are believed to have existed on the South Norfolk Rise [66,67] and in the Greater New Caledonia region [4,14]. In either case, our data do reject a 'museum' type hypothesis for New Caledonian *Araucaria* and implicate at least some over water dispersal, given that the New Caledonian endemic clade has arisen entirely after the estimated timing of the Palaeocene-Eocene marine transgressions and has radiated in the Neogene (c. 19-3 Ma; Table 2, Figure 1).

Supporting Information

Table S1 Accession details of individuals used in the phylogenetic analyses. Country abbreviations in the 'Location' column are AU…Australia, AR…Argentina, CL…Chile, NC…New Caledonia. Superscripts denote cultivated material: 1 Royal Botanic Garden Edinburgh (UK), 2 Mount Lofty Botanic Garden (AU), 3 Koishikawa Botanic Garden (Japan), 4 Adelaide Botanic Garden (AU). Voucher numbers refer to herbarium specimens deposited in Edinburgh (E), University of Adelaide (ADU) or Allan Herbarium, Christchurch (CHR).

Table S2 Genbank accession numbers of all samples used for the phylogenetic analyses.

Acknowledgments

We are grateful to IRD Nouméa – in particular G. Dagostini, T. Jaffré, J. Munzinger and F. Rigault – for logistical support in the field, and to the Province Sud and Province Nord authorities of New Caledonia for granting collection permits. We thank the Adelaide Botanic Garden of South Australia for access to living collections. We also thank Bob Hill, Richard Winkworth and an anonymous reviewer for feedback on earlier versions of the manuscript. The Royal Botanic Garden Edinburgh is supported by the Scottish Government's Rural and Environment Science and Analytical Services Division.

Author Contributions

Conceived and designed the experiments: PMH RAE MK. Performed the experiments: MK AC MLH EB. Analyzed the data: MK EB PMH. Contributed reagents/materials/analysis tools: MK AC MLH RAE MFG MG PT RRM AJL. Wrote the paper: EB PMH MR RRM.

References

1. Myers N, Mittermeier RA, Mittermeier CG, da Fonseca GAB, Kent J (2000) Biodiversity hotspots for conservation priorities. Nature 403: 853–858.
2. Grandcolas P, Murienne J, Robillard T, Desutter-Grandcolas L, Jourdan H, et al. (2008) New Caledonia: A very old Darwinian island? Philos Trans R Soc Lond B Biol Sci 363: 3309–3317.
3. Morat P (1993) Our knowledge of the flora of New Caledonia: endemism and diversity in relation to vegetation types and substrates. Biodiv Lett 1: 72–81.
4. Pelletier B, editor (2006) Geology of the New Caledonia region and its implications for the study of the New Caledonian biodiversity. Nouméa, New Caledonia: Documents Scientifiques et Techniques, Vol. II 7, 2nd ed. IRD.
5. Murienne J (2009) Testing biodiversity hypotheses in New Caledonia using phylogenetics. J Biogeogr 36: 1433–1434.
6. Pillon Y (2012) Time and tempo of diversification in the flora of New Caledonia. Bot J Linn Soc 170: 288–298.
7. Gaudeul M, Rouhan G, Gardner MF, Hollingsworth PM (2012) AFLP markers provide insights into the evolutionary relationships and diversification of New Caledonian *Araucaria* species (Araucariaceae). Am J Bot 99: 68–81.
8. Setoguchi H, Asakawa Osawa T, Pintaud J-C, Jaffré T, Veillon J-M (1998) Phylogenetic relationships within Araucariaceae based on rbcL gene sequences. Am J Bot 85: 1507–1516.
9. Kershaw P, Wagstaff B (2001) The southern conifer family Araucariaceae: History, status, and value for paleoenvironmental reconstruction. Annu Rev Ecol Syst 32: 397–414.

10. Kunzmann L (2007) Araucariaceae (Pinopsida): Aspects in palaeobiogeography and palaeobiodiversity in the Mesozoic. Zool Anz 246: 257–277.
11. Escapa IH, Catalano SA (2013) Phylogenetic analysis of Araucariaceae: integrating molecules, morphology, and fossils. Int J Plant Sci 174: 1153–1170.
12. Heads M (2010) Biogeographical affinities of the New Caledonian biota: a puzzle with 24 pieces. J Biogeogr 37: 1179–1201.
13. Bromham L, Penny D (2003) The modern molecular clock. Nat Rev Genet 4: 216–224.
14. Ladiges PY, Cantrill D (2007) New Caledonia-Australian connections: biogeographic patterns and geology. Aust Syst Bot 20: 383–389.
15. Biffin E, Hill RS, Lowe AJ (2010) Did Kauri (*Agathis*: Araucariaceae) really survive the Oligocene drowning of New Zealand? Syst Biol 59: 594–602.
16. Crisp MD, Trewick SA, Cook LG (2011) Hypothesis testing in biogeography. Trends Ecol Evol 26: 66–72.
17. Knapp M, Mudaliar R, Havell D, Wagstaff SJ, Lockhart PJ (2007) The drowning of New Zealand and the problem of *Agathis*. Syst Biol 56: 862–870.
18. Leslie AB, Beaulieu JM, Rai HS, Crane PR, Donoghue MJ, et al. (2012) Hemisphere-scale differences in conifer evolutionary dynamics. Proc Natl Acad Sci USA 109: 16217–16221.
19. Liu N, Zhu Y, Wei ZX, Chen J, Wang QB, et al. (2009) Phylogenetic relationships and divergence times of the family Araucariaceae based on the DNA sequences of eight genes. Chin Sci Bull 54: 2648–2655.
20. Hollingsworth ML, Clark AA, Forrest LL, Richardson J, Pennington RT, et al. (2009) Selecting barcoding loci for plants: evaluation of seven candidate loci with species-level sampling in three divergent groups of land plants. Mol Ecol Res 9: 439–457.
21. Grivet D, Heinze B, Vendramin GG, Petit RJ (2001) Genome walking with consensus primers: application to the large single copy region of chloroplast DNA. Mol Ecol Notes 1: 345–349.
22. Shaw J, Lickey EB, Beck JT, Farmer SB, Liu W, et al. (2005) The tortoise and the hare II: relative utility of 21 noncoding chloroplast DNA sequences for phylogenetic analysis. Am J Bot 92: 142–166.
23. Demesure B, Sodzi N, Petit RJ (1995) A set of universal primers for amplification of polymorphic noncoding regions of mitochondrial and chloroplast DNA in plants. Mol Ecol 4: 129–131.
24. Hamilton MB (1999) Four primer pairs for the amplification of chloroplast intergenic regions with intraspecific variation. Mol Ecol 8: 521–523.
25. Chen S, Yao H, Han J, Liu C, Song J, et al. (2010) Validation of the ITS2 Region as a novel DNA barcode for identifying medicinal plant species. PloS ONE 5: e8613.
26. Zwickl D (2006) Genetic algorithm approaches for the phylogenetic analysis of large biological sequence datasets under the maximum likelihood criterion: PhD dissertation, The University of Texas, Austin.
27. Huelsenbeck JP, Ronquist F (2001) MrBAYES: Bayesian inference of phylogenetic trees. Bioinformatics 17: 754–755.
28. Ronquist F, Huelsenbeck JP (2003) MrBayes 3: Bayesian phylogenetic inference under mixed models. Bioinformatics 19: 1572–1574.
29. Rambaut A, Drummond A (2007) Tracer Version 1.5. http://tree.bio.ed.ac.uk/software/tracer/
30. Hill RS, Lewis T, Carpenter RJ, Whang SS (2008) *Agathis* (Araucariaceae) macrofossils from Cainozoic sediments in south-eastern Australia. Aust Syst Bot 21: 162–177.
31. Axsmith BJ, Escapa IH, Huber P (2008) An araucarian conifer bract-scale complex from the lower Jurassic of Massachusetts: implications for estimating phylogenetic and stratigraphic congruence in the Araucariaceae. Palaeontologia Electronica 11: 13A.
32. Stockey RA (1982) The Araucariaceae - an evolutionary perspective. Rev Palaeobot Palynol 37: 133–154.
33. Hill RS, Brodribb TJ (1999) Turner Review No. 2 - Southern conifers in time and space. Aust J Bot 47: 639–696.
34. Pole M, Vajda V (2009) A new terrestrial Cretaceous-Paleogene site in New Zealand-turnover in macroflora confirmed by palynology. Cretac Res 30: 917–938.
35. Hill RS, Carpenter RJ (1991) Evolution of *Acmopyle* and *Dacrycarpus* Podocarpaceae foliage as inferred from macrofossils in south-eastern Australia. Aust Syst Bot 4: 449–480.
36. Sinclair WT, Mill RR, Gardner MF, Woltz P, Jaffré T, et al. (2002) Evolutionary relationships of the New Caledonian heterotrophic conifer, *Parasitaxus usta* (Podocarpaceae), inferred from chloroplast trnL-F intron/spacer and nuclear rDNA ITS2 sequences. Plant Syst Evol 233: 79–104.
37. Escapa I, Cuneo R, Axsmith B (2008) A new genus of the Cupressaceae (sensu lato) from the Jurassic of Patagonia: Implications for conifer megasporangiate cone homologies. Rev Palaeobot Palynol 151: 110–122.
38. Dornburg A, Beaulieu JM, Oliver JC, Near TJ (2011) Integrating fossil preservation biases in the selection of calibrations for molecular divergence time estimation. Syst Biol 60: 519–527.
39. Marshall CR (2008) A simple method for bracketing absolute divergence times on molecular phylogenies using multiple fossil calibration points. Am Nat 171: 726–742.
40. Drummond AJ, Rambaut A (2007) BEAST: Bayesian evolutionary analysis by sampling trees. BMC Evol Biol 7: 214.
41. Drummond AJ, Ho SYW, Phillips MJ, Rambaut A (2006) Relaxed phylogenetics and dating with confidence. PLoS Biol 4: 699–710.
42. Cantrill DJ, Raine JI (2006) *Wairarapaia mildenhallii* gen. et sp nov., a new Araucarian cone related to *Wollemia* from the Cretaceous (Albian-Cenomanian) of New Zealand. Int J Plant Sci 167: 1259–1269.
43. McLean D, Owens B, Pendleton JL, Bodman D (2013) Pennsylvanian (Namurian and Westphalian) miospore assemblages from the west coast of Scotland. Rev Palaeobot Palynol 190: 1–14.
44. van Waveren IM, Abbink OA, van Hoof TB, van Konijnenburg-van Cittert JHA (2008) Revision of the Late Carboniferous megaflora from the De Lutte-06 well (Twente, the Netherlands), and its stratigraphical implications. Geol Mijnbouw-N J G 87: 339–352.
45. Axsmith BJ, Taylor TN, Taylor EL (1998) Anatomically preserved leaves of the conifer *Notophytum krauselii* (Podocarpaceae) from the Triassic of Antarctica. Am J Bot 85: 704–713.
46. Miller CN (1999) Implications of fossil conifers for the phylogenetic relationships of living families. Bot Rev 65: 239–277.
47. Stockler K, Daniel IL, Lockhart PJ (2002) New Zealand Kauri (*Agathis australis* (D. Don) Lindl., Araucariaceae) survives Oligocene drowning. Syst Biol 51: 827–832.
48. Ruane S, Pyron RA, Burbrink FT (2011) Phylogenetic relationships of the Cretaceous frog Beelzebufo from Madagascar and the placement of fossil constraints based on temporal and phylogenetic evidence. J Evol Biol 24: 274–285.
49. Rutschmann F, Eriksson T, Abu Salim K, Conti E (2007) Assessing calibration uncertainty in molecular dating: the assignment of fossils to alternative calibration points. Syst Biol 56: 591–608.
50. Kenrick P, Crane PR (1997) The origin and early evolution of plants on land. Nature 389: 33–39.
51. Chambers TC, Drinnan AN, McLoughlin S (1998) Some morphological features of Wollemi pine (*Wollemia nobilis*: Araucariaceae) and their comparison to Cretaceous plant fossils. Int J Plant Sci 159: 160–171.
52. Ohsawa T, Nishida H, Nishida M (1995) *Yezonia* a new section of *Araucaria* (Araucariaceae) based on permineralized vegetative and reproductive-organs of *A. vulgaris* comb. nov. from the Upper Cretaceous of Hokkaido, Japan. J Plant Res 108: 25–39.
53. Cantrill DJ (1992) Araucarian foliage from the Lower Cretaceous of southern Victoria, Australia. Int J Plant Sci 153: 622–645.
54. Del Fueyo GM, Archangelsky A (2002) *Araucaria grandifolia* Feruglio from the Lower Cretaceous of Patagonia, Argentina. Cretac Res 23: 265–277.
55. Macphail M, Carpenter RJ (2013) New potential nearest living relatives for Araucariaceae producing fossil Wollemi Pine-type pollen (*Dilwynites granulatus* W.K. Harris, 1965). Alcheringa 38: 135–139.
56. Lee DE, Bannister JM, Lindqvist JK (2007) Late Oligocene-Early Miocene leaf macrofossils confirm a long history of *Agathis* in New Zealand. N Z J Bot 45: 565–578.
57. Pole M (2008) The record of Araucariaceae macrofossils in New Zealand. Alcheringa 32: 405–426.
58. Dettmann ME, Clifford HT, Peters M (2012) *Emwadea microcarpa* gen. et sp nov.-anatomically preserved araucarian seed cones from the Winton Formation (late Albian), western Queensland, Australia. Alcheringa 36: 217–237.
59. van der Ham RWJM, Jagt JWM, Renkens S, van Konijnenburg-van Cittert JHA (2010) Seed-cone scales from the upper Maastrichtian document the last occurrence in Europe of the Southern Hemisphere conifer family Araucariaceae. Palaeogeogr Palaeoclimatol Palaeoecol 291: 469–473.
60. McLoughlin S, Carpenter RJ, Jordan GJ, Hill RS (2008) Seed ferns survived the end-Cretaceous mass extinction in Tasmania. Am J Bot 95: 465–471.
61. Sanmartin I, Ronquist F (2004) Southern Hemisphere biogeography inferred by event-based models: Plant versus animal patterns. Syst Biol 53: 216–243.
62. Lawver LA, Gahagan LM (2003) Evolution of Cenozoic seaways in the circum-Antarctic region. Palaeogeogr Palaeoclimatol Palaeoecol 198: 11–37.
63. Veevers JJ, Powell CM, Roots SR (1991) Review of seafloor spreading around Australia. I. Synthesis of the patterns of spreading. Aust J Earth Sci 38: 373–389.
64. Francis J, Ashworth A, Cantrill D, Crame J, Howe J, et al. (2008) 100 million years of Antarctic climate evolution: Evidence from fossil plants. In: Cooper A, Barrett P, Stagg H, Storey B, Stump E, Wise W, editors. Antarctica: A Keystone in a Changing World. Proceedings of the 10th International Symposium on Antarctic Earth Sciences. Washington, DC: The National Academies Press. pp. 19–27.
65. Hall R (2001) Cenozoic reconstructions of SE Asia and the SW Pacific: changing patterns of land and sea. In: Metcalfe J, Smith M, Morwood I, Davidson A, editors. Faunal and Floral Migrations and Evolution in SE Asia-Australasia. Lisse: Swets and Zeitlinger. pp. 35–56.
66. Heads M (2008) Panbiogeography of New Caledonia, south-west Pacific: basal angiosperms on basement terranes, ultramafic endemics inherited from volcanic island arcs and old taxa endemic to young islands. J Biogeogr 35: 2153–2175.
67. Meffre S, Crawford A, Quilty P (2006) Arc-continent collision forming a large island between New Caledonia and New Zealand in the Oligocene. ASEG Extended Abstracts, Australian Earth Sciences Convention 2006 Melbourne 2006: 1–3.

The Dinoflagellate *Lingulodinium polyedrum* Responds to N Depletion by a Polarized Deposition of Starch and Lipid Bodies

Steve Dagenais Bellefeuille, Sonia Dorion, Jean Rivoal, David Morse*

Institut de Recherche en Biologie Végétale, Département de Sciences Biologiques, Université de Montréal, Montréal, Québec, Canada

Abstract

Dinoflagellates are important contributors to the marine phytoplankton and global carbon fixation, but are also infamous for their ability to form the spectacular harmful algal blooms called red tides. While blooms are often associated with high available nitrogen, there are instances where they are observed in oligotrophic environments. In order to maintain their massive population in conditions of nitrogen limitation, dinoflagellates must have evolved efficient adaptive mechanisms. Here we report the physiological responses to nitrogen deprivation in *Lingulodinium polyedrum*. We find that this species reacts to nitrogen stress, as do most plants and microalgae, by stopping cell growth and diminishing levels of internal nitrogen, in particular in the form of protein and chlorophyll. Photosynthesis is maintained at high levels for roughly a week following nitrate depletion, resulting in accumulated photosynthetic products in the form of starch. During the second week, photosynthesis rates decrease due to a reduction in the number of chloroplasts and the accumulation of neutral lipid droplets. Surprisingly, the starch granules and lipid droplets are seen to accumulate at opposite poles of the cell. Lastly, we observe that cells acclimated to nitrogen-depleted conditions resume normal growth after addition of inorganic nitrogen, but are able to maintain high cell densities far longer than cells grown continuously in nitrogen-replete conditions.

Editor: Senjie Lin, University of Connecticut, United States of America

Funding: Financial support by the Natural Sciences and Engineering Research Council of Canada (NSERC) for discovery grants to DM and JR and by the Fond de Recherche du Québec- Nature et Technologies (FQRNT) for a studentship (SD-B). The funders had no role in study design, data collection and analysis, decision to publish, or preparation of the manuscript.

Competing Interests: The authors have declared that no competing interests exist.

* Email: david.morse@umontreal.ca

Introduction

Nitrogen (N) is an essential nutrient for all living organisms as it is required for biosynthesis of proteins and nucleic acids. While N is highly abundant in the atmosphere as dinitrogen gas (N_2), this chemical form is inaccessible to most organisms, which thus depend on diazotrophs, prokaryotes able to transform N_2 to bioavailable N forms. As a result, demand for N often exceeds its supply, which limits growth in many ecosystems. This is particularly severe in the open oceans where diverse phytoplanktonic species must compete for small amounts of bioavailable N and be able to survive periods of natural N depletion [1].

Dinoflagellates are important members of the phytoplankton that have developed a variety of strategies to cope with N stress [2]. These strategies have a particular importance for those dinoflagellate species able to form harmful algal blooms (HAB), or "red tides". Some HAB can produce toxins, which can have a strong negative impact on other parts of the ecosystem and on human health. Strategies allowing blooms to deal with N stress might appear counterintuitive, as natural and anthropogenic N addition rather than N deprivation is a factor often associated with HAB [3,4]. However, in some cases dense blooms can form and persist in oligotrophic environments where the measured inorganic N supply would seem to be insufficient to support their biomass, as was observed for *Karenia brevis* in the West Florida Shelf [5]. In this particular example it was proposed that organic N coming from decaying fish killed by the HAB explained the persistence of the dinoflagellate populations [5,6]. However, another possibility is that acclimation mechanisms triggered by N deprivation during a bloom could allow the dinoflagellates to maintain bloom-density population levels until new N becomes available. Unfortunately, testing this hypothesis in the environment would be difficult, because of the complex interplay between biotic and abiotic factors involved in bloom dynamics [4].

Lingulodinium polyedrum is a marine dinoflagellate usually studied as a model system in circadian biology, but is known to form HAB in various regions of the world, particularly along the coast of Southern California [7,8,9]. Laboratory cultures of this species can be readily transferred to media of different composition after filtration, which makes it a good candidate to test the effects of controlled nutrient conditions on its biochemistry and physiology. Interestingly, *L. polyedrum* grown in N-deprived artificial sea water was reported to survive up to 4 times longer than when grown in the normal nitrate (NO_3^-) enriched f/2-medium [10]. These findings implied that this species was able to acclimate to N limitation, but the underlying mechanisms were not evaluated. In this study, we have addressed this issue by monitoring the physiological responses of *L. polyedrum* to N stress. In particular, we note that cells acclimated to N limitation rose to high cell densities when new inorganic N was supplied and these for durations normally observed in *L. polyedrum* blooming populations in the environment. This is in sharp contrast to the behavior of non acclimated cells, which abruptly die after reaching their maximum cell density. We also note an accumulation of reduced carbon (C) in the form of starch granules and lipid bodies in N-limited cells. Curiously, these two forms accumulated in different regions of the cell.

Materials and Methods

Cell culture

Unialgal but not axenic *Lingulodinium polyedrum* (CCMP 1936, previously *Gonyaulax polyedra*) was obtained from the Provasoli-Guillard National Center for Marine Algae and Microbiota (East Boothbay, ME, USA). Cell cultures were either grown in normal f/2 medium prepared using Instant Ocean (termed day 0) or in f/2 lacking added N (f/2-N) for one or two weeks (termed day 7 or day 14). Day 0 cells acted as control during all experiments. To transfer cells to f/2-N (N-depleted medium), cultures grown in f/2 were filtered on Whatman 541 paper and washed with 200 ml of f/2-N medium before resuspension in f/2-N medium. Normal f/2 medium is made from Instant Ocean supplemented with 0.88 mM NO_3, and Instant Ocean alone contains 1 μM NO_3, 10 μM NH_3 and 3 μM dissolved organic N [11]. All cultures were grown under a 12 h light (40 μmol photons m^{-2} s^{-1} cool white fluorescent light) and 12 h darkness at a temperature of 18±1°C. This light/dark regime is termed LD, with LD 0 corresponding to lights on and LD 12 to lights off. Cells used for elemental analysis, the quantification of total proteins, amino acids, starch and neutral lipids were harvested by filtration and stored at −80°C until use.

Cell density measurements

Cells in multiple 10 μl aliquots of f/2 and f/2-N cultures were placed on microscope slides and were counted every 3 or 4 days using a Leica Wild M3B stereo microscope. After 17 days, a time when the cultures had typically reached their maximum cell density, 880 μM of $NaNO_3^-$ was added to each culture. Counts were continued until no cells remained swimming in the medium.

Elemental analysis

Measurements of total C and N contents were performed using a Carlo Erba NC2500 Elemental Analyzer (at Geotop-UQAM, Montreal, PQ, Canada). The analytical step was preceded by a preparative step, where cells were harvested, lyophilized, weighed and inserted into tin capsules. Results are reported as percent of total dry weight.

Protein and amino acid quantification

For total protein quantification, cultures grown in f/2 were split into f/2 and f/2-N cultures, and samples were taken before, as well as 3, 7, 10 and 14 days after the split. Protein from 30 mg wet weight of cells was extracted following a slightly modified Trizol (Life technologies) protocol as described previously [12]. Protein pellets were resuspended in 7 M urea, 2 M thiourea, 4% CHAPS and 20 mM dithiotreitol, and incubated overnight to completely solubilize the pellets. Protein concentrations were measured using Bio-Rad (Bradford) Protein Assay following the manufacturer's protocol.

Amino acid quantification was performed by HPLC using a protocol previously optimized for plant tissues [13]. Samples were lysed mechanically in 80% (v/v) ethanol for 2 min at 4°C in a bead beater (BioSpec products). They were then extracted at 70°C in 80% (v/v) ethanol and fractionated into neutral, anionic and cationic fractions by preparative ion exchange purification [14]. The cationic fractions were aliquoted and evaporated to dryness before derivatization with the AccQ Fluor reagent kit (Waters). The amino acids in the aliquots were analyzed on a Waters HPLC system controlled by the Empower Pro software and equipped with a 600 controller, a 717 Plus refrigerated automatic sample injector, a 2996 Diode Array Detector and an AccQ.Tag Amino Acid Analysis Column (Waters). HPLC parameters were set according to manufacturer's recommendations. Quantification was done using standard curves from commercial amino acids.

Photosynthetic measurements

Chlorophyll *a* content was measured after extraction of 50 mg wet weight of cells in 100% acetone. The pigment-containing supernatants were collected after centrifugation and cell residues were re-extracted with small volumes of acetone until colorless. Supernatants were combined and chlorophyll optical densities were measured using a Cary 100 Spectrophotometer (Varian). Concentrations were calculated using the equations presented by [15].

Photosynthetic C fixation rates in normal and N-depleted cultures were calculated from incorporation of ^{14}C to an acid-soluble form. For each assay, 1.48 kBq (= 0.04 μCi) of radiolabelled bicarbonate ($NaH^{14}CO_{3(aq)}$, ICN, 310 MBq $mmol^{-1}$) was added to 5 mL of cell culture with dissolved inorganic carbon ≈2 mM [16] and the culture incubated for 90 min under normal lighting (40 μmol photons m^{-2} s^{-1} cool white fluorescent light). The ^{14}C labeling was quenched by adding HCl to a final concentration of 1 N and gaseous radioactivity was allowed to escape from unsealed samples vials during an overnight incubation in a fume hood. Samples were transferred into scintillation vials with 2 mL Scintiverse (Fisher Scientific) then counted in a TriCarb 2800TR scintillation counter (Perkin Elmer). Values were reported as the number of disintegrations per minute (DPM) from light-induced C fixation after subtraction of the number of DPM for identical subcultures incubated in the dark. Samples were normalized to cell number.

Starch quantification

For starch analysis, cells from all cultures were harvested at both LD 0 and at LD 12 and were lysed mechanically in 80% (v/v) ethanol for 2 min at 4°C in a bead beater (BioSpec products). Starch was assayed enzymatically as described [17]. Briefly, the procedure included the removal of soluble sugars with 80% (v/v) ethanol washes at 70°C, the solubilization of starch in 1 M KOH and its conversion into glucose by amyloglucosidase. The glucose concentration was determined using hexokinase (HK) to phosphorylate the glucose and glucose-6-phosphate dehydrogenase (G6PDH) to convert the glucose-6-phosphate to 6-phosphogluconolactone and simultaneously reduce NAD^+ to NADH. This reduction was monitored spectrophotometrically at 340 nm with the amount of NADH produced proportional to the amount of glucose in the samples.

Nile red quantification of neutral lipid

Neutral lipids were quantified with Nile red using a fluorometric assay previously optimized for various microalgae [18]. Two mg wet weight cell pellets were resuspended in 25% dimethyl sulfoxide (DMSO) in 1.5 mL tubes with glass beads (0.5 mm, BioSpec products) and were lysed mechanically for 2 min at 4°C in a bead beater (BioSpec products). Small aliquots of these cell solutions were added to a black 96-microplate and the volume was adjusted to 297 μL with 25% DMSO. Fluorescence was recorded using a SpectraMax M5 Microplate Reader (Molecular Devices) under high scan control and high PMT detector voltage mode, using emission and excitation wavelengths of 530 nm and 575 nm, respectively. These wavelengths were selected based on the emission/excitation spectra of the triolein standard used for quantification. Three μL of a 50 μg mL^{-1} Nile red solution was added to each well to a final concentration of 0.75 μg mL^{-1} and the plate was incubated at 37°C for 10 min. Fluorescence values after Nile red-staining were recorded and, after subtraction of the pre-stain fluorescence values, were used for quantification using a triolein standard curve.

Microscopy

Starch was observed in cells from the a culture taken at both LD 0 or LD 12, after fixation in 70% (w/v) ethanol, and staining using a solution containing 0.5% (w/v) I_2 and 1% (w/v) KI. Stained cells were visualized with a bright-field Axio Imager M1 microscope (Zeiss).

The contrast of all images was adjusted with Adobe Photoshop 8.0 using the same tonal curve to provide better contrast between the black-stained starch and the dark brown cell background.

Chloroplasts and lipid bodies of day phase living cells were visualized with a LSM 5 Duo confocal microscope (Zeiss). Chlorophyll autofluorescence was observed using a UV-diode (405 nm) excitation and a long pass filter (>575 nm) emission. For lipid bodies, 1 mL aliquots were incubated at 18°C in the dark for 20 minutes with Nile red at a final concentration of 10 μg mL^{-1}. Stained lipid bodies were observed using an argon laser (excitation at 514 nm) and emissions at 575 and 615 nm. Differential interference contrast (DIC) images were taken at the same time as the fluorescent images. Cells with the ventral side up were selected for imaging so that the optical slices would have a similar appearance with respect to the location of the C-shaped nucleus and the pyrenoids. All microscope parameters used were the same between samples. Nile red images were modified for contrast only using Adobe Photoshop 8.0.

For transmission electron microscopy, cells were harvested at LD 0, fixed with 2% glutaraldehyde in 0.4 M NaCl, 0.05 M PBS (pH 7.2) for 2 h and then washed 3 times in 0.4 M NaCl, 0.05 M PBS (pH 7.2). Fixed cells were dehydrated following standard procedures and embedded in Spurr's resin as recommended by the manufacturer. Thin sections were contrasted with a saturated solution of uranyl acetate in 50% (w/v) ethanol and immediately observed with a JEOL JEM-1010 electron microscope operating at a voltage of 80 keV.

Statistical analyses

All experiments were run in triplicate ($n = 3$) and results were presented as means ± SE. Analysis of variance or Student's t-tests were used for all data. Statistical tests were performed using the JMP software (SAS).

Results

To examine the acclimation mechanisms of *Lingulodinium* to N stress, we first determined the physiological responses after transferring N-replete cultures (in standard f/2 medium) to a medium lacking added nitrate (f/2-N). Cultures grown in f/2 medium generally show robust growth for 10–14 days after which the cell densities remain high for several days then drop precipitously (Fig. 1). In some cases, where average growth rates decrease at earlier times, the cultures can survive for slightly longer (Fig. 1B). In contrast, cell growth in f/2-N medium stalled immediately after transfer (Fig. 1A). This inhibition of cell growth is a direct consequence of N depletion, as the addition of nitrate to f/2-N cultures on the 17th day restarted cell growth (Fig. 1B). Interestingly, some cultures acclimated to N depletion demonstrated an ability to survive at high cell densities far longer than cells grown in the N-replete f/2 medium (Fig. 1B). A similar robustness in culture lifetime was also observed for cases where cell densities declined to intermediate values. The cultures acclimated to low N thus appear able to avoid the precipitous collapse of cell densities typically observed for cultures grown in N replete conditions.

Elemental analysis revealed that proportion of cell dry weight as N for cultures grown in f/2-N decreased to almost half their original value after 7 or 14 days (Fig. 2A), while the proportion of cell dry weight as C remained unchanged (Fig. 2B). As these values represent the fractional dry weight allocated to a particular element rather than the absolute amount of that element, the smaller percentage of N observed in f/2-N-grown cells implies that the absolute amounts of other elements are increasing while absolute cellular N levels remain constant.

The bulk of cellular N is normally sequestered in proteins, and so a reduction in the proportion of N should be accompanied by a reduction in the proportion of protein. To test this, protein levels were measured as a function of time for cultures growing in f/2 and f/2-N media. After 7 days in f/2-N, the protein concentration declined to half the value of the f/2-cultures and remained roughly constant for the following days (Fig. 3A), consistent with the decrease observed for elemental N. The decrease in protein

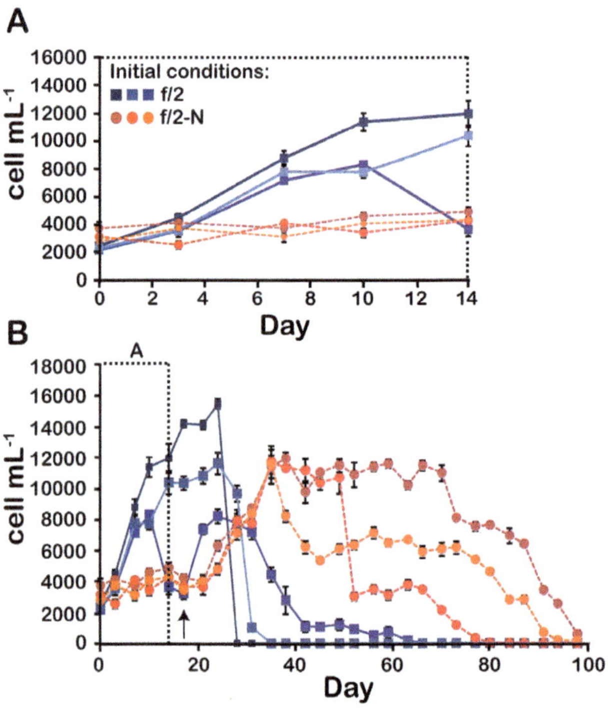

Figure 1. Growth of N-replete and N stressed cultures. The cell density in 6 independent cultures, 3 supplemented with 880 μM $NaNO_3^-$ (f/2) and 3 without added N (f/2-N), was assessed until no swimming cells remained in the culture flasks. The first 14 days are shown on an expanded time scale (A) or over the 100 days of the experiment. The time at which 880 μM $NaNO_3^-$ was added to all cultures is shown by a black arrow. Results are mean ± SE of 3 technical replicates made on cell counts.

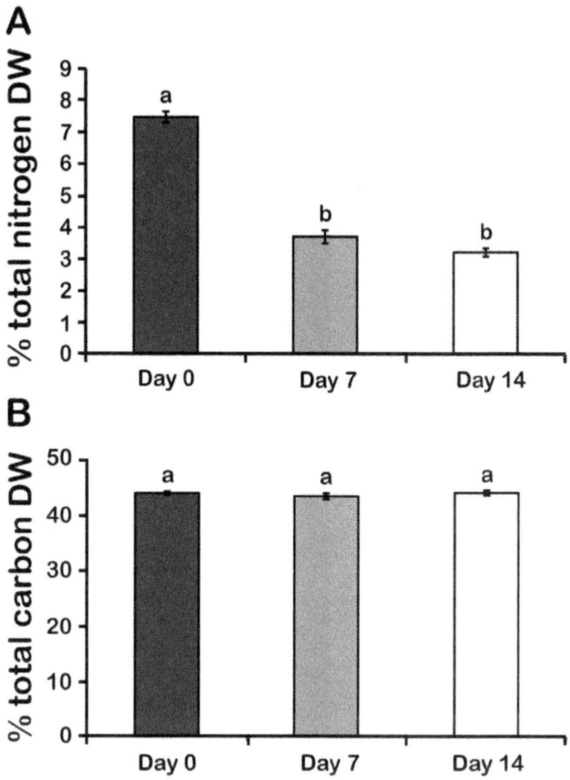

Figure 2. Elemental analysis shows a decreased N content in N stressed cells without a change in their C content. (A) Dry weight percent of total N and B) total C. Results are mean ± SE (n = 3). Statistically different results ($p<0.05$) are marked with a different letter (Analysis of variance). DW; Dry weight.

observed for cells grown in normal f/2 medium after 14 days may involve a decrease in protein synthesis rates such as observed in stationary phase cells of *Escherichia coli* [19]. A consequence of this, should protein degradation rates remain unchanged, would be a decrease in protein levels. However, we do not know what signals or environmental factors cause the transition to stationary phase in *Lingulodinium*.

To test if the changes in protein levels observed for f/2-N cultures were also reflected in the free amino acid (FAA) pools, we next characterized the FAA profile of N-depleted cultures using a previously described HPLC-based protocol [13]. Of the 14 FAAs that were quantified by this method, only Gln and Arg levels were found to decrease after N deprivation consistent with the requirement of NH_3^+ for their biosynthesis. Most of the FAAs measured, in particular the most highly abundant ones, did not vary significantly during N stress. However, levels of several of the less abundant FAAs, including Tyr, Met, Ile and Leu, were observed to increase after 7 or 14 days in f/2-N. When the FAA data are considered as a whole, the total N in FAA shows no significant change between f/2 and f/2-N cultures (Fig. S1A, S1B). However, the calculated Gln/Glu ratio, considered a useful indicator of N assimilation rates [20] was ~4 times lower in N-deplete than in N-replete cells (Fig. S1C). Thus, despite a constant total FAA pool, N-depleted cultures show the effect of a severe decrease in N assimilation.

When compared to the f/2-cultures, chlorophyll *a* concentrations gradually diminished by two- and six-fold by 7 and 14 days of incubation in f/2-N, respectively (Fig. 4A). However, a concomitant decrease in the rate of photosynthesis, as measured by ^{14}C-incorporation in acid-soluble compounds, was only observed after 14 days in N deprivation (Fig. 4B), with CO_2 fixation rates actually showing an increase after 7 days N deprivation. One possible explanation for this is that the initial decrease in chlorophyll aids C fixation by decreasing O_2 evolution rates [21], but by 14 days the ability of the cells to produce photosynthetic reductant is no longer sufficient to maintain high CO_2 fixation rates. Interestingly, observation of chlorophyll autofluorescence in living cells by confocal microscopy (Fig. 4C–K) supports the steady decrease in chlorophyll levels during N- stress from 7 (Fig. 4G) to 14 days (Fig. 4J). Cells observed after 14 days also show a decreased number of chloroplasts (Fig.4J).

The ability of the cells to maintain high levels of C fixation for at least one week in N-deplete conditions suggested that reduced C might accumulate in these cells. To test this, we assessed the levels of starch and neutral lipids, two C pools previously reported to increase during N stress in other dinoflagellates and microalgae [22,23,24,25]. *L. polyedrum* cultures show a clear increase in starch levels after 7 and 14 days of N depletion (Fig. 5A). In the leaves of higher plants, starch is synthesized during the day and degraded at night [26], and the data show that a similar rhythm is found in *L. polyedrum* grown in the presence of N. In these cells, starch levels were substantial at dusk (LD 12) yet almost undetectable at dawn (LD 0). However, this daily variation was abolished in N-deplete cells after 14 days, suggesting the reduced C stores are not being used for metabolism. Curiously, a higher variation in starch levels was repeatedly observed in cells grown for 7 days in f/2-N and measured at LD 0 indicative of a greater heterogeneity in starch metabolism at that stage. Starch accumulation can also be visualized microscopically using iodine to stain the starch granules (Fig. 5B–G). Most of the cells at day 0 did not contain starch at the end of the night (LD 0), but contained substantial amounts at the end of the day (LD 12) (Fig. 5B, C). In addition, higher levels of starch accumulate in N-deplete cultures (Fig. 5D–G). Intriguingly, starch granules always accumulated at the posterior end of the cells. This suggests that starch is not localized within chloroplasts, as these organelles are distributed over the entire cell. A cytosolic localization of starch granules has been reported in other dinoflagellates, in contrast to the plastidic localization seen in land plants and green algae [27,28,29].

Neutral lipids, particularly triacylglycerols (TAGs), also accumulate in N-deplete *Lingulodinium* (Fig. 6). TAG levels in cell extracts, measured using Nile red, were ~2 times and ~10 times higher than in cells at day 0 after 7 and 14 days in f/2-N, respectively (Fig. 6A). Again, this form of C storage can be visualized in cells microscopically (Fig. 6B–J). Abundance and size of lipid bodies clearly increased with the duration of N stress (Fig. 6C, F, I). However, in sharp contrast to the starch granules, TAGs accumulated preferentially at the anterior end of the cell (Fig. 6D, G, J).

To confirm the asymmetrical distribution of TAGs and starch granules, cells were examined using transmission electron microscopy to visualize both types of C stores in the same cell (Fig 7). Cells at day 0 and day 14 cells were fixed at LD 0, a time at which neither lipid bodies nor starch were observed in the day 0 cell (Fig. 7A). However, the two types of C stores have clearly accumulated at opposite ends of the cell by 14 days in f/2-N medium (Fig. 7B). Lipid bodies are located at the anterior end and appear dark due to the lipophilic nature of the osmium tetroxide stain used to contrast the sections (Fig. 7C) while starch granules appear white and are located at the posterior end of the cell (Fig. 7D). The white striations observed in the lipid droplets are likely due to a sectioning artifact, as their orientation is the same

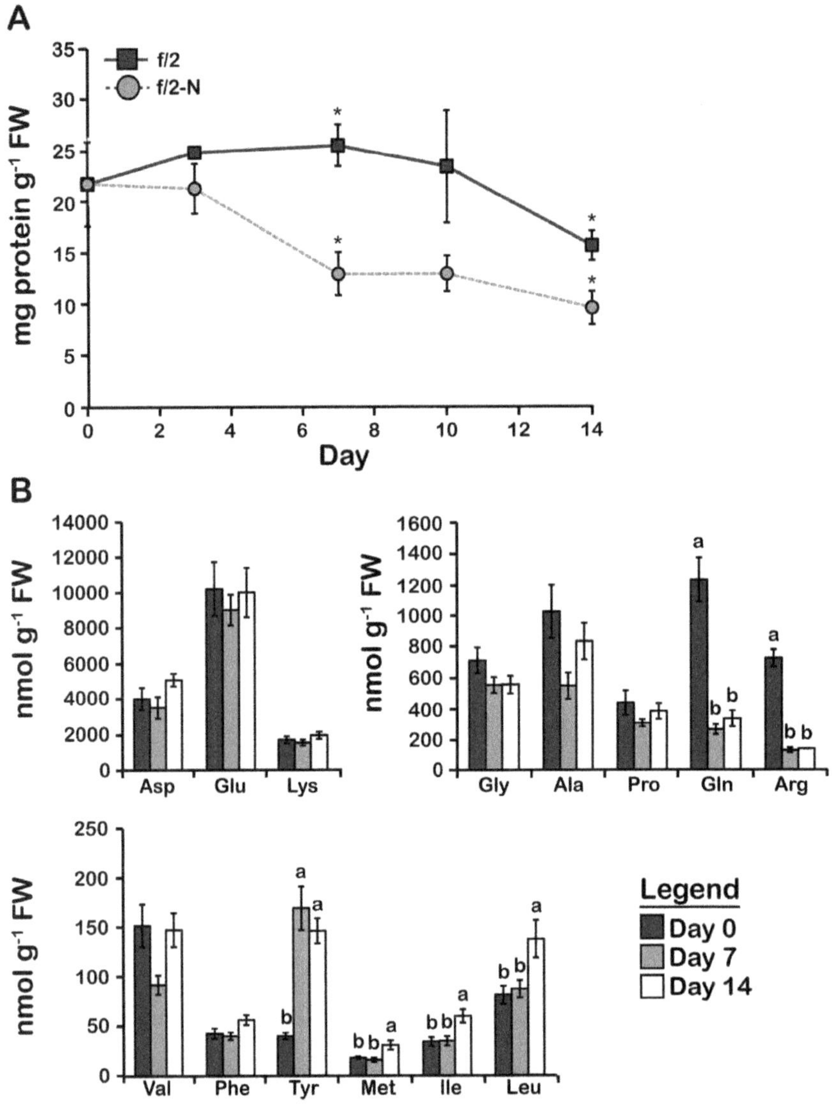

Figure 3. Changes in the total protein content and free amino acid profile in N stressed cells are consistent with a decrease in N assimilation. A) Total protein content. B) Free amino acids, classified into three groups based on their relative abundance. Results are mean ± SE (n = 3). Statistically different results ($p<0.05$) are marked using either a different letter (Analysis of variance) or an asterisk (Student's t-test). FW; Fresh weight.

for all cells in a section independent of how the individual cells are orientated in the section. Lastly, we also note that chloroplasts appear smaller and are less abundant in the day 14 cell when compared to the day 0 cell, in agreement with the confocal images (Fig. 4D, G, J).

Discussion

In this study, we show that *Lingulodinium polyedrum* cultures with cell densities similar to those observed in blooming populations in the field ($\sim 10^7$ cells L^{-1}), are able to survive and maintain these elevated cell densities for more than 2 weeks if previously acclimated to N-deplete conditions (Fig. 1) [30]. The cells cease cell division upon transfer to the N-depleted medium, a response to N limitation commonly observed in other dinoflagellates, plants and algae [31,32,33,34,35,36]. The arrest of growth in f/2-N was immediate, not gradual, and suggests that *L. polyedrum* grown in a NO_3^- replete medium does not store N in a form that could have sustained growth and division under N stress. This differs from the N storage observed in *Symbiodinium* spp. where uric acid crystals accumulate following increases in environmental N [37]. The crystalline inclusions rapidly dissolve, but it has been suggested that the resulting N is remobilized for long-term storage in other forms within the cytosol. Either *L. polyedrum* does not share this long-term storage ability with *Symbiodinium* or its storage capacity is insufficient to support cell division under N limitation. Another possibility is that sensors at the *Lingulodinium* plasma membrane, similar to the CHL1 nitrate transporter in *Arabidopsis*, detected an absence of extracellular NO_3^- and signaled a stress response, which included an early arrest of the cell cycle [38].

C and N metabolism are normally coupled because synthesis of amino acids and nucleotides is dependent upon C skeletons and energy provided by photosynthesis [35]. During N depletion, *L. polyedrum* was restricted in its ability to assimilate N, while its

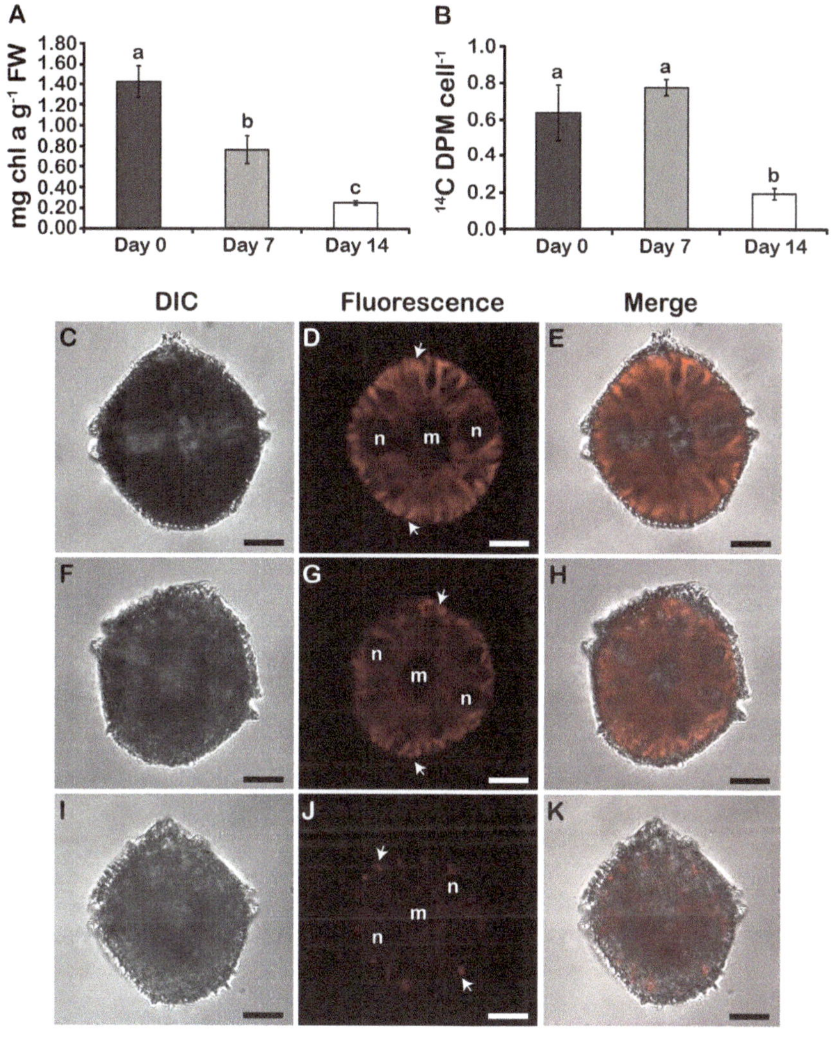

Figure 4. Photosynthesis decreases in N stressed cells. A) Chlorophyll *a* levels. FW; Fresh weight. B) Photosynthetic rates of ^{14}C fixation. Results are mean ± SE (n = 3). Statistically different results ($p<0.05$) are marked with a different letter (Analysis of variance). (C–K) DIC, chlorophyll autofluorescence and merged images of day phase cells at day 0 (C–E), day 7 (F–H) and day 14 (I–K). Chlorophyll intensity is highest in the periphery of day phase cells cells (white arrows) due to a separation of Rubisco and the light harvesting peridinin-chlorophyll *a*-protein within individual chloroplasts (Nassoury et al., 2001). All cells were pictured from a ventral view. In this orientation, two ends of the C-shaped nucleus (n) surround the centrally located ER and Golgi membranes (m). Scale bars are 10 μm.

photosynthetic machinery was, for at least a week, still fully functional. This is indicated by the ^{14}C fixation rates, which only showed a significant decrease after 14 days of N deprivation (Fig. 4B). We propose that during their first week of N-limitation, the cells divert their excess C toward the synthesis of starch and neutral lipids (Fig. 5; 6). Because neither of these storage compounds contain N, this would result in the decrease in the proportion of N in these N-limited cells that we have observed (Fig. 2A). Furthermore, as the proportion of C in sugar and TAG is roughly 40% C and 75% C, respectively, it seems reasonable that a mixture of both could provide a means of increasing the mass of the cell through new C fixation while at the same time maintaining a constant proportion of C. However, the situation after 14 days in N-limited condition differs. Importantly, the rates of C fixation have decreased markedly (Fig. 4B), in agreement with the small increase in starch observed (Fig. 5A). Despite this, we observe a pronounced increase in TAG (Fig. 6A). We suggest that the accumulation of TAG may result from a recycling of the membranes associated with the chloroplast, whose size and numbers have both decreased in 14 days N-limited-cells (Fig 4J). Therefore, the acclimation of cells to N-limited conditions appears to show a biphasic metabolic response, with an initial maintenance of photosynthesis producing primarily starch as accumulated C, and a subsequent recycling of chloroplast membranes resulting in an increased TAG accumulation.

The proportional reduction in elemental N content is mirrored by the decrease in cellular protein (Fig. 3A) since protein is the most important pool of cellular N in the cell. In contrast, the total FAA levels remained unchanged (Fig. S1A). Thus, since the mass of the cells increases as a result of ongoing photosynthesis, this implies that the absolute amount of FAA in the culture increases during N limitation in order to maintain the same levels relative to the cell mass. One possibility is that the extra FAA result from protein degradation. However, it seems unlikely that degradation of proteins would change the relative abundance of the FAA pools, suggesting that some of the variations may reflect other causes. For

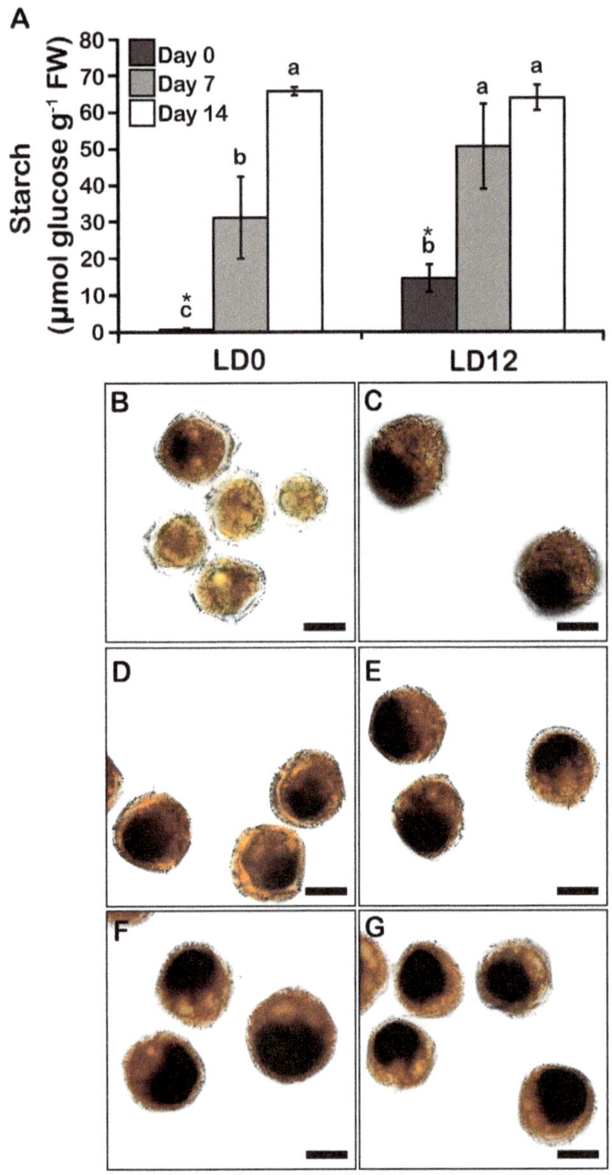

Figure 5. Starch accumulates in N stressed cells. A) Starch levels in cells harvested at LD 0 and LD 12. Results are mean ± SE (n = 3). Statistically different results ($p<0.05$) are marked using either a different letter (Analysis of variance) or an asterisk (Student's t-test). FW; Fresh weight. Bright-field microscopy of iodine-stained starch granules at day 0 (B–C), day 7 (D–E) and day 14-cells (F–G) at either LD 0 (B, D, F) or LD 12 (C, E, G). Starch is localized at the posterior part of the cells.

example, the lack of available N for ammonium synthesis seems the likely explaination for the decrease observed in Gln and Arg levels (Fig. 3B). In addition, the levels of some FAA, in particular those derived from pyruvate (such as Leu and Ile) or phospho-*enol*pyruvate (such as Tyr or Phe) might be augmented using excess C skeletons originating from glycolysis (Fig. 3B). An increase in the levels of these glycolytic intermediates might in turn result from a decreased activity of the tricarboxylic acid cycle under N limitation where energy requirements are likely to be lower than in actively dividing cells. However, additional studies investigating levels of both glycolytic and TCA cycle intermediates would be required to test these possibilities.

N limitation in *L. polyedrum* causes a gradual decrease in chlorophyll *a*, in agreement to what is observed in other photosynthetic eukaryotes [35,39]. It is unlikely that the dinoflagellate had difficulties in synthesizing new chlorophyll molecules in N-deplete conditions, since glutamate, the precursor for the porphyrin moiety in chlorophyll, remained stable under N stress (Fig. 3B). However, many different N-limited algal cells are shown to be more susceptible to photoinhibition in normal light conditions than N-replete cells [40]. In fact, levels of the D1 protein are markedly lower when the cells are N-limited than when N-replete. D1 is a component of the photosynthetic reaction center in PSII and is proposed to be the site of photodamage. It has been suggested that photoinhibition occurred at lower irradiance in N-limited cells, because the lower rates of protein synthesis are unable to keep pace with degradation of damaged D1 [40]. In our study, it is likely that *L. polyedrum* diminishes its chlorophyll content during N depletion to prevent photooxidative damage, because a similar susceptibility to photoinhibition has been previously reported for this species [41].

The reduction in chlorophyll levels, and the resulting protection against oxidative stress, thus appears distinct from the reduction in the size and number of the chloroplasts observed by 14 days in N limiting conditions (Fig. 4J). The decrease in chloroplast number and size during N stress is suggestive of autophagy, the process by which an organism degrades its own cellular components in order to recycle nutrients or eliminate damaged material. Interestingly, autophagy is important for the survival of *Arabidopsis* and *Chlamydomonas* in conditions of N limitation [42,43]. In the latter organism, N stress causes the polar lipids of plastid membranes to be recycled for the production of TAGs [44,45]. We propose that initially, neutral lipids could be produced *de novo*, fueled by C fixation, while at latter stages during N-limitation, TAG accumulation reflects autophagy of chloroplast membranes.

Interestingly, a recycling of plastid membranes might also explain the polarized distribution of lipid bodies in *L. polyedrum*. This species typically contains a single PAS (Periodic acid-Schiff) body, thought to be a dinoflagellate lysosome, in the posterior end of the cell where the lipid droplets are rare [46]. Thus, if the PAS body was involved in neutral lipid degradation, this might favor an anterior location for the accumulation of lipid droplets. TEM observations show small lipid bodies surrounding the cell center in addition to the larger droplets that accumulate predominantly at the anterior end (Fig. 7B). It is possible that the smaller droplets are associated with lipid bodies that have recently budded off from the ER membranes, centrally localized in this dinoflagellate [47]. This would agree with the ER membranes being the site for the last reactions of TAG biosynthesis. The restricted cellular location of the large lipid droplets could result from the fusion of smaller lipid bodies that would, because of the PAS body, strictly accumulate at the anterior end of the cell. Another intriguing possibility could be that enzymes for TAG biosynthesis are polarized within the cell. Independent of the mechanism for localizing the lipid body, however, their location may be functionally important, as lipids were shown to contribute to cell buoyancy in various planktonic species [48,49]. For example, an accumulation of lipid droplets at the anterior pole at the beginning of the day could help *L. polyedrum* migration to the water surface.

Starch accumulation is a common response to N limitation, both in plants and green algae, as well as in *L. polyedrum*. However, it is unlikely that these different organisms share a common regulatory pathway, because they synthesize starch in different cellular compartments. In the plastids of plants and green algae, glucose-1-phosphate is activated to ADP-glucose using an

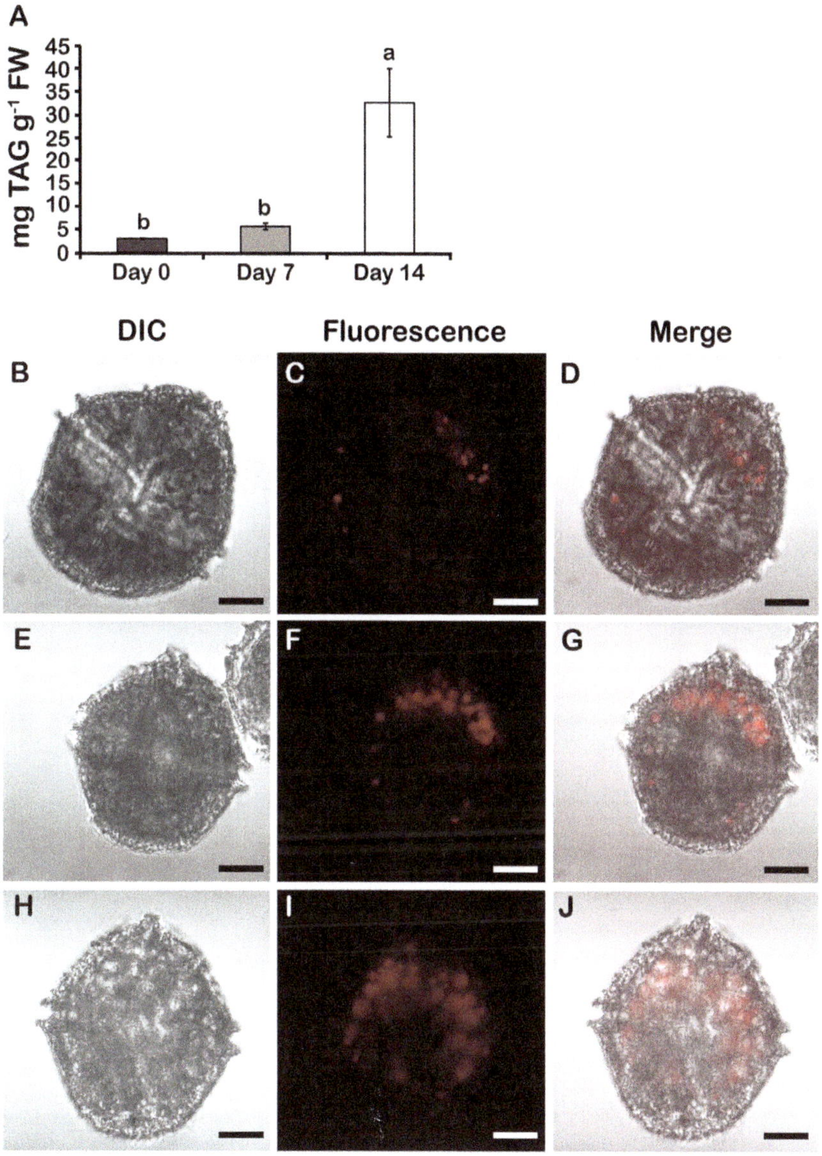

Figure 6. TAGs accumulate in N stressed cells. A) Neutral lipid levels. Results are mean ± SE (n = 3). Statistically different results ($p<0.05$) are marked with a different letter (Analysis of variance). FW; Fresh weight. DIC, Nile red-stained lipid bodies and merged images of day 0 (B–D), day 7 (E–G) and day 14 cells (H–J). All cells were pictured from a ventral view. Lipid bodies were most predominant in the anterior part of the cells.

ADP-glucose pyrophosphorylase before being transferred by starch synthase to a terminal glucose residue for elongation of a glucan chain [50]. The ADP-glucose pyrophosphorylase is inhibited by inorganic phosphate and activated by the Calvin cycle intermediate, 3-phosphoglycerate [50]. Cytosolic P_i is produced by sucrose synthesis and is exchanged with chloroplastic triose-phosphate by the triose phosphate-phosphate translocator [50]. In contrast, if starch is synthesized in the cytosol of dinoflagellates, then no exchange of metabolites across the plastid membranes is required. Furthermore, multiple soluble starch synthases and one granule-bound starch synthase of the heterotrophic dinoflagellate *Crypthecodinium cohnii,* display a marked substrate preference for UDP-glucose rather than ADP-glucose [29]. A similar preference is observed in all organisms storing starch in the cytosol, such as the rhodophytes, glaucophytes and cryptophytes [51,52,53]. Thus, starch synthesis in dinoflagellates might proceed through a cytosolic UDP-glucose pyrophosphorylase, but characterization of this enzyme and its regulators will be required to validate this suggestion.

Localization of the enzymes involved in *L. polyedrum* starch biosynthesis is of particular interest, considering the observation that starch granule accumulation is polarized in these cells (Fig 5B-G; Fig. 7B). The asymmetrical distribution is not simply a result of N limitation, because N-replete cells at LD12 accumulate granules at the same position as in N-limited cells. It also seems unlikely that starch could be transported to the posterior end by motor-proteins traveling on cytoskeletal filaments, because we were unable to detect any membrane surrounding the granules (Fig. 7D), in agreement with a previous TEM study of this species [47]. We suggest that starch synthases and starch branching enzymes might be tethered to membranes at the posterior end of *L. polyedrum* and these localized enzymes might provide a scaffold for polarized

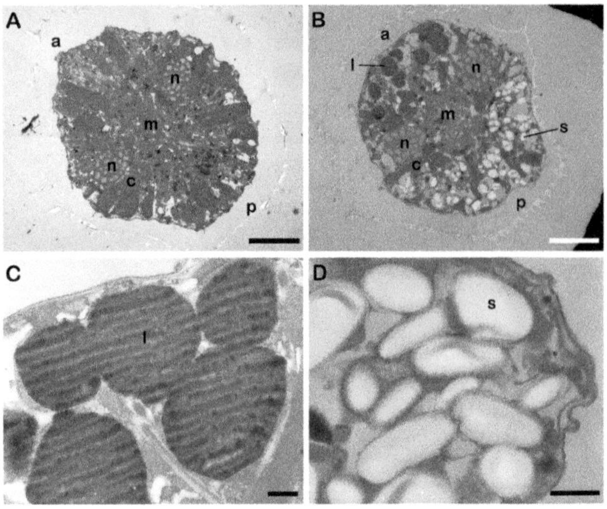

Figure 7. Polarized localization of lipid bodies and starch granules visualized by transmission electron microscopy. Cross-sections of a cell at day 0 (A) and at day 14 in f/2-N medium (B). Both are in a ventral orientation and scale bars are 10 μm. Lipid bodies (l) are located predominantly at the anterior (a) end of the cell, while starch granules (s) are localized at the posterior (p) end. The ends of the C-shaped nucleus (n) surround a central Golgi/ER membrane region (m). Chloroplasts (c) are less abundant in day 14 cells. Higher magnification images of lipid bodies (C) and starch granules (D) in a cell at day 14. Scale bars are 1 μm.

starch synthesis. Protein immunolocalization could assess this hypothesis, particularly now that sequences of starch metabolic enzymes are likely to be found in the complete transcriptome of *L. polyedrum* [54].

The adaptation to N-deplete conditions allow *L. polyedrum* cells that are re-exposed to N to survive at high cell densities for up to a month instead of the few days that N-replete cultures normally tolerate (Fig. 1B). Thus, during the period of N depletion, the cells have acquired a capacity to resist N stress that N-replete cells do not have. Interestingly, conditions of N stress are more likely to be encountered by this species in the field than continuous exposure to high N such as is the case for usual laboratory culture conditions. This increased endurance could help blooming populations survive and maintain their density for long periods of N stress in the environment. Further studies will be required to determine which molecular adaptations accompany this physiological adaptation.

Supporting Information

Figure S1 Sum of all free amino acids (A), sum of N in amino acids (B), and Gln/Glu ratio (C). The amino acid data were compiled from Fig. 3B.

Acknowledgments

We thank L. Pelletier for technical assistance. We also gratefully acknowledge the financial support of the Natural Sciences and Engineering Research Council of Canada (NSERC) for discovery grants to DM and JR and of the Fond de Recherche du Québec- Nature et Technologies (FQRNT) for a studentship (SD-B).

Author Contributions

Conceived and designed the experiments: SDB DM. Performed the experiments: SDB SD. Analyzed the data: SDB JR DM. Contributed reagents/materials/analysis tools: SDB SD JR. Wrote the paper: SDB SD JR DM.

References

1. Vitousek P, Howarth R (1991) Nitrogen limitation on land and in the sea: How can it occur? Biogeochemistry 13: 87–115.
2. Dagenais-Bellefeuille S, Morse D (2013) Putting the N in dinoflagellates. Front Microbiol 4: 369.
3. Anderson DM, Burkholder JM, Cochlan WP, Glibert PM, Gobler CJ, et al. (2008) Harmful algal blooms and eutrophication: Examining linkages from selected coastal regions of the United States. Harmful Algae 8: 39–53.
4. Anderson DM, Cembella AD, Hallegraeff GM (2012) Progress in understanding harmful algal blooms: paradigm shifts and new technologies for research, monitoring, and management. Ann Rev Mar Sci 4: 143–176.
5. Vargo GA, Heil CA, Fanning KA, Dixon LK, Neely MB, et al. (2008) Nutrient availability in support of *Karenia brevis* blooms on the central West Florida Shelf: What keeps *Karenia* blooming? Cont Shelf Res 28: 73–98.
6. Walsh JJ, Weisberg RH, Lenes JM, Chen FR, Dieterle DA, et al. (2009) Isotopic evidence for dead fish maintenance of Florida red tides, with implications for coastal fisheries over both source regions of the West Florida shelf and within downstream waters of the South Atlantic Bight. Progress in Oceanography 80: 51–73.
7. Kudela RM, Cochlan WP (2000) Nitrogen and carbon uptake kinetics and the influence of irradiance for a red tide bloom off southern California. Aquatic Microbial Ecology 21: 31–47.
8. Hastings JW (2007) The Gonyaulax clock at 50: translational control of circadian expression. Cold Spring Harb Symp Quant Biol 72: 141–144.
9. Moorthi SD, Countway PD, Stauffer BA, Caron DA (2006) Use of quantitative real-time PCR to investigate the dynamics of the red tide dinoflagellate *Lingulodinium polyedrum*. Microb Ecol 52: 136–150.
10. Roenneberg T, Rehman J (1996) Nitrate, a nonphotic signal for the circadian system. FASEB J 10: 1443–1447.
11. Atkinson M, Bingman C (1994) Elemental composition of commercial seasalts. J Aquari & Aqua Sci 8: 39–43.
12. Liu B, Lo SC, Matton DP, Lang BF, Morse D (2012) Daily changes in the phosphoproteome of the dinoflagellate Lingulodinium. Protist 163: 746–754.
13. Dorion S, Clendenning A, Jeukens J, Salas JJ, Parveen N, et al. (2012) A large decrease of cytosolic triosephosphate isomerase in transgenic potato roots affects the distribution of carbon in primary metabolism. Planta 236: 1177–1190.
14. Rivoal J, Hanson AD (1993) Evidence for a Large and Sustained Glycolytic Flux to Lactate in Anoxic Roots of Some Members of the Halophytic Genus *Limonium*. Plant Physiol 101: 553–560.
15. Jeffrey S, Humphrey G (1975) New spectrophotometric equations for determining chlorophylls a, b, c1 and c2 in higher plants, algae and natural phytoplankton. Biochem Physiol Pflanz: 191–194.
16. Mackenzie TD, Morse D (2011) Circadian photosynthetic reductant flow in the dinoflagellate Lingulodinium is limited by carbon availability. Plant Cell Environ 34: 669–680.
17. Trethewey RN, Geigenberger P, Riedel K, Hajirezaei MR, Sonnewald U, et al. (1998) Combined expression of glucokinase and invertase in potato tubers leads to a dramatic reduction in starch accumulation and a stimulation of glycolysis. Plant J 15: 109–118.
18. Chen W, Zhang C, Song L, Sommerfeld M, Hu Q (2009) A high throughput Nile red method for quantitative measurement of neutral lipids in microalgae. J Microbiol Methods 77: 41–47.
19. Kolter R, Siegele DA, Tormo A (1993) The stationary phase of the bacterial life cycle. Annu Rev Microbiol 47: 855–874.
20. Foyer CH, Parry M, Noctor G (2003) Markers and signals associated with nitrogen assimilation in higher plants. J Exp Bot 54: 585–593.
21. Nassoury N, Fritz L, Morse D (2001) Circadian changes in ribulose-1,5-bisphosphate carboxylase/oxygenase distribution inside individual chloroplasts can account for the rhythm in dinoflagellate carbon fixation. Plant Cell 13: 923–934.
22. Hicks GR, Hironaka CM, Dauvillee D, Funke RP, D'Hulst C, et al. (2001) When Simpler Is Better. Unicellular Green Algae for Discovering New Genes and Functions in Carbohydrate Metabolism. Plant Physiol 127: 1334–1338.
23. Hu Q, Sommerfeld M, Jarvis E, Ghirardi M, Posewitz M, et al. (2008) Microalgal triacylglycerols as feedstocks for biofuel production: perspectives and advances. Plant J 54: 621–639.
24. Jiang PL, Pasaribu B, Chen CS (2014) Nitrogen-deprivation elevates lipid levels in *Symbiodinium* spp. by lipid droplet accumulation: morphological and compositional analyses. PLoS One 9: e87416.
25. Pleissner D, Eriksen NT (2012) Effects of phosphorous, nitrogen, and carbon limitation on biomass composition in batch and continuous flow cultures of the

heterotrophic dinoflagellate *Crypthecodinium cohnii*. Biotechnol Bioeng 109: 2005–2016.
26. Graf A, Smith AM (2011) Starch and the clock: the dark side of plant productivity. Trends Plant Sci 16: 169–175.
27. Ball S, Colleoni C, Cenci U, Raj JN, Tirtiaux C (2011) The evolution of glycogen and starch metabolism in eukaryotes gives molecular clues to understand the establishment of plastid endosymbiosis. J Exp Bot 62: 1775–1801.
28. Dauvillée D, Deschamps P, Ral JP, Plancke C, Putaux JL, et al. (2009) Genetic dissection of floridean starch synthesis in the cytosol of the model dinoflagellate *Crypthecodinium cohnii*. Proc Natl Acad Sci U S A 106: 21126–21130.
29. Deschamps P, Guillebeault D, Devassine J, Dauvillee D, Haebel S, et al. (2008) The heterotrophic dinoflagellate *Crypthecodinium cohnii* defines a model genetic system to investigate cytoplasmic starch synthesis. Eukaryot Cell 7: 872–880.
30. Holmes R, Williams P, Eppley R (1967) Red water in La Jolla Bay, 1964–1967. Limnol Oceanogr 12: 503–512.
31. Olson RJ, Chisholm SW (1986) Effect of light and nitrogen limitation on the cell cycle of the dinoflagellate *Amphidinium carteri*. J Plankton Res 8: 785–793.
32. Lin S, Mulholland MR, Zhang H, Feinstein TN, Jochem FJ, et al. (2004) Intense grazing and prey-dependent growth of *Pfiesteria piscicida* (Dinophyceae) J Phycol 40: 1062–1073.
33. Lei QY, Lu SH (2011) Molecular ecological responses of dinoflagellate, *Karenia mikimotoi* to environmental nitrate stress. Mar Pollut Bull 62: 2692–2699.
34. Morey JS, Monroe EA, Kinney AL, Beal M, Johnson JG, et al. (2011) Transcriptomic response of the red tide dinoflagellate, Karenia brevis, to nitrogen and phosphorus depletion and addition. BMC Genomics 12: 346.
35. Turpin DH (1991) Effects of inorganic N availability on algal photosynthesis and carbon metabolism J Phycol 27: 14–20.
36. Vanucci S, Pezzolesi L, Pistocchi R, Ciminiello P, Dell'Aversano C, et al. (2012) Nitrogen and phosphorus limitation effects on cell growth, biovolume, and toxin production in *Ostreopsis cf. ovata*. Harmful Algae 15: 78–90.
37. Kopp C, Pernice M, Domart-Coulon I, Djediat C, Spangenberg JE, et al. (2013) Highly Dynamic Cellular-Level Response of Symbiotic Coral to a Sudden Increase in Environmental Nitrogen. mBio 4.
38. Ho C-H, Lin S-H, Hu H-C, Tsay Y-F (2009) CHL1 Functions as a Nitrate Sensor in Plants. Cell 138: 1184–1194.
39. Evans J (1989) Photosynthesis and nitrogen relationships in leaves of C3 plants. Oecologia 78: 9–19.
40. Kolber Z, Zehr J, Falkowski P (1988) Effects of Growth Irradiance and Nitrogen Limitation on Photosynthetic Energy Conversion in Photosystem II. Plant Physiol 88: 923–929.
41. Prézelin BB, Samuelsson G, Matlick HA (1986) Photosystem II photoinhibition and altered kinetics of photosynthesis during nutrient-dependent high-light photoadaptation in *Gonyaulax polyedra*. Marine Biology 93: 1–12.
42. Pérez-Pérez ME, Florencio FJ, Crespo JL (2010) Inhibition of Target of Rapamycin Signaling and Stress Activate Autophagy in *Chlamydomonas reinhardtii*. Plant Physiol 152: 1874–1888.
43. Thompson AR, Doelling JH, Suttangkakul A, Vierstra RD (2005) Autophagic nutrient recycling in *Arabidopsis* directed by the ATG8 and ATG12 conjugation pathways. Plant Physiol 138: 2097–2110.
44. Davey MP, Horst I, Duong GH, Tomsett EV, Litvinenko AC, et al. (2014) Triacylglyceride production and autophagous responses in *Chlamydomonas reinhardtii* depend on resource allocation and carbon source. Eukaryot Cell 13: 392–400.
45. Siaut M, Cuine S, Cagnon C, Fessler B, Nguyen M, et al. (2011) Oil accumulation in the model green alga *Chlamydomonas reinhardtii*: characterization, variability between common laboratory strains and relationship with starch reserves. BMC Biotechnol 11: 7.
46. Zhou J, Fritz L (1994) The PAS/ACCUMULATION bodies in *Prorocentrum lima* and *Prorocentrum maculosum* (Dinophyceae) are dinoflagellate lysosomes J Phycol 30: 39–44.
47. Schmitter RE (1971) The fine structure of *Gonyaulax polyedra*, a bioluminescent marine dinoflagellate. J Cell Sci 9: 147–173.
48. Eroglu E, Okada S, Melis A (2011) Hydrocarbon productivities in different Botryococcus strains: comparative methods in product quantification. J Appl Phycol 23: 763–775.
49. Campbell RW, Dower JF (2003) Role of lipids in the maintenance of neutral buoyancy by zooplankton. Marine Ecology Progress Series 263: 93–99.
50. Heldt H-W, Piechulla B (2011) 9 - Polysaccharides are storage and transport forms of carbohydrates produced by photosynthesis. In: Heldt H-W, Piechulla B, editors. Plant Biochemistry (Fourth Edition). San Diego: Academic Press. pp. 241–271.
51. Viola R, Nyvall P, Pedersen M (2001) The unique features of starch metabolism in red algae. Proc Biol Sci 268: 1417–1422.
52. Deschamps P, Haferkamp I, Dauvillee D, Haebel S, Steup M, et al. (2006) Nature of the periplastidial pathway of starch synthesis in the cryptophyte *Guillardia theta*. Eukaryot Cell 5: 954–963.
53. Plancke C, Colleoni C, Deschamps P, Dauvillee D, Nakamura Y, et al. (2008) Pathway of cytosolic starch synthesis in the model glaucophyte *Cyanophora paradoxa*. Eukaryot Cell 7: 247–257.
54. Beauchemin M, Roy S, Daoust P, Dagenais-Bellefeuille S, Bertomeu T, et al. (2012) Dinoflagellate tandem array gene transcripts are highly conserved and not polycistronic. Proc Natl Acad Sci U S A 109: 15793–15798.

19

Vacuolar-Iron-Transporter1-Like Proteins Mediate Iron Homeostasis in Arabidopsis

Julia Gollhofer[1], Roman Timofeev[1], Ping Lan[2], Wolfgang Schmidt[3,4,5], Thomas J. Buckhout[1]*

1 Institute of Biology, Humboldt University Berlin, Berlin, Germany, **2** State Key Laboratory of Soil and Sustainable Agriculture, Institute of Soil Science, Chinese Academy Sciences, Nanjing, PR China, **3** Academia Sinica, Institute of Plant and Microbial Biology, Taipei, Taiwan, **4** Graduate Institute of Biotechnology, National Chung Hsing University, Taichung, Taiwan, **5** Genome and Systems Biology Degree Program, College of Life Science, National Taiwan University, Taipei, Taiwan

Abstract

Iron deficiency is a nutritional problem in plants and reduces crop productivity, quality and yield. With the goal of improving the iron (Fe) storage properties of plants, we have investigated the function of three Arabidopsis proteins with homology to Vacuolar Iron Transporter1 (AtVIT1). Heterologous expression of *Vacuolar Iron Transporter-Like1* (*AtVTL1*; At1g21140), *AtVTL2* (At1g76800) or *AtVTL5* (At3g25190) in the yeast vacuolar Fe transport mutant, *Δccc1*, restored growth in the presence of 4 mM Fe. Isolated vacuoles from yeast expressing either of the *VTL* genes in the *Δccc1* background had a three- to four-fold increase in Fe concentration compared to vacuoles isolated from the untransformed mutant. Transiently expressed GFP-tagged AtVTL1 was localized exclusively and AtVTL2 was localized primarily to the vacuolar membrane of onion epidermis cells. Seedling root growth of the Arabidopsis *nramp3/nramp4* and *vit1-1* mutants was decreased compared to the wild type when seedlings were grown under Fe deficiency. When expressed under the 35S promoter in the *nramp3/nramp4* or *vit1-1* backgrounds, *AtVTL1*, *AtVTL2* or *AtVTL5* restored root growth in both mutants. The seed Fe concentration in the *nramp3/nramp4* mutant overexpressing *AtVTL1*, *AtVTL2* or *AtVTL5* was between 50 and 60% higher than in non-transformed double mutants or wild-type plants. We conclude that the VTL proteins catalyze Fe transport into vacuoles and thus contribute to the regulation of Fe homeostasis *in planta*.

Editor: Stephan Neil Witt, Louisiana State University Health Sciences Center, United States of America

Funding: This work was supported by the Deutsche Forschungsgemeinschaft to TJB (BU 713: 4-1). The funders had no role in study design, data collection and analysis, decision to publish,or preparation of the manuscript.

Competing Interests: The authors have declared that no competing interests exist.

* Email: thomas.buckhout@rz.hu-berlin.de

Introduction

Regulation of the cellular Fe concentration poses the problem of balancing Fe deficiency against toxicity [1]. Equilibrium is maintained by strict regulation of Fe uptake and storage in the cell. Uptake of soil Fe in Arabidopsis is catalyzed by the Fe^{2+} transporter AtIRT1 [2,3]. Transcription of *AtIRT1* is greatly increased under conditions of Fe deficiency [2], and AtIRT1 is also post-translationally modified to regulate its partitioning between the trans-Golgi network and plasma membranes [4]. Prior to transport, soil Fe^{3+}-chelates are reduced by the plasma membrane ferric reductase, AtFRO2 [5]. The coupling of uptake to reduction is a hallmark of strategy I plants, and an analogous mechanism is also found in *Chlamydomonas* [6] and *Saccharomyces* [7] among other organisms. The reduction-based acquisition of Fe is aided by secretion of Fe-binding compounds and H^+-ATPase-mediated acidification of the rhizosphere [8,9,10,11].

Once in the cell, Fe^{2+} can be incorporated into proteins or stored in cellular compartments. In chloroplast and presumably mitochondria, several thousand Fe atoms are stored per ferritin molecule [12]. However, in Arabidopsis only 5% of Fe is stored in ferritin, and vacuoles appear to be the major compartment for seed Fe storage [13]. Vacuolar Fe uptake was shown to be catalyzed by the ferroportin homologue, AtFPN2 [14,15], and by the CCC1-like protein AtVIT1 (16); however, the primary substrates of AtFPN2 have been reported to be Ni and Co in addition to Fe [14,15]. AtVIT1 has a specific function in the vacuolar transport of Fe into xylem parenchyma of developing embryos, and the *vit1-1* mutant shows misdistribution of Fe in seeds; although, *vit1-1* mutant seeds have unchanged Fe concentration compared with the wild type [16]. Efflux of Fe from the vacuole is catalyzed by two NRAMP proteins, NRAMP3 and NRAMP4. The double mutant *nramp3/nramp4* shows decreased Fe mobilization from the vacuole in germinating seeds [17].

Two rice orthologs of AtVIT1, OsVIT1 and OsVIT2 [18], and one in tulip, TgVIT1 [19], have been shown to catalyze vacuolar Fe transport. TgVIT1 catalyzes the transport of Fe into the proximal perianth cell vacuole, which was shown to be essential for blue color development in tulips. Both OsVIT1 and OsVIT2 complemented the yeast *Δccc1* mutant, and vacuoles that were isolated from complemented cells had increased Fe and Mn concentrations. In addition, both rice genes complemented the Zn transport mutant, *Δzrc1*, and vacuoles isolated from these cells also had an increased Zn concentration [18]. The transcript abundance of *OsVIT2* increased within hours in roots and shoots grown under conditions of excess Fe (4 mM) but decreased in roots and shoots when rice was grown under Fe-deficient conditions. In contrast, *OsVIT1* expression responded only weakly to changes in the Fe status. *OsVIT1* and *OsVIT2* were expressed in flag leaf blades and sheaths, respectively. Consistent with the localization of expression, the *osvit1-1* and *osvit2-1* T-DNA knockout mutants had decreased Fe and Zn content in flag leaves with no change in Mn. Seeds of *osvit1-1* and *osvit2-1* had correspondingly increased Fe and Zn content. OsVIT1 and OsVIT2 were shown to regulate the partitioning of Fe and Zn in rice between source and sink tissues [18].

In an analysis of the transcriptional response of Arabidopsis roots to Fe deficiency, we and others have identified three genes whose mRNA abundance decreased in Fe-deficient roots and whose putative amino acid sequences showed significant homology to AtVIT1 and consequently also to yeast Ccc1p [20,21,22]. These genes belong to a small, five-membered family that has been annotated as nodulin or nodulin-like in databases. These genes will be subsequently referred to as *Vacuolar Iron Transporter-Like* (*VTL*). The *VTL* family was found both in mono- and dicotyledon plants, as well as *Chamydomonas* and *Physcomitrella*. Promoter-β-glucuronidase (GUS) assays showed expression of *AtVTL1* in roots, hypocotyls, and cotyledons of seedlings with the greatest activity associated with the vascular bundle and the root stele [21]. The promoter activity was greatly reduced in Fe-deficient seedlings, consistent with the transcriptional analysis. In the present report, we show that AtVTL1 (At1g21140) and AtVTL2 (At1g76800) are localized to the vacuolar membrane in plants, and that AtVTL1, AtVTL2 and AtVTL5 (At3g25190) complement the *Δccc1* mutation in *Saccharomyces*. Over-expression of *AtVTL1*, *AtVTL2* and *AtVTL5* also complemented the Fe deficiency-dependent root growth phenotype in the *nramp3/nramp4* double mutation and the *vit1-1* mutation in Arabidopsis seedlings. These results indicate that the three members of the *VTL* family are involved in regulation of cellular Fe homeostasis, likely by acting as vacuolar Fe transporters.

Materials and Methods

Arabidopsis Growth Conditions

Arabidopsis seeds were surface-sterilized, vernalized and grown hydroponically as described by Gollhofer et al. [21]. Seeds of the accession Columbia (Col-0) were obtained from the Arabidopsis Biological Resource Center (Ohio State University, Columbus, Ohio). Plants were grown hydroponically according to the method described by Buckhout et al. [20], and the hydroponic nutrient solution was composed of KNO_3 (3 mM), $MgSO_4$ (0.5 mM), $CaCl_2$ (1.5 mM), K_2SO_4 (1.5 mM), NaH_2PO_4 (1.5 mM), H_3BO_3 (25 µM), $MnCl_2$ (1 µM), $ZnSO_4$ (0.5 µM), $(NH_4)_6Mo_7O_{24}$ (0.05 µM) $CuSO_4$ (0.3 µM), Fe-EDTA (40 µM) with the pH adjusted to 6.0 with KOH. For growth on Petri dishes, the method of Santi and Schmidt [11] was used. The medium contained: KNO_3 (5 mM), $Ca(NO_3)_2$ (2 mM), $MgSO_4$ (2 mM), KH_2PO_4 (2.5 mM), $MnCl_2$ (14 µM), H_3BO_3 (70 µM), $ZnSO_4$ (1 µM), $CuSO_4$ (0.5 µM), Na_2MoO_4 (0.2 µM), $CoCl_2$ (0.01 µM), NaCl (10 µM), and Fe-EDTA (40 µM). Sucrose (44 mM) and 5 mM MES (2-[N-morpholino]ethanesulfonic acid) were included, and the pH was adjusted to 5.5. The medium was solidified with 0.8% agar (Fluka, Taufkirchen, Germany). For determination of root growth seedlings were photographed at the appropriate time, the photographs enlarged and root length measured.

Analysis of the Fe Concentration in Arabidopsis Seeds and in Yeast Vacuoles

The Fe content in seeds was determined by the BPDS method [23]. Dried and powdered samples (4–8 mg) were mixed with glass beads (~20; ø 425–600 µm) in 2 ml Eppendorf tubes and incubated at 95°C in 75 µl nitric acid (65%) for 6 h to digest the plant material. Fifty µl of H_2O_2 (30%) was added and the solution incubated at 56°C for 2 h. The volume was adjusted to 200 µl with water. Twenty µl of this solution were diluted in 980 µl of BPDS buffer (1 mM bathophenanthroline disulfonic acid, 0.6 M sodium acetate and 0.48 M hydroxylammonium chloride). The concentration of Fe-BPDS was determined photometrically at 535 nm. A standard calibration curve was prepared by dilution of a stock $FeSO_4$ solution dissolved in 0.1 N HCl. Samples were measured in triplicate and the experiments were conducted at least three times.

Yeast Growth, Complementation and Isolation of Vacuoles

The yeast strain used in this study was *Δccc1* (*ura3, leu2, his3, ade2, can1, CCC1::HIS3*) in DY150. Yeast transformation was carried out by the Li-acetate method [24]. Yeast growth was determined under selection on agar plates with 10^0–10^5 cells per dot.

Vacuoles were isolated as described by Li et al. [25]. Briefly, cells harvest at $OD_{600} = 0.4$ were suspended in 30 ml of 0.1 M Tris-HCl (pH 9.3) buffer containing 10 mM dithiothreitol and incubated for 10 min at 30°C. The cells were washed once with spheroplast buffer (1.2 M sorbitol, 20 mM K-phosphate, pH 7.4) and incubated with 20 µg/ml lyticase for 30 min at 30°C. Spheroplasts were collected by centrifugation at 3,500 *g* for 5 min., and the pellet was resuspended in 10 ml of 15% Ficoll buffer (15% Ficoll (Sigma) in 0.2 M sorbitol and 10 mM PIPES-KOH, pH 6.8). Fifty µg/ml DEAE-Dextran (Amersham Pharmacia Biotech) were added to the spheroplasts, and the sample was incubated for 3 min. on ice and for an additional 5 min. at 30°C. The lysate (10 ml) was transferred to SW28 centrifuge tubes (Beckman Instruments) and overlaid with 10 ml of 8% Ficoll, 10 ml of 4% Ficoll, and 10 ml of 0% Ficoll. The tubes were centrifuged at 110,000 *g* for 3 h. The vacuolar fraction was collected from the 0%/4% interphase. Fe concentration of yeast vacuoles was determined by the BPDS method [23].

Transformation and Transient Expression of *GFP::VTL*

Arabidopsis wild-type Col-0 was transformed using the floral dip method of Clough and Bent [26]. Transgenic plants were selected for BASTA resistance on potting soil.

Onion epidermal cells were co-bombarded with 2.5 µg of plasmids bearing free GFP or GFP-tagged *VTL* genes (35S:*GFP::AtVTL1* or 35S:*GFP::AtVTL2*) and the vacuole marker plasmid vac-rk CD3 975 [27]. Plasmids were coated on 1 µm gold particles and delivered into onion epidermal cells at a pressure of 900 psi by a PDS 1000/He particle delivery system (BioRad, U.S.A). After bombardment, onion slices (2 cm^2) were placed in a Petri dish containing Murashige and Skoog (MS) salts, 30 g/l sucrose and 1.5% agar (pH = 5.7). Following a minimum of 24 h, the epidermis cells were observed under a confocal laser scanning microscope (Zeiss LSM510 Meta) using a ×63 water objective. Vac-rk CD3 975 images were captured in the 560 to 615 nm range after excitation at 543 nm with a HeNe laser beam. The GFP images were captured in the 505 to 530 nm range after excitation at 488 nm with an argon laser beam. Image overlay was carried out by Z-stack analysis at 0.8 µm intervals and further processed with the projection function under LSM510-Expert Mode software.

qRT-PCR

Total RNA was isolated using TRIsure (Bioline) and treated with DNase using DNAse Kit (Fermentas) as suggested by the manufacturer. cDNA was synthesized using DNA-free RNA with oligo-dT(18) primer and RevertAid reverse transcriptase (Thermo scientific). The cDNA was used as a PCR template in a 10 µl reaction system using the SensiMix SYBR No-ROX Kit (Bioline) with programs recommended by the manufacturer in a CFX96 Realtime system (BioRAD). Three biological replicates were performed for each sample. The $\Delta\Delta C_T$ method was used to determine the relative transcript abundance

Results

Complementation of the yeast mutant *Δccc1* by *AtVTL1*, *AtVTL2* and *AtVTL5*

As in Arabidopsis, the yeast vacuole is the major compartment for Fe storage but also protects the cell from Fe toxicity. The yeast gene *CCC1* encodes a vacuolar Fe^{2+}/Mn^{2+} transporter that catalyzes Fe^{2+} uptake into the vacuole. The *Δccc1* mutant is hypersensitive to Fe toxicity, not growing in the presence of greater than ca. 3 mM Fe [25]. Because of the homology of the VTL family member to yeast Ccc1p and Arabidopsis AtVIT1 and because of the transcriptional repression of *AtVTL1*, *AtVTL2* and *AtVTL5* under conditions of Fe deficiency, the ability of the three *AtVTL* genes to complement the *Δccc1* yeast mutant was investigated. When expressed in yeast, the VTL proteins were effective in complementing of the *Δccc1* phenotype in the order *AtVTL1*> *AtVTL2*> *AtVTL5* (Fig. 1). These results are consistent with an AtVTL-dependent decreased the cytosolic Fe in the *Δccc1* mutant, presumably by transport of Fe into the vacuole.

To test this hypothesis, the Fe concentration in isolated yeast vacuoles was investigated. Intact yeast vacuoles were isolated by floatation through a Ficoll step gradient as described by Li et al. [25]. We confirmed the composition of the vacuolar fraction by re-centrifugation the fraction on a continuous sucrose gradient followed by marker enzyme analysis (Fig. S1). As expected, the vacuole marker, bafilomycin-sensitive ATPase, co-localized with membrane proteins and no detectable plasma membrane (vanadate-sensitive ATPase) or endoplasmic reticulum (NADPH cytochrome c reductase) activities were detected in the gradient. Vacuoles, isolated from log-phase *Δccc1* cells or from *Δccc1* cells transformed with the control vector grown in liquid SD medium containing 1 mM $FeSO_4$, had approximately 2 nmoles Fe per µg protein, whereas *Δccc1* cells expressing *AtVTL1*, *AtVTL2* or *AtVTL5* had a greater than 3-fold higher Fe content (Fig. 2A).

Localization of AtVTL1 was demonstrated by transforming *Δccc1* cells with *AtVTL1* fused to a 3′ His tag (*AtVTL1-H₆*). *Δccc1* cells that were transformed with *AtVTL1-H₆* complemented the Fe-sensitive phenotype (Fig. S2). Using a His-tag antibody, AtVTL1-H_6 protein was localized to the vacuolar fraction at the 0/4% Ficoll interface (Fig. 2B). Iron was also concentrated in this fraction compared to fractions in other regions of the Ficoll gradient. The immunological signals in the pellet and 15%/8% Ficoll interface were most like the result of vacuolar membranes from ruptured vacuoles that that did not float through the gradient. We demonstrated that AtVTL1 was localized on the yeast vacuolar membrane, where it presumably catalyzed transport of Fe into the yeast vacuole.

Localization of AtVTL1 and AtVTL2 *in planta*

The localization of AtVTL1 and AtVTL2 was investigated in onion epidermis cells. Attempts to localize AtVTL5 in onion or tobacco leaf cells have not been successful. Onion cells were transformed by particle bombardment with 35S:*GFP::VTL1* or 35S:*GFP::VTL2* reporter gene constructs. In the case of AtVTL1, the GFP fluorescence co-localized with the vacuolar marker vac-rk CD3-975 (Fig. 3A, [27]). When cells were plasmolyzed in 0.8 M mannitol, the GFP signal remained associated with the vacuole (Fig. 3B). The localization of AtVTL2 in epidermis cells was predominately associated with the vacuolar membrane (Fig. 3A); however, GFP fluorescence was also found to be associated with the plasma membrane and the cytoplasm upon plasmolysis (Fig. 3B). The localization of the two AtVTL proteins on the onion vacuolar membrane was consistent with the complementation results in *Saccharomyces* and with the increased Fe concentration in vacuoles isolated from *VTL*-expressing yeast cells.

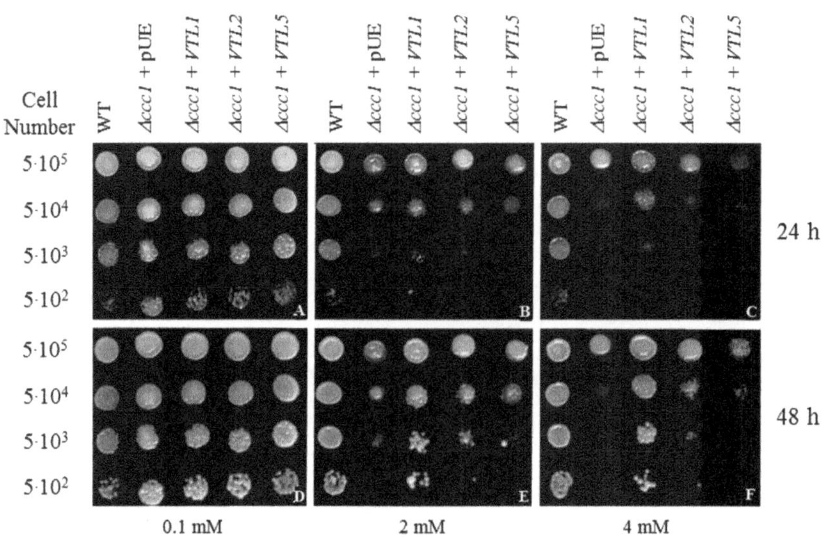

Figure 1. Complementation of the *Δccc1* by heterologous expression of the *VTL* genes. *Δccc1* (vacuolar Fe^{2+}/Mn^{2+} transporter) cells were transformed with each of the three VTL genes or the empty vector (pUE) under the control of the PGK promoter and grown on SD medium containing $FeSO_4$ at the concentrations indicated for 24 or 48 h at 30°C. Cells were plated at the densities indicated in the figure.

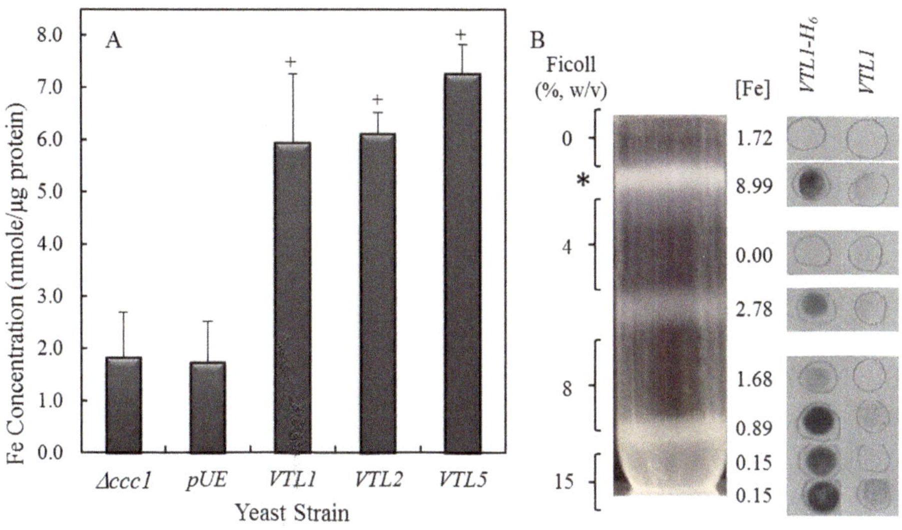

Figure 2. Determination of the Fe concentration in *Δccc1* cells transformed with *AtVTL1, AtVTL2* or *AtVTL5*. (**A**) Shown is the Fe concentration in yeast vacuoles isolated from the *Δccc1* mutant, *Δccc1* transformed with an empty plasmid (pUE), *AtVTL1, AtVTL2* or *AtVTL5*. Error bars indicate the standard error of the mean. Bar graphs labeled with "+" are significantly different for the *Δccc1* mutant ($p<0.05$). (**B**) Vacuoles were isolated from *Δccc1* yeast cells transformed with His-tagged *AtVTL1* (*VTL1-H₆*) or with untagged *AtVTL1* (*VTL1*) as a control. AtVTL1-H_6 was localized by dot-blots using an anti-His antibody. The positions of the dot-blots are approximate. The Fe concentrations (nmole/µg protein) in the fraction corresponding to the dot-blot are given.

Complementation of the *nramp3/nramp4* and the *vit1-1* mutants

Over-expression of *AtVTL1* in Arabidopsis did not greatly alter Fe content of shoots and roots [21], nor were changes observed in the kinetics of the response to Fe deficiency, as determined by the root plasma membrane Fe^{3+}-chelate reductase and chlorophyll content (Figs. S3A and B). In an attempt to characterize their functions, we have ectopically expressed the *AtVTL* genes in the Arabidopsis *nramp3/nramp4* double mutant background (17). Over-expression was confirmed by semi-quantitative RT-PCR (Fig. S4). Both NRAMP3 and NRAMP4 have been shown to be Fe vacuolar efflux carriers. The double mutant had no obvious phenotype when grown on soil; however, growth of mutant seedlings was retarded shortly after germination on an Fe-deficient substrate [17]. The phenotype was not persistent and disappeared after a few days of growth or with supplemental Fe in the media. Under Fe deficiency root growth of the *nramp3/nramp4* double mutant and of the double mutant transformed with the empty vector (GL1), was greatly inhibited compared to the wild-type control at 5 d following germination (Fig. 4 and Fig. S5; $p<0.001$) [17]. Importantly, over-expression of *AtVTL1*, *AtVTL2* or *AtVTL5* in *nramp3/nramp4* complemented this mutant growth phenotype. The early seedling root length was significantly increased (Fig. 4 and Fig. S5; $p<0.001$) compared to the double mutant; however, the roots were still significantly shorter than those of the wild type ($p<0.001$). Thus, over-expression of *AtVTL1*, *AtVTL2* or *AtVTL5* partially restored the wild-type phenotype in the *nramp3/nramp4* double mutant.

The proposed function of NRAMP3 and NRAMP4 was to mobilize Fe from the vacuole of young seedlings [17]. We hypothesized that over-expression of the AtVTLs might have increased total seed Fe, and thereby increased the Fe supply to the germinating seedling. To test this, we analyzed the Fe concentration in *nramp3/nramp4* seeds that over-expressed the *AtVTL* genes. The results of these analyses demonstrated that the *AtVTL1*, *AtVTL2* or *AtVTL5* over-expressing lines had between 40 and 60% increased Fe concentration compared to either the wild type or the *nramp3/nramp4* double mutant (Fig. 5; $p<0.01$). These results supported the concept that an increased content of Fe in the seeds could have supplied the embryo with sufficient Fe even when efflux from the vacuole through the NRAMP transporters was compromised.

As mentioned above, the CCC1-like protein, AtVIT1, is localized on the vacuolar membrane and functions in the transport of Fe into the parenchyma cell of the provascular strands in developing embryos. The *vit1-1* mutant showed misdistribution of Fe; although, *vit1-1* mutant seeds had unchanged Fe content compared to the wild type (Fig. 5) [16]. *vit1-1* had a severely chlorotic phenotype when grown on alkaline soil [16]. In addition, we observed that *vit1-1* displayed a short-root phenotype similar to the *nramp3/nramp4* mutant when grown on Fe-deficient media in the presence of the Fe^{2+} chelator Ferrozine (Fig. 6 and Fig. S6). Over-expression of *AtVTL1*, *AtVTL2* or *AtVTL5* restored root growth to greater than the wild-type length ($p>0.001$), thus complementing the *vit1-1* mutation.

Regulation of *AtVTL1*, *AtVTL2* and *AtVTL5* Gene Expression

Finally, we investigated the response of *AtVTL1*, *AtVTL2* and *AtVTL5* to nutrient supply using quantitative qRT-PCR. Previously, we reported that the transcriptional activity of these genes positively correlated with the Fe supply [20]. Quantitative RT-PCR analyses confirmed these results (Fig. 7). Compared to Fe-sufficient controls (40 µM Fe), the transcript abundance of *AtVTL1*, *AtVTL2* and *AtVTL5* was decreased under Fe deficiency and increased in plants grown on media containing 120 µM Fe. Similarly, transcript abundance also decreased under Zn deficiency for *AtVTL1*, *AtVTL2* and *AtVTL5* (Fig. 7). However, when the Zn concentration was increased to 5 µM, the transcript abundance for *AtVTL1*, *AtVTL2* and *AtVTL5* was not significantly different from that of control plants grown on standard media (Fig. 7). The *VTL* genes were unable to restore growth in the Zn sensitive *Δzrc1*

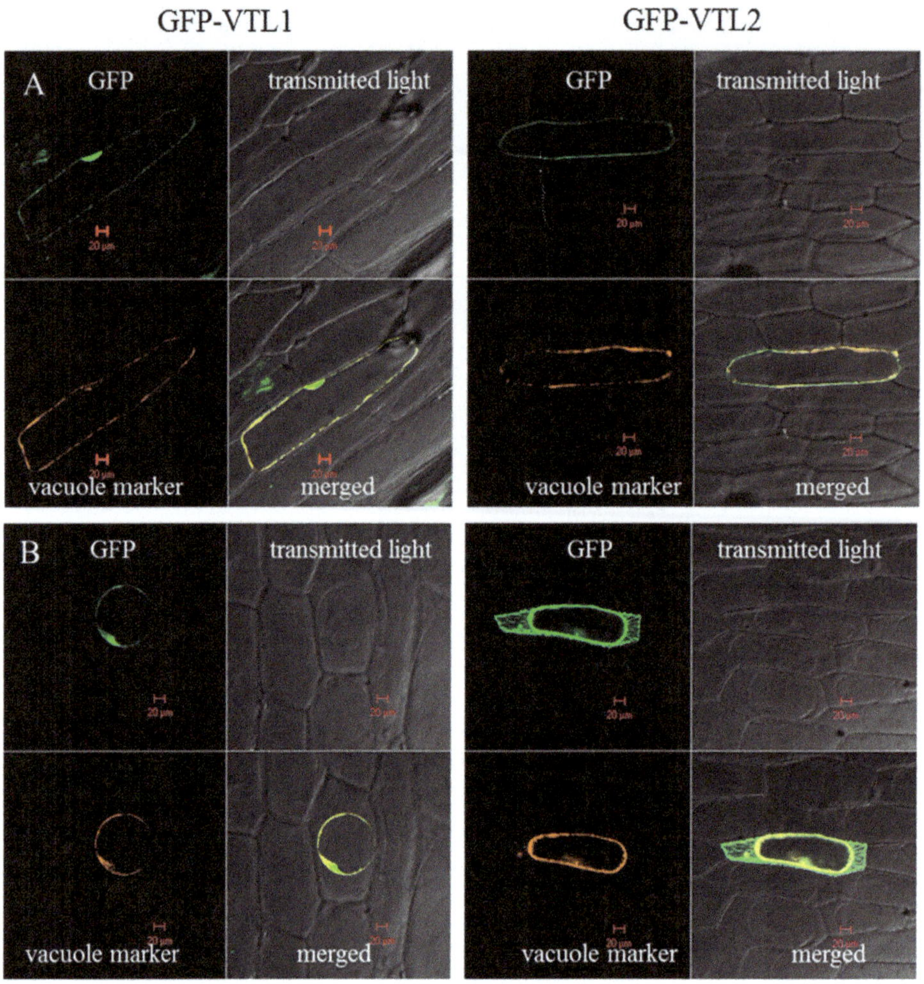

Figure 3. Transient 35S:GFP::*VTL1* and 35S:GFP::*VTL2* expression in onion epidermis cells. The GFP-VTL1 fluorescent signal co-localized with the vacuolar marker, vac-rk CD3-975, in turgescent (**A**) and plasmolyzed (**B**) cells. The GFP-VTL2 signal was also observed on the plasma membrane and cytoplasma (**B**).

mutant (Fig. S7). Thus, a function of AtVTL1, AtVTL2 or AtVTL5 in vacuolar Zn transport is unlikely. Since *AtVTL1, AtVTL2* and *AtVTL5* responded positively to Fe supply and since their over-expression complemented vacuolar Fe transport mutations in yeast and Arabidopsis mutants, we conclude that *VTL* genes encode protein involved in Fe homeostasis, presumably as Fe vacuolar transporters.

Discussion

Heterologous expression of *AtVIT1, OsVIT1, OsVIT2* and *TgVIT1* in the *Δccc1* mutant has been shown to complement the *Δccc1* mutation by restoring Fe uptake into the vacuole, and thus, to protect the cell from the deleterious effects of high cytosolic Fe ([6,18,19]. Vacuolar Fe transporter activity has been reported for FPN2, NRAMP3 and NRAMP4. Whereas AtVIT1 and FPN2 transported Fe into vacuoles, the two NRAMPs were shown to be efflux carriers. Previously, we have identified five genes in Arabidopsis, which encoded proteins with significant sequence homology to AtVIT1 [20]. The transcript abundance of *AtVTL1, AtVTL2* and *AtVTL5* was rapidly decreased in roots under conditions of Fe deficiency [21,28]. We have shown in this report, that heterologous expression of *AtVTL1, AtVTL2* and *AtVTL5* in yeast *Δccc1* cells restored their ability to grow in the presence of 4 mM Fe. Since the heterologous expression of these genes in yeast correlated with increased Fe in the vacuolar fraction, we proposed that these three proteins increased Fe transport into the vacuole and thereby reduced toxic Fe in the cytoplasm of *Δccc1* cells.

VTLs are Localized to the Arabidopsis Vacuolar Membrane

Transient expression of *GFP::VTL1* and *GFP::VTL2* chimeric proteins in onion epidermis cells resulted in a specific co-localization of GFP::VTL1 with the vacuolar marker vac-rk CD3-975. In the case of *GFP::VTL2* transient expression, a prominent signal was co-localized with the vacuole marker, but GFP fluorescence was also found in the cytoplasm and associated with the plasma membrane, indicating possible localization of AtVTL2 on other membranes in addition to the vacuole. Attempts at localizing AtVTL5 have not been successful.

Evidence for a function of AtVTL1, 2 and 5 in Fe homeostasis was obtained by complementation of the *nramp3/nramp4* and *vit1-1* mutations. The NRAMP3 and NRAMP4 proteins act redundantly as vacuolar Fe efflux carriers. The abundance of both

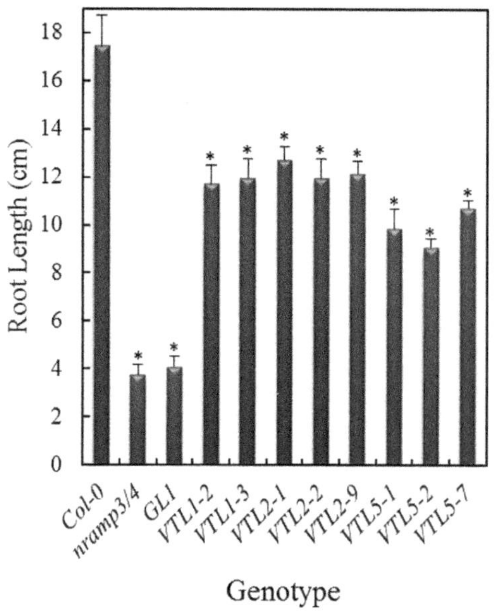

Figure 4. Complementation of the *nramp3/nramp4* double mutant by over-expression of AtVTLs. The three VTL genes were over-expressed in the *nramp3/nramp4* double mutant background under the control of the 35S promoter and root length was determined after 5 days of growth. Seedlings were grown without added Fe. Seedlings were grown on agarose Petri plates as described in the Materials and Methods. Asterisks indicate significant differences from the wild-type ($p<0.001$).

NRAMPs was increased during Fe deficiency, and in the double mutant, greening of cotyledons and early root growth of seedlings were delayed compared to controls when seeds were germinated under Fe deficiency [17]. This inhibition of root growth could be largely eliminated by supplying Fe to the growth medium. The root growth phenotype was complemented by over-expression of *AtVTL1, 2* or *5* in the *nramp3/nramp4* mutant. In addition the total Fe content of seeds from over-expressing *AtVTL1, 2* or *5* plants was increased compared to the wild type and the mutant. It might have seemed paradoxical that a mutation in a vacuole efflux transport system could be complemented by a putative vacuolar uptake transporter; however, an increased vacuolar Fe concentration might have been sufficient to compensate for a decreased efflux caused by the mutations in *NRAMP3* and *NRAMP4*.

The *vit1-1* mutant also displayed a Fe deficiency-dependent inhibition of seedling root growth that is likely the result of a mis-localization of Fe in the seed radicle. Over-expression of *AtVTL1, 2* or *5* might have restored the Fe supply to the growing root. In summary and based on the complementation activity in yeast, the increased vacuolar Fe concentration in yeast cells expressing AtVTL1, 2 or 5, the localization of AtVTL1 and 2 on the vacuolar membrane in onion epidermis cells and the restoration of seedling root growth in the *nramp3/nramp4* and *vit1-1* mutants, a critical role for the VTL proteins in the subcellular distribution of Fe is apparent.

AtVTLs and Metal Homeostasis

Low substrate specificity is commonly found in cation transports. For example, AtIRT1 is the predominant plasma membrane Fe^{2+} transporter and is essential for Fe uptake from the soil [3]. In the absence of Fe or in the presence of excess divalent cations, AtIRT1 can also catalyze the transport of Zn, Mn and Cd [3,29]. Both NRAMP3 and NRAMP4 have been shown also to be vacuolar Mn exporters [30]; although, transcription was correlated with the Fe and not the Mn nutrient status. Lanquar et al. [30] envision a passive function of NRAMP3 and NRAMP4 in cycling of Mn from the vacuole to the plastid in mesophyll cells. The *nramp3/nramp4* double mutant is hypersensitive to Zn and Cd, and as Sinclair and Krämer [31] suggested, the cause of the hypersensitivity appeared to be an indirect effect related to

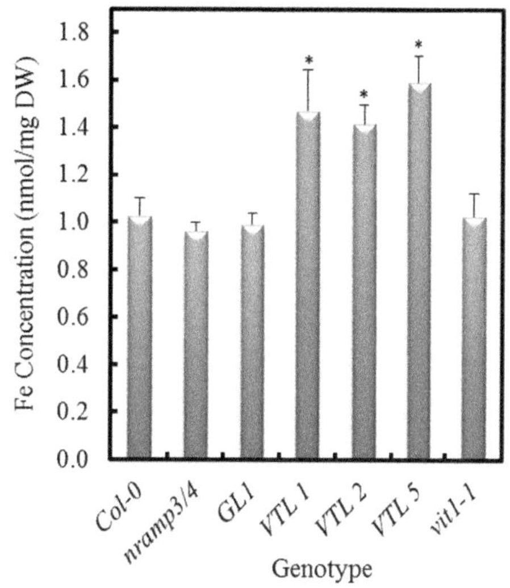

Figure 5. Fe concentration of Arabidopsis seeds. Fe was determined in the wild-type (Col-0), the *nramp3/nramp4* double mutant, the vector control (GL1), the *vit1-1* mutant and in seeds from the double mutant expressing one of the three VTL genes under the control of the 35S promoter. Asterisks indicate significant differences from the wild-type ($p<0.01$).

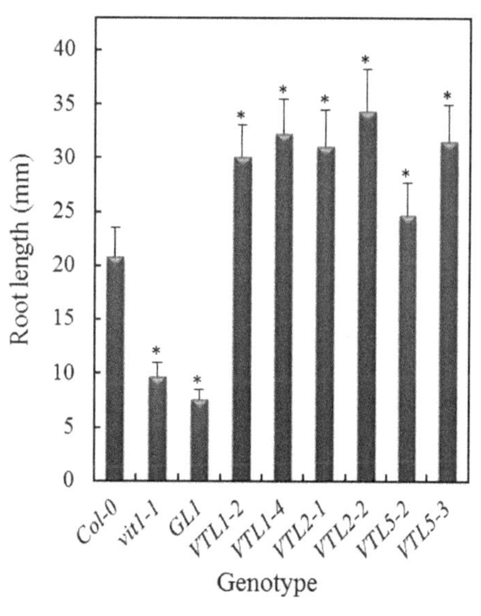

Figure 6. Complementation of the *vit1-1* mutant by over-expression of *AtVTL1, AtVTL2* or *AtVTL5*. The three VTL genes were over-expressed under the control of the 35S promoter in the *vit1-1* background. Root length was determined after 13 days of growth on agarose Petri plates as described in the Materials and Methods ion ES medium without added Fe and in the presence of the Fe^{2+} chelator, Ferrozine. Asterisks indicate significant differences from the wild-type ($p<0.001$). Shown are results from 2 independent transformants.

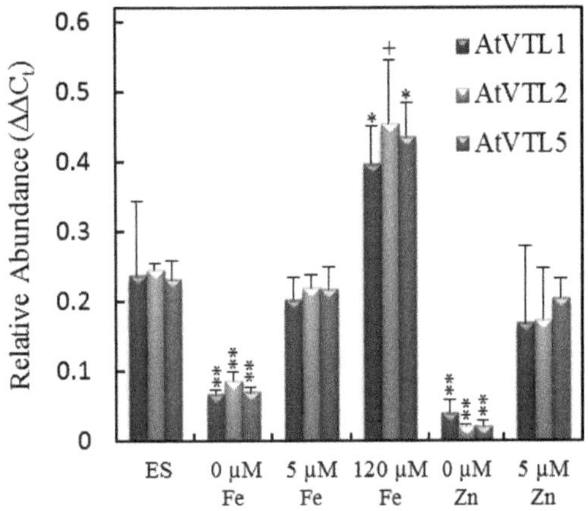

Figure 7. RT-qPCR of At*VTL1*, At*VTL2* and *AtVTL5* in response to Fe and Zn supply. Seedlings were grown on agarose petri plates as described in the Materials and Methods. Growth conditions were: ES medium (ES); "0", ES without Fe or Zn or at the concentrations indicated. RT-qPCR was standardized to the expression of the *ACTIN2*. Symbols indicate significant differences to expression in the ES medium in three biological replicates: **, $p<0.001$; *, $p<0.01$; and +, $p<0.05$.

impaired Fe and Mn transport. Although the transcript data do support a role for the VTLs in Fe homeostasis, a role in homeostasis of other divalent cations analogous to the roles of NRAMP3 and NRAMP4 cannot be excluded.

Five Zn transport genes are among the Arabidopsis genes that response to Fe deficiency [32]. Four of these genes that transport Zn into the vacuole are strongly induced under Fe deficiency. The remaining gene, *AtZIP3*, was repressed and shown to catalyzed Zn uptake into cells. Zn has been shown to accumulate in roots grown under Fe-deficient conditions, likely the result of the Zn transport activity of IRT1 [3,31]. It is noteworthy that the expression of *AtVTL1*, *AtVTL2* and *AtVTL5* was greatly decreased under Zn deficiency but unchanged when grown under Zn excess (Fig. 7). Expression of *AtVTL1*, *AtVTL2* and *AtVTL5* in the yeast *Δzrc1*, a vacuolar Zn transport mutant, did not complement this mutation (Fig. S7).

Although a direct role of the VTL proteins in Zn transport is lacking, an involvement of Zn in Fe homeostasis has been clearly demonstrated. Transcriptional regulation in strategy I plants to Fe deficiency is mediated in part by the bHLH transcription factors FIT (bHLH029) [33,34,35] and POPEYE (bHLH047, PYE) [36]. The central components of PYE regulation include PYE (At3g47640), IAA–LEUCINE-RESISTANT3 (ILR3, bHLH105, At5g54680) and BRUTUS (BTS, At3g47640). *PYE* is weakly and *BTS* is strongly induced under Fe deficiency [20,36], and PYE and BTS both interact but with ILR3 in a yeast two-hybrid assay but not with each other [36]. Either a ternary complex of PYE-ILR3-BTS or the competition between PYE and BTS for ILR3 has been proposed as the regulatory switch in the PYE network [36]. PYE directly regulated genes that were associated with the response to Fe deficiency including: *FRO3*, *NAS4* and *ZIF1*. The transcription of *AtVTL1*, *2* and *5* was increased 1.9, 16.9 and 2.9-fold, respectively, in the *pye-1* knockout mutant [36], and roots of the *pye-1* mutant had an elevated Fe content compared with controls. These observations are consistent with the positive correlation between the root Fe concentration and transcript abundance of *AtVTL1*, *2* and *5* (Fig. 7) [21].

In the *ilr3-1* gain-of-function mutant, the transcript abundance of *AtVTL1*, *2* and *5* was decreased between 3- and 4-fold; an observation that can be explained by a gain-of-function in the PYE-ILR3-BTS repression [22]. In general, BTS and the rice homologs OsHRZs negatively regulate the Fe deficiency response in Arabidopsis and rice, and thus plants with decreased expression of these genes showed tolerance to Fe deficiency [36,37]. Recombinant BTS has been shown to bind both Fe and Zn, primarily at the hemerythrin domain of the protein [36]. In a recent model, Kobayshi and Nishizawa [38] have proposed a mechanism for transcriptional regulation through the BTS/OsHRZ proteins based on binding of Zn and Fe to the hemerythrin domain of these proteins. In their model, Fe binding to BTS/OsHRZ in the absence of Zn would lead to repression PYE-regulated gene expression as seen in Fig. 7. The implications of their model are germane to the repression of *VTL* transcription under Zn deficiency observed in our experiments. Taken together, we have identified a group of putative Fe transporters that participate in the Fe homeostasis in Arabidopsis. As vacuolar Fe transporters these proteins may have importance in biofortification efforts in the future.

Supporting Information

Figure S1 Sucrose density gradient isolation of membrane in the yeast vacuolar fraction. The vacuolar fraction was isolated as described in the Materials and Methods, and the vacuoles ruptured by repeated pipetting. The membranes were layered onto a continuous, 10 to 60% sucrose gradient and centrifuged at 110,000xg in a swing-out rotor over-night. The gradient was fractionated into 1 ml fractions and marker enzymes for the vacuole (bafilomycin-sensitive ATPase), plasma membrane (vanadate-sensitive ATPase) and endoplasmic reticulum (cytochrome *c* reductase) were determined by the method of Luster and Buckhout (Plant Physiol. 1989; 91(3): 1014-9). The activity of the vanadate-sensitive ATPase and the cytochrome c reductase were below the limits of detection.

Figure S2 Complementation of the yeast *Δccc1* (vacuolar Fe^{2+}/Mn^{2+} transporter) mutant with his-tagged *AtVTL1* and *AtVTL2* genes. Cells were transformed with the empty vector (pUE) or the *VTL* gene containing a H_6 tag under the control of the PGK promoter and grown on YPD medium containing 7.5 mM $FeSO_4$.

Figure S3 A. Chlorophyll content in wild-type (Col-0) and wild-type plants over-expressing *AtVTL1* grown in the Fe concentrations indicated. Shown are chlorophyll a and b and the chlorophyll a/b ratio. **B. Analysis of the Fe^{3+}-chelate reductase activity in Col-0 and Col-0 plants over-expressing *AtVTL1*.** Bars are standard error of the mean.

Figure S4 Semi-quantitative PCR of of *AtVTL1*, *AtVTL2* and *AtVTL5*. Expression was determined in Col-0 (WT), the *nramp3/nramp4* double mutant and in the double mutant over-expressing each of the *VTL1*, *VTL2* or *VTL5* genes. Expression was standardized to the level of *ACTIN2* expression.

Figure S5 Root growth in the *nramp3/nramp4* double mutant transformed with *AtVTL1*, *AtVTL2* or *AtVTL5*. Seedlings were grown for 5 days on standard media lacking Fe (see Materials and Methods). Shown are results from two to three independent transformants taken from one repetition of the experiment reported in Fig. 4.

Figure S6 Root growth in the *vit1-1* mutant transformed with *AtVTL1*, *AtVTL2* or *AtVTL5*. Seedlings were grown for 13 days on standard media (see Materials and Methods) lacking Fe and in the presence of the Fe^{2+} chelator, Ferrozine. Shown are results from an experiment similar to that reported in Fig. 6.

Figure S7 Complementation of the *Δzrc1* by heterologous expression of the *VTL* genes. *Δzrc1* (vacuolar Zn^{2+} transporter) cells were transformed with each of the three VTL genes or the empty vector (pUE) under the control of the PGK promoter and grown on SD medium containing $ZnSO_4$ at the concentrations indicated for 24 or 48 h at 30°C. Cells were plated at the densities indicated in the figure.

Acknowledgments

We thank Drs. Mary-Lou Guerinot, Jerry Kaplan and Sabastian Thomine for kindly providing the *Saccharomyes* and Arabidopsis mutants used in this study and Christin Schläwicke for determination of seed Fe and leaf chlorophyll. We gratefully acknowledge the excellent technical support of Marion Dewender and Susanne Olstowski-Jacoby.

Author Contributions

Conceived and designed the experiments: JG WS TJB. Performed the experiments: JG RT PL. Analyzed the data: JG RT PL WS TJB. Contributed reagents/materials/analysis tools: JG RT PL. Wrote the paper: WS TJB.

References

1. Kosman DJ (2010) Redox cycling in iron uptake, efflux, and trafficking. J Biol Chem 285: 26729–26735. doi: 10.1074/jbc.R110.113217.
2. Eide D, Broderius M, Fett J, Guerinot ML (1996) A novel iron-regulated metal transporter from plants identified by functional expression in yeast. Proc Natl Acad Sci U S A 93: 5624–5628.
3. Vert G, Grotz N, Dedaldechamp F, Gaymard F, Guerinot ML, et al. (2002) IRT1, an Arabidopsis transporter essential for iron uptake from the soil and for plant growth. Plant Cell 14: 1223–1233.
4. Barberon M, Zelazny E, Robert S, Conéjéro G, Curie C, et al. (2011) Monoubiquitin-dependent endocytosis of the iron-regulated transporter 1 (IRT1) transporter controls iron uptake in plants. Proc Natl Acad Sci USA 108: E450–458. doi: 10.1073/pnas.1100659108.
5. Robinson NJ, Procter CM, Connolly EL, Guerinot ML (1999) A ferric-chelate reductase for iron uptake from soils. Nature 397: 694–697. doi: 10.1038/17800.
6. Eckhardt U, Buckhout TJ (1998) Iron assimilation in Chlamydomonas reinhardtii involves ferric reduction and is similar to Strategy I higher plants. Journal of Experimental Botany 49: 1219–1226.
7. Kosman DJ (2003) Molecular mechanisms of iron uptake in fungi. Mol Microbiol 47: 1185–1197.
8. Rodriguez-Celma J, Lin W-D, Fu G-M, Abadia J, Lopez-Millan A-F, et al. (2013) Mutually exclusive alterations in secondary metabolism are critical for the uptake of insoluble iron compounds by Arabidopsis and Medicago truncatula. Plant Physiol 162: 1473–1485. doi: 10.1104/pp.113.220426.
9. Fourcroy P, Siso-Terraza P, Sudre D, Saviron M, Reyt G, et al. (2014) Involvement of the ABCG37 transporter in secretion of scopoletin and derivatives by Arabidopsis roots in response to iron deficiency. New Phytol 201: 155–167. doi: 10.1111/nph.12471.
10. Schmid NB, Giehl RFH, Doll S, Mock H-P, Strehmel N, et al. (2014) Feruloyl-CoA 6′-Hydroxylase1-dependent coumarins mediate iron acquisition from alkaline substrates in arabidopsis. Plant Physiol 164: 160–172. doi: 10.1104/pp.113.228544.
11. Santi S, Schmidt W (2009) Dissecting iron deficiency-induced proton extrusion in Arabidopsis roots. New Phytol 183: 1072–1084. doi: 10.1111/j.1469-8137.2009.02908.x.
12. Briat J-F, Duc C, Ravet K, Gaymard F (2010) Ferritins and iron storage in plants. Biochim Biophys Acta 1800: 806–814. doi: 10.1016/j.bbagen.2009.12.003.
13. Ravet K, Touraine B, Boucherez J, Briat J-F, Gaymard F, et al. (2009) Ferritins control interaction between iron homeostasis and oxidative stress in Arabidopsis. Plant J 57: 400–412. doi: 10.1111/j.1365-313X.2008.03698.x.
14. Morrissey J, Baxter IR, Lee J, Li L, Lahner B, et al. (2009) The ferroportin metal efflux proteins function in iron and cobalt homeostasis in Arabidopsis. Plant Cell 21: 3326–3338. doi: 10.1105/tpc.109.069401.
15. Schaaf G, Honsbein A, Meda AR, Kirchner S, Wipf D, et al. (2006) AtIREG2 encodes a tonoplast transport protein involved in iron-dependent nickel detoxification in Arabidopsis thaliana roots. J Biol Chem 281: 25532–25540. doi: 10.1074/jbc.M601062200.
16. Kim SA, Punshon T, Lanzirotti A, Li L, Alonso JM, et al. (2006) Localization of iron in Arabidopsis seed requires the vacuolar membrane transporter VIT1. Science 314: 1295–1298. doi: 10.1126/science.1132563.
17. Lanquar V, Lelievre F, Bolte S, Hames C, Alcon C, et al. (2005) Mobilization of vacuolar iron by AtNRAMP3 and AtNRAMP4 is essential for seed germination on low iron. EMBO J 24: 4041–4051. doi: 10.1038/sj.emboj.7600864.
18. Zhang Y, Xu Y-H, Yi H-Y, Gong J-M (2012) Vacuolar membrane transporters OsVIT1 and OsVIT2 modulate iron translocation between flag leaves and seeds in rice. Plant J 72: 400–410. doi: 10.1111/j.1365-313X.2012.05088.x.
19. Momonoi K, Yoshida K, Mano S, Takahashi H, Nakamori C, et al. (2009) A vacuolar iron transporter in tulip, TgVit1, is responsible for blue coloration in petal cells through iron accumulation. Plant J 59: 437–447. doi: 10.1111/j.1365-313X.2009.03879.x.
20. Buckhout TJ, Yang TJW, Schmidt W (2009) Early iron-deficiency-induced transcriptional changes in Arabidopsis roots as revealed by microarray analyses. BMC Genomics 10. doi: 10.1186/1471-2164-10-147.
21. Gollhofer J, Schlawicke C, Jungnick N, Schmidt W, Buckhout TJ (2011) Members of a small family of nodulin-like genes are regulated under iron deficiency in roots of Arabidopsis thaliana. Plant Physiol Biochem 49: 557–564. doi: 10.1016/j.plaphy.2011.02.011.
22. Rampey RA, Woodward AW, Hobbs BN, Tierney MP, Lahner B, et al. (2006) An Arabidopsis basic helix-loop-helix leucine zipper protein modulates metal homeostasis and auxin conjugate responsiveness. Genetics 174: 1841–1857. doi: 10.1534/genetics.106.061044.
23. Blair D, Diehl H (1961) Bathophenanthrolinedisulphonic acid and bathocuproinedisulphonic acid, water soluble reagents for iron and copper. Talanta 7: 163–174.
24. Rose M, Winston F, Hieter P (1990) Methods in yeast genetics: A laboratory course manual. Cold Spring Harbor, New York: Cold Spring Harbor Press.
25. Li L, Chen OS, McVey Ward D, Kaplan J (2001) CCC1 is a transporter that mediates vacuolar iron storage in yeast. J Biol Chem 276: 29515–29519. doi: 10.1074/jbc.M103944200.
26. Clough SJ, Bent AF (1998) Floral dip: a simplified method for Agrobacterium-mediated transformation of Arabidopsis thaliana. Plant J 16: 735–743.
27. Nelson BK, Cai X, Nebenfuhr A (2007) A multicolored set of in vivo organelle markers for co-localization studies in Arabidopsis and other plants. Plant J 51: 1126–1136. doi: 10.1111/j.1365-313X.2007.03212.x.
28. Yang TJW, Lin W-D, Schmidt W (2010) Transcriptional profiling of the Arabidopsis iron deficiency response reveals conserved transition metal homeostasis networks. Plant Physiol 152: 2130–2141. doi: 10.1104/pp.109.152728.
29. Cohen CK, Fox TC, Garvin DF, Kochian LV (1998) The role of iron-deficiency stress responses in stimulating heavy-metal transport in plants. Plant Physiol 116: 1063–1072.
30. Lanquar V, Ramos MS, Lelievre F, Barbier-Brygoo H, Krieger-Liszkay A, et al. (2010) Export of vacuolar manganese by AtNRAMP3 and AtNRAMP4 is required for optimal photosynthesis and growth under manganese deficiency. Plant Physiol 152: 1986–1999. doi: 10.1104/pp.109.150946.
31. Sinclair SA, Kramer U (2012) The zinc homeostasis network of land plants. Biochim Biophys Acta 1823: 1553–1567. doi: 10.1016/j.bbamcr.2012.05.016.
32. Schmidt W, Buckhout TJ (2011) A hitchhiker's guide to the Arabidopsis ferrome. Plant Physiol Biochem 49: 462–470. doi: 10.1016/j.plaphy.2010.12.001.
33. Colangelo EP, Guerinot ML (2004) The essential basic helix-loop-helix protein FIT1 is required for the iron deficiency response. Plant Cell 16: 3400–3412. doi: 10.1105/tpc.104.024315.
34. Jakoby M, Wang H-Y, Reidt W, Weisshaar B, Bauer P (2004) FRU (BHLH029) is required for induction of iron mobilization genes in Arabidopsis thaliana. FEBS Lett 577: 528–534. doi: 10.1016/j.febslet.2004.10.062.
35. Yuan Y, Wu H, Wang N, Li J, Zhao W, et al. (2008) FIT interacts with AtbHLH38 and AtbHLH39 in regulating iron uptake gene expression for iron homeostasis in Arabidopsis. Cell Res 18: 385–397. doi: 10.1038/cr.2008.26.
36. Long TA, Tsukagoshi H, Busch W, Lahner B, Salt DE, et al. (2010) The bHLH transcription factor POPEYE regulates response to iron deficiency in Arabidopsis roots. Plant Cell 22: 2219–2236. doi: 10.1105/tpc.110.074096.
37. Kobayashi T, Nagasaka S, Senoura T, Itai RN, Nakanishi H, et al. (2013) Iron-binding Haemerythrin RING ubiquitin ligases regulate plant iron responses and accumulation. Nat Commun 4. doi: 10.1038/ncomms3792.
38. Kobayashi T, Nishizawa NK (2014) Iron sensors and signals in response to iron deficiency. Plant Sci 224C. doi: 10.1016/j.plantsci.2014.04.002.

20

Saharan Dust Deposition May Affect Phytoplankton Growth in the Mediterranean Sea at Ecological Time Scales

Rachele Gallisai[1]*, Francesc Peters[1], Gianluca Volpe[2], Sara Basart[3], José Maria Baldasano[3,4]

1 Departament de Biologia Marina i Oceanografia, Institut de Ciències del Mar, CSIC, Barcelona, Spain, **2** Istituto di Scienze dell'Atmosfera e del Clima, Roma, Italy, **3** Earth Sciences Department, Barcelona Supercomputing Center-Centro Nacional de Supercomputación, BSC-CNS, Barcelona, Spain, **4** Environmental Modelling Laboratory, Technical University of Catalonia, Barcelona, Spain

Abstract

The surface waters of the Mediterranean Sea are extremely poor in the nutrients necessary for plankton growth. At the same time, the Mediterranean Sea borders with the largest and most active desert areas in the world and the atmosphere over the basin is subject to frequent injections of mineral dust particles. We describe statistical correlations between dust deposition over the Mediterranean Sea and surface chlorophyll concentrations at ecological time scales. Aerosol deposition of Saharan origin may explain 1 to 10% (average 5%) of seasonally detrended chlorophyll variability in the low nutrient-low chlorophyll Mediterranean. Most of the statistically significant correlations are positive with main effects in spring over the Eastern and Central Mediterranean, conforming to a view of dust events fueling needed nutrients to the planktonic community. Some areas show negative effects of dust deposition on chlorophyll, coinciding with regions under a large influence of aerosols from European origin. The influence of dust deposition on chlorophyll dynamics may become larger in future scenarios of increased aridity and shallowing of the mixed layer.

Editor: Tomoya Iwata, University of Yamanashi, Japan

Funding: The work was supported by the following: JAE-Predoc fellowship from the Spanish Scientific Research Council (CSIC), RG; Aerosol deposition and ocean plankton dynamics (CTM2011-23458), Spanish Ministry of Science and Innovation project, http://www.idi.mineco.gob.es/, RG and FP; Estructura de la materia orgánica en el océano costero: implicaciones biogeoquímicas y ecológicas (CTM2009-09352), Spanish Ministry of Science and Innovation project, http://www.idi.mineco.gob.es/, FP; Grup d'Oceanografia Mediterrània (2009/SGR/588), Generalitat de Catalunya project, http://www10.gencat.cat/agaur_web/AppJava/catala/index.jsp, RG and FP; La ricerca italiana per il mare, Italian Ministry of Education, University and Research, http://www.ritmare.it/, GV; Acoplamiento on-line de un modulo completo de aerosoles multicomponente al modelo atmosferico global regional NMMB (CGL2010-19652), Spanish Ministry of Science and Innovation project, http://www.idi.mineco.gob.es/, SB and JMB; Supercomputación and e-ciencia project (CSD2007-0050) from the Consolider-Ingenio 2010 program of the Spanish Ministry of Economy and Competitivity, http://www.idi.mineco.gob.es/, SB and JMB; and Severo Ochoa program (SEV-2011-00067) from Spanish Ministry of Economy and Competitivity, http://www.idi.mineco.gob.es/, SB and JMB. The funders had no role in study design, data collection and analysis, decision to publish, or preparation of the manuscript.

Competing Interests: The authors have declared that no competing interests exist.

* Email: gallisai@icm.csic.es

Introduction

Aerosols have major impacts on weather and climate regulations [1,2] and even on crop production [3]. Atmospheric desert dust may travel large distances from its source and has been proposed to have ocean production regulation effects over geological times scales [4]. The Mediterranean Sea (hereafter Med) atmosphere is subject to the continuous injection of Saharan and Middle East mineral dust particles [5]. The deposition of these mineral particles supply numerous macro and micro- nutrients to the ocean surface [6–14] and some authors consider it as the major source of "new" nutrients [15] for system production.

Calculations show that the atmospheric input of nutrients in the Med is of the same magnitude as riverine inputs [16–18], thus playing a significant role in the regulation of the nutrient balance of the basin at decadal or longer time scales [19,20]. The contribution of atmospheric deposition can be especially important and efficient in oligotrophic environments such as the Med, which has a marked stratification period and a pronounced nutrient limitation [21]. The deposition of some of these soluble compounds on surface waters may influence biological production, at least during certain events [9,22,23]. Dust deposition spreads over vast areas and dilutes into the water column often preventing the potential effects on system production to be unequivocally detected at ecological time scales. Experiments and observations in low nutrient – low chlorophyll areas have so far shown mixed results [24–26]. Reasons may include a tremendous variability in dust nutrient bioavailability content [27–29] and a relatively small increase of the background nutrient concentration when vertical mixing is active and represents the major source of nutrients [21] as well as a rapid transfer of increased primary production to other trophic levels and a variety of plankton community structures and physiological states.

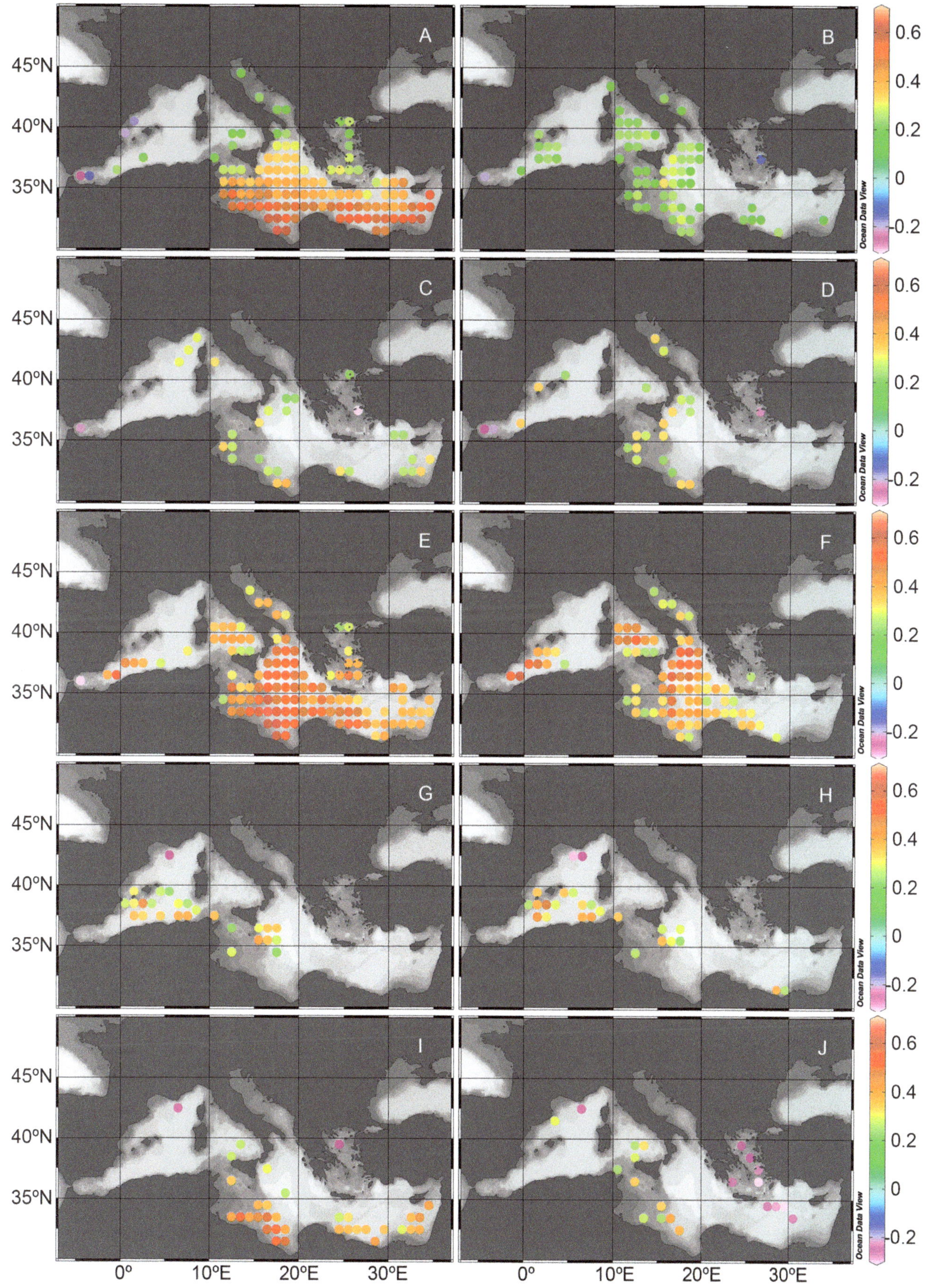

A
B
C
D
E
F
G
H
I
J
45ºN
40ºN
35ºN
0º
10ºE
20ºE
30ºE
0.6
0.4
0.2
0
-0.2
Ocean Data View

Figure 1. Correlation between chlorophyll concentration and dust deposition. Statistically significant ($p<0.05$) correlation coefficient (r) between chlorophyll concentration and dust deposition (left panels) and between the seasonally detrended chlorophyll concentration and the seasonally detrended dust deposition (right panels) for the whole time series and for different seasons. Panels: a, b) annual; c, d) winter (January to March); e, f) spring (April to June); g, h) summer (July to September) and i, j) autumn (October to December).

Given the episodic nature of dust events, an additional complication may reside in the human capacity of detecting the dust event deposition with sufficient space-time resolution in order to build a statistically significant dust event database. Previous attempts used satellite-derived aerosol optical thickness (AOT) as a proxy of dust in the atmosphere to infer the deposition events [26,30,31]. Dust generally travels from several hundreds to thousands of meters high in the atmosphere, making this approach not quantitatively adequate for discerning between transport and deposition. Deposition is measured *in situ* at a few terrestrial (mainland and islands) sites, which are extremely valuable for ground truth validation but are dependent on local conditions, making generalizations hard to draw especially towards the open ocean. Here, we employ a state-of-art atmospheric transport and deposition model, the BSC-DREAM8b model [32], which has been validated [33–35] and gives the power of having aerosol deposition data over the whole Med basin with daily temporal resolution. A previous study showed the potential positive effects of dust deposition on SeaWiFS-derived chlorophyll (Chl) in the Med [32]. However, in the Med, the used NASA OC4v4 algorithm falls far short to retrieve Chl with accuracy smaller than 100%, casting doubts on the relationships found. Here we extend this approach by relating deposition to SeaWiFS Chl using the Med-specific algorithm MedOC4 [36]. When we think of dust deposition, we tend to think about very large events, those that are obvious in true color images or that we recognize because we find our cars covered with red dust, but the truth is that, to some extent, there is Saharan dust in the atmosphere over the Mediterranean almost continuously and deposition does not occur only during large events but also when atmospheric aerosol concentrations are not so high. Thus, rather than focusing on single events or experiments, we take a correlational approach using an 8-year data time series in order to find relationships between Chl dynamics and dust deposition over the Mediterranean Sea.

Methods

Chlorophyll data

SeaWiFS HRPT Level-1A data (2000–2007) were collected at the Istituto di Scienze dell'Atmosfera e del Clima of Rome, Italy, and processed up to Level-3 using the MedOC4 regional algorithm (http://www.myocean.eu/web/69-myocean-interactive-catalogue.php) [36]. This algorithm takes into account the peculiar blue-green ratio of Med waters. Level-3 Chl data, with a native 1 km resolution, were $\log_{10}$-transformed averaged, over a period of eight days, and regridded over the 1° resolution grid of the basin (179 cells, see Table S1). A previous study showed that it is recommended not to use 8-d averages when computing correlation analysis between Chl and dust events [26]. To account for the possible contamination by atmospheric dust mimicking chlorophyll, here, before averaging over the period of eight days, the quality of the entire Chl dataset was carefully checked by i) applying all the SeaDAS Level-2 processing masks and flags (http://oceancolor.gsfc.nasa.gov/VALIDATION/flags.html), ii) removing all isolated pixels, iii) removing all pixels exceeding 3 standard deviations within a moving box of 3×3 pixels, and iv) by applying a median filter over all remaining good pixels. This procedure increases the confidence level on data quality, with the only shortcoming of reducing the number of observations with respect to the NASA standard processing. The time series of daily observations was temporally binned into periods of 8 days. This results into 45 bins up to the 360th day of the year. The last bin was computed with the remaining 5 days, and in the case of leap years, with the remaining 6 days. The climatic mean is then calculated across years for each of the natural 8-d time periods.

Dust deposition

For the present study, a dust deposition simulation from the BSC-DREAM8b model (http://www.bsc.es/earth-sciences/mineral-dust/catalogo-datos-dust) model [33,37] was used for the period between 1 January 2000 and 31 December 2007, over the Med basin. BSC-DREAM8b tracks mineral dust particles from their sources in the Sahara and Middle East regions. Output, after being $\log_{10}$-transformed, was provided for the same space and time resolution as for chlorophyll. A low cut-off threshold (10^{-8} Kg m^{-2} d^{-1}) is applied to the numerical deposition output from BSC-DREAM8b since the dataset showed numerically correct but physically unrealistic low value spikes [32]. The model main features were described in detail in Pérez et al. [37] and Basart et al. [35]. It has been used for dust forecasting and as a dust research tool in North Africa and the Med [32,38–40]. Several studies have checked its performance [33,41], concerning both the horizontal and vertical extent of the dust plumes in the Med Basin. The model daily evaluation with near-real time observations is conducted at the Barcelona Supercomputing Center, and includes satellite data (MODIS and MSG) and AERONET sun photometers. BSC-DREAM8b has also been validated and tested over longer time periods in the European region [34,42,43] and against measurements at source regions [44].

Aerosol Optical Thickness

AOT at 865 nm data were derived from SeaWiFS radiometer measurements and they were downloaded from the Giovanni database (http://gdata1.sci.gsfc.nasa.gov/daac-bin/G3/gui.cgi?instance_id=ocean_8day). We acquired 8-d averaged, 9 km resolution product from 2000 to 2007. Similarly to Chl and deposition data, AOT data were $\log_{10}$-transformed and regridded over the 1° resolution grid of the basin, with the same temporal binning. This was done for the same 179 1°×1° cells as for chlorophyll. It should be noted that AOT contains information of total aerosol particles in the atmosphere, not only of particles from Saharan origin. However, over much of the Mediterranean Sea most particles are indeed of Saharan origin [45].

Statistical analyses

Pearson's correlation coefficient (*r*) was calculated between chlorophyll concentration, modeled dust deposition and AOT time series for each grid cell. Significance was considered at $p<0.05$ using Student's t-test. In addition, the degrees of freedom used for significance testing were adjusted to take into account the possible presence of autocorrelation in the time series. The number of effective independent observations, N*, were calculated as described in Pyper el al. [46]. Correlations were computed both for the entire series and for each season. The same analyses were performed after seasonally detrending the data by subtracting the

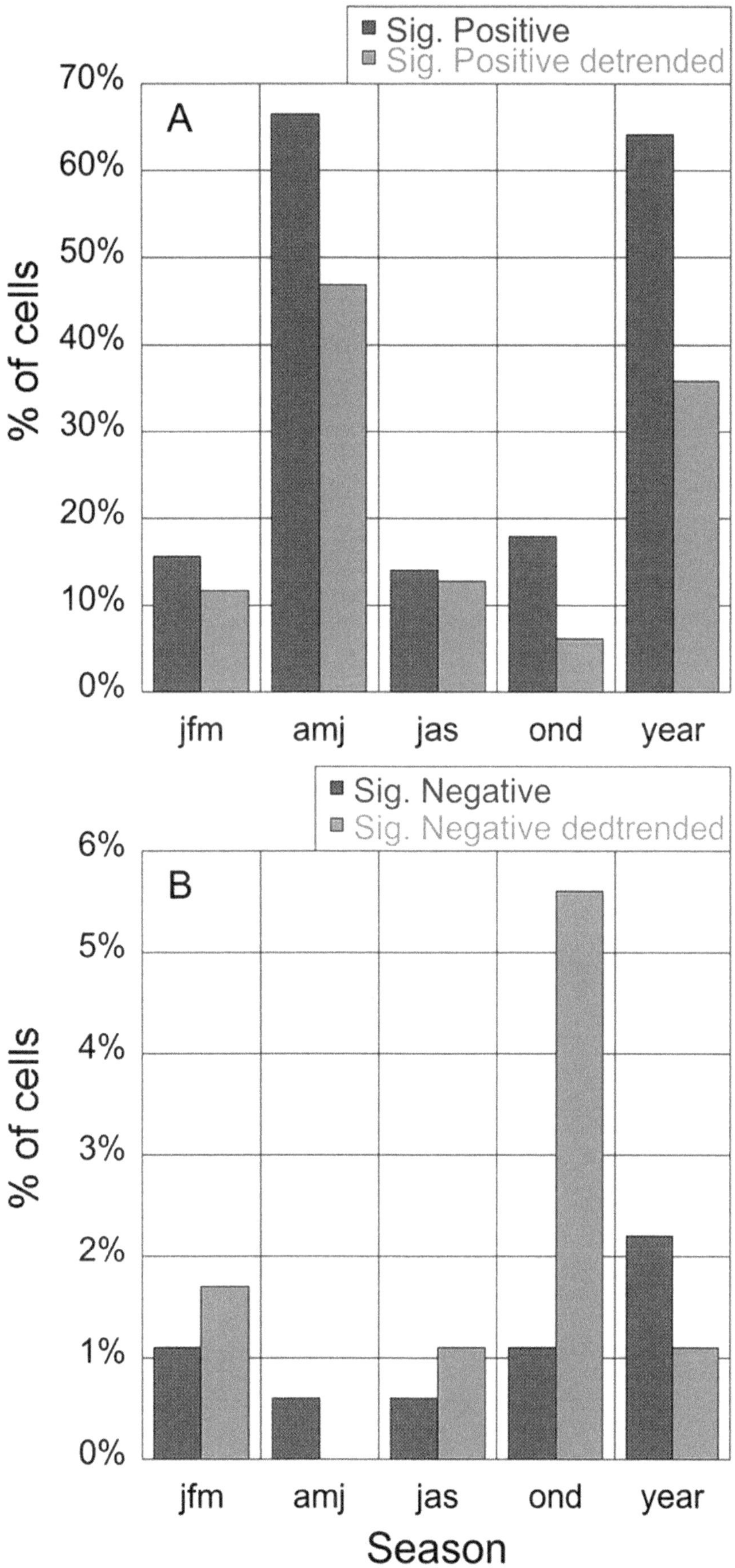
Sig. Positive
Sig. Positive detrended
A
70%
60%
50%
40%
30%
20%
10%
0%
% of cells
jfm
amj
jas
ond
year
Sig. Negative
Sig. Negative dedtrended
B
6%
5%
4%
3%
2%
1%
0%
% of cells
jfm
amj
jas
ond
year
Season

Figure 2. Percentage of cells showing significant correlations between chlorophyll and deposition. Left panel: positive correlations. Right panel: negative correlations. Red bars represent non seasonally detrended data and blue bars seasonally detrended data.

climatic mean at each time series data point. The r^2 of the correlation in a cell is the variance explained by the correlation in that cell. The minimum, maximum and average variance-explained values (expressed in %variability) were calculated for the population of cells with a $p<0.05$.

Results and Discussion

We have found statistically significant positive correlations between surface chlorophyll and mineral dust deposition in large areas of the Med, covering 64% of the analyzed surface and located mainly in the Central and Eastern basins (Fig. 1) and with a clear south to north gradient in correlation intensity from 0.63 to 0.12. Significant negative correlations (r from −0.15 to −0.25) are observed in only 4 cells located in the Alboran Sea and in the eastern coast of Spain. Positive correlations can be found during all seasons, although it is in spring when we see the largest effects with correlations ranging from 0.22 to 0.65 mainly in the Central, Eastern and Southwestern Med. The Western and Central Med also show regions with positive correlations in summer, while in autumn there are some areas affected in the Central and Eastern Med. Most of the Med phytoplankton variability (>80%) is well explained by the variability of the mixed layer depth [47], and especially the winter-spring mixing bringing nutrient-rich deep waters to the surface. Thus, at least part of the explained variability between our deposition and chlorophyll time series must be due to the partial matching of the annual cycles of both variables.

The relationship between the seasonally detrended data of chlorophyll and dust deposition, that represents more of a response of short-term chlorophyll peaks to dust outbreaks, is somewhat weaker in intensity and in area covered. Largest positive correlations are found in the Central Med (from 0.13 to 0.32) (Fig. 1). Again, it is in spring where the largest impacted area is found, mainly in the Central Med and extending into the Eastern Med and Southwestern Med with r ranging from 0.24 to 0.58. The Western Med shows the largest area affected in summer. This is not surprising given that the seasonal dust event frequency peaks during spring in the Central-Eastern Med, and during summer over the Central-Western basin [26]. Seasonally detrended data tend to slightly increase the number of cells showing significant negative correlations and decrease the number of significantly positive correlated cells (Fig. 2). It is in autumn when we see the largest number of negatively correlated cells (6% of analyzed surface) and located mainly in the Aegean Sea and extending southeasterly of Crete.

For the seasonally detrended data, we checked that the correlation values were not caused by chance. We generated synthetic seasonally detrended chlorophyll time series with the observed mean and standard deviation for each cell. Correlations were computed with dust deposition model outputs, and the process repeated 100 times (Fig. 3). The observed significant correlations were compared to the distribution of the synthetic correlations for each cell, and in all cases they were statistically different with an $\alpha<0.001$ and a power (1-β) undistinguishable from 1. This confirmed the non-spurious nature of the relationships between dust deposition and non-seasonal chlorophyll time series.

Bulk Saharan dust deposition over the Med is not straightforwardly related to dust travelling in the atmosphere (Fig. S1). Meteorological conditions and wind patterns at different times of the year often have large amounts of dust (AOT) travelling at altitude with little deposition [48–50]. AOT and dust deposition

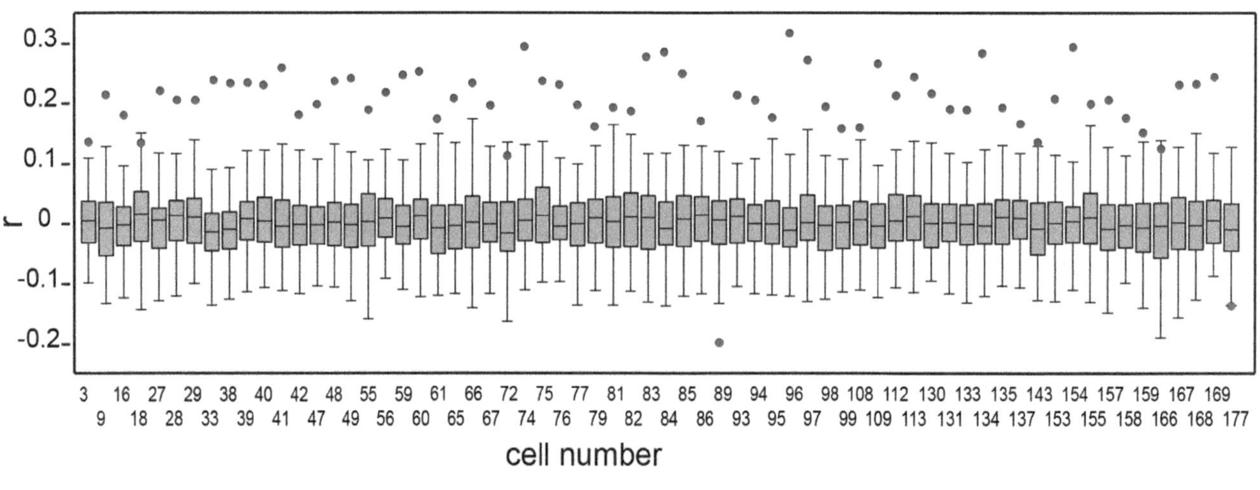

Figure 3. Analysis of the chance of significant correlations being spurious. Comparison between the box plots of the distribution of correlation coefficients between synthetic seasonally detrended chlorophyll time series and seasonally detrended dust deposition model outputs (N = 100) and the actual observed correlation between the seasonally detrended chlorophyll and the seasonally detrended dust deposition model outputs (dots). Data is shown only for those cells showing significant ($p<0.05$) observed correlations. Dots in blue represent significantly positive correlations and red significantly negative correlations. Box plots show the median, the grey box englobing all data between the 25 and 75 percentiles, and the range between the smallest and largest values that are not outliers. Starting from the detrended data of the cells that show statistically significant correlations between detrended chlorophyll and deposition (Fig. 1b), synthetic Chl time series with the same mean and standard deviation (normal distribution) as the original detrended chlorophyll time series, were computed for each cell. The correlation between these synthetic Chl time series and the modelled dust deposition were computed. For each cell, this process was repeated 100 times, and the probability distribution functions (PDFs) of the correlations were then obtained and presented as box plots.

Table 1. Percentage of observed chlorophyll variability explained by modeled dust deposition.

Season	Non-detrended data				Seasonally detrended data			
	N	Min	Max	Average	n	Min	Max	Average
Annual	115	1.4	40.2	16.4	64	1.3	10.1	4.7
Winter	28	4.3	15.1	7.8	21	4.4	12.6	8.5
Spring	119	5.0	41.6	19.1	84	5.7	33.3	15.0
Summer	25	4.2	19.6	10.2	23	4.7	21.3	11.1
Autumn	32	6.5	26.9	15.6	11	5.4	16.5	9.6

Number of cells (n) with significantly ($p<0.05$) positive correlations. Minimum (min), maximum (max) and average percentage of chlorophyll variability explained in the significantly positive cells.

show positive correlation in the Western Med (Fig. S2.) especially in spring and summer with correlated areas shifting depending on the season. The Eastern sub-basin presents the highest correlations in spring (from 0.23 to 0.51) and the Central Med (Tyrrhenian Sea, Sicily channel and Dardanelle strait) in autumn. Once the data are seasonally detrended, deposition events are more related to AOT events, both when the whole series is considered and when the data are analyzed for the different times of the year (Fig. S2). With respect to non-detrended data, seasonally detrended data show main increases in correlation and correlated area in the Central Med for most of the year, as well as in the Eastern Med in autumn. Some overall hotspots appeared in the Alboran and in the Tyrrhenian Sea and around Crete, where the correlations ranged from 0.36 to 0.48.

The annual cycles of chlorophyll and AOT do not match (Fig. S1). The maximum chlorophyll concentrations occur in winter and minima coincide with the summer months. On the contrary, the highest AOT is found in summer and the minimum in autumn. Overall, AOT and chlorophyll (Fig. S3) show no significant correlations in the Med, except for some areas near the African coasts, where the correlation is negative (from -0.28 to -0.36). While no correlations are evident between AOT and chlorophyll there are significant correlations between seasonally detrended AOT and chlorophyll data (Fig. S3). A plume of higher correlation, with r-values between 0.33 and 0.39, appear in the northern part of Cyrenaica region with an extension up to the south of Italy. The best match between both series was found in summer (Fig. S3). Volpe et al. [36] ground truthed the chlorophyll satellite estimates with in situ measurements and concluded that the atmospheric correction was appropriate. In addition, we compared the data from the chlorophyll measurements at the DYFAMED station (1998–2007) with SeaWiFS estimates corrected with a regional algorithm giving a slope of ~ 1 ($\log DYF = 0.0129 + 1.0497 \cdot \log SW$; Adjusted $R^2 = 0.68$; $N = 91$; $p<0.001$). Moreover DYFAMED chlorophyll was unrelated to AOT, providing further evidence of the independence between satellite measurements of chlorophyll and AOT. Aerosols travelling over a certain area are not necessarily depositing. When a deposition event is occurring, it should coincide with high aerosol content in the air (AOT), thus if we find relationships between dust deposition events and non-seasonal chlorophyll peaks it is also logical to expect that chlorophyll is related to AOT, while the non-detrended AOT data show little or no relationship.

As mentioned before, the largest positive correlations between dust deposition and chlorophyll occur around the Central and Eastern Med. Calculations [8,20,51] and experiments [11,20,24] tell us that aerosol deposition effects on primary production should be small in most situations and thus we do not expect African dust deposition in general to explain a large portion of chlorophyll variability. Accordingly, positive significant correlations between mineral dust deposition of Saharan origin and chlorophyll do explain only a 1 to 10% (average 5%) of chlorophyll variability for seasonally detrended and a 1 to 40% (average 16%) for non-detrended data although it may be higher for certain seasons (Table 1). It should be noted that the explained variability does not provide direct information of the magnitude of chlorophyll impacted.

Winter shows overall the lowest significantly positive correlations, while spring presents the highest. This is to be expected since the entrance of new nutrients should be mostly due to seasonal winter overturning and mixing of nutrient-rich deep waters with upper ocean surface waters, through a number of physical processes that increase vertical diffusion at certain moments. But even at times when nutrient concentrations are expected to be relatively high in the water, low concentrations and strong imbalances between N and P are often observed [52,53], opening windows of opportunity for the nutrients from atmospheric deposition to have an impact in the sustainment of phytoplankton production. We can only speculate on the positive cause-effect relationship between aerosol deposition and chlorophyll in the Med at certain times. Terrestrial inputs through major rivers occur mainly in the Western Med [54], and atmospheric inputs may dominate nutrient supply at certain times [10,55]. Phosphorus (P) limitation alleviation has often been invoked [8,56] as the surface waters of the Med are among the most P-limited in the world [57]. Although aerosols show a disproportionally large ratio of nitrogen to phosphorus [18], potentially only exacerbating P-limitation, they do carry an amount of P that could be used by phytoplankton and bacteria, especially in spring and summer when the concentrations of this element in surface waters of the open Med are at their lowest. Guieu et al. [10] calculated that, if P is considered the limiting element for phytoplankton growth, atmospheric deposition could account for chlorophyll increases of ca. 0.2 $\mu g\ L^{-1}$ in the upper mixed layer for a single large deposition event or for the average total deposition during the summer-stratified period. The Central and Eastern Med do not show the typical spring phytoplankton bloom and have been defined as no blooming areas [58]. The ultra-oligotrophic conditions [59] found in these areas should make them most responsive to external nutrient supplies. As is the case for high nutrient – low chlorophyll areas, micronutrients such as iron from aerosols have also been proposed to stimulate Med phytoplankton production under certain situations [60], albeit addition experiments have not shown a direct increase in dissolved iron (Fe) [61]. Fe in the mixed layer of the Mediterranean is found at concentrations from 0.13 to 2.7 nM [62,63]. It seems though that Fe is, relative to the needs of plankton, in excess with respect to P in the Mediterranean [64]. Nevertheless, in a system where all elements are relatively scarce, responses to the combination of elements arriving through aerosol deposition, may be very complex, with elements becoming successively limiting in a chained reaction. Ridame et al. [65] found stimulation of nitrogen (N) fixation in dust pulse experiments, in general related to a primary alleviation of P-limitation. In their Central Med experiment though, they found high N-fixation stimulation unrelated to P- or Fe-limitation, further showing the complexity of the processes involved and the potential spatial and temporal variability. An initial stimulation of heterotrophic bacteria [25,66] should not be discarded since these organisms have a potential advantage at low nutrient concentrations owing to their high surface to volume ratio. Secondarily, released nutrients from recycling could then stimulate phytoplankton processes. Contrary to the Eastern Med showing the lowest nutrient concentrations in the Mediterranean [59], the Central Med was found somewhat more responsive to dust deposition in the present study. Pey et al. [5] mention the Central Med as a transitional area, receiving a higher frequency of dust outbreaks than similar latitudes in the Western and Eastern Med. Additionally, the dust source areas are not homogeneous. The Libyan Desert is the main source of dust for the Central Med while the Eastern Med receives dust from Libya and from the Middle East [67]. Thus, positive correlations between dust deposition and surface chlorophyll seem to arise from the combination of areas of low nutrient concentrations with the right nature, timing and frequency of dust outbreaks.

Negative relationships between dust deposition and chlorophyll have been related to metal (mainly Cu but also Al) inhibition of phytoplankton growth [13,68]. The toxicity of Cu in reducing phytoplankton growth rate has been shown in laboratory experiments (see [68] and references therein). A recent correlation study between chlorophyll and metals from onshore-measured aerosols in the Northwest Med shows negative relationships in the area under northerly wind (Tramontane) conditions [13]. These winds favor the transport of anthropogenic aerosols from Europe to the Med. Although most Cu pulses are anthropogenically derived, pulses originating in Africa showed effects on chlorophyll undistinguishable from those originating locally [13]. A reduction in chlorophyll growth of up to 20% can be seen along the French and Spanish coasts. In addition, Jordi et al. [13] argue that since Cu toxicity seems to be taxon specific, the summer phytoplankton community with a predominance of nanoflagellates over the less sensitive diatoms, is more vulnerable to atmospheric deposition. This is an area where we also see some negative correlations between the modeled deposition and chlorophyll. We only track Saharan mineral dust, while some of the high load of metals may be more related to local anthropogenic sources. Most of the large deposition events in the Northwest Med come in the form of wet deposition [69]. In our model, the deposition field only originates from Saharan and Middle East dust transport and does not account for local anthropogenic aerosol sources, but rain washes out the entire atmospheric column aerosol loading, no matter the origin. Results from our correlation analysis agree with previous more detailed local studies [13]. We also see a negative relationship between deposition and chlorophyll, both seasonally detrended, mainly in autumn in the Aegean region (Fig. 1). This area is affected by long-range transport of air pollutants from Eastern Europe [70] but it is also heavily impacted by anthropogenic emissions generated in Athens and Istanbul [71,72]]. A high-density population together with a massive number of vehicles, many of them still using non-catalytic or old technology diesel engines, contributed to exceed the EU annual aerosol limit. The amount of Cu in these aerosols is high with an annual mean concentration between 0.013 and 0.22 $\mu g\ m^{-3}$ ([73] and references therein). An estimated dry deposition flux of Cu over the sea ranges then between 22 and 380 μg Cu $m^{-2}\ d^{-1}$ surpassing the threshold limit for Cu to inhibit phytoplankton growth rate according to [13] and [68].

Conclusion

Desert dust storm events seem to be increasing in frequency and intensity [14,69,72–76] in the last decades, due to human activities and climate forcing. This means that the presence of aerosols over the Med is likely to increase with future aridity. Thus, it is important to understand basin level patterns in the response of Med biogeochemistry to aerosol deposition. Only a few studies [26,30] have tried to analyze the potential links between aerosols in the air column and chlorophyll for the entire Med basin, with non definitive results. In this study, we use a modeled actual aerosol deposition product and show both positive and negative significant correlations with chlorophyll dynamics in certain areas and times of the year. Mineral dust from North Africa and the Middle East correlates to chlorophyll in large areas of the Med Sea. This is especially true for the Central and Eastern Med sub-basins, where Saharan dust deposition dynamics matches that of chlorophyll, particularly during spring. Here the atmospheric input may be an intrinsic part of the annual ecosystem dynamics. In terms of large dust outbreaks, chlorophyll best relates to aerosol deposition in the Central Med, extending both into the Eastern and Southwestern Med (Fig. S3). Some areas of the Western Med and Aegean Sea show negative correlations between chlorophyll and deposition in accordance with some recent findings of toxicity brought by metals in aerosols. As expected, dust deposition does not explain an overall large amount of chlorophyll variability since the main ecosystem production driver in the Med is the vertical mixing of nutrients from deep waters. Variability related to carbon to chlorophyll ratios, the consumption of biomass with a varying degree of coupling and the variable settling of primary production, are all additional sources of surface chlorophyll variability that we could not account for in our correlations and thus add to the noise. No matter how small significant correlations are, they are not distributed randomly in space and coincide with independent estimates that follow the same trend. Thus, albeit the mechanisms that affect chlorophyll through aerosol deposition cannot be pinpointed and may be indirect, non-unique, and dependent on local spatio-temporal conditions, our study shows a clear potential for effects at ecological scales. These effects should become more important in a future scenario with increased aerosols over the Med and a shallower mixed layer depth, owing to increased temperatures, over which aerosols may leach out nutrients.

Supporting Information

Figure S1 Seasonal average values of chlorophyll concentration, dust deposition and aerosol optical thickness. Average chlorophyll concentration (left panels). Average dust deposition (central panels) and average aerosol optical thickness (right panels) for different seasons. Winter (a, b, c), spring (d, e, f), summer (g, h, i) and autumn (j, k, l).

Figure S2 Correlation between dust deposition and aerosol optical thickness. Statistically significant ($p<0.05$) correlation coefficient (r) between dust deposition and aerosol optical thickness (left panels) and between seasonally detrended dust deposition and seasonally detrended aerosol optical thickness (right panels) for the whole time series and for different seasons. Panels: a, b) annual; c, d) winter (January to March); e, f) spring (April to June); g, h) summer (July to September) and i, j) autumn (October to December).

Figure S3 Correlation between chlorophyll concentration and aerosol optical thickness. Same as Fig. S2 but for chlorophyll concentration versus aerosol optical thickness.

Table S1 Geographical coordinates for the 1°×1° grid cells analyzed in this study. Coordinates refer to the central point of the cell. Latitudes are all North. Positive longitudes are East and negative longitudes West.

Acknowledgments

The Goddard Earth Sciences Data and Information Services Center (GES DISC), the SeaWiFS mission scientists and associated NASA personnel for the production of the data used in this study and for the development and maintenance of SeaDAS software. The Barcelona Supercomputing Center (BSC) for hosting the BSC-DREAM8b simulations on the Mare Nostrum Supercomputer. The DYFAMED (CNRS-INSU) Observatoire Océanologique de Villefranche-sur-mer (http://www.obs-vlfr.fr/dyfBase/) for providing *in situ* chlorophyll measurements. The developers of Ocean Data View (http://odv.awi.de) for their useful tool.

Author Contributions

Conceived and designed the experiments: RG FP. Performed the experiments: RG FP. Analyzed the data: RG FP GV SB JMB. Contributed reagents/materials/analysis tools: RG FP GV SB JMB. Contributed to the writing of the manuscript: RG FP GV SB JMB.

References

1. Booth BBB, Dunstone NJ, Halloran PR, Andrews T, Bellouin N (2012) Aerosols implicated as a prime driver of twentieth-century North Atlantic climate variability. Nature 484: 228–232.
2. Creamean JM, Suski KJ, Rosenfeld D, Cazorla A, DeMott PJ, et al. (2013) Dust and Biological Aerosols from the Sahara and Asia Influence Precipitation in the Western U.S. Science 339: 1572–1578.
3. Liu X, Zhang Y, Han W, Tang A, Shen J, et al. (2013) Enhanced nitrogen deposition over China. Nature 494: 459–462.
4. Jaccard SL, Hayes CT, Martinez-Garcia A, Hodell DA, Anderson RF, et al. (2013) Two Modes of Change in Southern Ocean Productivity Over the Past Million Years. Science 339: 1419–1423.
5. Pey J, Querol X, Alastuey A, Forastiere F, Stafoggia M (2013) African dust outbreaks over the Mediterranean Basin during 2001–2011: PM10 concentrations, phenomenology and trends, and its relation with synoptic and mesoscale meteorology. Atmospheric Chemistry and Physics 13: 1395–1410.
6. Bonnet S, Guieu C, Chiaverini J, Ras J, Stock A (2005) Effect of atmospheric nutrients on the autotrophic communities in a low nutrient, low chlorophyll system. Limnology and Oceanography 50: 1810–1819.
7. Bergametti G, Remoudaki E, Losno R, Steiner E, Chatenet B, et al. (1992) Source, transport and deposition of atmospheric phosphorus over the northwestern Mediterranean. Journal of Atmospheric Chemistry 14: 501–513.
8. Ridame C, Guieu C (2002) Saharan input of phosphate to the oligotrophic water of the open western Mediterranean sea. Limnology and Oceanography 47: 856–869.
9. Markaki Z, Oikonomou K, Kocak M, Kouvarakis G, Chaniotaki A, et al. (2003) Atmospheric deposition of inorganic phosphorus in the Levantine Basin, eastern Mediterranean: Spatial and temporal variability and its role in seawater productivity. Limnology and Oceanography 48: 1557–1568.
10. Guieu C, Loÿe-Pilot MD, Benyahya L, Dufour A (2010) Spatial variability of atmospheric fluxes of metals (Al, Fe, Cd, Zn and Pb) and phosphorus over the whole Mediterranean from a one-year monitoring experiment: Biogeochemical implications. Marine Chemistry 120: 164–178.
11. Pulido-Villena E, Rerolle V, Guieu C (2010) Transient fertilizing effect of dust in P-deficient LNLC surface ocean. Geophysical Research Letters 37: L01603.
12. Herut B, Krom MD, Pan G, Mortimer R (1999) Atmospheric input of nitrogen and phosphorus to the Southeast Mediterranean: Sources, fluxes, and possible impact. Limnology and Oceanography 44: 1683–1692.
13. Jordi A, Basterretxea G, Tovar-Sanchez A, Alastuey A, Querol X (2012) Copper aerosols inhibit phytoplankton growth in the Mediterranean Sea. Proceedings of the National Academy of Sciences of the United States of America 109: 21246–21249.
14. Goudie AS, Middleton NJ (2001) Saharan dust storms: nature and consequences. Earth-Science Reviews 56: 179–204.
15. Ternon E, Guieu C, Ridame C, L'Helguen S, Catala P (2011) Longitudinal variability of the biogeochemical role of Mediterranean aerosols in the Mediterranean Sea. Biogeosciences 8: 1067–1080.
16. Guieu C, Martin JM, Thomas AJ, Elbazpoulichet F (1991) Atmospheric versus river inputs of metals to the Gulf of Lions: Total concentrations, partitioning and fluxes. Marine Pollution Bulletin 22: 176–183.
17. Ludwig W, Bouwman AF, Dumont E, Lespinas F (2010) Water and nutrient fluxes from major Mediterranean and Black Sea rivers: Past and future trends and their implications for the basin-scale budgets. Global Biogeochemical Cycles 24: GB0A13.
18. Markaki Z, Loÿe-Pilot MD, Violaki K, Benyahya L, Mihalopoulos N (2010) Variability of atmospheric deposition of dissolved nitrogen and phosphorus in the Mediterranean and possible link to the anomalous seawater N/P ratio. Marine Chemistry 120: 187–194.
19. Bethoux JP, Morin P, Ruiz-Pino DP (2002) Temporal trends in nutrient ratios: chemical evidence of Mediterranean ecosystem changes driven by human activity. Deep-Sea Research Part Ii-Topical Studies in Oceanography 49: 2007–2016.
20. Herut B, Zohary T, Krom MD, Mantoura RFC, Pitta P, et al. (2005) Response of East Mediterranean surface water to Saharan dust: On-board microcosm experiment and field observations. Deep-Sea Research Part Ii-Topical Studies in Oceanography 52: 3024–3040.
21. Estrada M (1996) Primary production in the northwestern Mediterranean. Scientia Marina 60: 55–64.
22. Guerzoni S, Chester R, Dulac F, Herut B, Loÿe-Pilot MD, et al. (1999) The role of atmospheric deposition in the biogeochemistry of the Mediterranean Sea. Progress in Oceanography 44: 147–190.
23. Morales-Baquero R, Pulido-Villena E, Reche I (2006) Atmospheric Inputs of Phosphorus and Nitrogen to the Southwest Mediterranean Region: Biogeochemical Responses of High Mountain Lakes. Limnology and Oceanography 51: 830–837.

24. Romero E, Peters F, Marrasé C, Guadayol T, Gasol JM, et al. (2011) Coastal Mediterranean plankton stimulation dynamics through a dust storm event: An experimental simulation. Estuarine, Coastal and Shelf Science 93: 27–39.
25. Lekunberri I, Lefort T, Romero E, Vázquez-Domínguez E, Romera-Castillo C, et al. (2010) Effects of a dust deposition event on coastal marine microbial abundance and activity, bacterial community structure and ecosystem function. Journal of Plankton Research 32: 381–396.
26. Volpe G, Banzon VF, Evans RH, Santoleri R, Mariano AJ, et al. (2009) Satellite observations of the impact of dust in a low-nutrient, low-chlorophyll region: Fertilization or artifact? Global Biogeochem Cycles 23: GB3007.
27. Carbo P, Krom MD, Homoky WB, Benning LG, Herut B (2005) Impact of atmospheric deposition on N and P geochemistry in the southeastern Levantine basin. Deep-Sea Research Part II: Topical Studies in Oceanography 52: 3041–3053.
28. Herut B, Collier R, Krom MD (2002) The role of dust in supplying nitrogen and phosphorus to the Southeast Mediterranean. Limnology and Oceanography 47: 870–878.
29. Baker AR, French M, Linge KL (2006) Trends in aerosol nutrient solubility along a west-east transect of the Saharan dust plume. Geophysical Research Letters 33: L07805.
30. Cropp RA, Gabric AJ, McTainsh GH, Braddock RD, Tindale N (2005) Coupling between ocean biota and atmospheric aerosols: Dust, dimethylsulphide, or artifact? Global Biogeochemical Cycles 19: GB4002.
31. Gabric AJ, Cropp R, Ayers GP, McTainsh G, Braddock R (2002) Coupling between cycles of phytoplankton biomass and aerosol optical depth as derived from SeaWiFS time series in the Subantarctic Southern Ocean. Geophysical Research Letters 29: 1112.
32. Gallisai R, Peters F, Basart S, Baldasano JM (2012) Mediterranean basin-wide correlations between Saharan dust deposition and ocean chlorophyll concentration. Biogeosciences Discussion 9: 28.
33. Pérez C, Nickovic S, Baldasano JM, Sicard M, Rocadenbosch F, et al. (2006) A long Saharan dust event over the western Mediterranean: Lidar, Sun photometer observations, and regional dust modeling. Journal of Geophysical Research-Atmospheres 111: D15214.
34. Basart S, Pay MT, Jorba O, Pérez C, Jinez-Guerrero P, et al. (2012) Aerosols in the CALIOPE air quality modelling system: Evaluation and analysis of PM levels, optical depths and chemical composition over Europe. Atmospheric Chemistry and Physics 12: 3363–3392.
35. Basart S, Perez C, Nickovic S, Cuevas E, Baldasano JM (2012) Development and evaluation of the BSC-DREAM8b dust regional model over Northern Africa, the Mediterranean and the Middle East. Tellus Series B-Chemical and Physical Meteorology 64: 18539.
36. Volpe G, Santoleri R, Vellucci V, d'Alcala MR, Marullo S, et al. (2007) The colour of the Mediterranean Sea: Global versus regional bio-optical algorithms evaluation and implication for satellite chlorophyll estimates. Remote Sensing of Environment 107: 625–638.
37. Pérez C, Nickovic S, Pejanovic G, Baldasano JM, Özsoy E (2006) Interactive dust-radiation modeling: A step to improve weather forecasts. Journal of Geophysical Research-Atmospheres 111: D16206.
38. Amiridis V, Kafatos M, Perez C, Kazadzis S, Gerasopoulos E, et al. (2009) The potential of the synergistic use of passive and active remote sensing measurements for the validation of a regional dust model. Annales Geophysicae 27: 3155–3164.
39. Alonso-Perez S, Cuevas E, Perez C, Querol X, Baldasano JM, et al. (2011) Trend changes of African airmass intrusions in the marine boundary layer over the subtropical Eastern North Atlantic region in winter. Tellus, Series B: Chemical and Physical Meteorology 63: 255–265.
40. Pay MT, Jiménez-Guerrero P, Jorba O, Basart S, Querol X, et al. (2012) Spatio-temporal variability of concentrations and speciation of particulate matter across Spain in the CALIOPE modeling system. Atmospheric Environment 46: 376–396.
41. Papanastasiou DK, Poupkou A, Katragkou E, Amiridis V, Melas D, et al. (2010) An Assessment of the Efficiency of Dust Regional Modelling to Predict Saharan Dust Transport Episodes. Advances in Meteorology 2010: 154368.
42. Jiménez-Guerrero P, Pérez C, Jorba O, Baldasano JM (2008) Contribution of Saharan dust in an integrated air quality system and its on-line assessment. Geophysical Research Letters 35: L03814.
43. Pay MT, Piot M, Jorba O, Gassó S, Gonçalves M, et al. (2010) A full year evaluation of the CALIOPE-EU air quality modeling system over Europe for 2004. Atmospheric Environment 44: 3322–3342.
44. Haustein K, Pérez C, Baldasano JM, Müller D, Tesche M, et al. (2009) Regional dust model performance during SAMUM 2006. Geophysical Research Letters 36: L03812.
45. Barnaba F, Gobbi GP (2004) Aerosol seasonal variability over the Mediterranean region and relative impact of maritime, continental and Saharan dust particles over the basin from MODIS data in the year 2001. Atmos. Chem. Phys. 4: 2367–2391.
46. Pyper BJ, Peterman RM (1998) Comparison of methods to account for autocorrelation in correlation analyses of fish data. Can. J. Fish. Aquat. Sci. 55: 2127–2140.
47. Volpe G, Nardelli BB, Cipollini P, Santoleri R, Robinson IS (2012) Seasonal to interannual phytoplankton response to physical processes in the Mediterranean Sea from satellite observations. Remote Sens. Environ. 117: 223–235.
48. Papayannis A, Amiridis V, Mona L, Tsaknakis G, Balis D, et al. (2008) Systematic lidar observations of Saharan dust over Europe in the frame of EARLINET (2000–2002). J. Geophys. Res. (Atmos.) 113: D10204.
49. Mona L, Amodeo A, Pandolfi M, Pappalardo G (2006) Saharan dust intrusions in the Mediterranean area: Three years of Raman lidar measurements. J. Geophys. Res. (Atmos.) 111: D16203.
50. Gobbi GP, Angelini F, Barnaba F, Costabile F, Baldasano JM, et al. (2013) Changes in particulate matter physical properties during Saharan advections over Rome (Italy): a four-year study, 2001–2004. Atmos. Chem. Phys. 13: 7395–7404.
51. Eker-Develi E, Kideys AE, Tugrul S (2006) Role of Saharan dust on phytoplankton dynamics in the northeastern Mediterranean. Mar. Ecol. Prog. Ser. 314: 61–75.
52. Diaz F, Raimbault P, Boudjellal B, Garcia N, Moutin T (2001) Early spring phosphorus limitation of primary productivity in a NW Mediterranean coastal zone (Gulf of Lions). Mar. Ecol. Prog. Ser. 211: 51–62.
53. Rahav E, Herut B, Levi A, Mulholland MR, Berman-Frank I (2013) Springtime contribution of dinitrogen fixation to primary production across the Mediterranean Sea. Ocean Sci. 9: 489–498.
54. Struglia MV, Mariotti A, Filograsso A (2004) River discharge into the Mediterranean Sea: Climatology and aspects of the observed variability. J. Clim. 17: 4740–4751.
55. Durrieu de Madron X, Guieu C, Sempéré R, Conan P, Cossa D, et al. (2011) Marine ecosystems' responses to climatic and anthropogenic forcings in the Mediterranean. Prog. Oceanogr. 91: 97–166.
56. Izquierdo R, Benítez-Nelson CR, Masqué P, Castillo S, Alastuey A, et al. (2012) Atmospheric phosphorus deposition in a near-coastal rural site in the NE Iberian Peninsula and its role in marine productivity. Atmos. Environ. 49: 361–370.
57. Marty JC, Chiaverini J, Pizay MD, Avril B (2002) Seasonal and interannual dynamics of nutrients and phytoplankton pigments in the western Mediterranean Sea at the DYFAMED time-series station (1991–1999). Deep-Sea Res. Part II. 49: 1965–1985.
58. D'Ortenzio F, Ribera D'Alcalà M (2009) On the trophic regimes of the Mediterranean Sea: A satellite analysis. Biogeosciences 6: 139–148.
59. Pujo-Pay M, Conan P, Oriol L, Cornet-Barthaux V, Falco C, et al. (2011) Integrated survey of elemental stoichiometry (C, N, P) from the western to eastern Mediterranean Sea. Biogeosciences 8: 883–899.
60. Bonnet S, Guieu C (2006) Atmospheric forcing on the annual iron cycle in the western Mediterranean Sea: A 1-year survey. J. Geophys. Res. (Oceans) 111: C09010.
61. Wagener T, Guieu C, Leblond N (2010) Effects of dust deposition on iron cycle in the surface Mediterranean Sea: results from a mesocosm seeding experiment. Biogeosciences 7: 3769–3781.
62. Sarthou G, Jeandel C (2001) Seasonal variations of iron concentrations in the Ligurian Sea and iron budget in the Western Mediterranean Sea. Mar. Chem. 74: 115–129.
63. Guieu C, Bozec Y, Blain S, Ridame C, Sarthou G, et al. (2002) Impact of high Saharan dust inputs on dissolved iron concentrations in the Mediterranean Sea. Geophys. Res. Lett. 29: 1911.
64. Guieu C, Dulac F, Desboeufs K, Wagener T, Pulido-Villena E, et al. (2010) Large clean mesocosms and simulated dust deposition: a new methodology to investigate responses of marine oligotrophic ecosystems to atmospheric inputs. Biogeosciences 7: 2765–2784.
65. Ridame C, Le Moal M, Guieu C, Ternon E, Biegala IC, et al. (2011) Nutrient control of N-2 fixation in the oligotrophic Mediterranean Sea and the impact of Saharan dust events. Biogeosciences 8: 2773–2783.
66. Pulido-Villena E, Wagener T, Guieu C (2008) Bacterial response to dust pulses in the western Mediterranean: Implications for carbon cycling in the oligotropic ocean. Global. Biogeochem. Cy. 22: GB1020.
67. Gaetani M, Pasqui M (2012) Synoptic patterns associated with extreme dust events in the Mediterranean Basin. Reg. Environ. Change.: 1–14.
68. Paytan A, Mackey KRM, Chen Y, Lima ID, Doney SC, et al. (2009) Toxicity of atmospheric aerosols on marine phytoplankton. Proc. Natl. Acad. Sci. USA. 106: 4601–4605.
69. Avila A, Peñuelas J (1999) Increasing frequency of Saharan rains over northeastern Spain and its ecological consequences. Sci. Total Environ. 228: 153–156.
70. Lelieveld J, Berresheim H, Borrmann S, Crutzen PJ, Dentener FJ, et al. (2002) Global air pollution crossroads over the Mediterranean. Science 298: 794–799.
71. Kanakidou M, Mihalopoulos N, Kindap T, Im U, Vrekoussis M, et al. (2011) Megacities as hot spots of air pollution in the East Mediterranean. Atmos. Environ. 45: 1223–1235.
72. Querol X, Alastuey A, Pey J, Cusack M, Perez N, et al. (2009) Variability in regional background aerosols within the Mediterranean. Atmos. Chem. Phys. 9: 4575–4591.
73. Theodosi C, Grivas G, Zarmpas P, Chaloulakou A, Mihalopoulos N (2011) Mass and chemical composition of size-segregated aerosols (PM1, PM2.5,

PM10) over Athens, Greece: local versus regional sources. Atmos. Chem. Phys. 11: 11895–11911.
74. Ganor E, Osetinsky I, Stupp A, Alpert P (2010) Increasing trend of African dust, over 49 years, in the eastern Mediterranean. J. Geophys. Res. (Atmos.) 115: D07201.
75. Goudie AS (2009) Dust storms: Recent developments. J. Environ. Manage. 90: 89–94.
76. Mahowald NM, Kloster S, Engelstaedter S, Moore JK, Mukhopadhyay S, et al. (2010) Observed 20th century desert dust variability: impact on climate and biogeochemistry. Atmos. Chem. Phys. 10: 10875–10893.

The Photoprotective Role of Spermidine in Tomato Seedlings under Salinity-Alkalinity Stress

Lipan Hu[1,2]ᵠ, Lixia Xiang[1,2]ᵠ, Li Zhang[1,2], Xiaoting Zhou[1,2], Zhirong Zou[1,2]*, Xiaohui Hu[1,2]*

1 College of Horticulture, Northwest Agricultural & Forest University, Yangling, China, **2** Key Laboratory of Protected Horticultural Engineering in Northwest, Ministry of Agriculture, Shaanxi Yangling, China

Abstract

Polyamines are small, ubiquitous, nitrogenous compounds that scavenge reactive oxygen species and stabilize the structure and function of the photosynthetic apparatus in response to abiotic stresses. Molecular details underlying polyamine-mediated photoprotective mechanisms are not completely resolved. This study investigated the role of spermidine (Spd) in the structure and function of the photosynthetic apparatus. Tomato seedlings were subjected to salinity-alkalinity stress with and without foliar application of Spd, and photosynthetic and morphological parameters were analyzed. Leaf dry weight and net photosynthetic rate were reduced by salinity-alkalinity stress. Salinity-alkalinity stress reduced photochemical quenching parameters, including maximum photochemistry efficiency of photosystem II, quantum yield of linear electron flux, and coefficient of photochemical quenching (qP). Salinity-alkalinity stress elevated nonphotochemical quenching parameters, including the de-epoxidation state of the xanthophyll cycle and nonphotochemical quenching (NPQ). Microscopic analysis revealed that salinity-alkalinity stress disrupted the internal lamellar system of granal and stromal thylakoids. Exogenous Spd alleviated the stress-induced reduction of leaf dry weight, net photosynthetic rate, and qP parameters. The NPQ parameters increased by salinity-alkalinity stress were also alleviated by Spd. Seedlings treated with exogenous Spd had higher zeaxanthin (Z) contents than those without Spd under salinity-alkalinity stress. The chloroplast ultrastructure had a more ordered arrangement in seedlings treated with exogenous Spd than in those without Spd under salinity-alkalinity stress. These results indicate that exogenous Spd can alleviate the growth inhibition and thylakoid membrane photodamage caused by salinity-alkalinity stress. The Spd-induced accumulation of Z also may have an important role in stabilizing the photosynthetic apparatus.

Editor: Keqiang Wu, National Taiwan University, Taiwan

Funding: This work was supported by grants from the National Key Technology R&D Program in the 12th Five-year Plan of China (No. 2011BAD29B01), the Scientific Research Special Fund of Northwest Agriculture & Forestry University (QN2013018), and the China Agriculture Research System (No. CARS-25-D-02). The funders had no role in study design, data collection and analysis, decision to publish, or preparation of the manuscript.

Competing Interests: The authors have declared that no competing interests exist.

* Email: zouzhirong2005@ hotmail.com (ZZ); hxh1977@163.com (X-HH)

ᵠ These authors contributed equally to this work.

Introduction

Salinity-alkalinity stress is becoming one of the most serious abiotic stresses affecting crop growth and yield [1], and causes considerable economic losses. Our previous studies clearly show that plants are more sensitive to salinity-alkalinity stress than to neutral salt stress [1,2]. Previous reports show that salinity-alkalinity stress strongly inhibits assimilation of photosynthetic carbon [3–5], which leads to an unfavorable dissipation of excess light energy. When photosystem II (PSII) is exposed to excess light, reactive oxygen species (ROS) such as singlet oxygen ($^1O_2$), hydroxyl ions (OH•), and H_2O_2 are formed by the interaction of molecular oxygen and triplet chlorophyll [6,7], which causes photoinhibition and photodamage [8]. ROS directly affects chloroplasts and inhibits the repair of photodamaged PSII [7]. Salinity stress reduces photosynthesis and electron transport activity, thereby exacerbating problems arising from photoinhibition and photodamage of the photosynthetic apparatus [9].

Polyamines (PAs) are organic, low molecular weight, nitrogen-containing compounds; they include spermidine (Spd), spermine (Spm), and their diamine obligate precursor putrescine (Put) [10]. PAs function as plant growth regulators or intracellular messengers that regulate plant growth, development, and responses to abiotic stresses [11]. They also function as antioxidant, scavengers of ROS and free radicals [12], inhibit lipid peroxidation reactions [13] due to a combination of their anion- and cation-binding properties [14], and respond to a wide range of abiotic stresses. PAs also stabilize the structure and function of the photosynthetic apparatus in response to unfavorable environment factors [15–17]. PAs protect the structure of chloroplasts by keeping the thylakoid membranes in an orderly arrangement and maintaining high photosynthetic efficiency [18]. Exogenous Spd may regulate the synthesis of xanthophyll components and promote the conversion of violaxanthin (V) to zeaxanthin (Z), which can protect LHCII reaction centers against salinity-induced oxidative damage and injury from excess light energy [19].

The xanthophyll cycle pigment Z is formed from V via the intermediate anthoxanthin (A) during abiotic stress. Z is thought to be involved in photoprotective thermal dissipation and nonphotochemical quenching (NPQ) [20,21]. The thermal dissipation process protects PSII against photodamage by decreasing excess redox potential of PSII and the electron transfer chain [22]. Z has an important antioxidant function in the lipid phase of the membrane and protects chloroplasts from photo-oxidative stress [23]. Z reduces the formation of ROS, or detoxifies already-formed ROS, to minimize photo-oxidative damage of the photosynthetic apparatus [21].

Previous studies showed that exogenous application of PAs partially alleviated the salinity-induced decline in photosynthetic efficiency [24]. However, the molecular mechanism of Spd-mediated photoprotection of the photosynthetic apparatus and the relationship between Spd and the xanthophyll cycle in response to salinity-alkalinity stress were not completely resolved. Therefore, we sought to determine whether exogenous Spd could protect tomato seedlings from photoinhibition and photodamage and interact with the endogenous xanthophyll cycle under salinity-alkalinity conditions.

Materials and Methods

Plant Materials and Growth Conditions

Tomato (*Solanum lycopersicum* cv. Jinpengchaoguan) seeds were germinated at 27°C in Petri plates lined with two layers of filter paper moistened with sterile distilled water. The germinated seeds were then sown in commix medium (Xintiandi Co., Yangling, Shanxi, China). Seedlings were grown in a controlled environmental greenhouse at Northwest A & F University, where the air temperature was maintained at 25±1°C/15±1°C (day/night). A 12-h light/12-h dark photoperiod was imposed with a daylight intensity of 800 $\mu mol \cdot m^{-2}s^{-1}$. Seedlings were transferred into 40-L containers of one-half-strength Hoagland solution (pH 6.5±0.1; electrical conductivity, 1.4−1.8 $ds \cdot m^{-1}$; dissolved oxygen, 6.0±0.2 $mg \cdot L^{-1}$) when the third true leaves were fully expanded.

Salinity-alkalinity Conditions and Spermidine Treatment

When seedlings had five true leaves, they were treated with 75 mM complex neutral and alkali salts [$NaCl:Na_2SO_4:NaHCO_3:Na_2CO_3$ = 1:9:9:1] and the foliage was sprayed with 0.25 mM Spd. The experiments included the following four treatments: 0 mM salinity-alkalinity plus 0 mM Spd (CK); 0 mM salinity-alkalinity plus 0.25 mM Spd (CS); 75 mM salinity-alkalinity plus 0 mM Spd (S); and 75 mM salinity-alkalinity plus 0.25 mM Spd (SS). Each treatment group included 48 plants and four replicates. Plant containers were arranged in a randomized block design. The solutions were renewed every 3 days. All experiments were performed at least three times with similar results.

Plant Growth Measurements

Four days after the salinity-alkalinity treatment, plants were washed with sterile distilled water and dissected into leaves, stems, and roots. The fresh weights of dissected tissues were determined. The leaves, stems, and roots were dried at 105°C for 15 min and then at 75°C for 72 h to constant mass, and their dry masses were recorded. The shoot weight was equal to the sum of the leaf weight and the stem weight. The root/shoot ratio (R/S) was calculated as (R/S) = root weight/shoot weight. Four plants were measured per treatment.

Gas-Exchange Measurements

Gas-exchange parameters were measured using a portable photosynthesis system (LI-6400, LI-COR Inc, USA). The photosynthetic rate was measured at 380±10 $\mu mol \cdot mol^{-1}$ CO_2, 25°C, 70% relative humidity, and 800 $\mu mol \cdot m^{-2}s^{-1}$ light intensity. Stomatal limitation (L_s) was calculated as $L_s = 1 - C_i/C_a$ (C_i and C_a represent the intercellular and ambient CO_2 concentration, respectively) according to Farquhar et al. [25].

Chlorophyll Fluorescence Parameters

Chlorophyll fluorescence was measured using a modulated fluorometer (PAM-2500; Walz, Effeltrich, Germany). The minimal (F_o) and maximal (F_m) Chl fluorescence emissions were determined after 30 min of dark adaptation, and F_o', F_s, and F_m' were measured after light adaption at 600 $\mu mol \cdot m^{-2}s^{-1}$. Fluorescence parameters were calculated as follows: maximum quantum yield of PSII, $(F_v/F_m) = (F_m - F_o)/F_m$; maximum photochemistry efficiency of PSII, $(F_v'/F_m') = (F_m' - F_o')/F_m'$; PSII operating efficiency, $\Phi PSII = (F_m' - F_s)/F_m'$; coefficient of photochemical quenching, $Qp = (F_m' - F_s)/(F_m' - F_o')$; and nonphotochemical quenching, $NPQ = (F_m - F_m')/F_m'$ [26].

Pigments Analysis

Pigment composition was determined according to the procedure of Chen et al. with some minor modifications [27]. Xanthophyll pigments were extracted with 100% acetone under weak light. The extract was centrifuged at 12,000×*g* for 10 min. The supernatant was passed through a 0.2-μm syringe filter into an amber HPLC vial for immediate analysis. Twenty-five microliters of the filtered supernatant was analyzed with a Shimadzu LC-20A HPLC system (Kyoto, Japan). The Waters

Table 1. Effects of exogenous spermidine on tomato seedling fresh weight under salinity-alkalinity stress.

Treatment	Shoot fresh weight (g)	Root fresh weight (g)	Stem fresh weight (g)	Leaf fresh weight (g)	R/S fresh weight
CK	40.726±2.066[a]	6.743±0.276[a]	17.968±1.263[a]	22.759±0.946[a]	0.166±0.007[a]
CS	43.578±3.130[a]	7.400±0.650[a]	18.348±1.737[a]	25.231±1.543[ab]	0.170±0.007[a]
S	22.482±0.711[b]	3.718±0.338[b]	5.649±0.555[b]	16.834±1.232[c]	0.167±0.020[a]
SS	26.050±0.938[b]	3.911±0.272[b]	6.877±0.940[b]	19.172±0.906[bc]	0.150±0.008[a]

Note: Data were measured after salinity-alkalinity treatment for 4 days. Each value represents mean ± standard error of four independent experiments ($n = 4$). Different letters indicate significant differences between treatments ($P<0.05$).

Abbreviations: CK, 0 mM salinity-alkalinity plus 0 mM Spd; CS, 0 mM salinity-alkalinity plus 0.25 mM Spd; S, 75 mM salinity-alkalinity plus 0 mM Spd; SS, 75 mM salinity-alkalinity plus 0.25 mM Spd.

Table 2. Effects of exogenous spermidine on tomato seedling dry weight under salinity-alkalinity stress.

Treatment	Shoot dry weight (g)	Root dry weight (g)	Stem dry weight (g)	Leaf dry weight (g)	R/S dry weight
CK	3.049±0.312[a]	0.396±0.034[a]	0.978±0.139[a]	2.071±0.176[a]	0.131±0.009[b]
CS	3.018±0.250[a]	0.396±0.029[a]	0.979±0.086[a]	2.040±0.164[a]	0.132±0.004[b]
S	1.547±0.116[c]	0.289±0.037[b]	0.451±0.045[b]	1.097±0.084[c]	0.185±0.011[a]
SS	2.216±0.115[b]	0.264±0.023[b]	0.626±0.076[b]	1.591±0.046[b]	0.118±0.006[b]

Note: Data were measured after salinity-alkalinity treatment for 4 days. Each value represents mean ± standard error of four independent experiments ($n = 4$). Different letters indicate significant differences between treatments ($P < 0.05$).

Abbreviations: CK, 0 mM salinity-alkalinity plus 0 mM Spd; CS, 0 mM salinity-alkalinity plus 0.25 mM Spd; S, 75 mM salinity-alkalinity plus 0 mM Spd; SS, 75 mM salinity-alkalinity plus 0.25 mM Spd.

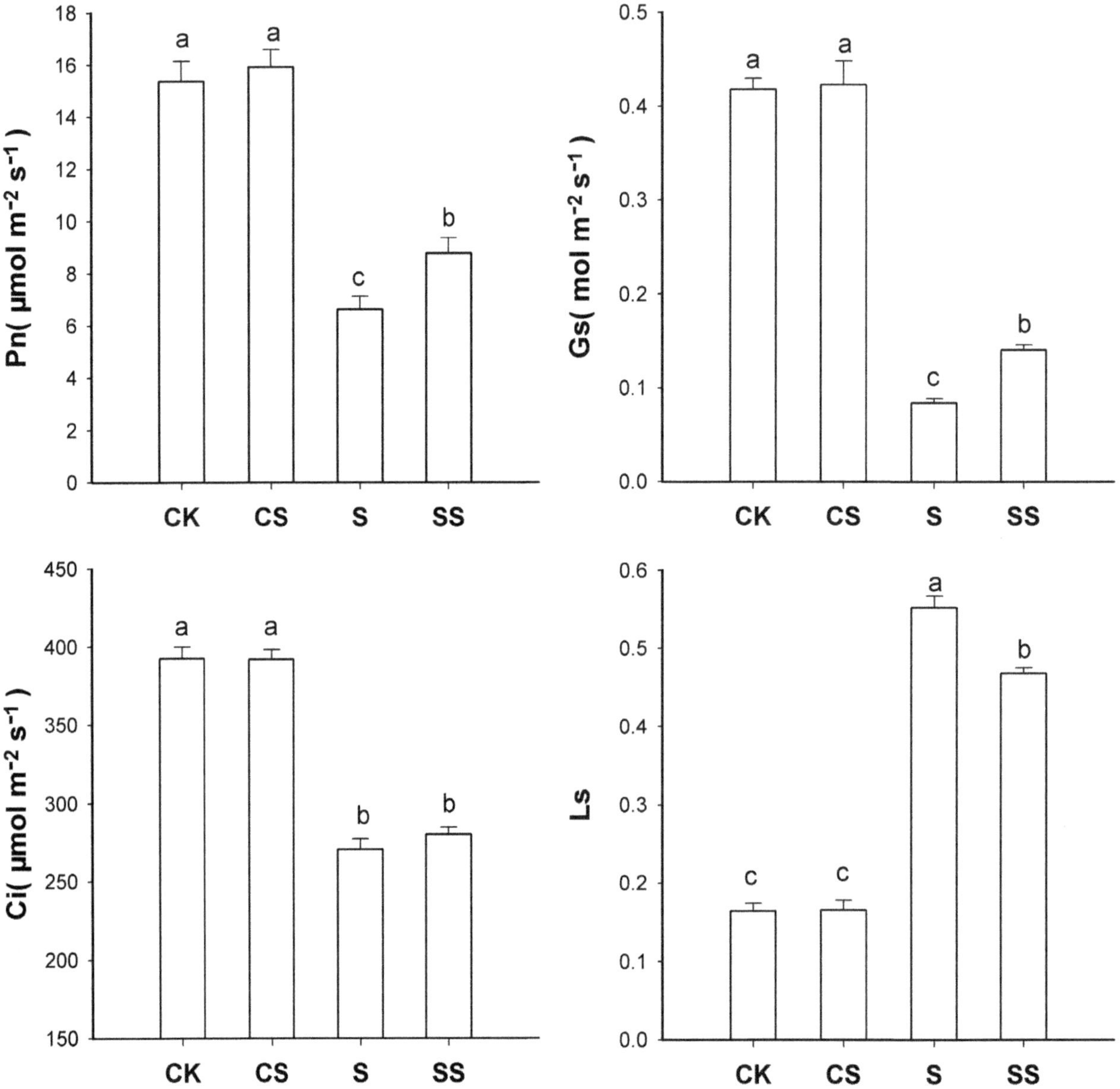

Figure 1. Effects of exogenous spermidine on gas-exchange parameters in tomato seedlings under salinity-alkalinity stress. Data were measured in the second expanded leaves (numbered basipetally) after salinity-alkalinity treatment for 4 days. Each histogram represents the mean ± standard error of four independent experiments ($n = 4$). Different letters indicate significant differences between treatments ($P < 0.05$).

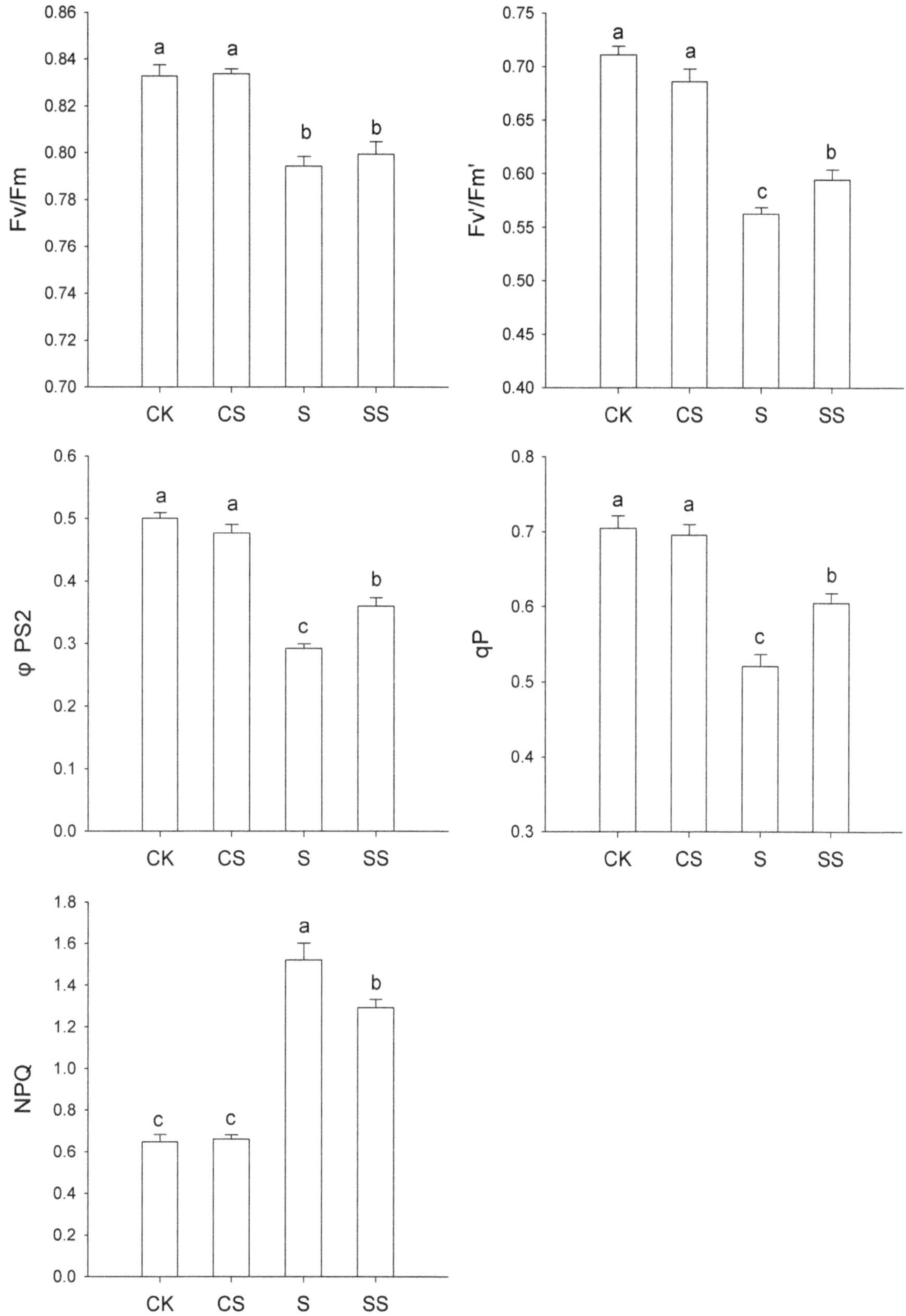

Figure 2. Effects of exogenous spermidine on chlorophyll fluorescence parameters in tomato seedlings under salinity-alkalinity stress. Data were measured in the second expanded leaves (numbered basipetally) after salinity-alkalinity treatment for 4 days. Each histogram represents the mean ± standard error of four independent experiments ($n=4$). Different letters indicate significant differences between treatments ($P<0.05$).

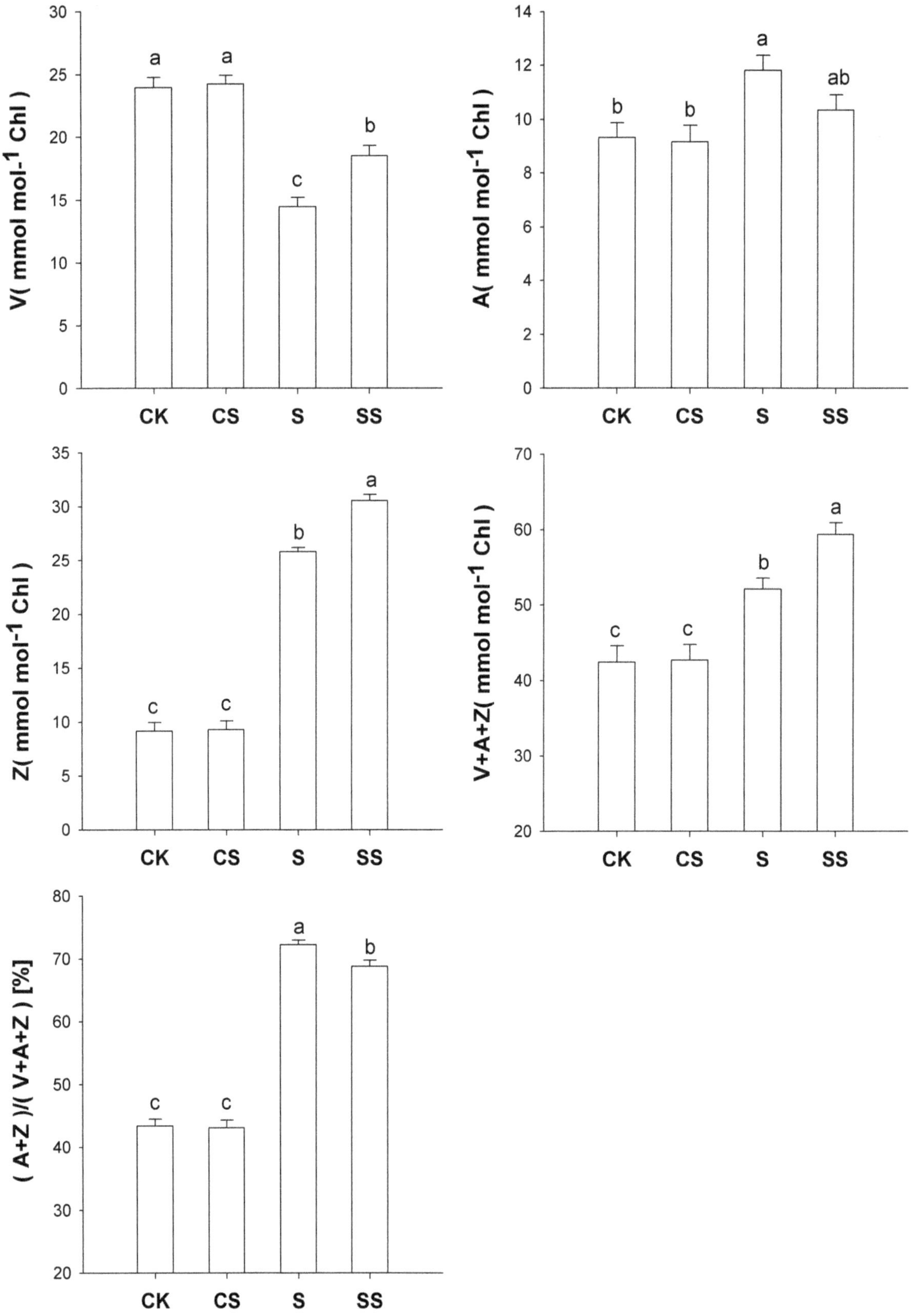

Figure 3. Effects of exogenous spermidine on xanthophyll cycle components in tomato seedlings under salinity-alkalinity stress. Data were measured in the second expanded leaves (numbered basipetally) after salinity-alkalinity treatment for 4 days. Each histogram represents the mean ± standard error of four independent experiments ($n = 3$). Different letters indicate significant differences between treatments ($P < 0.05$).

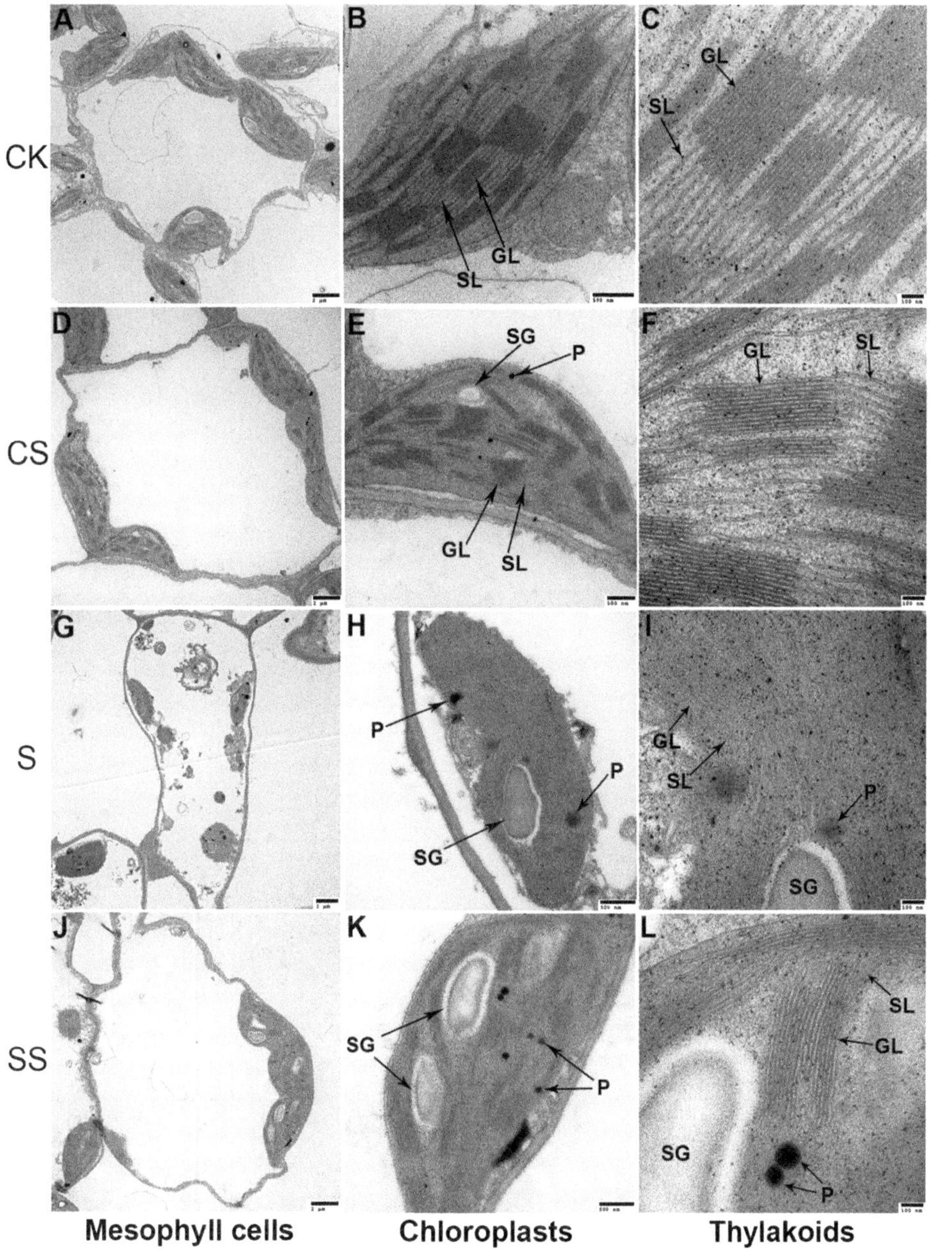

Figure 4. Effects of exogenous spermidine on chloroplast ultrastructure in tomato seedlings under salinity-alkalinity stress. Data were measured in the second expanded leaves (numbered basipetally) after salinity-alkalinity treatment for 4 days. SL, stroma lamellae; GL, grana lamellae; SG, starch grains; P, plastoglobuli. Scale bars for mesophyll cells, chloroplasts, and thylakoids are 2, 0.5, and 0.1 μm, respectively.

Spherisorb ODS1 column (5.0-μm particle size, 4.6×250 mm, Waters, USA) was used in the separation at 35°C. Solvent A [acetonitrile:methanol:50 mM Tris-HCl, 72:8:3 (v/v/v), pH 7.5] was used for the first 10 min, followed by a 2.5-min linear gradient to 100% solvent B [methanol:hexane, 5:1 (v/v)], with a run time of 30 min and a flow rate of 1.5 ml·min^{-1}. Pigments were detected by absorbance at 445 nm. The de-epoxidation state of the xanthophyll cycle was calculated as (Z+A)/(V+Z+A).

Transmission Electron Microscopy of Chloroplasts

The second fully expanded leaves from the top of the plants were randomly selected for microscopic examination. The leaf samples were cut into pieces of approximately 1 mm^2 and fixed overnight with 4% glutaraldehyde in 0.1 M phosphate-buffered saline (PBS, pH 7.4). The fixed samples were washed three times with the same solution for 10 min each. The samples were then post-fixed in 1% osmium tetroxide in cacodylate buffer for 2 h, and then washed three times in 0.1 M PBS (pH 7.4). The samples were dehydrated in a graded ethanol series (50%, 70%, 90%, and 100%), and then in absolute acetone for 15 min. The samples were embedded in Spurr's resin, and ultra-thin sections were cut and stained with uranium acetate and lead citrate, in series. The ultra-thin sections were mounted on copper grids and examined with a HITACHI HT7700 transmission electron microscope.

Statistical Analysis

All treatments and measurements were conducted at least in triplicate. All data were statistically analyzed with statistical software SAS (version 8.0, SAS Institute, Cary, NC) using Duncan's multiple range test at the *P* less than 0.05 level of significance.

Results

Plant Growth

The fresh and dry weights of tomato seedlings were significantly reduced by salinity-alkalinity stress compared with controls after 4 days of treatment (Tables 1 and 2). Reductions in dry leaf and shoot weights were alleviated by addition of exogenous Spd (Table 2). The S treatment had higher R/S than any other treatment (Table 2). There were no significant differences between S and SS treatments with respect to dry root weight and dry stem weight (Table 2). These results indicated that tomato seedling leaf growth was inhibited by salinity-alkalinity stress, and this inhibition was alleviated by exogenous Spd. However, exogenous Spd did not significantly affect the fresh weight measurements of tomato seedlings with or without salinity-alkalinity stress (Table 1).

Gas-Exchange Parameters

Salinity-alkalinity stress significantly reduced the net photosynthetic rate (P_n), stomatal conductance (G_s), and C_i, and increased L_s, compared with those of controls (Fig. 1). Exogenous Spd alleviated reductions in P_n and G_s, and increase in L_s. Exogenous Spd had no significant effect on C_i under salinity-alkalinity stress (SS treatment) compared with that of S treatment. No significant differences in P_n, G_s, C_i, or L_s were detected in CS and CK treatments.

Chlorophyll Fluorescence

F_v'/F_m', ΦPSII, and qP were substantially reduced under salinity-alkalinity conditions compared with controls (Fig. 2). The addition of exogenous Spd alleviated these effects (Fig. 2). The maximum quantum yield of PSII (F_v/F_m) was reduced in plants subjected to salinity-alkalinity stress. The addition of exogenous Spd did not protect the quantum yield of PSII. Plants subjected to salinity-alkalinity stress had significantly elevated NPQ, and this increase was alleviated by exogenous Spd. Exogenous Spd did not affect F_v/F_m, F_v'/F_m', ΦPSII, qP, or NPQ in the absence of salinity-alkalinity stress.

Xanthophyll Cycle Components

The Z content and the total V+A+Z pool size increased in response to salinity-alkalinity stress, and exogenous Spd induced an even higher increase in the contents of these metabolites. The V content markedly decreased in response to salinity-alkalinity stress, and this decrease was greater in leaves without exogenous Spd than in those with Spd. The (A+Z)/(V+Z+A) ratio increased in response to salinity-alkalinity stress in plants treated with exogenous Spd, but it was lower than in those without Spd. These results indicate that Spd increased the total V+A+Z pool size and the constitutive accumulation of Z in response to salinity-alkalinity stress.

Ultrastructural Changes in Chloroplasts

A typical elongated chloroplast was observed in controls (Fig. 3B), which had intact double membranes and a regular arrangement of granal and stromal thylakoids. Under salinity-alkalinity stress (Fig. 3G–I), chloroplasts were swollen and contained more plastoglobuli than those in controls (Fig. 3A–C). The chloroplast membrane structures became dilatant, undefined, and in some cases appeared disintegrated (Fig. 3H). The disoriented lamellar system of the granal and stromal thylakoids became swollen and unclear (Fig. 3I). The numbers of grana were considerably reduced. Exogenous Spd alleviated the salinity-alkalinity-mediated damage of the chloroplast photosynthetic apparatus, and more normal chloroplast ultrastructure was observed (Fig. 3J–L). No significant differences in chloroplast ultrastructure were observed in controls with or without the application of Spd (Fig. 3A–F).

Discussion

PAs are low molecular weight, ubiquitous, nitrogenous compounds that function in many cellular processes in numerous organisms [28]. Due to their polycationic nature and positive charge at physiological pH, PAs bind to macromolecules such as proteins and nucleic acids [14,29]. In this way, PAs are involved in regulating the physical and chemical properties of membranes, stabilizing nucleic acid structure and function, and modulating enzyme activities [30]. Chloroplasts are the primary sites of photosynthetic reactions in higher plants. Some types of stress reduce photochemical efficiency and electron transport activity, which might reflect changes in the structure of the photosynthetic apparatus [31]. Previous studies report that PAs stabilize the photosynthetic apparatus in response to stress [15].

In the present study, the deleterious effects of salinity-alkalinity stress on leaf and shoot dry weights in tomato seedlings were partly counteracted by the addition of exogenous Spd (Table 1). The improved dry matter accumulation in shoots and leaves might be associated with Spd-mediated improvement in P_n in tomato seedlings under salinity-alkalinity conditions (Fig. 1). A similar result was reported by Zhang et al. in salt-stressed cucumber plants sprayed with Put, which enhanced leaf area, plant height, and fresh and dry weight accumulation [24]. These authors demonstrated that Put likely enhanced photosynthetic production of salt-stressed plants, and this increased production appeared to effectively support the enhance growth. In tomato plants subjected to salinity-alkalinity stress and treated with exogenous Spd, our R/S results showed that most photosynthetic products were utilized to produce shoots rather than roots (Table 2).

Salinity-alkalinity stress reduced P_n and G_s, and increased L_s (Fig. 1). Previous studies evaluated the relationship between L_s and nonstomatal factors. If both C_i and G_s decreased, P_n was limited primarily by stomatal conductance. If G_s decreased but C_i did not change or increased, the P_n decrease could be ascribed to nonstomatal factors [19,24,32,33]. In the present study, salinity-alkalinity-induced reductions in P_n and G_s and increase in L_s were alleviated by exogenous Spd. However, exogenous Spd had no significant effect on C_i with or without stress (Fig. 1). These observed changes in G_s, C_i, and L_s imply that exogenous Spd alleviates both L_s and nonstomatal limitations under salinity-alkalinity conditions.

In general, F_v/F_m, F_v'/F_m', ΦPSII, and qP are parameters that reflect photochemical quenching, whereas NPQ reflects nonphotochemical quenching [26]. In this study, the values of F_v/F_m, F_v'/F_m', ΦPSII, and qP were reduced by salinity-alkalinity treatment, and these decreases were alleviated by exogenous Spd. A previous report suggested that salt-stress-induced reductions in F_v'/F_m' and ΦPSII might reflect damage to PSII electron transport [19]. These authors demonstrated that inhibition of electron transport may occur during transfer from the primary acceptor plastoquinone (QA) to the secondary acceptor plastoquinone (QB) at the acceptor side of PSII. Although exogenous Spd had no significant effect on F_v/F_m of tomato seedlings with or without salinity-alkalinity stress, it alleviated the stress-induced reductions in F_v'/F_m' and ΦPSII. This result indicates that exogenous Spd alleviates the stress-induced inhibition of photosynthetic electron transport.

Exogenous Spd alleviated the salinity-alkalinity-induced decrease in qP and increase in NPQ. This suggests that exogenous Spd may increase qP at the expense of NPQ under stress conditions, thereby mitigating the dissipation of excitation energy in the PSII antennae [34]. This is supported by the observation that exogenous Spd improved P_n and decreased the de-epoxidation status of the xanthophyll cycle in tomato seedlings under salinity-alkalinity stress (Fig. 2). The xanthophyll cycle serves as an essential photoprotective process in the dissipation of excess energy [35,36]. This thermal dissipation is an important function of NPQ to alleviate excess excitation energy on PSII reaction centers by converting excess excitation energy into heat [37]. Previous work showed that, to a certain extent, severe stress might result in higher levels of NPQ and (A+Z)/(V+Z+A) [38,39]. Consequently, under salinity-alkalinity conditions, leaves treated with Spd suffered less photoinhibition than those without Spd. Exogenous Spd may protect the photosynthetic apparatus against over-excitation, perhaps by preventing a loss of thylakoid membrane integrity [40]. The observed chloroplast ultrastructures support the proposal that exogenous Spd alleviates injury to the photosynthetic membrane caused by salinity-alkalinity stress (Fig. 4J–L). This result may be attributed to increases in the total V+A+Z pool size and the Spd-induced increases in Z content.

Z serves another important function as an antioxidant in the lipid phase of the thylakoid membrane to block photo-oxidative stress [21]. Photoinhibition occurs even under moderate light conditions if excess light energy cannot be appropriately dissipated [41]. $^1O_2$ is one of the most important ROS involved in photoinhibition and photo-oxidative damage [42]. Nonprotein-bound Z enhances photoprotection by blocking the formation of $^1O_2$ and $^1O_2$-related oxidation products [43,44], which reduces lipid peroxidation. ROS scavenging protects chloroplasts from direct injury and diminishes photon (electron)-related stress [7]. In our study, exogenous Spd increased the amount of Z under salinity-alkalinity stress, which stabilized thylakoid membrane structure and improved photosynthetic function.

Chloroplasts are crucial plant organelles because they are the site of photosynthesis. Chloroplast ultrastructure can be substantially disrupted by salinity stress [45]. In the present study, increased numbers of plastoglobuli were observed in mesophyll cells subjected to salinity-alkalinity stress, and some of these were larger than those in controls (Fig. 4G–I). Increases in plastoglobuli numbers may be caused by thylakoid membrane degradation [46], and individual plastoglobuli may be larger when chloroplasts are under oxidative stress [47]. Chloroplasts are a major source of ROS, including $^1O_2$, OH•, and H_2O_2 [7], which cause membrane lipid peroxidation and cellular abnormalities when plants are exposed to environmental stresses [48]. Tian et al. showed that exogenous Spd enhanced the activities of ascorbate peroxidase (APX), superoxide dismutase (SOD), peroxidase (POD), and catalase when cucumber seedlings were subjected to high temperatures [49]. Exogenous Spm alleviated salt-induced chloroplast membrane injury by increasing the levels of antioxidant metabolites (including dehydro-L-ascorbic acid, ascorbic acid, glutathione-S-S-glutathione, and glutathione-stimulating hormone) and enzyme activities (including SOD, POD, and APX) in chloroplasts [16]. PAs also bind membrane proteins, including polypeptides of the light-harvesting complex (LHC) [50]. Exogenous PAs with a high net positive charge can stabilize PSII proteins such as D1 and D2 under photoinhibition conditions [51]. PA binding to membrane proteins may provide stability for protein structure during stress, and consequently preserve photosynthetic activity. Our results indicate that Spd may serve a protective role in chloroplasts by enhancing ROS scavenging and stabilizing photosynthetic membrane proteins.

In conclusion, our results show that salinity-alkalinity stress induces photoinhibition and photodamage in tomato seedlings. Exogenous Spd alleviates the stress-induced inhibition of plant shoot growth and P_n by reducing ROS and stabilizing thylakoid membrane structure. Exogenous Spd indirectly improves total V+A+Z pool size, especially the amount of Z, which acts as an antioxidant in the lipid phase of the thylakoid membrane to block photo-oxidative stress. However, under salinity-alkalinity stress, Spd-induced increases in Z content did not increase NPQ compared with that in seedlings without Spd. This may be because Spd protects leaves from light-induced damage due to qP, which differs from heat dissipation.

Acknowledgments

The authors are grateful to Yanyan Zhao Ph.D. (NWSUAF) for advice regarding laboratory techniques, Fei Yan Ph.D. (NWSUAF) for insightful comments, and Hao Zhang (NWSUAF) for data analysis methods.

Author Contributions

Conceived and designed the experiments: X-HH ZZ. Performed the experiments: LH LZ XZ LX. Analyzed the data: LH LX. Contributed reagents/materials/analysis tools: LH LX. Contributed to the writing of the manuscript: LH XH.

References

1. Zhang Y, Hu XH, Shi Y, Zou ZR, Yan F, et al. (2013) Beneficial role of exogenous spermidine on nitrogen metabolism in tomato seedlings exposed to saline-alkaline stress. Journal of the American Society for Horticultural Science 138: 38–49.
2. Hu X, Zhang Y, Shi Y, Zhang Z, Zou Z, et al. (2012) Effect of exogenous spermidine on polyamine content and metabolism in tomato exposed to salinity-alkalinity mixed stress. Plant physiology and biochemistry: PPB/Societe francaise de physiologie vegetale 57: 200–209.
3. Yang J, Zheng W, Tian Y, Wu Y, Zhou D (2011) Effects of various mixed salt-alkaline stresses on growth, photosynthesis, and photosynthetic pigment concentrations of Medicago ruthenica seedlings. Photosynthetica 49: 275–284.
4. Li R, Shi F, Fukuda K, Yang Y (2010) Effects of salt and alkali stresses on germination, growth, photosynthesis and ion accumulation in alfalfa (Medicago sativa L.). Soil Science & Plant Nutrition 56: 725–733.
5. Deng C, Zhang G, Pan X (2011) Photosynthetic Responses in Reed (Phragmites australis (CAV.) TRIN. ex Steud.) Seedlings Induced by Different Salinity-Alkalinity and Nitrogen levels. Journal of Agricultural Science and Technology 13: 687–699.
6. Yadav DK, Pospíšil P (2012) Evidence on the formation of singlet oxygen in the donor side photoinhibition of photosystem II: EPR spin-trapping study. PloS one 7: e45883.
7. Asada K (2006) Production and scavenging of reactive oxygen species in chloroplasts and their functions. Plant physiology 141: 391–396.
8. Murata N, Takahashi S, Nishiyama Y, Allakhverdiev SI (2007) Photoinhibition of photosystem II under environmental stress. Biochimica et Biophysica Acta (BBA)-Bioenergetics 1767: 414–421.
9. Mittal S, Kumari N, Sharma V (2012) Differential response of salt stress on Brassica juncea: Photosynthetic performance, pigment, proline, D1 and antioxidant enzymes. Plant Physiology and Biochemistry 54: 17–26.
10. Bouchereau A, Aziz A, Larher F, Martintanguy J (1999) Polyamines and environmental challenges: recent development. Plant Science 140: 103–125.
11. Liu Y, Gu D, Wu W, Wen X, Liao Y (2013) The Relationship between Polyamines and Hormones in the Regulation of Wheat Grain Filling. PloS one 8: e78196.
12. Ha HC, Sirisoma NS, Kuppusamy P, Zweier JL, Woster PM, et al. (1998) The natural polyamine spermine functions directly as a free radical scavenger. Proceedings of the National Academy of Sciences 95: 11140–11145.

13. Bellé NAV, Dalmolin GD, Fonini G, Rubin MA, Rocha JBT (2004) Polyamines reduces lipid peroxidation induced by different pro-oxidant agents. Brain research 1008: 245–251.
14. Groppa MD, Benavides MP (2008) Polyamines and abiotic stress: recent advances. Amino acids 34: 35–45.
15. Demetriou G, Neonaki C, Navakoudis E, Kotzabasis K (2007) Salt stress impact on the molecular structure and function of the photosynthetic apparatus–the protective role of polyamines. Biochimica et Biophysica Acta (BBA)-Bioenergetics 1767: 272–280.
16. Shu S, Yuan LY, Guo SR, Sun J, Yuan YH (2013) Effects of exogenous spermine on chlorophyll fluorescence, antioxidant system and ultrastructure of chloroplasts in Cucumis sativus L. under salt stress. Plant physiology and biochemistry: PPB/Societe francaise de physiologie vegetale 63: 209–216.
17. Shu S, Guo SR, Sun J, Yuan LY (2012) Effects of salt stress on the structure and function of the photosynthetic apparatus in Cucumis sativus and its protection by exogenous putrescine. Physiologia plantarum 146: 285–296.
18. Li Y, Shi G, Wang H, Zhao J, Yuan Q (2009) Exogenous spermidine can mitigate the poison of cadmium in nymphoides peltatum. Chinese Bulletin of Botany 44: 571−577.
19. Sheng S (2012) Effects of exogenous spermidine on photosynthesis, xanthophyll cycle and endogenous polyamines in cucumber seedlings exposed to salinity. African Journal of Biotechnology 11: 6064–6074.
20. Demmigadams B, Adams III WW (2006) Photoprotection in an ecological context: the remarkable complexity of thermal energy dissipation. The New phytologist 172: 11–21.
21. Jahns P, Holzwarth AR (2012) The role of the xanthophyll cycle and of lutein in photoprotection of photosystem II. Biochimica et biophysica acta 1817: 182–193.
22. Demmigadams B, Adams III WW (1996) The role of xanthophyll cycle carotenoids in the protection of photosynthesis. Trends in Plant science 1: 21–26.
23. Zhang R, Kramer DM, Cruz JA, Struck KR, Sharkey TD (2011) The effects of moderately high temperature on zeaxanthin accumulation and decay. Photosynthesis research 108: 171–181.
24. Zhang RH, Li J, Guo SR, Tezuka T (2009) Effects of exogenous putrescine on gas-exchange characteristics and chlorophyll fluorescence of NaCl-stressed cucumber seedlings. Photosynthesis research 100: 155–162.
25. Farquhar GD, Sharkey TD (1982) Stomatal conductance and photosynthesis. Annual Review of Plant Physiology 33: 317–345.
26. Baker NR (2008) Chlorophyll fluorescence: a probe of photosynthesis in vivo. Annual Review of Plant Biology 59: 89–113.
27. Chen C, Zhang D, Li P, Ma F (2012) Partitioning of absorbed light energy differed between the sun-exposed side and the shaded side of apple fruits under high light conditions. Plant physiology and biochemistry: PPB/Societe francaise de physiologie vegetale 60: 12–17.
28. Hussain SS, Ali M, Ahmad M, Siddique KH (2011) Polyamines: natural and engineered abiotic and biotic stress tolerance in plants. Biotechnology advances 29: 300–311.
29. Childs AC, Mehta DJ, Gerner EW (2003) Polyamine-dependent gene expression. Cellular and molecular life sciences: CMLS 60: 1394–1406.
30. Galston AW, Sawhney RK (1990) Polyamines in plant physiology. Plant physiology 94: 406–410.
31. Parida A, Das A, Mittra B (2003) Effects of NaCl stress on the structure, pigment complex composition, and photosynthetic activity of mangrove Bruguiéra parviflora chloroplasts. Photosynthetica 41: 191–200.
32. Liu ZX, Bie ZL, Huang Y, Zhen A, Lei B, et al. (2012) Grafting onto Cucurbita moschata rootstock alleviates salt stress in cucumber plants by delaying photoinhibition. Photosynthetica 50: 152–160.
33. Begcy K, Mariano ED, Gentile A, Lembke CG, Zingaretti SM, et al. (2012) A novel stress-induced sugarcane gene confers tolerance to drought, salt and oxidative stress in transgenic tobacco plants. PloS one 7: e44697.
34. Horton P, Ruban A, Walters R (1996) Regulation of light harvesting in green plants. Annual review of plant biology 47: 655–684.
35. Bilger W, Björkman O (1990) Role of the xanthophyll cycle in photoprotection elucidated by measurements of light-induced absorbance changes, fluorescence and photosynthesis in leaves of Hedera canariensis. Photosynthesis research 25: 173–185.
36. Goss R, Jakob T (2010) Regulation and function of xanthophyll cycle-dependent photoprotection in algae. Photosynthesis research 106: 103–122.
37. Nilkens M, Kress E, Lambrev P, Miloslavina Y, Muller M, et al. (2010) Identification of a slowly inducible zeaxanthin-dependent component of non-photochemical quenching of chlorophyll fluorescence generated under steady-state conditions in Arabidopsis. Biochimica et biophysica acta 1797: 466–475.
38. Yin Y, Li S, Liao W, Lu Q, Wen X, et al. (2010) Photosystem II photochemistry, photoinhibition, and the xanthophyll cycle in heat-stressed rice leaves. Journal of plant physiology 167: 959–966.
39. Guo YP, Guo DP, Zhou HF, Hu MJ, Shen YG (2006) Photoinhibition and xanthophyll cycle activity in bayberry (Myrica rubra) leaves induced by high irradiance. Photosynthetica 44: 439–446.
40. Ogweno JO, Hu WH, Song XS, Shi K, Mao WH, et al. (2010) Photoinhibition-induced reduction in photosynthesis is alleviated by abscisic acid, cytokinin and brassinosteroid in detached tomato leaves. Plant Growth Regulation 60: 175–182.
41. Demmigadams B, Adams III WW (1992) Photoprotection and other responses of plants to high light stress. Annual review of plant biology 43: 599–626.
42. Triantaphylidès C, Krischke M, Hoeberichts FA, Ksas B, Gresser G, et al. (2008) Singlet oxygen is the major reactive oxygen species involved in photooxidative damage to plants. Plant physiology 148: 960–968.
43. Wrona M, Różanowska M, Sarna T (2004) Zeaxanthin in combination with ascorbic acid or α-tocopherol protects ARPE-19 cells against photosensitized peroxidation of lipids. Free Radical Biology and Medicine 36: 1094–1101.
44. Havaux M, DallOsto L, Bassi R (2007) Zeaxanthin has enhanced antioxidant capacity with respect to all other xanthophylls in Arabidopsis leaves and functions independent of binding to PSII antennae. Plant physiology 145: 1506–1520.
45. Paramonova N, Shevyakova N, Kuznetsov VV (2004) Ultrastructure of chloroplasts and their storage inclusions in the primary leaves of Mesembryanthemum crystallinum affected by putrescine and NaCl. Russian Journal of Plant Physiology 51: 86–96.
46. Naeem MS, Warusawitharana H, Liu H, Liu D, Ahmad R, et al. (2012) 5-aminolevulinic acid alleviates the salinity-induced changes in Brassica napus as revealed by the ultrastructural study of chloroplast. Plant physiology and biochemistry: PPB/Societe francaise de physiologie vegetale 57: 84–92.
47. Austin JR, Frost E, Vidi PA, Kessler F, Staehelin LA (2006) Plastoglobules are lipoprotein subcompartments of the chloroplast that are permanently coupled to thylakoid membranes and contain biosynthetic enzymes. The Plant Cell Online 18: 1693–1703.
48. Gill SS, Tuteja N (2010) Reactive oxygen species and antioxidant machinery in abiotic stress tolerance in crop plants. Plant Physiology and Biochemistry 48: 909–930.
49. Tian J, Wang LP, Yang YJ, Sun J, Guo SR (2012) Exogenous spermidine alleviates the oxidative damage in cucumber seedlings subjected to high temperatures. Journal of the American Society for Horticultural Science 137: 11–19.
50. Hamdani S, Yaakoubi H, Carpentier R (2011) Polyamines interaction with thylakoid proteins during stress. Journal of Photochemistry and Photobiology B: Biology 104: 314–319.
51. Hamdani S, Gauthier A, Msilini N, Carpentier R (2011) Positive charges of polyamines protect PSII in isolated thylakoid membranes during photoinhibitory conditions. Plant and cell physiology 52: 866–873.

22

Direct Chloroplast Sequencing: Comparison of Sequencing Platforms and Analysis Tools for Whole Chloroplast Barcoding

Marta Brozynska, Agnelo Furtado, Robert James Henry*

Queensland Alliance for Agriculture and Food Innovation, The University of Queensland, Brisbane, Queensland, Australia

Abstract

Direct sequencing of total plant DNA using next generation sequencing technologies generates a whole chloroplast genome sequence that has the potential to provide a barcode for use in plant and food identification. Advances in DNA sequencing platforms may make this an attractive approach for routine plant identification. The HiSeq (Illumina) and Ion Torrent (Life Technology) sequencing platforms were used to sequence total DNA from rice to identify polymorphisms in the whole chloroplast genome sequence of a wild rice plant relative to cultivated rice (cv. Nipponbare). Consensus chloroplast sequences were produced by mapping sequence reads to the reference rice chloroplast genome or by *de novo* assembly and mapping of the resulting contigs to the reference sequence. A total of 122 polymorphisms (SNPs and indels) between the wild and cultivated rice chloroplasts were predicted by these different sequencing and analysis methods. Of these, a total of 102 polymorphisms including 90 SNPs were predicted by both platforms. Indels were more variable with different sequencing methods, with almost all discrepancies found in homopolymers. The Ion Torrent platform gave no apparent false SNP but was less reliable for indels. The methods should be suitable for routine barcoding using appropriate combinations of sequencing platform and data analysis.

Editor: Hikmet Budak, Sabanci University, Turkey

Funding: The research was funded by Rural Industries Research and Development Corporation (Australia). The funders played no role in study design, data collection, decision to publish or preparation of the manuscript.

Competing Interests: This research was supported by a grant from Rural Industries Research and Development Corporation, Australian Government.

* Email: robert.henry@uq.edu.au

Introduction

A universal plant barcoding method based upon sequencing a specific gene or combination of specific genes has proven elusive. Analysis of different plant groups has required the use of different genetic loci. The emergence of next generation sequencing (NGS) technology allowing whole chloroplast genomic DNA sequencing has provided the opportunity to use the whole chloroplast genome sequence as a barcode [1]. The chloroplast has unique features found in all green plants, conserved sufficiently to be readily aligned for comparison of different samples and large enough to contain variation, which allow species or sub species distinction across the seed plants [2]. Analysis of complex mixtures of plants has applications in many ecological studies [3].

A generic method for rapid and cost effective DNA-based identification of plants at this level (species or sub-species) will find wide application in industry and research [4,5]. Industrial applications will include the identification of plant components in whole and processed foods and the management of food processing to ensure food safety and authenticity of labelling [6]. This will complement the use of next generation sequencing to screen for the presence of pathogenic microbes [7]. Protection of intellectual property rights associated with plant varieties will also be simplified by a standard approach to plant identification. These tools will also support protection of biosecurity and management of wild plant populations of rare or threatened plants.

Early analysis relied on either specific amplification of the chloroplast genome or separation of chloroplasts from nuclear and mitochondrial DNA before chloroplast DNA isolation and sequencing the amplified or cloned product derived from PCR amplification [8]. However, the universal amplification of chloroplast sequences from all species has proven difficult and chloroplast purification is laborious and not perfect [9,10]. Recently a simple approach involving NGS of a total genomic DNA preparation was proposed [11]. This method relies on successful extraction of chloroplast genome sequence reads from total genomic DNA reads and their assembly to form a consensus sequence for the sample that can be used as a barcode. Reference-guided mapping of sequence reads and *de novo* assembly have been used to generate these whole chloroplast barcodes [12].

DNA sequencing platforms differ in their throughput and accuracy [13]. Accuracy of a DNA sequence assembled using NGS reads is dependent on the read length, sequencing depth, sequence coverage or width and evenness of coverage but also on the accuracy of the sequencing platform and the assembly and analysis pipeline [14]. We compared two different sequencing technologies, the Illumina platform that uses sequencing-by-syntheses (SBS) chemistry and the Ion Torrent that is based on semiconductor sequencing technology. Sequencing on both platforms commences with library preparation, which involves fragmentation of total genomic DNA, purifying to uniform and desired fragment size and ligation to sequencing adapters specific to the platform [15,16]. In Illumina systems the fragments are subsequently denatured and fixed on the surface of a proprietary glass flowcell, followed by solid-phase amplification (bridge amplification). As a result clusters are created which contain clonally amplified DNA templates. Next the clusters are sequenced in parallel using four fluorescently labelled nucleotides with reversible dye terminators. After every sequencing cycle and base incorporation to the template the polymerization terminates, a CCD (charge-coupled device) captures the fluorescent signal and identifies the base. After the base call the dye is cleaved and the sequencing process continues. This technology was first implemented in an instrument called Genome Analyzer (GA) with subsequent releases of improved series up to GAIIx. The next major system launched by Illumina was the HiSeq2000 with improved output and read length. Data from both GAIIx and HiSeq was used in the current study.

In Ion Torrent systems the prepared libraries are immobilized to beads and amplified using emulsion PCR which takes place within microdroplets of aqueous solution and oil [17]. During the sequencing individual bases are incorporated by the action of a DNA polymerase. As a result of this reaction a proton is released and the resulting change in pH is measured. The reaction occurs in a proprietary chip, which acts as a pH meter. Unlike Illumina, this system does not use fluorescent dyes and light detection, shortening the sequencing time. The Ion Torrent instrument used in the study was a Personal Genome Machine (PGM).

We examined the reliability of these two different sequencing platforms coupled with different data analysis approaches when applied to obtaining the chloroplast consensus sequences of a wild rice genotype and a cultivated rice genotype. Although the Ion Torrent platform is widely used for sequencing and limited published data is available on its application for transcript [18] and amplicon [19] sequencing in plants, it has not previously been applied to plant genomic DNA analysis. The most common applications of the Ion Torrent system have been to small genomes (bacterial samples) and targeted sequencing of amplicons or transcripts, whereas the Illumina has been widely used for applications including whole human genome re-sequencing, *de novo* plant and animal genome sequencing, exome, transcriptome (RNA-Seq) and metagenomics investigations. The chloroplast genome represents a specific sequencing challenge with the presence of an inverted repeat. This study compares total DNA sequencing with the Illumina and Ion Torrent platforms and options for analysis of the sequences generated to produce a whole chloroplast genome sequence as a barcode for plant identification. Whole genome NGS was carried out on two rice genotypes: *Oryza sativa* spp. *japonica* var. Nipponbare referred to here as the reference-rice-genotype (R-rice-genotype) as its chloroplast sequence (reference-sequence) is publicly available and an Australian wild rice genotype sample (W-rice-genotype).

Materials and Methods

Plant materials

Seeds of rice (*Oryza sativa* cv. Nipponbare), referred to as reference-rice-genotype (R-rice-genotype), were germinated and entire seedlings were used for DNA extraction. Leaf tissue from a wild rice plant, referred to as wild-rice-genotype (W-rice-genotype) collected from a field containing a population of perennial wild rice (*Oryza rufipogon*-like) at Abattoir Creek in North Queensland, was used for DNA extraction [20].

DNA extraction

DNA from leaf tissue of the W-rice-genotype was extracted using modification of the CTAB method [21]. DNA from seedlings of the R-rice-genotype was extracted as described for the W-rice-genotype but with further purification as described by [11].

Sequencing

The Illumina and the Ion Torrent sequencing platforms were used for the shotgun sequencing of total genomic DNA of both rice samples. The wild rice sample was sequenced on the Ion Torrent (200 bp reads) and the Illumina HiSeq platform (100 bp paired end reads). The cultivated rice genotype was subjected to sequencing using the Ion Torrent (200 bp reads) platform while Illumina reads (36 bp paired-end), generated on the GAII, were sourced from available archived data [11]. Sequencing on the Illumina HiSeq 2000 (Illumina, San Diego, CA, USA), to generate 100 bp paired end reads with an average library insert size of 500–600 bp, was outsourced to the Australian Genome Research Facility (AGRF, Melbourne, Australia). For sequencing using the Ion Torrent platform, genomic DNA was sheared using the Covaris S220 instrument (www.covarisinc.com) and used for preparing sequencing libraries according to the standard Ion Torrent PGM protocol. The resulting individual DNA libraries were quality checked and quantified on the Agilent 2100 Bioanalyzer using the High Sensitivity DNA kit (Agilent). Following template amplification and enrichment on the Ion OneTouch2 (Ion OneTouch 200 Template Kit v2; #4478316) and OneTouch2 ES instruments, each sample was loaded onto one PGM #318 chip and sequenced using Ion PGM Sequencing 200 Kit v.2 (#4482006) according to manufacturer's protocol.

Data analysis – consensus

Raw reads from both sequencing platforms, PGM Ion Torrent and Illumina, were imported to CLC Genomics Workbench 6.0 (CLC-GW) (www.clcbio.com) and read statistics assessed using sequencing data quality control, followed by read trimming for quality, length and presence of ambiguous bases. Ion Torrent reads were trimmed with the quality score limit set to 0.05 (which corresponded to PHRED quality value >15) and a minimum read length of 30 bp. Reads from both Illumina platforms, GAII and HiSeq, were trimmed with a quality score limit of 0.01 (which corresponded to PHRED >22) and the same minimum read length of 30 bp. Ion Torrent data was trimmed at a slightly lower quality value due to the lower average quality compared to the Illumina reads (Table S1; avg. PHRED for Ion Torrent reads: 25 and 26, and avg. PHRED for Illumina: 29 and 32). The Illumina HiSeq platform generates sequencing data fifty to sixty times more in order of magnitude as compared to the Ion Torrent data. Thus, in order to compare Ion Torrent and Illumina HiSeq platforms at similar coverage, a subset of HiSeq Illumina reads for the wild rice sample and matching the Ion Torrent read numbers for that same sample, was randomly extracted from the whole reads set. Ion Torrent raw reads were also alternatively trimmed using the Torrent Suite Software version 3.6 using default analysis settings with the following modifications: the quality-based trimming of the sequencing reads was adjusted to predicted-PHRED-score 17 over the sliding window of 20 bp (default values 15, and 30 respectively) and the minimal length filter for reported reads was set to 20 (default 8).

Consensus sequence generation using the mapping tool was carried out as follows. The CLC-GW was used to map trimmed reads to the chloroplast genome sequence of *Oryza sativa* spp. *japonica* var. Nipponbare (GenBank accession – GU592207) used as the reference-sequence. The mapping process involved the following parameters: mismatch cost 2, insertion cost 3, deletion cost 3, length fraction 0.9, similarity fraction 0.9, and global alignment setting. Non-specific matches were also mapped randomly to the reference genome. Consensus sequences were extracted from all mapping runs giving five sequences in total; two consensus sequences for *O. sativa* Nipponbare each derived from the Illumina GAII reads and the Ion Torrent reads, and three consensus sequences for the wild rice sample each derived from the Illumina HiSeq sequence reads, Illumina HiSeq sequence subset-reads and the Ion Torrent reads. Conflicts between reads were resolved by voting for the majority of the reads at the given position. Consensus sequence were aligned to the reference sequence and analysed for SNPs and indels.

Consensus sequence generation using the assembly tool was carried out as follows using both the CLC-GW and the Torrent Suite Software. The CLC trimmed reads from both platforms were assembled in CLC-GW using the *de novo* assembly tool with the following parameters: automatic word and bubble sizes with 200 bp for minimum contig length. Moreover the reads were mapped back to contigs (mismatch cost 2, insertion cost 3, deletion cost 3, length fraction 0.9, and similarity fraction 0.9) and contigs were updated based on the results. The trimmed Ion Torrent reads were also assembled in the Torrent Suite Software using the Assembler plugin v.3.4.2 utilizing the MIRA assembly algorithm (http://sourceforge.net/projects/mira-assembler/). The assembly was performed in the reference-assisted mode using the reference-sequence. All contigs generated were aligned to the reference-sequence using NUCmer alignment generator from MUMmer package [22]. Owing to the chloroplast genome structure (two long inverted repeats) the option to use all matches regardless of the uniqueness was enabled in the script. All other parameters were left as default. Based on the MUMer output the longest contigs with the best mapping results (query coverage, precent identity and alignment length) were chosen as scaffolds and manually joined together by overlapping the remaining short contigs. In cases when there was not enough contigs to reconstruct the chloroplast genome sequence, *de novo* assembly was repeated with different parameter for word size in the CLC-GW and subjected again to the alignment tool. Consensus sequences from contigs mapping and joining were also screened for SNPs and indels based on the reference-sequence.

Data analysis – variant detection

The consensus sequences generated by both mapping and *de novo* assembly of the reads were imported into Geneious 6.1.6 (www.geneious.com) and each of them was separately mapped to reference-sequence. To generate the alignments the Geneious Map to Reference tool was used with the default parameters. Subsequently the alignments were screened to find variations/SNPs using the tool available in Geneious package.

Validation of the most probable variant

The wild rice chloroplast consensus sequences derived from reads mapping from Illumina and Ion Torrent platforms were aligned and scanned for SNPs and indels (as described above). If a variant was found in one consensus but was not present in the other it was reported as a discrepancy and closely studied. To confirm which of the disagreeing variants was more likely to be true the following procedure was adopted: (1) visual investigation of mapping results in CLC-GW and (2) aligning the conflicting regions from mapping derived consensus with contigs sequences originated from *de novo* assemblies.

Data deposition

The chloroplast genome for Australian wild rice genotype sample (W-rice-genotype) has been deposited in GenBank with the sequence accession number KF428978.

Results

Mapping and *de novo* assembly of sequence reads

The CLC Genomics Workbench was used to trim sequence reads from both the Illumina and Ion Torrent platforms and then processed, using the mapping and *de novo* assembly mode, to generate a mapping-consensus or *de novo*-consensus chloroplast sequences, respectively. The Torrent Suite Software, a dedicated analysis tool for Ion Torrent reads, was used to trim sequence reads from the Ion Torrent platform and then processed using the reference-guided *de novo* assembly mode (the only available mode) to generate a reference-guided *de novo*-consensus chloroplast sequence. The published chloroplast sequence of the R-rice-genotype (*Oryza sativa* spp. *japonica* var. Nipponbare) was used as a reference chloroplast genome sequence (reference-sequence) for mapping of reads or contigs. Sequencing data of the two rice genotypes, generated by both sequencing platforms, was of good quality with average PHRED scores ranging above 25 (Table S1).

In case of the R-rice-genotype, CLC trimmed-reads from both sequencing platforms mapped to the entire reference-sequence indicating complete reference-sequence coverage. However, the mapping-consensus chloroplast sequences varied in length based on sequencing platform reads. Similar results were obtained for the W-rice-genotype (Table S2).

Using the *de novo* mode in the CLC-GW and reads from both sequencing platforms, a number of contigs were generated with a subset of these mapping to the reference-sequence (Table S3). Similar results were obtained with the Ion Torrent reads when using the reference-assisted *de novo* mode on the Torrent Suite Software (Table S3). For both rice genotypes, contigs, which mapped to the reference-sequence, were checked if contiguous/overlapping to generate a *de novo*-consensus sequence. Using the *de novo* mode in the CLC-GW, respective *de novo*-consensus chloroplast sequences were generated for the reference-rice-genotype and the wild-rice-genotype but only from their respective Ion Torrent reads and not their Illumina reads. Using the reference-assisted *de novo* mode, reference-guided *de novo*-consensus chloroplast sequences were generated for both genotypes.

Thus, all contigs generated from Ion Torrent reads using both assembly algorithms (in the CLC-GW and Torrent Suite Software) could be joined together based on their position and overlapping information creating full length consensus sequences for the chloroplast genome (Table S3). Consensus sequences could not be created from contigs generated in the CLC-GW from Illumina reads for either of the rice genotypes, possibly because of the limitations of read length (Table S3). Although these contigs were well mapped to the reference, there were gaps between some of them that prevented their joining.

Analysis of reference-rice-genotype

The mapping-consensus chloroplast genome sequence (134,551 bp) generated by mapping the Illumina reads from [11] to the reference-sequence in the CLC-GW was, as expected, identical to the reference-sequence (Table 1). The mapping-

Table 1. Comparison of chloroplast consensus sequences of the cultivated reference rice genotype (*Oryza sativa* Nipponbare).

Sequencing platform	Source	Sequence length (bp)	Variants	Deletions	Insertions	Mismatches
Used available reference	GenBank (GU592207)	134,551	–	–	–	–
Illumina [11]	mapping- consensus	134,551	0	0	0	0
Ion Torrent	mapping- consensus	134,520	31	30	1	0
	de novo- consensus (CLC-GW	134,503	54	48	6	0
	reference-assisted *de novo*- consensus (Torrent Suite)	134,581	50	9	41	0

consensus sequence generated by mapping of the Ion Torrent reads to the reference-sequence in the CLC-GW was shorter than the reference-sequence by 31 nucleotides with no base mismatches, but diverged at 31 positions comprising of 30 deletions and 1 insertion and where all but one of the polymorphisms were found in homopolymer regions of length from 2 to 9 bases. The *de novo*-consensus sequence built by mapping and joining the contigs from *de novo* assembly of Ion Torrent reads (performed in CLC-GW) was shorter than the reference-sequence by 48 bp with no base mismatches, but diverged at 54 positions comprising of 48 deletions and 6 insertion and where all but two of the polymorphisms (indels) were found in homopolymer regions of length from 2 to 17 identical bases.

The reference-guided *de novo*-consensus chloroplast sequence obtained from contigs generated in Torrent Suite Software using Ion Torrent reads was longer than the reference-sequence by 30 bases with no base mismatches, but diverged at 50 positions comprising of 9 deletions and 41 insertions and where all polymorphisms (indels) were found in homopolymer regions of length from 2 to 12 identical bases.

The variants in the mapping-derived consensus sequences generated by the CLC-GW and using the Illumina or Ion Torrent reads were predominantly deletions, which is in contrast to the reference-guided *de novo*-consensus chloroplast sequence generated by the Torrent Suite Software using Ion Torrent reads, where the variants were predominantly insertions. Overall, 135 variants were detected using these different approaches. Among them only 8 were common for all of the methods (the same position, length and type). The rest of the variants were either present in two of the approaches or were unique to only one of them.

The analysis of regions in the chloroplast genome where the indels were found in the Ion Torrent data (both mapping and assembly consensus) revealed biases in terms of polymorphism nature, nucleotide type and location. Data from all approaches were merged and duplicated regions were removed. The number of indels that involved C/G bases was slightly higher than A/T bases (53 and 44, respectively). It is worth noting that the GC content of the chloroplast genome is 39%. Also, more deletions than insertions were observed (except for the consensus from the Torrent Suite Software assembly) and the majority were of G/C nucleotides in short homopolymeric regions (2–5 bp long with the vast majority of these variants occurring in 3 bp long homopolymer) (Figure 1). On the other hand, insertions were found mostly in longer homopolymers (5–17 bp long) and involved almost only A/T nucleotides.

Analysis of wild-rice-genotype

The reference-sequence was used to generate consensus chloroplast sequences using mapping or *de novo* approaches. For variant analysis, consensus sequences were compared to the reference-sequence unless indicated otherwise. The mapping-consensus sequence generated by mapping the Illumina reads of the wild rice plant using the CLC-GW was shorter than the reference-sequence by 20 bases (Table 2). In comparison, there were 128 variants found between the two sequences, which included 18 deletions, 13 insertions and 97 mismatches. Among mismatches 92 were single-nucleotide variants (SNPs) (48 transitions and 44 transversions) and 5 were multi-nucleotide variants (MNVs). The mapping-consensus obtained by mapping a subset of Illumina reads differed from the previous sequence by one deletion of 2 bases (Table 2), although, this variation was not found in the reference-sequence.

The consensus sequence generated by mapping Ion Torrent reads of the wild-rice-genotype using the CLC-GW was shorter than the reference-sequence by 26 bases and shorter by 6 bases than the consensus sequence from Illumina sequencing. The two sequences differed by 139 variants, which included 30 deletions, 14 insertions and 95 mismatches. Among the mismatches 90 were SNPs (47 transitions and 43 transversions) and 5 were MNVs.

As was observed for the R-rice-genotype, the de novo-consensus sequence differed from the reference-sequence at more positions than the mapping-consensus sequence (Table 1 and Table 2). In addition, a higher number of deletions were observed in consensus sequences obtained from the CLC-GW analysis as compared to the Torrent Suite Software analysis where more insertions were observed (Table 1 and Table 2).

Comparison of sequencing platforms

The mapping-consensus chloroplast sequences of the wild-rice-genotype generated in the CLC-GW using the Illumina and Ion Torrent reads were used to compare sequence platforms. When comparing variations found in the wild-rice-genotype by both sequencing technologies, 20 discrepancies (variant in one consensus not present in the other) were observed (Table 3). However, in the Ion Torrent consensus sequence there were fifteen variants and in the Illumina consensus there were only three. Moreover, two variants found at the same position differed in nucleotide composition in both consensus sequences.

Read mapping files derived from both mapping-consensus sequences were visually inspected to check for reads mapped to all 20 positions to determine sequence read error or for mapping error. The region in the Ion Torrent mapping-consensus sequence with the substitution of TT → AA (Table 3, variant #1) had several mapped reads with many mismatches. Reads from this region when extracted and blasted against the nucleotide sequence collection (nr/nt) at NCBI website (http://blast.ncbi.nlm.nih.gov) gave hits to the chloroplast but also to the nuclear DNA of *Oryza* species. In addition, this variant was not observed in the corresponding region of the contig sequences obtained from *de novo* assembly (from both analysis software), indicating that the

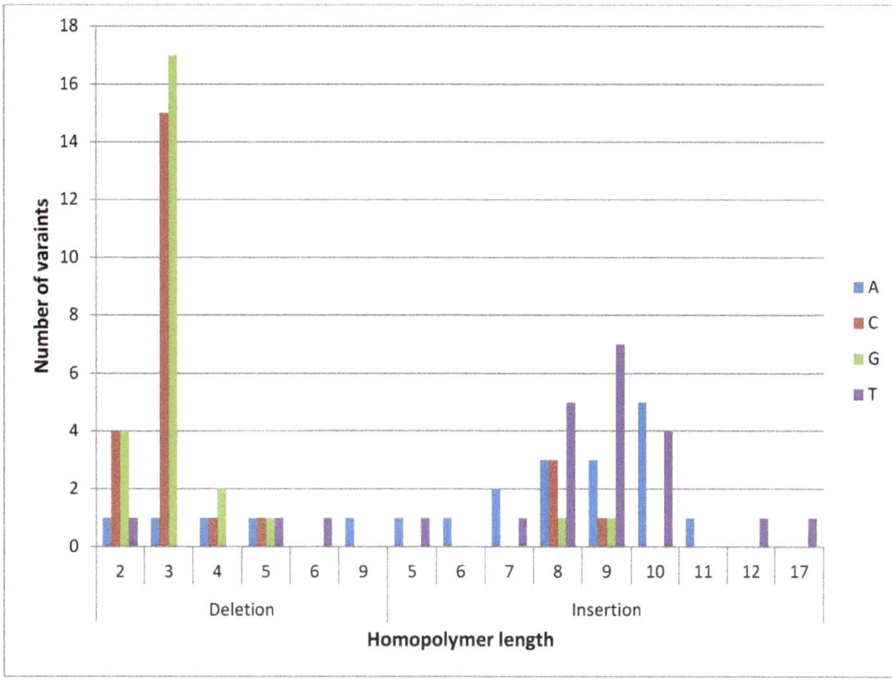

Figure 1. Variants in indels in cultivated (cv. Nipponbare) rice chloroplast consensus. Sequences generated by mapping and assembly of Ion Torrent reads to the available chloroplast sequence in GenBank for this genotype. The number of variants is shown with respect to its type (deletion or insertion) and position (the length of homopolymer region where the variants were found).

variant was not due to sequencing error but due to mapping artefact. Similarly, mapping error and not sequence read error was the reason for variants detected at several positions in one or the other mapping-consensus sequences (Table 3: for # 2 see Figure 2, for #3 and #4 see Figure 3), as these variants were not observed in the corresponding region of the contig sequences obtained from *de novo* assembly (from both analysis software). One of the variants (#7), an insertion of an A in the Ion Torrent consensus and AA in the Illumina consensus, was found in a long homopolymer stretch of 10 A's. The location of the variation suggested that in both cases it could be an error. Variations at this position in contigs sequences varied from 2 to 3 A's insertions. This polymorphism was not called in the consensus from the subset of Illumina data. Comparison of other chloroplast genomes, known to have been sequenced on the Illumina platform (GAII) [11], showed some to have this insertion, namely Australian *Oryza rufipogon* (GenBank accession – JN005833), Asian *Oryza rufipogon* (GenBank accession – JN005832) and *Oryza meridionalis* (GenBank accession – JN005831). Interestingly, the insertion was not present in the chloroplast genome sequence of *Oryza sativa* spp. *japonica* var. Nipponbare (reference-sequence), *Oryza sativa* ssp. *indica* isolate 93-11 (GenBank accession – AY522329) and *Oryza nivara* (GenBank accession – AP006728) which were sequenced by Sanger technology. Thus for this variant we cannot conclude with certainty if this discrepancy is due to mapping artefact or due to read error. However, all other errors found in Ion Torrent sequence (Table 3, #8 to #20) were not due mapping errors but read errors from either deletions or insertions in homopolymer regions.

Based on the analysis and all the findings using various approaches outlined above, an amended wild-rice-genotype chloroplast consensus sequence was created incorporating the most probable variants from Table 3. To assess the correctness of this amended chloroplast sequence paired-end reads from Illumina

Table 2. Comparison of chloroplast consensus sequences of the wild rice (*Oryza rufipogon*-like).

Sequencing platform	Source	Sequence length (bp)	Variants	Deletions	Insertions	Mismatches
Used available reference	GenBank (GU592207)	134,551	–	–	–	–
Illumina	mapping- consensus	134,531	128	18	13	97
	mapping-consensus (subset of reads)	134,529	129	19	13	97
Ion Torrent	mapping consensus	134,525	139	30	14	95
	de novo- consensus (CLC-GW)	134,521	155	43	17	95
	reference-assisted *de novo*- consensus (Torrent Suite)	134,554	147	23	28	96

Table 3. Inconsistent variations found in wild rice chloroplast mapping-consensus sequences and their validation.

Variations		Reference-sequence		Mapping-consensus sequence		The most probable variant
No	Type	Position	Allele	HiSeq	Ion Torrent	
#1	MNV	57,036	TT	TT	AA	TT
#2	ins	65,465^65,466	–	–	TCCTATTTAATA	TTCCTATTTAATA
#3	MNV	66,897	CGAT	TAGA	CGAT	TAGAAATAAAAAATTCTAA
#4	SNP	66,902	C	A	C	TAGAAATAAAAAATTCTAA
#5	SNP	17,366	T	A	T	T
#6	SNP/del	17,368	C	A	–	–
#7	ins	3,545^3,546	–	AA	A	–
#8	del	21,808	C	C	–	C
#9	del	57,027	T	T	–	T
#10	del	81,342	G	G	–	G
#11	del	91,427	C	C	–	C
#12	del	91,589	C	C	–	C
#13	del	97,135	G	G	–	G
#14	del	111,639	G	G	–	G
#15	del	116,139	C	C	–	C
#16	del	118,025	C	C	–	C
#17	del	119,245	G	G	–	G
#18	del	122,914	C	C	–	C
#19	del	123,568	G	G	–	G
#20	del	133,816	C	C	–	C

Variations derived by Illumina and Ion Torrent sequencing.
SNP – single-nucleotide variant, MNV – multi-nucleotide variant, ins – insertion, del – deletion.

sequencing platform were re-mapped to this new wild rice chloroplast genome. Illumina reads as against Ion Torrent reads of the W-rice-genotype were selected for remapping as even short Illumina reads (36 bases) of the R-rice-genotype provided a consensus matching the reference-sequence (Table 1) [11], as against an inaccurate consensus when Ion Torrent reads were used (Table 1) due to indel-associated errors. The new mapping results for the wild rice show only one discrepancy described earlier (Table 3, #7) in a long homopolymer region. All other sites were identical giving an additional assurance of the correctness of the final consensus.

Discussion

Various NGS technologies are now available for the rapid sequencing of whole genomes [17,23,24]. The choice of selecting one or more NGS technologies depends on the yield of data required, read length required, cost per data point and accuracy of the sequence data [25]. Systematic as against random errors can compromise the use of sequence data [26,27]. In this study we analysed NGS reads of two rice genotypes obtained from the Ion Torrent platform and compared it for accuracy to those obtained from the Illumina platform. A comparison of the three sequencing systems is presented in Table 4.

The Ion Personal Genome Machine (PGM) is characterised by long read length, very short run times and inexpensive consumables [28]. The drawbacks include low output, high cost per Gb (gigabase) of sequence [25] and the longer time required for the library preparation relative to that required for the sequencing run. Moreover, the PGM produces biased sequences from high homopolymer regions, which result in indel errors [27]. However, this is a developing platform with high potential for improvement in accuracy, read length and cost per read and per Gb. As the machine is the chip, it can be easily upgraded by the release of new chips. The PGM was the first next-generation sequencer, which price that dropped below $100 K. At the time of the study the PGM cost was around $80 K, which included the instrument, temple preparation system, enrichment system and server. The main advantages of HiSeq2000 system are the high throughput, large output data volumes per run and low cost of reagents per Gb [29]. The shortcomings of the platform embrace short read length and the need for advanced computational resources to process and store the enormous data volumes created from each run. Furthermore, the run time is considerably longer than that for the PGM, and the Illumina has a much higher initial capital cost. The specification of the GAIIx system was very similar to the HiSeq2000 with smaller output per run, higher cost per Gb and significantly lower instrument cost. The sequencing accuracy on both Illumina platforms is high with the most commonly encountered errors being substitutions. A wide range of Illumina instruments are now available with differing data volumes, running times and costs.

We selected the cultivated rice genotype (cv. Nipponbare) as one of the samples as its chloroplast sequence, derived by Sanger sequencing and considered as accurate, was available and could be used as reference sequence to check for errors in consensus sequences derived from NGS sequencing on both platforms. The aim was to generate both mapping and *de novo*-consensus sequences to determine read accuracy of both platforms.

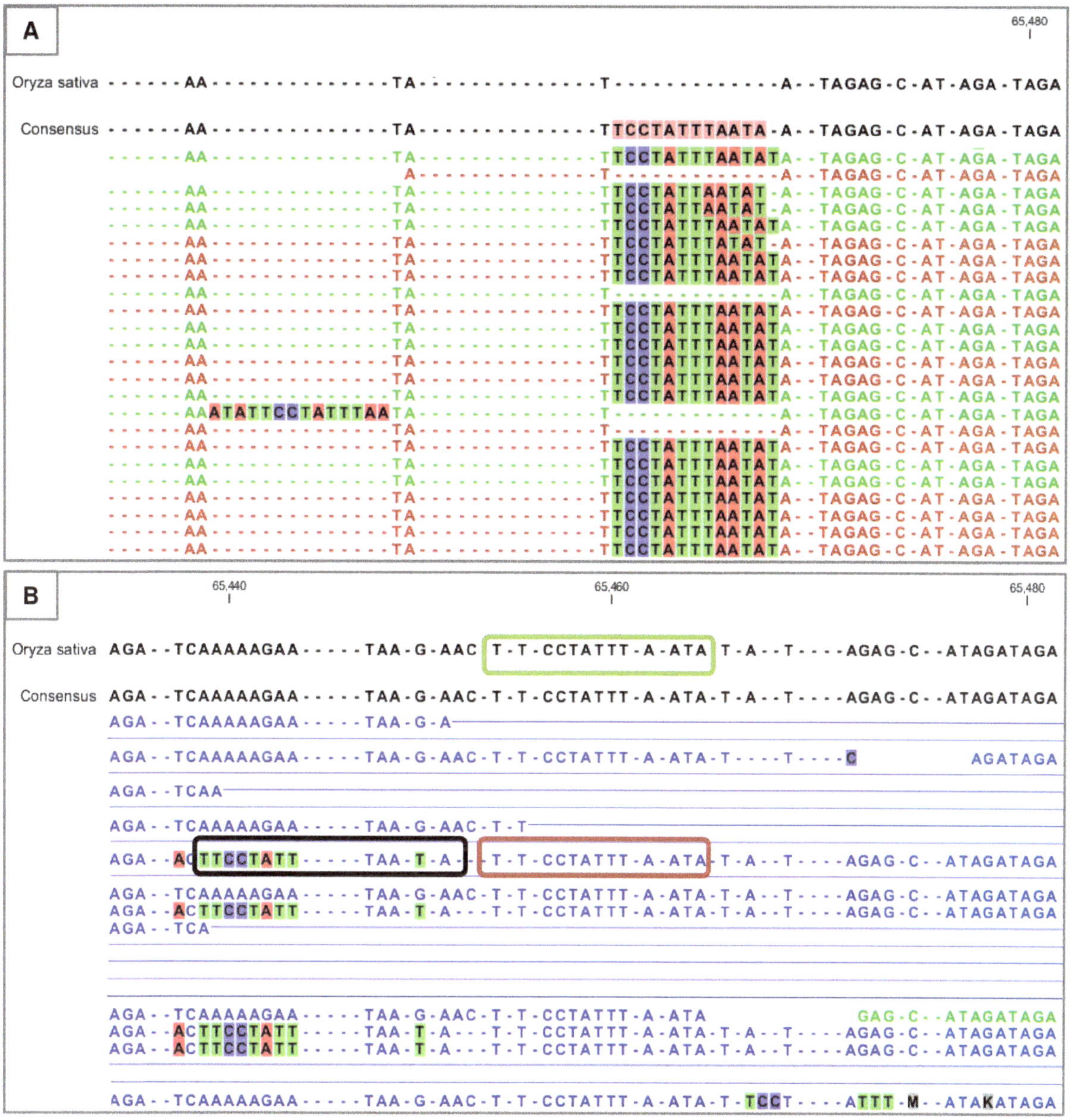

Figure 2. Snapshot of mapping results of wild rice Ion Torrent (A) and Illumina (B) reads. Reads were mapped to the chloroplast reference of *Oryza sativa* cv. Nipponbare. In the mapping of Ion Torrent reads there was a long insertion (TCCTATTTAATA) reported in the consensus sequence of wild rice chloroplast ((A), marked with orange background colour). This insertion was missed in the mapping of Illumina reads, although it was present in the reads ((B), example of the read sequence marked in black rectangle). The nucleotides in the insertion were duplicated in wild rice (sequence marked in red rectangle), and not in the reference genome where only one copy of these nucleotides was present (marked in green rectangle). The duplicated region was a probable cause of the misalignment of reads. Oryza sativa – fragment of chloroplast sequence of *Oryza sativa* spp. *japonica* var. Nipponbare; Consensus – consensus sequence of wild rice chloroplast sequence derived by mapping reads from Illumina (A) and Ion Torrent (B) platforms to the reference. Nucleotides with background colours represent the mismatches between reads and the reference sequence; paired end reads are shown in blue; single reads are shown in green and red (in forward and reverse orientation, respectively).

We were unable to obtain *de novo*-consensus sequences for both genotypes with Illumina reads when processed by CLC-GW as the contigs generated were non-contiguous when mapped to the reference-sequence. In the case of reference-genotype, read coverage depth and read length contributed to non-contiguous contigs as the read depth in regions corresponding to gaps was between 7 and 30, when examined for base calling and read mapping depth to the mapping-consensus, which is well below the average mapping read coverage depth of 127.7 (Table S2). In the case of wild-rice-genotype with longer read length (100 bases), read length but not read coverage depth may have contributed to non-contiguous contigs as the read depth in regions corresponding to the gaps was between 7,000 to 11,000, when examined for base calling and read mapping depth to the in the mapping-consensus, which is well above the average mapping read coverage depth of 8,143.2 (Table S2). Interestingly, *de novo*-consensus, although not accurate (Table 1) was generated for the reference-genotype using CLC-GW and the longer (200 bases) Ion Torrent reads. We can

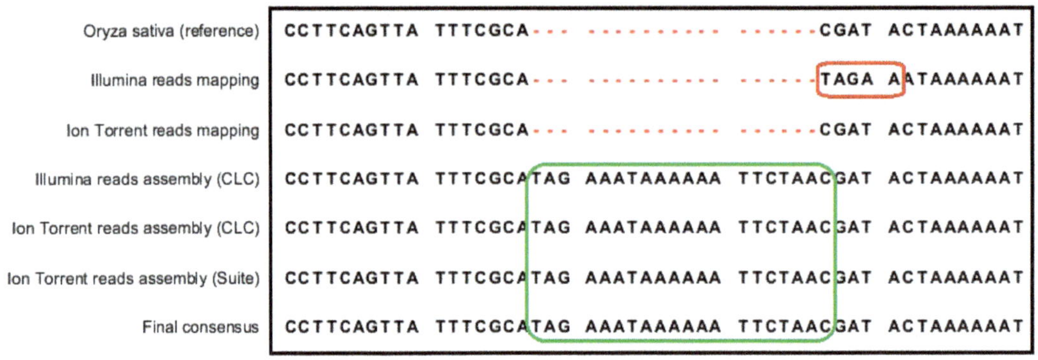

Figure 3. Alignment of regions #3 and #4 from Table 3 showing discrepancies in consensus sequences. The fragment circled in red shows false called SNPs (#3 and #4, Table 3, Illumina consensus); these SNPs were incorrect because of the long insertion present in wild rice sequence but not in the reference. The fragments circled in green illustrate this long insertion found in wild rice chloroplast genome by means of reads assembly from both platforms and both assembly tools. Final sequence was created based on this information. Oryza sativa (reference) – region 66860.66940 from chloroplast sequence of *Oryza sativa* spp. *japonica* var. Nipponbare; Illumina reads mapping and Ion Torrent reads mapping – regions from consensus sequence generated by mapping wild rice Illumina and Ion Torrent reads, respectively, to the reference sequence; Illumina reads assembly and Ion Torrent reads assembly – regions from contigs generated by assembly of reads from Illumina and Ion Torrent platforms, respectively; CLC – assembly performed in CLC Genomic Workbench; Suite – assembly performed in Torrent Suite Software; Final consensus – final wild rice chloroplast genome sequence (GenBank accession – KF428978).

thus conclude that read lengths of more than 100 bases would be required to successfully obtain a *de novo*-consensus of the chloroplast genome.

In case of the reference-genotype, mapping-consensus sequences were generated using reads from both platforms, with no mismatches. However, mapping-consensus sequences showed several indels due to inaccuracies in the Ion Torrent reads mainly in homopolymer regions. In case of the short Illumina reads (36 bases) of the R-rice-genotype, neither the read length (36 bases) nor the sequence yield (minimum reference-sequence read mapping coverage of 7) was a limitation in generating an accurate mapping-consensus when processed by the CLC-GW, indicating the accuracy of the Illumina sequencing platform. In case of the longer Ion Torrent reads (200 bases), when processed by the CLC-GW and the Torrent Suite Software, consensus-sequences with no mismatches were generated but with several variants all of which were indels and predominantly in homopolymer regions (Table 1). These results indicate the actual presence of these anomalous indels in the Ion Torrent reads which contributed to the inaccuracy in the consensus sequence generated and to the unreliability of any indels calls made when using these reads. It is known that the Ion Torrent sequencing platform generates anomalous indels [27]. Interestingly, the preference of the CLC-GW and the Torrent Suite analysis tool in filtering out predominantly either insertions or deletion, respectively, as observed in both rice genotypes (Tables 1 and 2), indicates the

Table 4. The comparison between the tree sequencing systems utilised in the study.

	Illumina GAIIx	Illumina HiSeq2000	Ion Torrent PGM –318 chip
Sequencing method	Synthesis (light detection)	Synthesis (light detection)	Synthesis (proton detection)
Amplification method	Bridge PCR	Bridge PCR	Emulsion PCR
Read length[a]	Up to 2×150 bp	1×50 bp, 2×50 bp, 2×100 bp	~200 bp, ~400 bp
Paired reads	Yes	yes	Yes
Insert size	Up to 700 bp	Up to 700 bp	Up to 250 bp
Output data/run	30 Gb	600 Gb	Up to 2 Gb
Time/run[b]	10–14 days	8–11 days	4–7 hours
Cost/Gb[c]	$148	$41	$1000
Instrument cost	$256 K	$654 K	$80 K
Accuracy	>99.9%	>99.9%	99%
Error rate[d]	≥0.1%	≥0.1%	~1%
Primary errors	Substitutions	Substitutions	Indels
DNA requirements	0.05–1 ug	0.05–1 ug	0.1–1 ug

Gb – gigabase, bp – base pair, K – thousand, uq - microgram.
[a] Annotation '2 x' refers to paired end reads and '1 x' to single reads.
[b] Run time from minimum to maximum read lengths.
[c] Includes one sample and one sequencing kit per run.
[d] percentage of errors per base in single read.

possibility of these analysis tools to be amended to deal with anomalous indels in the Ion Torrent reads. We also observed a bias for indels in homopolymer regions, where deletions occurred in shorter regions while insertions occurred in longer regions. These findings are inconsistent with a recent study on Ion Torrent sequencing bias where it was reported that an increased deletion rate was positively correlated with increased lengths of homopolymer regions and that insertions were independent of homopolymer length [27,30]. This is probably because of the different genomes used or that the chloroplast genome is AT-rich and the longest C/G stretches found are 9 bp long (2 regions), 8 bp (4 regions) and 7 bp (9 regions).

The systematic pattern of bias, specific types of nucleotide variants in long versus short repeats, in the single base indels in homopolymers (Figure 1) is likely to be a feature of chloroplasts with their specific patterns of homopolymer repeat frequency. However, systematic errors reported in the Ion Torrent data in earlier studies [27] were not found in this analysis of the chloroplast possibly because we used a more recent version of the technology. We can conclude that both sequencing platforms produce high quality data with relatively low rates of discrepancy in the calling of polymorphisms, especially SNPs, but the Ion Torrent data would be less reliable for indel calls.

We compared the two sequencing platforms in generating an accurate consensus chloroplast sequence by comparing the mapping-consensus sequences generated for both genotypes. For the R-rice genotype, no mismatches were observed indicating the two platforms to be comparable for SNP calls as has been reported for microbial genomes [25]. When comparing the mapping-consensus sequences of the W-rice-genotype we identified twenty discrepancies. It must be noted that some of the discrepancies were caused by misalignment of reads and as has been previously recognized [31] there is an important issue with mismatches in close proximity of indel events. Some read mapping tools can have problems with these regions resulting in false SNPs calling. However, for both the rice genotypes we observed erroneous indel calls with Ion Torrent reads mainly in homopolymer regions.

The importance of obtaining a *de novo* sequence was clear from our study as discrepancies observed in the mapping sequences of the W-rice-genotype were curated using the *de novo* contigs. The *de novo* assembly of chloroplast genomes using the tools applied in this study was easier with the longer Ion Torrent data than with the Illumina data. Although increasing number of reference genome sequences are being generated to support chloroplast genome analysis, the ability to generate whole chloroplast genome sequences *de novo* will find wide application. The first report of total plant DNA analysis for chloroplast sequencing [11] used short Illumina reads (36 bp) and relied on a reference genome for successful assembly. More recent studies based upon longer reads [12,23] have reported more success with *de novo* assembly.

Conclusions

This study demonstrates that *de novo* assembly of an accurate whole chloroplast genome sequence will be possible for routine plant barcoding. Analysis of plants based upon appropriate and careful analysis of shotgun sequencing of total DNA promises to provide a barcode that will have wide application. Having a well curated and reliable consensus chloroplast sequence for a plant sample provides greater certainty of obtaining reliable results in many critical studies such as plant identification, purity assessment, phylogenetic analysis and heteroplasmy analysis. Continuing improvements in sequencing platforms and analysis tools will make this method more reliable and cost effective for a wide range of research and industrial applications. The significant advances in these competing sequencing platforms that have been foreshadowed by the manufactures promise dramatic reductions in cost in the near future. This would make barcoding by sequencing the whole chloroplast in this way the preferred option for plant identification in many research and industrial applications. A comparison of the platforms at the same level of coverage is useful in providing a guide to likely comparative performance as the number of reads on these platforms increases in the future. Costs will change for each platform as read length and read volume are increased.

However cost is not the only advantage of this approach. The use of total DNA from the sample without amplification simplifies the analysis but also provides greater opportunity to measure in an unbiased way the contribution of different genotypes in complex mixtures. The risks of preferential amplification or enrichment of chloroplasts from a specific genotype that complicated earlier analysis will be avoided.

Supporting Information

Table S1 Summary statistics of raw reads obtained from PGM Ion Torrent and Illumina platforms for rice (*Oryza sativa* spp. *japonica* var. Nipponbare) and wild rice (*Oryza rufipogon*-like plant from Australia). Quality distribution is represented as PHRED scores.

Table S2 Mapping statistics of CLC trimmed reads from Illumina and PGM Ion Torrent platforms.

Table S3 Assembly results performed on CLC Genomic Workbench and Torrent Suite Software of reads from Illumina and Ion Torrent platforms and mapping results of generated contigs to chloroplast reference-sequence.

Acknowledgments

We thank the local support team from Life Technologies for technical assistance with this project.

Author Contributions

Conceived and designed the experiments: MB AF RH. Performed the experiments: MB. Analyzed the data: MB AF. Contributed to the writing of the manuscript: MB AF RH.

References

1. Straub SCK, Parks M, Weitemier K, Fishbein M, Cronn RC, et al. (2012) Navigating the tip of the genomic iceberg: Next-generation sequencing for plant systematics. Am J Bot 99: 349–364.
2. Parks M, Cronn R, Liston A (2009) Increasing phylogenetic resolution at low taxonomic levels using massively parallel sequencing of chloroplast genomes. BMC Biol 7: 84–84.
3. Taberlet P, Coissac E, Pompanon F, Brochmann C, Willerslev E (2012) Towards next-generation biodiversity assessment using DNA metabarcoding. Mol Ecol 21: 2045–2050.
4. Henry RJ (2001) Plant genotyping: the DNA fingerprinting of plants. New York: CABI Pub.
5. Henry RJ (2013) Molecular Markers in Plants. Hoboken, NJ: John Wiley & Sons.

6. Galimberti A, De Mattia F, Losa A, Bruni I, Federici S, et al. (2013) DNA barcoding as a new tool for food traceability. Food Res Int 50: 55–63.
7. Diaz-Sanchez S, Hanning I, Pendleton S, D'Souza D (2013) Next-generation sequencing: the future of molecular genetics in poultry production and food safety. Poult Sci 92: 562–572.
8. Jansen RK, Raubeson LA, Boore JL, dePamphilis CW, Chumley TW, et al. (2005) Methods for Obtaining and Analyzing Whole Chloroplast Genome Sequences. In: Elizabeth AZ, Eric HR, editors. Methods in Enzymology: Academic Press. pp. 348–384.
9. Lang BF, Burger G (2007) Purification of mitochondrial and plastid DNA. Nat Protoc 2: 652–660.
10. Cronn R, Liston A, Parks M, Gernandt DS, Shen R, et al. (2008) Multiplex sequencing of plant chloroplast genomes using Solexa sequencing-by-synthesis technology. Nucleic Acids Res 36: e122.
11. Nock CJ, Waters DLE, Edwards MA, Bowen SG, Rice N, et al. (2011) Chloroplast genome sequences from total DNA for plant identification. Plant Biotechnol J 9: 328–333.
12. McPherson H, Rossetto M, van der Merwe M, Delaney SK, Edwards MA, et al. (2013) Capturing chloroplast variation for molecular ecology studies: a simple next generation sequencing approach applied to a rainforest tree. BMC Ecol 13: 8–18.
13. Loman NJ, Misra RV, Dallman TJ, Constantinidou C, Gharbia SE, et al. (2012) Performance comparison of benchtop high-throughput sequencing platforms. Nat Biotechnol 30: 434–434.
14. Sims D, Sudbery I, Ilott NE, Heger A, Ponting CP (2014) Sequencing depth and coverage: key considerations in genomic analyses. Nat Rev Genet 15: 121–132.
15. Illumina Inc website. Available: http://res.illumina.com/documents/products/techspotlights/techspotlight_sequencing.pdf. Accessed 2014 September 8.
16. Illumina Inc website. Available: http://res.illumina.com/documents/products/illumina_sequencing_introduction.pdf. Accessed 2014 September 8.
17. Rothberg JM, Edwards M, Hoon J, Simons JF, Marran D, et al. (2011) An integrated semiconductor device enabling non-optical genome sequencing. Nature 475: 348–352.
18. Lee KH, Park J, Williams DS, Xiong Y, Hwang I, et al. (2013) Defective chloroplast development inhibits maintenance of normal levels of abscisic acid in a mutant of the Arabidopsis RH3 DEAD-box protein during early post-germination growth. Plant J 73: 720–732.
19. Hartwig B, James GV, Konrad K, Schneeberger K, Turck F (2012) Fast isogenic mapping-by-sequencing of ethyl methanesulfonate-induced mutant bulks. Plant Physiol 160: 591–600.
20. Sotowa M, Ootsuka K, Kobayashi Y, Hao Y, Tanaka K, et al. (2013) Molecular relationships between Australian annual wild rice, Oryza meridionalis, and two related perennial forms. Rice 6: 26.
21. Furtado A (2014) DNA extraction from vegetative tissue for next-generation sequencing. In: Henry RJ, Furtado A, editors. Cereal Genomics. Methods in Molecular Biology. New York: Springer Science+Business Media. pp. 1.
22. Kurtz S, Phillippy A, Delcher AL, Smoot M, Shumway M, et al. (2004) Versatile and open software for comparing large genomes. Genome Biol 5: 12–R12.
23. Ferrarini M, Cestaro A, Sargent DJ, Moretto M, Ward JA, et al. (2013) An evaluation of the PacBio RS platform for sequencing and de novo assembly of a chloroplast genome. BMC Genomics 14: 670–670.
24. Shendure J, Ji H (2008) Next-generation DNA sequencing. Nat Biotechnol 26: 1135–1145.
25. Quail MA, Smith M, Coupland P, Otto TD, Harris SR, et al. (2012) A tale of three next generation sequencing platforms: comparison of Ion Torrent, Pacific Biosciences and Illumina MiSeq sequencers. BMC Genomics 13: 341–341.
26. Meacham F, Boffelli D, Dhahbi J, Martin DIK, Singer M, et al. (2011) Identification and correction of systematic error in high-throughput sequence data. BMC Bioinformatics 12: 451–451.
27. Bragg LM, Stone G, Butler MK, Hugenholtz P, Tyson GW (2013) Shining a light on dark sequencing: characterising errors in Ion Torrent PGM data. Plos Comput Biol 9: e1003031.
28. Liu L, Li Y, Li S, Hu N, He Y, et al. (2012) Comparison of next-generation sequencing systems. J Biomed Biotechnol 2012: 251364.
29. Glenn TC (2011) Field guide to next-generation DNA sequencers. Molecular ecology resources 11: 759–769.
30. Ross MG, Russ C, Costello M, Hollinger A, Lennon NJ, et al. (2013) Characterizing and measuring bias in sequence data. Genome Biol 14: R51.
31. Albers CA, Lunter G, MacArthur DG, McVean G, Ouwehand WH, et al. (2011) Dindel: accurate indel calls from short-read data. Genome Res 21: 961–973.

Light Affects the Chloroplast Ultrastructure and Post-Storage Photosynthetic Performance of Watermelon (*Citrullus lanatus*) Plug Seedlings

Qingqing Duan[1,2], Wu Jiang[1,2], Ming Ding[3], Ye Lin[1,2], Danfeng Huang[1,2]*

1 Department of Plant Science, School of Agriculture & Biology, Shanghai Jiao Tong University, Shanghai, China, **2** Key Laboratory of Urban Agriculture (South), Ministry of Agriculture, Shanghai, China, **3** Department of Horticulture, Northwest A&F University, Yangling, China

Abstract

Watermelon [*Citrullus lanatus* (Thunb.) Matsum. and Nakai] plug seedlings were stored at 15°C in the light at a photosynthetic photon flux density of 15 $\mu mol \cdot m^{-2} \cdot s^{-1}$ or in darkness for 6 days, to evaluate their chloroplast ultrastructure, and associated photosynthetic characteristics. Storage in the dark caused swelling, disordered granal arrangement, and starch grain disappearance in the chloroplasts. In contrast, the chloroplasts stored in the light were relatively normal. As a result, the light-stored seedlings had a significantly higher chlorophyll content, Fv/Fm, and Pn than did dark-stored seedlings. Regardless of whether the seedlings were stored in light or darkness, the Gs and Ls of the seedlings significantly decreased, while the Ci obviously increased when the Pn decreased after 6 days of storage. This result suggests that the decreased Pn is not solely a stomatal effect, as the effects on the chloroplasts contributed to this photosynthetic inhibition. Six days after transplanting, seedlings that were stored in the light or darkness for 2 or 4 days showed complete recovery of chloroplast ultrastructure, chlorophyll content, Fv/Fm, Gs and Pn. When the storage period increased to 6 days, the dark-stored seedlings had a significantly lower Fv/Fm and Pn than the light-stored and control seedlings 6 days after transplanting, which was mainly ascribed to incomplete recovery of chloroplast ultrastructure. Furthermore, the light-stored seedlings exhibited a significantly higher shoot dry weight during storage and a higher percentage dry weight increase after transplanting than the dark-stored seedlings. These effects were enhanced by prolonged storage (4 to 6 days). This study demonstrated that dim light during storage is beneficial for maintaining chloroplast ultrastructure as well as photosynthetic efficiency in watermelon seedlings, thus contributing to the rapid recovery of post-storage photosynthetic performance, which ensures the transplant quality of the seedlings after removal from storage.

Editor: John R Battista, Louisiana State University and A & M College, United States of America

Funding: This work was supported by funds from the National High-tech Research & Development Program of China (Grant No. 2012AA101903), National Natural Science Foundation of China (No. 61233006), the National Science-tech Support Plan Project (Grant No. 2011BAD43B02), and Shanghai Graduate Education and Innovation Program (Horticulture). The funders had no role in study design, data collection and analysis, decision to publish, or preparation of the manuscript.

Competing Interests: The authors have declared that no competing interests exist.

* Email: hdf@sjtu.edu.cn

Introduction

Watermelon [(*Citrullus lanatus* (Thunb.) Matsum. and Nakai)] is a major cucurbit crop worldwide. China is the largest watermelon producer, providing almost two-thirds of the world supply [1]. In China, watermelon is mainly produced in industrial seedling nurseries. To ensure an adequate supply of seedlings, the plants are occasionally stored (e.g., when waiting for optimal weather for transplantation, during shipping, or during labor shortages). Seedlings are normally stored at low temperature in darkness [2–4], which can decrease seedling quality. When the plants are stored in darkness, photosynthesis ceases, but respiration continues, depleting the carbohydrate reserves and triggering a series of changes in cellular events, resulting in dry weight loss, decreased chlorophyll content [5,6], reduced soluble sugars [7,8], and enhanced lipid peroxidation [9,10].

The provision of light, even at a very low intensity, during low temperature storage can improve the quality of many horticultural species [8,11–13]. Kubota and Kozai found that the photosynthetic ability of broccoli plantlets that were stored in the light was preserved [5]. Similar results were also observed in seedlings of *Phalaenopsis* [14] and radiata pine [15]. The recovery of post-storage photosynthetic performance is crucial for plants [4,16], and the post-storage photosynthetic rate is positively associated with new root growth of radiata pine seedlings [15] and survival of Douglas fir seedlings in a greenhouse [17].

However, minimal attention has been focused on how light affects the photosynthetic apparatus of plants during storage and the subsequent recovery period. Plants photosynthetically acclimate to various light conditions through both anatomical and physiological changes [18–20]. Chloroplasts are the sites of photosynthesis, and the chloroplast ultrastructure provides a structural framework for photosynthetic performance. Decreases

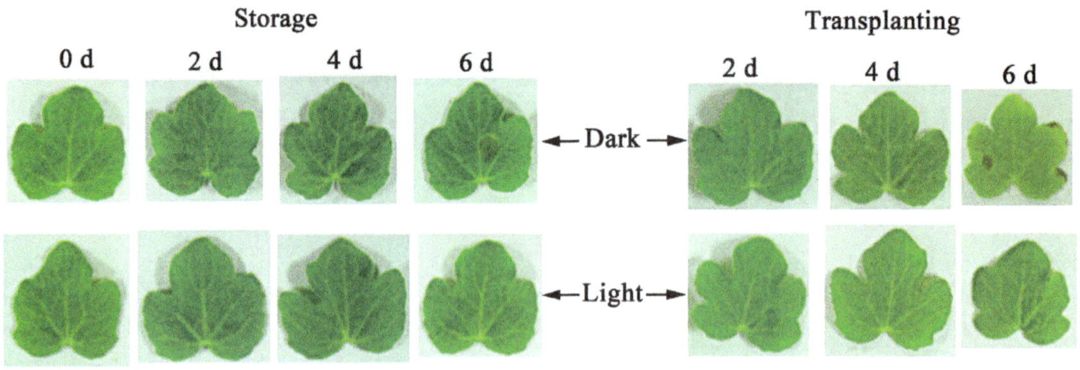

Figure 1. Photographs of the first true leaves of watermelon (*Citrullus lanatus*) seedlings during storage in light or darkness at 15°C and subsequent transplanting. (The seedlings were stored for 2, 4, or 6 d. After removal from storage, the seedlings were transplanted for 6 days.).

in photosynthesis mainly correspond with ultrastructural alterations in the chloroplast [21–23].

We performed experiments to quantify how short-term storage in light or darkness at 15°C and durations from 0 to 6 d influence the chloroplast ultrastructure and photosynthetic status during storage and the subsequent recovery of watermelon seedlings. The specific objectives of this study were (1) to identify the effect of the storage duration and condition (light or darkness) on the chloroplast ultrastructure, chlorophyll fluorescence parameters and gas exchange measurements during storage and (2) to evaluate how storage influences the photosynthetic performance and dry weight accumulation of seedlings after transplanting. Accurately assessing these effects is essential for maintaining watermelon seedling quality during storage and for improving the capacity of photosynthetic recovery after removal from storage.

Materials and Methods

Plant Material and Growth Conditions

Seeds of watermelon [*Citrullus lanatus* (Thunb.) Matsum. & Nakai] cv. Zaojia 84–24 were sown in 72-cell plastic plug trays (50 cm^3 per cell) that were filled with a substrate consisting of peat, perlite, and vermiculite (9:3:1, v/v/v) in a greenhouse under a natural photoperiod with day/night temperatures of 24/17°C ±3°C. At 20 d after sowing, uniform seedlings with one fully expanded true leaf and one small true leaf were selected and transferred to growth chambers (Model HP GS1000-D, Wuhan Ruihua Instrument & Equipment Co., Ltd., Wuhan, China). The conditions in the growth chambers were as follows: 300 $\mu mol \cdot m^{-2} \cdot s^{-1}$ photosynthetic photon flux density (PPFD) under a 12-h photoperiod, 25/18°C ±2°C day/night, and 70±10% relative humidity. The plants were supplied daily with half-strength Hoagland's solution.

Storage and Transplant Conditions

The storage period began when two true leaves had fully expanded and when one small true leaf was in the process of expanding. Six chambers, including twelve plug trays of seedlings, were used in this experiment. There were six trays in three chambers per treatment, providing 432 seedlings per treatment. Two trays in one chamber with 144 seedlings were considered as replications for each treatment. The air temperature inside the chambers was set at 15±1°C, which was recommended as the optimal storage temperature for watermelon plug seedlings [24]. The seedlings were stored for 6 d in darkness or under light at an average PPFD of 15±0.5 $\mu mol \cdot m^{-2} \cdot s^{-1}$. Light was provided during storage by cool white light lamps (Philips LIFEMAX, TLD 30w/865, Royal Philips Electronics, Amsterdam, the Netherlands) that were positioned approximately 30 cm above the seedling canopy. The light compensation point is the optimum PPFD for the low-temperature storage of seedlings [12]. Because there is no reported recommended light intensity for cucurbit storage, the light compensation point for watermelon was estimated as ~15 $\mu mol \cdot m^{-2} \cdot s^{-1}$ PPFD according to the results of Yongjian [25]. Three additional trays were left under normal growth conditions without storage and served as the non-stored controls. Every 2 d, the seedling trays were taken from the chambers and subirrigated with the same nutrient solution as that used before storage for 5 min under dim light conditions (< 1 $\mu mol \cdot m^{-2} \cdot s^{-1}$) at ambient room temperatures. Then, the trays were relocated to different chambers to minimize environmental differences.

Upon removal from storage at 2, 4, or 6 d, the seedlings from two trays in each treatment were transplanted into 20-cm-diameter plastic pots and allowed to recover in growth chambers under normal culture conditions (25/18°C ±2°C (day/night), PPFD = 300 $\mu mol \cdot m^{-2} \cdot s^{-1}$) for 6 d. The seedlings were supplied with half-strength Hoagland's solution once per day.

The seedlings were sampled just before the start of storage (0 d) and 2, 4, and 6 d later. After storage in light or darkness, the seedlings were allowed to recover for 6 d and sampled again. In this experiment, measurements were obtained using the first true leaves that were considered mature.

Electron Microscopy Studies

To avoid variation due to the differential structure in different parts of the leaves, the middle part of the leaves without the midrib was used and cut into small pieces (approximately 12 mm). These small pieces were placed in a bottle with 2.5% glutaraldehyde in phosphate buffer (pH 7.2), and the air was pumped out of the bottle with a syringe to ensure that the leaves would become fully soaked in the buffer solution according to the method of Chen [26] with some modifications. The leaves were fixed at 4°C for 24 h and post-fixed in 2% OsO_4 (w/v) for 2 h with the same buffer at 4°C. The fixed samples were washed in buffer (three times for 15 min each) and dehydrated in an ascending alcohol series at 4°C: 50% for 15 min, 70% (containing 2% uranyl acetate) for 12 h, 70% for 15 min, and 90% for 15 min. The samples were subsequently washed in steps of 20 min in 90% alcohol/90% acetone (1:1), 90% acetone at 4°C and in 100% acetone (three times) at room temperature. Then, the samples were immersed in

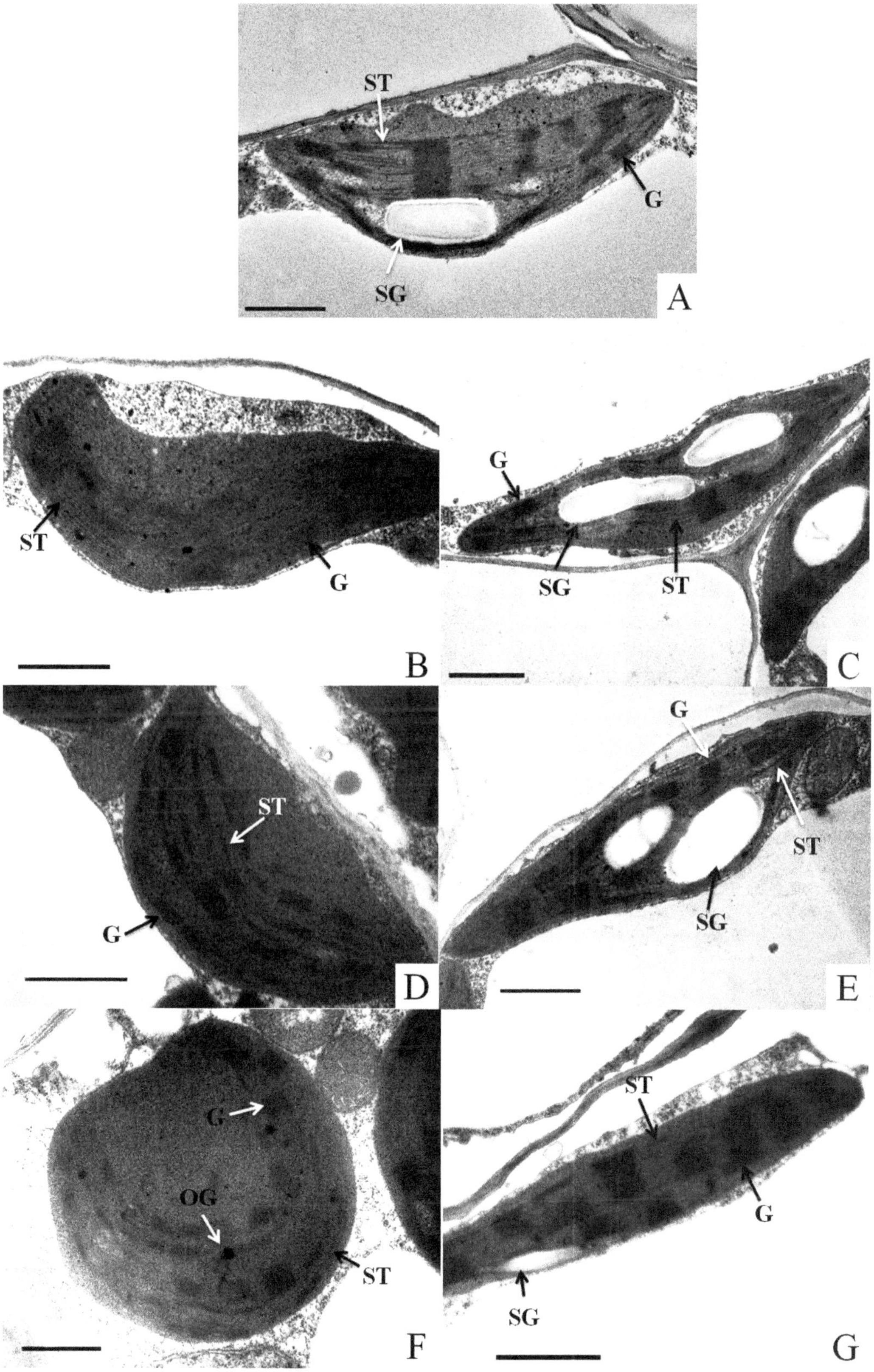

Figure 2. Changes in the ultrastructure of chloroplasts from the leaves of watermelon (*Citrullus lanatus*) seedlings stored in light (C, E, G) or darkness (B, D, F) at 15°C for 0 (A), 2 (B, C), 4 (D, E), or 6 (F, G) days. Bar = 1 μm. Abbreviations: ST, stroma thylakoid; G, grana thylakoid; SG, starch grain; OG, osmiophilic globule.

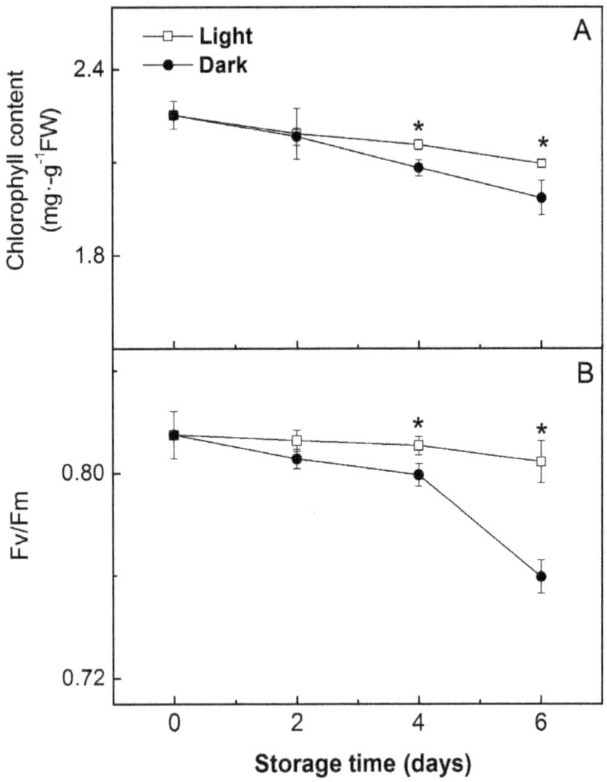

Figure 3. Changes in the chlorophyll content (A) and maximal photochemical efficiency of PSII (Fv/Fm) (B) in the leaves of watermelon (*Citrullus lanatus*) seedlings stored in the light (□) or in the dark (•) at 15°C. Data in A are the means of five replicates, and data in B are the means of ten replicates. Standard errors are shown with a vertical bar. Asterisks indicate significant differences between treatments on a given day according to the Tukey test ($P<0.05$).

acetone-Epon 812 resin at a ratio of 1:1 for 4 h and subsequently at a ratio of 1:2 overnight followed by immersion in 100% resin (three times for 4 h each) and polymerization at 60°C for 48 h. Thin sections (approximately 70 nm thickness) were obtained using a Leica EM UC6 ultramicrotome (Leica Co., Austria) and double-stained with uranium acetate-lead citrate before being examined with a Tecnai G2 Spirit BioTWIN (FEI Company, Hillsboro, Oregon, USA) transmission electron microscope operating at 120 kV.

Pigment Analysis and Chlorophyll Fluorescence Measurements

The middle part of the first true leaves without the midrib was used to measurechlorophyll content. The samples were collected by punching small disks (0.75 cm diameter) from five plants, and ten disks for each treatment were used for the measurements. Chlorophyll was extracted by incubation in 10 mL of 80% acetone for 48 h in darkness. The chlorophyll content was determined by measuring the absorbance at 663 and 646 nm using a UV–vis spectrophotometer (Model: U-2900, Hitachi Co., Tokyo, Japan) and calculated according to Lichtenthaler [27].

The chlorophyll fluorescence parameters were monitored using a portable modulated chlorophyll fluorometer (FMS-2, Hansatech, Norfolk, UK). The initial fluorescence (Fo) and maximum fluorescence (Fm) were measured after dark adaptation for 30 min. The maximum photochemical efficiency of PSII (Fv/Fm) was calculated as Fv/Fm = (Fm-Fo)/Fm. Ten plants from each treatment were measured.

Photosynthetic Gas Exchange

The net photosynthesis rate (Pn), stomatal conductance (Gs), and intercellular CO_2 concentration (Ci) were measured using a Ciras-2 portable photosynthesis system (PP Systems, Amesbury, MA, USA) with an LED light source at 800 $\mu mol \cdot m^{-2} \cdot s^{-1}$ (PPFD). The stomatal limitation (Ls) was calculated according the following equation: Ls = 1 - Ci/Ca, where Ci is the intercellular CO_2 concentration and Ca is the ambient CO_2 concentration. The leaf area clipped by the chamber was 2.5 cm^2. The temperature, relative humidity, and CO_2 concentration were maintained at 25°C, 45±5%, and 375 $\mu mol \cdot mol^{-1}$, respectively, during the measurements. The measurements were performed on leaves from at least three plants for each treatment during storage and after transplanting.

Plant Growth Parameters and Tissue Biomass After Transplanting

Twelve plants from each treatment were sampled. The sampled plants were washed carefully to remove the medium, and the shoots were cut from the plants. The washing procedure was performed at room temperature within a few minutes for each plant. Subsequently, the shoots were dried at 80 °C for 48 h in folded paper envelopes to obtain the shoot dry weight. The percentage increase in shoot dry weight for each treatment was measured on the 1st day after removal from storage and the 6th day after transplanting.

Table 1. The net photosynthesis rate (Pn), stomatal conductance (Gs), intercellular CO_2 concentration (Ci), stomatal limitation (Ls) of mature leaves, and shoot dry weight of watermelon (*Citrullus lanatus*) seedlings before and after 6 days of storage.

Treatments	Pn ($\mu mol \cdot m^{-2} \cdot s^{-1}$)	Gs ($mmol \cdot m^{-2} \cdot s^{-1}$)	Ci ($\mu mol \cdot mol^{-1}$)	Ls	Shoot dry weight ($g \cdot plant^{-1}$)
Before storage					
	20.0 a	120.1 a	254.7 c	0.56 a	0.114 a
After 6 days of storage					
Light storage	7.5 b	62.8 b	266.9 b	0.47 b	0.105 a
Dark storage	2.7 c	49.7 c	439.8 a	0.14 c	0.087 b

Means in columns followed by different letters are significantly different by Tukey test at $P<0.05$.

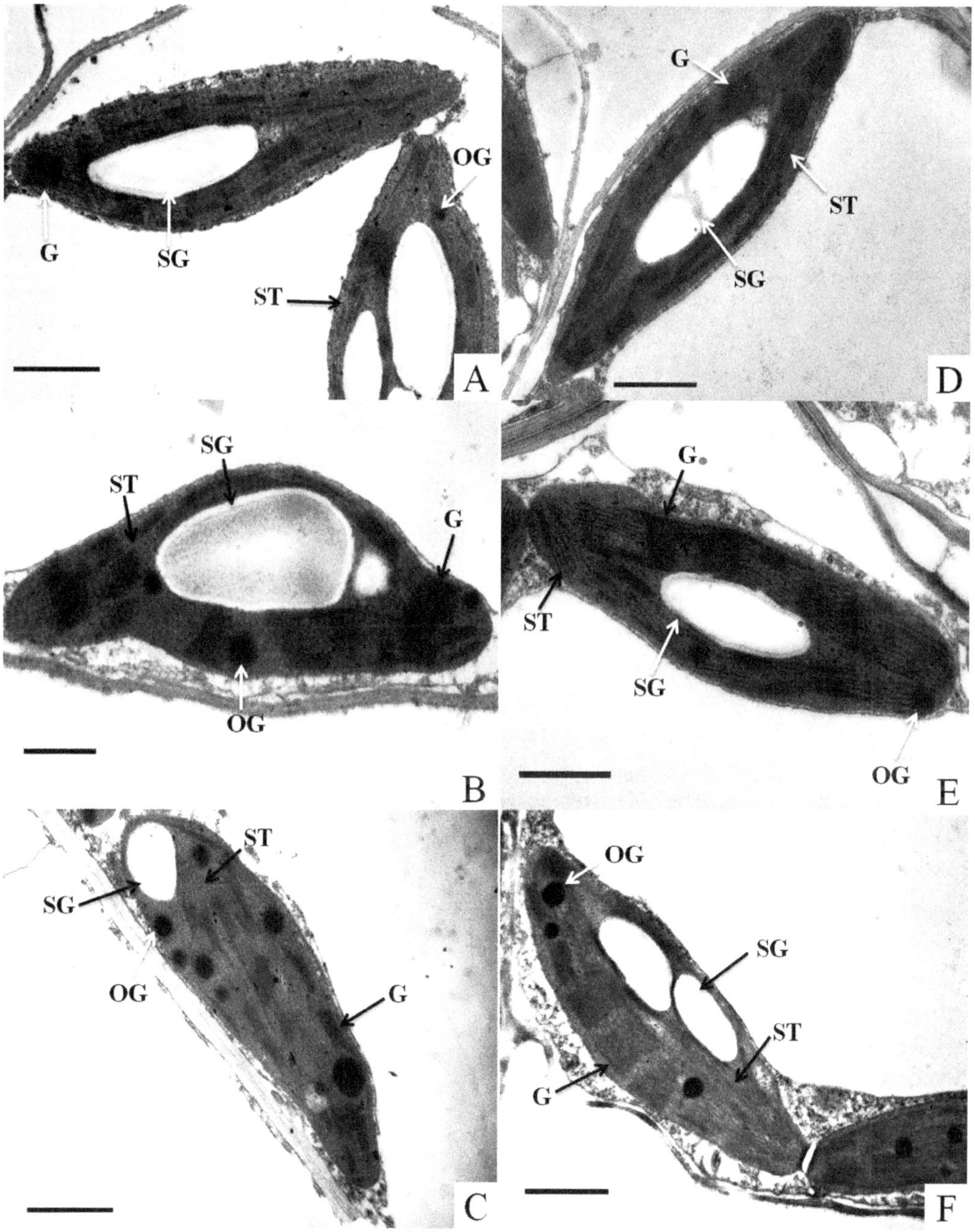

Figure 4. Effect of storage duration on the post-storage ultrastructure of chloroplasts from the leaves of watermelon (*Citrullus lanatus*) seedlings. Seedlings were transplanted for 6 days after being stored in the dark (A, B, C) or in the light (D, E, F) for 2 (A, D), 4 (B, E), or 6 (C, F) days at 15°C. Bar = 1 μm. Abbreviations: ST, stroma thylakoid; G, grana thylakoid; SG, starch grain; OG, osmiophilic globule.

Statistical Analysis

The statistical analyses were performed using SPSS 16.0 for Windows (SPSS, Chicago, IL, USA). The data were subjected to analysis of variance (ANOVA), and the mean values were compared by Tukey's test ($P<0.05$) when a significant difference was detected.

Results

Changes in Chloroplast Ultrastructure, Chlorophyll Content, Chlorophyll Fluorescence Measurements, and Photosynthesis During Storage in Light or Darkness

Changes in The Appearance of Leaves and Ultrastructure of Chloroplasts. No obvious differences in leaf appearance were observed between the light and dark treatments within 4 d of storage. However, after 6 d of storage, dark-stored seedlings exhibited some necrosis on the leaves, whereas light-stored leaves

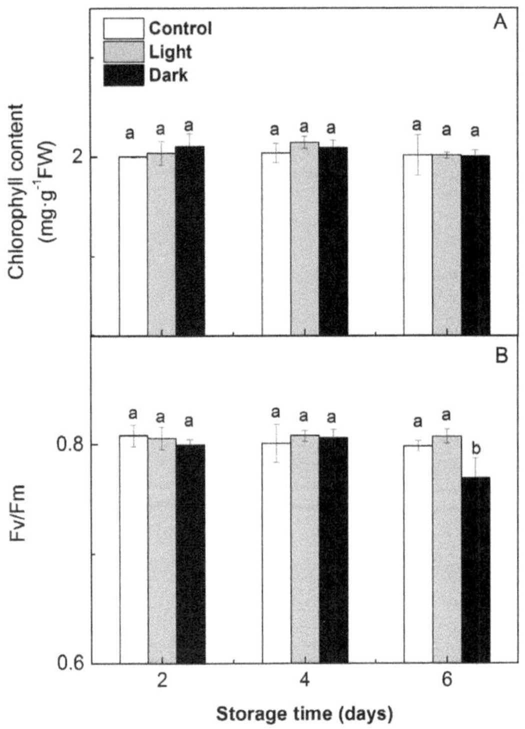

Figure 5. Effect of storage duration on the post-storage chlorophyll content (A) and maximal photochemical efficiency of PSII (Fv/Fm) (B) from the leaves of watermelon (*Citrullus lanatus*) seedlings. Seedlings were transplanted for 6 days after being stored at 15°C in the dark (*black*) or in the light (*gray*) for 2–6 days. The controls (*white*) were taken from seedlings that had never been stored. Data in A are the means of five replicates, and data in B are the means of ten replicates. Standard errors are shown with a vertical bar. Small letters indicate significant differences between treatments on a given day according to the Tukey test (P<0.05).

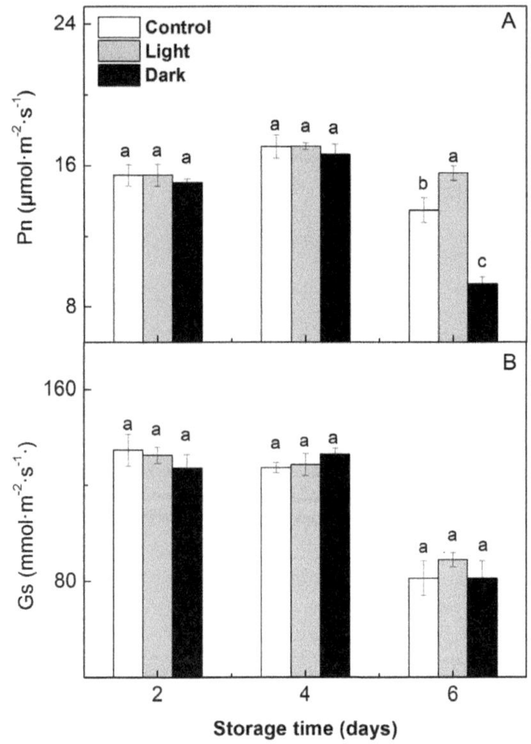

Figure 6. Effect of storage duration on the post-storage net photosynthesis rate (Pn) (A) and stomatal conductance (Gs) (B) in the leaves of watermelon (*Citrullus lanatus*) seedlings. Seedlings were transplanted for 6 days after being stored at 15°C in the dark (*black*) or in the light (*gray*) for 2–6 days. The controls (*white*) were taken from seedlings that have never been stored. Data are the means of nine replicates; standard errors are shown with a vertical bar. Small letters indicate significant differences between treatments on a given day according to the Tukey test (P<0.05).

appeared vigorous relative to dark-stored leaves (Fig. 1). In normal leaves (0 d), the chloroplasts were ellipsoid in shape, and the grana and stroma lamellae of the chloroplasts were well developed; the thylakoid was arranged densely along the long axis of the chloroplasts, and starch grains were observed (Fig. 2 A). No substantial differences in the chloroplasts were found between light-stored and non-stored seedlings until the 4th day of storage (Fig. 2 C, E). On the 6th day of light storage, the starch grains of the chloroplasts decreased in number and size, and the chloroplasts appeared elongated in shape, with an increased thickness and an increased number of granal thylakoids (Fig. 2 G). In contrast, chloroplasts from the dark-stored leaves showed remarkable differences, with the chloroplast appearing to be swollen during storage (Fig. 2 B, D) and exhibiting a round shape on the 6th day of storage. In addition, some osmiophilic globules were observed in the chloroplasts, and the granal arrangement was distorted, with an obscure boundary between the grana thylakoids and stroma thylakoids (Fig. 2 F). Furthermore, dark storage caused the disappearance of starch grains.

Chlorophyll Content and Chlorophyll Fluorescence Measurements. As shown in Fig. 3 A, the chlorophyll content of the leaves decreased with the duration of the storage period under dark conditions and significantly decreased after 6 d of storage in the dark, while the chlorophyll content was maintained at a relatively higher level in the light compared with the darkness. The Fv/Fm is the maximum photochemical efficiency of PSII and

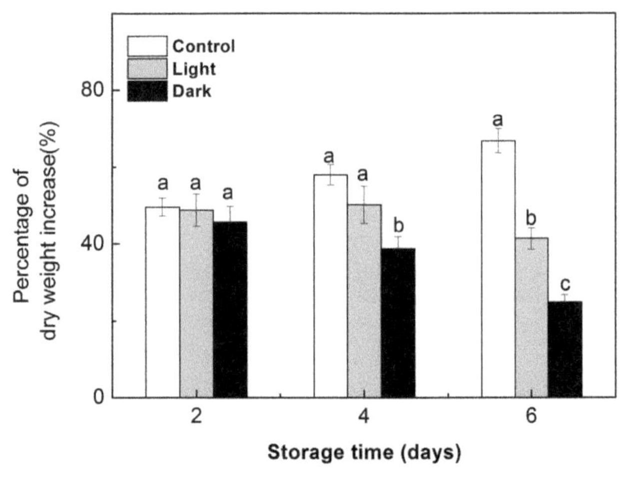

Figure 7. Effect of storage duration on the post-storage percentage dry weight increase in shoots of watermelon (*Citrullus lanatus*) seedlings. Seedlings were transplanted for 6 days after being stored at 15°C in the dark (*black*) or in the light (*gray*) for 2–6 days. The controls (*white*) were taken from seedlings that have never been stored. Data are the means of twelve replicates; standard errors are shown with a vertical bar. Small letters indicate significant differences between treatments on a given day according to the Tukey test (P<0.05).

represents the function of PSII. Similar to chlorophyll, the Fv/Fm of the seedlings remained relatively unchanged during storage under light conditions but decreased under dark conditions (Fig. 3 B). Moreover, the seedlings that were stored in the light had a higher Fv/Fm as well as a higher chlorophyll content during storage compared with those that were stored in darkness.

Photosynthesis. During storage, the Pn, Gs, and Ci of the light-stored and dark-stored seedlings decreased gradually over time; however, the Ls increased until the 4th day of storage (data not shown). In contrast, after 6 d of storage, the seedlings showed a marked decrease in the Pn accompanied by a significant decrease in the Gs and Ls and a remarkable increase in Ci in the leaves compared with the seedlings before storage (Table 1). The seedlings that were stored in the light showed a higher Pn and Gs during storage than those that were stored in darkness.

Shoot Dry Weight. The shoot dry weight per seedling (the total dry weight of the leaf and stem) after storage in the light for 6 d was not significantly different from that before storage (Table 1), while the shoot dry weight of the seedlings that were stored in darkness for 6 d decreased significantly compared to that of the seedlings before storage and was obviously lower than that of the light-stored seedlings.

Recovery of Chloroplast Ultrastructure, Chlorophyll Fluorescence, and Photosynthesis After Transplanting

Chloroplast Ultrastructure and Leaf Appearance. Six days after transplanting following dark storage for 2 or 4 d, the chloroplasts in the leaves returned to a normal oval shape, the thylakoids were orderly and dense, and the starch grains reaccumulated (Fig. 4 A, B). Moreover, there were no differences in the leaf appearance (Fig. 1) or chloroplast ultrastructure compared to light-stored seedlings (Fig. 3 D, E). However, the seedlings that were stored for 6 d in the dark did not recover normal leaf phenotypes or chloroplast ultrastructure by the 6th day of transplanting, showing greater necrosis on the leaves, numerous osmiophilic globules and disordered grana thylakoids in the chloroplasts (Fig. 4 C). In contrast, 6 d after transplantation, the seedlings that were stored in the light for 6 d exhibited healthier leaves (Fig. 1) and had more starch grains and fewer osmiophilic globules in the chloroplasts than dark-stored seedlings (Fig. 4 C). Moreover, no differences were observed in the chloroplast shape (Fig. 4 F) or the thylakoid structure of light-stored chloroplasts compared with normal chloroplasts (Fig. 2 A).

Chlorophyll Content and Chlorophyll Fluorescence Measurements. As shown in Fig. 5, the chlorophyll content and the Fv/Fm of the seedlings that were stored in the light for 2, 4, or 6 d had recovered by 6 d after removal from storage. When the seedlings were stored in the darkness for 2 or 4 d, the chlorophyll content and Fv/Fm also recovered to control (non-stored seedling) levels by 6 d after transplanting. When the duration of dark storage was extended to 6 d, the Fv/Fm of the seedlings was significantly lower than that of the control and light-stored seedlings, but the chlorophyll content recovered to control levels by the 6th day after transplanting.

Photosynthesis. The post-storage recovery of the Pn varied depending on the duration of storage and the light conditions during storage. As shown in Fig. 6, at 6 d after transplantation, the seedlings that were stored under dark or light conditions for 2 or 4 d showed no differences in Pn or Gs; however, as the storage duration increased to 6 d, the post-storage Pn was significantly higher in the light-stored plants than in the control and dark-stored plants. The Gs of the seedlings at 6 d after transplanting was not influenced by 6 d of storage in light or darkness.

Shoot Dry Weight Increase. The seedlings that were stored under light or dark conditions for 2 d exhibited no obvious differences in the percent dry mass increase compared to the control seedlings at 6 d after transplanting (Fig. 7). The percentage dry mass increase of the seedlings that were transplanted after 6 d following storage in the light for 6 d or in darkness for 4 or 6 d was significantly lower relative to that observed in the respective control seedlings. Moreover, the percent dry mass increase in the seedlings that were stored in darkness for 4 or 6 d was significantly lower than that of the light-stored seedlings at 6 d after transplantation.

Discussion

Light Preserves the Chloroplast Ultrastructure and Photosynthetic Performance of Watermelon Seedlings During Storage

Chloroplast development depends on light, and light affects chloroplast structure and photosynthetic changes in plants [28,29]. In this study, dark storage significantly altered the chloroplast ultrastructure of the leaves: the chloroplasts gradually swelled to a rounded shape, and the granal thylakoids were disordered with a low stacking degree (Fig. 2 B, D, F). Similar results were also found in detached barley leaves that were cultured in the dark for 4 d, showing a distorted granal arrangement and an increase in the number and size of plastoglobuli [30]. Furthermore, dark storage caused the disappearance of starch grains, mainly because starch grains serve as a reserve material that can be used to allow plants to survive starvation induced by darkness [31]. In contrast, the chloroplasts from the light-stored leaves were relatively normal, with no differences compared to the chloroplasts observed before storage until the 4th day after storage (Fig. 2 A, C, E). However, on the 6th day, it was noted that the thickness of the granal thylakoids was increased compared with that before storage (Fig. 2 G). This result agrees with earlier reports, which showed that leaves that were grown under low light had thicker grana lamellae and an increased number of granal thylakoids in the chloroplasts than those under normal light intensity; these alterations would improve the light-harvesting ability of the thylakoids, serving as an adaptation to low-light conditions [28,32,33]. These results indicate that the starch grains in the chloroplast and the chloroplast ultrastructure of the leaves, including a regular thylakoid arrangement and dense stacking of grana, are maintained during light storage but are destroyed during dark storage.

During dark storage, the chlorophyll in seedlings and leafy vegetables is susceptible to degradation [5,6]. The degradation of chlorophyll involves alteration of the chloroplast ultrastructure [34]. The considerable alteration of the chloroplast ultrastructure in dark-stored leaves might be the result of their rapid loss of chlorophyll (Fig. 3 A). When the chlorophyll of PSII is destroyed, the Fv/Fm decreases [35]. The present data indicate that the seedlings that were stored in the light had a higher chlorophyll content and higher Fv/Fm (Fig. 3 B) during storage than those that were stored in darkness, indicating that the chlorophyll remaining in the leaves of the seedlings that were stored in the light could facilitate photosynthesis in a relatively efficient manner.

After 6 d of storage, the Pn, Gs and Ls decreased significantly compared to their levels before storage, but the Ci increased remarkably in the leaves of the seedlings that were stored in light or darkness (Table 1); however, the seedlings that were stored in the light had a higher Pn and Gs than those that were stored in darkness. This result indicates that non-stomatal limitations contribute to the storage-induced decrease in Pn, regardless of the light conditions during storage. Chlorophyll breakdown and damage of the chloroplast structure are always accompanied by a decrease in photosynthesis [34]. In this study, the loss of chlorophyll and the alteration of the chloroplast ultrastructure in the leaves might be responsible for the reduction in photosynthesis during storage.

Dry mass is an important factor in plant storage because it indicates how environmental conditions affect the plant growth rate and dry mass accumulation [11]. Illumination during storage helps to maintain the dry weight of plantlets better than dark storage. Compared to that before storage, the shoot dry weight of the seedlings remained relatively unchanged under light storage for 6 d but decreased significantly under dark storage due to the continuous respiration that occurs in darkness (Table 1). Moreover, the seedlings that were stored in the light had a significantly higher shoot dry mass than those that were stored in darkness. The stability of the dry weight during light storage indicates that the PPFD of 15 $\mu mol \cdot m^{-2} \cdot s^{-1}$ in this study was close to the light compensation point for watermelon seedlings at 15°C.

Post-storage Photosynthetic Performance is Associated with Recovery of the Chloroplast Ultrastructure and the Duration of Storage

Recovery of the chloroplast ultrastructure in light- or dark-stored seedlings varies among plant species. In conifer trees, including larch, pine, and spruce, 2 d of illumination was adequate for the restoration of chloroplast ultrastructure when seedlings were grown in the dark for 13 d [36], while chloroplasts from the cotyledons of *Cucurbita pepo* that were treated in the dark for 5 d required 5 d of photoactivation to restore their normal shape and thylakoid membrane system [37]. In this study, the seedlings that were stored in the dark for 2 or 4 d recovered a normal chloroplast shape and thylakoid structure 6 d after transplanting, while the seedlings that were stored in the dark for 6 d did not recover their thylakoid structure, with many osmiophilic globules and disordered grana thylakoids (Fig. 4). In contrast, the seedlings that were stored in the light for 6 d completely recovered their chloroplast structure and had fewer osmiophilic globules and more ordered grana in the chloroplasts than the seedlings that were stored in darkness (Fig. 4).

When the storage duration increased to 6 d, the post-storage photosynthesis was less dependent on the recovery of the chlorophyll content and Gs. On the 6th day after transplanting, the chlorophyll content (Fig. 5 A) and Gs (Fig. 6 B) in the seedlings that were stored in light or darkness did not differ from those of the control seedlings, regardless of the storage duration. Although their chlorophyll concentrations and Gs were the same, the light-stored seedlings had a significantly higher Pn than the dark-stored and control seedlings (Fig. 6 A). These data suggest that the pigment concentration and stomatal factors are not the main elements affecting photosynthesis during this stage.

The structure and function of chloroplasts are important for the growth of plants and influence plants' physiological and ecological responses [38]. It is commonly accepted that a decrease in photosynthesis is related to disturbance of the chloroplast structure [34,39]. Simultaneously with the change in the chloroplast structure, 6 d after transplanting, the Fv/Fm and Pn of the seedlings that were stored in the light for 2, 4 or 6 d and the seedlings that were stored in darkness for 2 or 4 d were completely restored to control levels (Fig. 5, Fig. 6). However, on the 6th day after removal from storage, the seedlings that were stored in the dark for 6 d had a significantly lower Fv/Fm and Pn than the light-stored and control seedlings, indicating that the Fv/Fm and Pn could not be recovered in the dark-stored seedlings in this time period, mainly due to the incomplete restoration of the chloroplast ultrastructure in these seedlings.

The high photosynthetic ability during storage contributes to the subsequent growth of seedlings after storage [15,40]. The seedlings that were stored in the light for 6 d appeared vigorous and survived, while the seedlings that were stored in darkness deteriorated due to necrosis or chlorosis on old leaves or cotyledons, resulting in the death of 30% of the seedlings after transplanting (data not shown). The percentage dry weight increase of the seedlings decreased as the storage time increased from 2 to 6 d for the stored seedlings; in contrast, the percentage dry weight increase of the control seedlings continuously increased over time (Fig. 7). These data indicate that the storage duration affects the regrowth potential of seedlings, with a lengthened storage period causing a decrease in the percentage dry mass increase after transplanting, regardless of whether the seedlings were stored under light or dark conditions. However, the light-stored seedlings had a higher regrowth ability than the dark-stored seedlings after removal from storage.

The purpose of seedling storage is to stop or suppress the growth and development of seedlings while preserving their quality and not adversely affecting their future growth. Photosynthetic ability is a good indicator of the visual quality of seedlings during storage [15,41] and the growth potential of seedlings after transplanting [40]. In this experiment, the watermelon seedlings that were stored in the light had a higher Pn and Fv/Fm than those that were stored in darkness (Table 1 and Fig. 3), which contributed to the preservation of the shoot dry mass of seedlings (Table 1). After transplanting, the light-stored seedlings had a higher percentage dry weight increase (Fig. 7) and exhibited faster recovery of their photosynthetic ability than the dark-stored seedlings; 6 d was adequate for the recovery of the Pn and Fv/Fm, even when the storage time was extended to 6 d (Fig. 5 and Fig. 6). In this study, the quality of the light-stored watermelon seedlings might have been higher because 1) the maintenance of the chloroplast ultrastructure with densely stacked grana in the leaves during light storage contributed to photosynthesis; 2) the higher chlorophyll concentration and the higher Fv/Fm of PSII in the seedlings that were stored in the light reduced damage to PSII reaction centers; 3) the rapid recovery of the chloroplast ultrastructure and photochemical activities had beneficial effects on photosynthesis after removal from storage; or 4) the increased vigor of the seedlings that were stored in the light increased the dry weight during storage and enhanced the regrowth potential after transplanting.

Conclusions

This study demonstrated the effects of light during storage and the storage duration on the photosynthetic apparatus and its ability in watermelon seedlings. The current data indicate that the seedlings that were stored at 15°C in the light exhibited a normal chloroplast ultrastructure with well-organized grana thylakoids and showed an improvement in photosynthetic performance with a significantly higher Pn and Fv/Fm compared with the dark-stored seedlings. Six days might be adequate for the post-storage recovery of Pn and Fv/Fm, but the effect depends on the light condition and the duration of storage. Thus, the seedlings that were stored in the darkness for 6 d appeared to require a longer time period to recover their photosynthetic ability. Furthermore, the percentage dry mass increase was greatly reduced when the seedlings were stored in the dark, and this effect was worsened by prolonged storage (4 to 6 d). This study indicates that dim light (PPFD = 15 $\mu mol \cdot m^{-2} \cdot s^{-1}$) during storage is beneficial for maintaining the chloroplast ultrastructure and the photosynthetic efficiency in watermelon seedlings, thus contributing to the rapid post-storage recovery of photosynthetic performance, which ensures the transplant quality of the seedlings after removal from storage.

Author Contributions

Conceived and designed the experiments: QD. Performed the experiments: QD WJ MD YL. Analyzed the data: QD. Contributed reagents/materials/analysis tools: QD DH. Contributed to the writing of the manuscript: QD. Revised this manuscript: DH.

References

1. FAO (04 August 2014) FAOSTAT Domains/Production/Crops. Available: http://faostat3.fao.org/faostat-gateway/go/to/browse/Q/QC/E. Accessed 23 Sept 2014.
2. Bessembinder J, Staritsky G, Zandvoort E (1993) Long-term in vitro storage of *Colocasia esculenta* under minimal growth conditions. Plant Cell Tissue Organ Cult 33: 121–127.
3. Kaczperski MP, Armitage AM, Lewis PM (1996) Performance of plug-grown geranium seedlings preconditioned with nitrogen fertilizer or low-temperature storage. HortScience 31: 361–363.
4. Sato F, Yoshioka H, Fujiwara T, Higashio H, Uragami A, et al. (2004) Physiological responses of cabbage plug seedlings to water stress during low-temperature storage in darkness. Sci Hortic 101: 349–357.
5. Kubota C, Kozai T (1994) Low-temperature storage for quality preservation and growth suppression of broccoli plantlets cultured in vitro. HortScience 29: 1191–1194.
6. Wilson SB, Iwabuchi K, Rajapakse NC, Young RE (1998) Responses of broccoli seedlings to light quality during low-temperature storage in vitro: I. Morphology and survival. HortScience 33: 1253–1257.
7. Wilson SB, Iwabuchi K, Rajapakse NC, Young RE (1998) Responses of broccoli seedlings to light quality during low-temperature storage in vitro: II. Sugar content and photosynthetic efficiency. HortScience 33: 1258–1261.
8. Kubota C, Rajapakse NC, Young RE (1997) Carbohydrate status and transplant quality of micropropagated broccoli plantlets stored under different light environments. Postharvest Biol Technol 12: 165–173.
9. Ding M, Bie B, Jiang W, Duan Q, Du H, et al. (2011) Physiological advantages of grafted watermelon (*Citrullus lanatus*) seedlings under low-temperature storage in darkness. HortScience 46: 993–996.
10. Jiang W, Ding M, Duan Q, Zhou Q, Huang D (2012) Exogenous glucose preserves the quality of watermelon (*Citrullus lanatus*) plug seedlings for low-temperature storage. Sci Hortic 148: 23–29.
11. Justus I, Kubota C (2010) Effects of low temperature storage on growth and transplant quality of non-grafted and grafted cantaloupe-type muskmelon seedlings. Sci Hortic 125: 47–54.
12. Kubota C, Seiyama S, Kozai T (2002) Manipulation of photoperiod and light intensity in low-temperature storage of eggplant plug seedlings. Sci Hortic 94: 13–20.
13. Park J, Fujiwara K (2009) Quality preservation of tomato seedlings affected by different wavelengths of LED-low light during storage. Europ J Hort Sci 74: 21–23.
14. Su V, Hsu BD, Chen WH (2001) The photosynthetic activities of bare rooted *Phalaenopsis* during storage. Sci Hortic 87: 311–318.
15. Mena-Petite A, Robredo A, Alcalde S, Duñabeitia M, González-Moro M, et al. (2003) Gas exchange and chlorophyll fluorescence responses of *Pinus radiata* D. Don seedlings during and after several storage regimes and their effects on post-planting survival. Trees 17: 133–143.
16. Hou JY, Setter TL, Chang YC (2010) Effects of simulated dark shipping on photosynthetic status and post-shipping performance in *Phalaenopsis* Sogo Yukidian 'V3'. J Am Soc Hortic Sci 135: 183–190.
17. Harper R, O'reilly C, Perks M (2010) Rapidly detecting the effects of warm storage stress on Douglas fir seedlings. Plant Biosyst 144: 769–777.
18. Oguchi R, Hikosaka K, Hirose T (2005) Leaf anatomy as a constraint for photosynthetic acclimation-differential responses in leaf anatomy to increasing growth irradiance among three deciduous trees. Plant Cell Environ 7: 916–927.
19. Peri PL, Moot DJ, Jarvis P, McNeil DL, Lucas RJ (2007) Morphological, anatomical, and physiological changes of orchardgrass leaves grown under fluctuating light regimes. Agron J 99: 1502.
20. Vidal D, Griera E, Marin P, Sabido J (1990) Anatomical and physiological acclimation of *Fatsia japonica* leaves to irradiance. Am J Bot 77: 1149–1158.
21. Migge A, Kahmann U, Fock H, Becker T (1999) Prolonged exposure of tobacco to a low oxygen atmosphere to suppress photorespiration decreases net photosynthesis and results in changes in plant morphology and chloroplast structure. Photosynthetica 36: 107–116.
22. Bondada BR, Oosterhuis D (1998) Decline in photosynthesis as related to alterations in chloroplast ultrastructure of a cotton leaf during ontogeny. Photosynthetica 35: 467–471.
23. Yang XS, Chen GX, Xie KB, Wei XD, Zhang MP, et al. (2010) Photosynthetic, biochemical, and ultra-structural changes in the chloroplasts of two ginkgo (*Ginkgo biloba* L.) cultivars during leaf development. J Hortic Sci Biotechnol 85: 323.
24. Holcomb EJ (1994) Bedding Plant Performance. In: Bedding plants IV 4th ed. Batavia, IL: Ball Publishing. 399–424.
25. Wang Y, Zhang H, Zhang F, Xu Y, Cao W, et al. (2001) Effects of low temperature and low light intensity stress on photosynthesis in seedlings of different cucumber varieties. Acta Horticulturae Sinica 28: 230–234. (in Chinese with English abstract).
26. Chen GX, Liu SH, Zhang CJ, Lu CG (2004) Effects of drought on photosynthetic characteristics of flag leaves of a newly-developed super-high-yield rice hybrid. Photosynthetica 42: 573–578.
27. Lichtenthaler HK (1987) Chlorophylls and carotenoids: pigments of photosynthetic biomembranes. Methods Enzymol 148: 350–382.
28. Huang J, Guo S, Wu Z, Li S (2007) Effects of weak light on photosynthetic characteristics and chloroplast ultrastructure of non-heading Chinese cabbage. Chinese Journal of Applied Ecology 18: 352–358. (in Chinese with English abstract).
29. Doncheva S, Vassileva V, Ignatov G (2008) Influence of nitrogen deficiency on photosynthesis and chloroplast ultrastructure of pepper plants (Research Note). Agricultural and Food Science 10: 59–64.
30. Špundová M, Popelková H, Ilík P, Skotnica J, Novotný R, et al. (2003) Ultrastructural and functional changes in the chloroplasts of detached barley leaves senescing under dark and light conditions. J Plant Physiol 160: 1051–1058.
31. Stitt M, Heldt HW (1981) Physiological rates of starch breakdown in isolated intact spinach chloroplasts. Plant Physiol 68: 755–761.
32. Yao YC, Wang SH, Kong Y (2007) Characteristics of photosynthesis machinism in different peach species under low light intensity. Scientia Agricultura Sinica 4: 855–863. (in Chinese with English abstract).
33. Ai X, Guo Y, Ma X, Xing Y (2003) Photosynthetic characteristics and ultrastructure of chloroplast of cucumber under low light intensity in solargreenhouse. Scientia Agricultura Sinica 37: 268–273. (in Chinese with English abstract).
34. Baka Z, Aldesuquy H (1991) Changes in ultrastructure and hormones of the fully senescent leaf of *Senecio aegyptius*. Beitr Biol Pflanzen 66: 271–281.
35. Kubota C, Kozai T (1995) Low-temperature storage of transplants at the light compensation point: air temperature and light intensity for growth suppression and quality preservation. Sci Hortic 61: 193–204.
36. Walles B, Hudák J (1975) A comparative study of chloroplast morphogenesis in seedlings of some conifers (*Larix decidua*, *Pinus sylvestris* and *Picea abies*). Studia Forestalia Suecica 127: 2–22.
37. Ananieva K, Malbeck J, Kaminek M, Van Staden J (2004) Changes in endogenous cytokinin levels in cotyledons of Cucurbita pepo (zucchini) during natural and dark-induced senescence. Physiol plantarum 122: 133–142.
38. Peng Q, Zhou Q (2009) Influence of lanthanum on chloroplast ultrastructure of soybean leaves under ultraviolet-B stress. J Rare Earth 27: 304–307.
39. Shao Q, Wang H, Guo H, Zhou A, Huang Y, et al. (2014) Effects of Shade Treatments on Photosynthetic Characteristics, Chloroplast Ultrastructure, and Physiology of *Anoectochilus roxburghii*. PloS one 9: e85996.
40. Hou JY, Miller WB, Chang YC (2011) Effects of simulated dark shipping on the carbohydrate status and post-shipping performance of *Phalaenopsis*. J Am Soc Hortic Sci 136: 364–371.
41. Wilson SB, Rajapakse NC, Young RE (2000) Use of low temperature to improve storage of in vitro broccoli seedlings under various light qualities. J veg crop prod 6: 51–67.

PERMISSIONS

All chapters in this book were first published in PLOS ONE, by The Public Library of Science; hereby published with permission under the Creative Commons Attribution License or equivalent. Every chapter published in this book has been scrutinized by our experts. Their significance has been extensively debated. The topics covered herein carry significant findings which will fuel the growth of the discipline. They may even be implemented as practical applications or may be referred to as a beginning point for another development.

The contributors of this book come from diverse backgrounds, making this book a truly international effort. This book will bring forth new frontiers with its revolutionizing research information and detailed analysis of the nascent developments around the world.

We would like to thank all the contributing authors for lending their expertise to make the book truly unique. They have played a crucial role in the development of this book. Without their invaluable contributions this book wouldn't have been possible. They have made vital efforts to compile up to date information on the varied aspects of this subject to make this book a valuable addition to the collection of many professionals and students.

This book was conceptualized with the vision of imparting up-to-date information and advanced data in this field. To ensure the same, a matchless editorial board was set up. Every individual on the board went through rigorous rounds of assessment to prove their worth. After which they invested a large part of their time researching and compiling the most relevant data for our readers.

The editorial board has been involved in producing this book since its inception. They have spent rigorous hours researching and exploring the diverse topics which have resulted in the successful publishing of this book. They have passed on their knowledge of decades through this book. To expedite this challenging task, the publisher supported the team at every step. A small team of assistant editors was also appointed to further simplify the editing procedure and attain best results for the readers.

Apart from the editorial board, the designing team has also invested a significant amount of their time in understanding the subject and creating the most relevant covers. They scrutinized every image to scout for the most suitable representation of the subject and create an appropriate cover for the book.

The publishing team has been an ardent support to the editorial, designing and production team. Their endless efforts to recruit the best for this project, has resulted in the accomplishment of this book. They are a veteran in the field of academics and their pool of knowledge is as vast as their experience in printing. Their expertise and guidance has proved useful at every step. Their uncompromising quality standards have made this book an exceptional effort. Their encouragement from time to time has been an inspiration for everyone.

The publisher and the editorial board hope that this book will prove to be a valuable piece of knowledge for researchers, students, practitioners and scholars across the globe.

LIST OF CONTRIBUTORS

Liming Lai and Steven X. Ge
Department of Mathematics and Statistics, South Dakota State University, Brookings, South Dakota, United States of America

Soledad Gonzalo
Departamento de Ingeniería Química, Universidad de Alcalá, Alcalá de Henares, Madrid, Spain

Veronica Llaneza and Jean Claude Bonzongo
Department of Environmental Engineering Sciences, University of Florida, Gainesville, Florida, United States of America

Gerardo Pulido-Reyes
Departamento de Ingeniería Química, Universidad de Alcalá, Alcalá de Henares, Madrid, Spain
Departamento de Biología, Facultad de Ciencias, Universidad Autónoma de Madrid, Madrid, Spain

Francisca Fernández-Piñas, Francisco Leganes and Ismael Rodea-Palomares
Departamento de Biología, Facultad de Ciencias, Universidad Autónoma de Madrid, Madrid, Spain

Roberto Rosal and Eloy García-Calvo
Departamento de Ingeniería Química, Universidad de Alcalá, Alcalá de Henares, Madrid, Spain
Instituto Madrileño de Estudios Avanzados (IMDEA) Agua, Alcalá de Henares, Madrid, Spain

Daryl Moorhead
Department of Environmental Sciences, University of Toledo, Toledo, Ohio, United States of America
INRA, UMR614 Fractionnement des AgroRessources et Environnement, Reims, France
Université Reims-Champagne Ardenne, UMR614 Fractionnement des AgroRessources et Environnement, Reims, France

Gwenaëlle Lashermes, Sylvie Recous and Isabelle Bertrand
INRA, UMR614 Fractionnement des AgroRessources et Environnement, Reims, France
Université Reims-Champagne Ardenne, UMR614 Fractionnement des AgroRessources et Environnement, Reims, France

Yoko Kimata-Ariga and Toshiharu Hase
Institute for Protein Research, Osaka University, Suita, Osaka, Japan

Jaime F. Martínez-García
Institució Catalana de Recerca i Estudis Avançats, Barcelona, Spain
Centre for Research in Agricultural Genomics (CRAG), Consortium CSIC-IRTA-UAB-UB, Barcelona, Spain

Marc¸al Gallemí, María José Molina-Contreras and Briardo Llorente
Centre for Research in Agricultural Genomics (CRAG), Consortium CSIC-IRTA-UAB-UB, Barcelona, Spain

Maycon R. R. Bevilaqua
Centre for Research in Agricultural Genomics (CRAG), Consortium CSIC-IRTA-UAB-UB, Barcelona, Spain
CAPES foundation, Ministry of Education of Brazil, Brasilia - DF, Brazil

Peter H. Quail
Department of Plant and Microbial Biology, University of California, Berkeley, California, United States of America
US Department of Agriculture/Agriculture Research Service, Plant Gene Expression Center, Albany, California, United States of America

Justin N. Vaughn, Srinivasa R. Chaluvadi and Jeffrey L. Bennetzen
Department of Genetics, University of Georgia, Athens, Georgia, United States of America

Tushar and Latha Rangan
Department of Biotechnology, Indian Institute of Technology Guwahati, Assam, India

Lori B. Huberman and Andrew W. Murray
Molecular and Cellular Biology, Harvard University, Cambridge, MA, United States of America
Faculty of Arts and Sciences Center for Systems Biology, Harvard University, Cambridge, MA, United States of America

Marco Lauteri
Institute of Agro-Environmental and Forest Biology, National Research Council, Porano, Italy

Matthew Haworth and Mauro Centritto
Trees and Timber Institute (IVALSA - Istituto per la valorizzazione del legno e delle specie arboree), National Research Council, Florence, Italy

Rachid Serraj
Consultative Group on International Agricultural Research, Independent Science and Partnership Council Secretariat, Food and Agriculture Organisation of the United Nations, Rome, Italy
International Rice Research Institute, Los Banos, Philippines

Maria Cristina Monteverdi
Forestry Research Centre, National Research Council, Arezzo, Italy

Rachel M. Mitchell
School of Environmental and Forest Sciences, University of Washington, Seattle, Washington, United States of America

Jonathan D. Bakker
School of Environmental and Forest Sciences, University of Washington, Seattle, Washington, United States of America
Smithsonian Environmental Research Center, Edgewater, Maryland, United States of America

Candida Vannini, Marcella Bracale, Paola Campomenosi and Milena Marsoni
Dipartimento di Biotecnologie e Scienze della Vita, Università degli Studi dell'Insubria, Varese, Italy

Rita Crinelli and Valerio Marconi
Dipartimento di Scienze Biomolecolari, Università di Urbino "Carlo Bo", Urbino, Italy

Valeria Scoccianti
Dipartimento di Scienze della Terra, della Vita e dell'Ambiente, Sezione di Biologia Vegetale, Universita` di Urbino "Carlo Bo", Urbino, Italy

Luhua Zhang, Haiwei Chen, Yanlan Li, Yanan Li, Defu Chen and Xiwen Chen
Laboratory of Molecular Genetics, College of Life Sciences, Nankai University, Tianjin, China

Shengjun Wang, Jinping Su and Xuejun Liu
Tianjin Crop Research Institute, Tianjin, China

Mounira Chaki, Izabella Kovacs and Christian Lindermayr
Institute of Biochemical Plant Pathology, Helmholtz Zentrum München-German Research Center for Environmental Health, Neuherberg, Germany

Manuel Spannagl
Institute of Bioinformatics and Systems Biology, Helmholtz Zentrum München-German Research Center for Environmental Health, Neuherberg, Germany

Chieko Sugita, Yoshihiro Komura, Korechika Tanaka, Kazuki Kometani and Mamoru Sugita
Center for Gene Research, Nagoya University, Nagoya, Japan

Hiroyuki Satoh
Department of Biomolecular Science, Toho University, Funabashi, Japan

Fan Yang, Yong Wang and Zhulong Chan
Key Laboratory of Plant Germplasm Enhancement and Specialty Agriculture, Wuhan Botanical Garden, Chinese Academy of Sciences, Wuhan, Hubei, PR China

Tingting Li, Yuanyuan Hu, Hui Tang, Chaohua Shen and Jiasheng Wu
Nurturing Station for the State Key Laboratory of Subtropical Silviculture, Zhejiang A & F University, Lin'an, Hangzhou, P. R. China
School of Forestry & Biotechnology, Zhejiang A & F University, Lin'an, Hangzhou, P. R. China

Xuhua Du
China National Bamboo Research Center, Hangzhou, P. R. China

Adam D. Idoine, Alix Boulouis, Jens Rupprecht and Ralph Bock
Max-Planck-Institut für Molekulare Pflanzen physiologie, Potsdam, Golm, Germany

Mai Lan Kranitz, Alexandra Clark, Michelle L. Hollingsworth, Markus Ruhsam,
Martin F. Gardner, Philip Thomas, Robert R. Mill and Peter M. Hollingsworth
Royal Botanic Garden Edinburgh, Edinburgh, United Kingdom

Edward Biffin and Andrew J. Lowe
Australian Centre for Evolutionary Biology and Biodiversity, Environment Institute, School of Earth and Environmental Science, University of Adelaide, Adelaide, Australia

Richard A. Ennos
University of Edinburgh, Institute of Evolutionary Biology, Edinburgh, United Kingdom

Myriam Gaudeul
UMR CNRS MNHN UPMC EPHE 7205 Institut de Systématique, Evolution, Biodiversité, Muséum National d'Histoire Naturelle, Paris, France

Steve Dagenais Bellefeuille, Sonia Dorion, Jean Rivoal and David Morse
Institut de Recherche en Biologie Végétale, Département de Sciences Biologiques, Université de Montréal, Montréal, Québec, Canada

Julia Gollhofer, Roman Timofeev and Thomas J. Buckhout
Institute of Biology, Humboldt University Berlin, Berlin, Germany

Ping Lan
State Key Laboratory of Soil and Sustainable Agriculture, Institute of Soil Science, Chinese Academy Sciences, Nanjing, PR China

Wolfgang Schmidt
Academia Sinica, Institute of Plant and Microbial Biology, Taipei, Taiwan
Graduate Institute of Biotechnology, National Chung Hsing University, Taichung, Taiwan
Genome and Systems Biology Degree Program, College of Life Science, National Taiwan University, Taipei, Taiwan

Rachele Gallisai and Francesc Peters
Departament de Biologia Marina i Oceanografia, Institut de Ciències del Mar, CSIC, Barcelona, Spain

Gianluca Volpe
Istituto di Scienze dell'Atmosfera e del Clima, Roma, Italy

Sara Basart
Earth Sciences Department, Barcelona Supercomputing Center-Centro Nacional de Supercomputación, BSC-CNS, Barcelona, Spain

José Maria Baldasano
Earth Sciences Department, Barcelona Supercomputing Center-Centro Nacional de Supercomputación, BSC-CNS, Barcelona, Spain
Environmental Modelling Laboratory, Technical University of Catalonia, Barcelona, Spain

Lipan Hu, Lixia Xiang, Li Zhang, Xiaoting Zhou, Zhirong Zou and Xiaohui Hu
College of Horticulture, Northwest Agricultural & Forest University, Yangling, China
Key Laboratory of Protected Horticultural Engineering in Northwest, Ministry of Agriculture, Shaanxi Yangling, China

Marta Brozynska, Agnelo Furtado and Robert James Henry
Queensland Alliance for Agriculture and Food Innovation, The University of Queensland, Brisbane, Queensland, Australia

Qingqing Duan, Wu Jiang, Ye Lin and Danfeng Huang
Department of Plant Science, School of Agriculture & Biology, Shanghai Jiao Tong University, Shanghai, China
Key Laboratory of Urban Agriculture (South), Ministry of Agriculture, Shanghai, China

Ming Ding
Department of Horticulture, Northwest A&F University, Yangling, China

Index